D0251296

A GUIDE TO BIRD FINDING

WEST OF THE MISSISSIPPI

Second Edition

A Guide to

Bird Finding WEST OF THE MISSISSIPPI

Second Edition

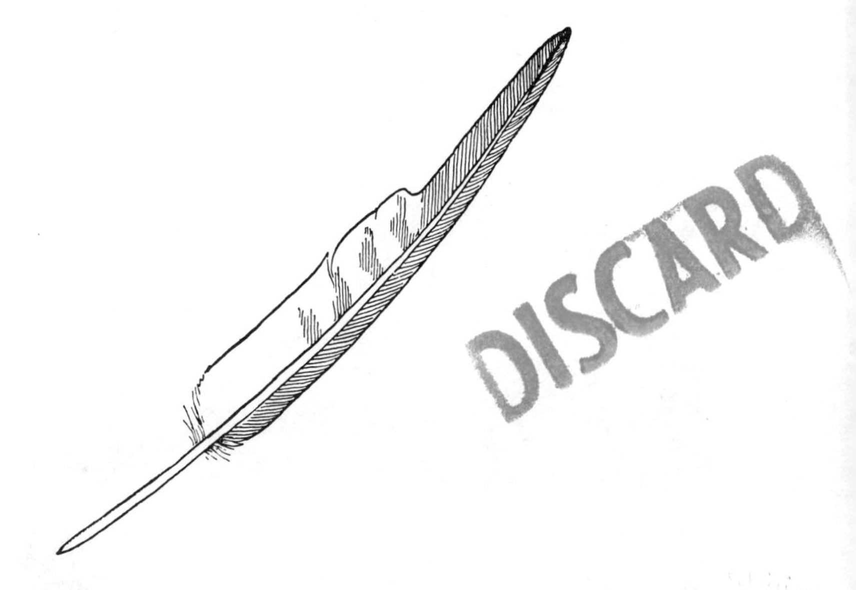

OLIN SEWALL PETTINGILL, JR.

with illustrations by GEORGE MIKSCH SUTTON

New York · OXFORD UNIVERSITY PRESS · 1981

Library of Congress Cataloging in Publication Data

Pettingill, Olin Sewall, Jr. 1907–
A guide to bird finding west of the Mississippi.

Includes index.
1. Birds—The West—Identification. 2. Bird watching—The West. 3. Birds—Identification.
I. Title. II. Title: Bird finding west of the Mississippi.
QL683.W4P45 1981 598'.97'23478 80-18666
ISBN 0-19-502818-X

PRINTED IN THE UNITED STATES OF AMERICA

FOR MY FATHER AND MOTHER

Preface to the First Edition

This volume is the counterpart of *A Guide to Bird Finding East of the Mississippi* (1951) and is identical with it in organization and style. Much of this text had, in fact, already been prepared when the first volume went to press and remains essentially unchanged except for minor revisions to bring the information up to date.

Like the first volume, this work attempts to cover each state with respect to physiographic regions, natural areas, and the principal ornithological attractions. For the somewhat specialized information I relied heavily on ornithological literature and the generous help of various ornithologists, amateur and professional.

In persuing books, pamphlets, and journals pertaining to plant and animal life in the states west of the Mississippi, I took the liberty of extracting the facts that would prove useful and incorporated many of them in the text. All of the books and pamphlets drawn upon extensively are included in the reference material at the end of this book. Among the journals consulted, the following contained papers of greatest value: *The Auk*, *The Condor*, *The Wilson Bulletin*, *Audubon Field Notes*, *Audubon Magazine*, *Journal of Wildlife Management*, *The American Midland Naturalist*, *The Murrelet*, *The Nebraska Bird Review*, *Iowa Bird Life*, *The Flicker*, and *South Dakota Bird Notes*.

Without the co-operation of over 300 persons, each one of whom contributed information, the preparation of this guide would have been impossible. When I received the information relating to a particular bird-finding area, I wrote it up in my own style, then returned it to the donor for checking. This was a time-consuming procedure and, needless to say, a tax on the generosity of the donor, who had already gone to the trouble of assembling the data. Listed at the end of each chapter are the individuals who helped supply the necessary facts, or who played some other part in the preparation of the chapter. To all of them I am immensely indebted.

Of the many persons who helped with the preparation of the manuscript, I wish to single out for my special thanks John W. Aldrich, Roger Tory Peterson, George Miksch Sutton, and Josselyn Van Tyne for their advice, or suggestions, on some of the common names chosen for certain bird species; Harry C. Oberholser not only for carefully going over the manuscript of the Texas chapter, but for giving me his computed total of bird species known in the state; Frederick M. and Marguerite Heydweiller Baumgartner for their critical reading of the Oklahoma chapter and for bringing the list of species in the chapter's introduction into conformity with the information in their forthcoming *Book of Oklahoma Bird Life;* the following for contributing substantially to the contents of a particular chapter and, later, reading the whole chapter: W. J. Breckenridge, Harvey L. Gunderson, and Dwain W. Warner (Minnesota), Ben B. Coffey, Jr. (Arkansas), Robert B. Lea and George H. Lowery, Jr. (Louisiana), C. J. Henry and Mrs. Robert T. (Ann M.) Gammell (North Dakota), Levi L. Mohler, R. Allyn Moser, and William F. Rapp, Jr. (Nebraska), George G. Williams (Texas), Clifford V. Davis (Montana), Robert J. Niedrach (Colorado), Allan R. Phillips (Arizona), Thomas D. Burleigh (Idaho), Ernest S. Booth and Earl J. Larrison (Washington), and Stanley G. Jewett (Oregon); the following for contributing substantially to three or more area accounts: M. Dale Arvey, L. Irby Davis, W. B. Davis, William R. Eastman, William H. Elder, Donald S. Farner, Reed W. Fautin, John E. and Margret Galley, Harold C. Hedges, Zell C. Lee, S. Walter Lesher, Karl H. Maslowski, Kenneth D. Morrison, Clifford C. Presnall, Alexander Sprunt, Jr., and John H. Wampole; and Edward F. and Doris Dana and G. Reeves Butchart for their timely assistance with editorial matters and proofreading.

In order to have the accounts of National Wildlife Refuges and National Parks and Monuments correct in all details, I made it a custom to send preliminary copies to Refuge managers and Park and Monument superintendents or naturalists with the request that they read them for accuracy. I was greatly impressed with their willingness to comply and am deeply grateful to them for their co-operation.

George Miksch Sutton's delightful pen-and-ink drawings enliven the pages of this volume as they did those of the first and will, I am confident, point up the ornithological attractions of the 22 western states.

Finally, I wish to express my everlasting gratitude to Robert M. Mengel for the fine account of Wyoming's Big Horn Mountains and to the men listed below, who were willing to prepare entire chapters, faithfully following my specifications in regard to organization and style: Gale Monson (Arizona), W. J. Baerg (Arkansas), Charles G. Sibley and Howard L. Cogswell (California), Jean M. Linsdale (Nevada), J. Stokley Ligon (New Mexico), Robert M. Storm (Oregon), William H. Behle (Utah).

The publication of this second volume marks the end of a project that began in 1946. Besides bringing me pleasant acquaintances with ornithologists the country over, the work has served to impress upon me the great diversity of bird-finding opportunities in all the 48 states. Whereas I once believed that there were only a few states with truly exciting places to find birds, I now know that every state encompasses at least several spots waiting to enrich a bird finder's quest for the interesting and the unusual. All in all, there are enough places in our nation, each particularly inviting, to keep the bird finder 'on the go' for a lifetime.

That some parts of this book will be out of date in a relatively short space of time is almost a certainty. The natural processes of plant growth and ecological succession, the commercial and agricultural 'development' of wild areas, the construction of expressways and other new roads—all will create changes and hence make necessary regular revisions or new editions. Users of this volume and its eastern counterpart will be of great service if they will notify me of any changes in environments (including resultant effects on birdlife) and routes that come to their attention.

OLIN SEWALL PETTINGILL, JR.

Carleton College
Northfield, Minnesota
May 1953

Preface to the Second Edition

Like its counterpart, *A Guide to Bird Finding East of the Mississippi* (1951, 1977), this work covers each state, chapter by chapter, describing its physiographic regions and principal ornithological attractions. Throughout I have strived for parallel coverage—that is, in the chapter introductions, to describe the regions in similar detail, and, in the area accounts that follow, to include at least one bird-finding site representative of each region. This was difficult, as I learned in preparing the first edition. There was no published source, or series of sources, that treated all the states with consistent thoroughness as to topography, geological characteristics, vegetation, and birdlife. For the chapter introductions my only recourse was to gather the needed information from scores of assorted books, journals, and pamphlets and bring it into suitable conformity. For the bird-finding sites in the area accounts, only a small fraction had been written about in sufficient detail and I lacked the personal familiarity with even a smaller fraction. Thus I turned to hundreds of persons, knowledgeable about the sites, for the information.

In this second edition, most chapter introductions are retained virtually unchanged except for corrections, updating, and otherwise slight alterations. But all bird-finding sites are either new, or, if repeated from the first edition, have been revised in accordance with changes in bird distribution during the past quarter-century, modifications of the natural environment by human creations and activities, and highway construction.

Throughout this second edition I have used the English names of bird species authorized by the Checklist Committee of the American Birding Association in the *A.B.A. Checklist: Birds of Continental United States and Canada* (1975), with the following exceptions. I have used Traill's Flycatcher for the two species, Willow and Alder, because in many instances I was unable to determine whether one species or the other occupied the locality in question. I have used Yellow-shafted Flicker, Red-shafted Flicker, and Gilded Flicker for the Common Flicker; Myrtle Warbler and Audubon's Warbler for the Yellow-rumped Warbler; Baltimore Oriole and Bullock's Oriole for the Northern Oriole; and Slate-colored Junco, White-winged Junco, and Oregon Junco for the Northern Junco. My reason is this. The populations of each of the three flickers, the two warblers, the two orioles, and the three juncos are perfectly distinct and identifiable, even though now regarded by some ornithologists as subspecies or races of one species because their populations hybridize where they come together. I have no quarrel with the reasoning for lumping the populations in one species, but in a handbook of this sort that covers a wide continental area, I deem it of interest to indicate the obvious differences in geographical populations by keeping the long-established names. Where populations—for example, of the Yellow-shafted and Red-shafted Flickers—meet and are known to produce hybrids, I have so indicated.

For information on bird-finding sites, as in the first edition, I resorted to available literature, relied upon my recent visits to certain localities, and, more especially, imposed upon many persons for their particular knowledge.

In preparing this edition, I drew useful information from articles and notes in *American Birds* and *Birding*. I also took information from my bird-finding columns in *Audubon Magazine* between 1956 and 1969 and from state and local publications on bird finding which have been appearing with increasing frequency since the first edition.

I have included the major National Parks, many of the National (Nature) Monuments, and a majority of the National Wildlife Refuges. All my accounts of them have been checked for accuracy, if not by officials—usually chief naturalists at the Parks and Monuments, managers of the Refuges—then by other persons familiar with them. All along I was greatly impressed by the willingness and promptness of the Park, Monument, and Refuge personnel in responding to my requests.

I cannot possibly thank every individual who has taken part in the preparation of this edition. Like its predecessor, it has been a cooperative undertaking from its start four years ago with the preparation of the first chapter to the proofreading and indexing. However, I must single out many individuals for my special thanks. First, those who authored the chapters on Arizona, Arkansas, Kansas, New Mexico, and Utah, conforming faithfully to the organization and style of the book and my use of English bird names. I owe each of them my everlasting gratitude. Second, those who contributed information and the authors of state or local publications on bird finding. I am indebted to each of them, as the case may be, for so generously meeting my exacting requests or granting me use of the information in their publications.

After each of the states listed below I have mentioned one or more persons for their role in the chapter. Included are those who have contributed comprehensive information—that is, information whose acquisition for me required the personal expenditure of much time and effort. All these individuals—except the authors of chapters—and many others who have been sources of information for this edition are named under 'Authorities' at the conclusion of the introduction to the appropriate chapters.

Arizona. Gale Monson completely revised his chapter that was published in the first edition.

Arkansas. Edith and Henry Halberg authored the chapter, which replaces the one written for the first edition by the late W. J. Baerg.

California. Charles G. Sibley and Howard L. Cogswell, authors of the chapter in the first edition, declined to revise the chapter in this edition. Consequently, I revised the chapter, retaining much of their introduction and parts of several area accounts. The chapter is otherwise updated and new, thanks to the generous help of many persons, particularly the following. Arnold Small carefully reviewed the introduction, contributing to it information on pelagic bird trips, and assisted me in sundry ways with the preparation of the area accounts. Jean Brandt wrote the account of the Los Angeles area, based largely on her bird-finding articles in *The Western Tanager* (vols. 42–45), and contributed to the information on the Salton Sea Basin, San Bernardino Mountains, and Yosemite National Park. Others contributing substantial information for specific areas: Stanley W. Harris, northwestern California; Laurence C. Binford, Theodore A. Chandik, and Richard A. Erickson, San Francisco Bay areas; William Reese, Monterey; Paul Lehman, Santa Barbara; Elizabeth Copper, San Diego area; Jeri Langham, Sacramento; E. Turner Biddle, Palm Springs.

Colorado. Ronald A. Ryder reviewed the chapter. Much useful information was drawn from *A Birder's Guide to Denver and Eastern Colorado* by James A. Lane and Harold R. Holt and *Birds in Western Colorado* (1969, out of print) by William A. Davis.

Idaho. Charles H. Trost helped in updating the entire chapter. He also contributed information on many areas in southern Idaho, as did Liven A. Peterson, Jr.

Iowa. The many bird-finding articles in *Iowa Bird Life*, since collected and published as *Birding Areas in Iowa*, provided useful data. Supplying much new information were Robert F. and Jean Vane for Cedar Rapids, Myrle M. Burk for Cedar Falls-Waterloo, Charles C. Ayers, Jr. for Ottumwa, Margaret Brooke for Des Moines, Darwin Koenig for northeastern Iowa, and Robert L. Nickolson for northwestern Iowa.

Kansas. Robert M. Mengel, with the collaboration of Charles A. Ely, Max C. Thompson, and Marvin Schwilling, wrote the chapter, which is entirely new.

Louisiana. Kathleen S. Harrington and Robert J. Newman collaborated in reviewing the chapter. Additionally, Mrs. Harrington, with the assistance of other authorities, provided extensive information for all the sites in southern Louisiana. Horace H. Jeter contributed the information for Shreveport.

Minnesota. I referred repeatedly to *Minnesota Birds: Where, When, and How Many* by Janet C. Green and Robert B. Janssen, and *A Birder's Guide to Minnesota* by K. R. Eckert. Giving me substantial information were Mrs. Green for northeastern Minnesota, Mr. Eckert for southwestern Minnesota, Frederick Z. Lesher for southeastern Minnesota, and David K. Weaver for Minneapolis–St. Paul.

Missouri. Richard A. Anderson and Paul E. Bauer gave me permission to take information from their *Guide to Finding Birds in the St. Louis Area*. Mr. Anderson reviewed the entire chapter and made many pertinent suggestions. William H. Elder advised me on bird-finding sites in central Missouri. Earl S. McHugh assisted me in revising the Kansas City area, Simon Rositzky, the St. Joseph area.

Montana. Clifford V. Davis reviewed the whole chapter, suggested numerous changes, and provided new information.

Nebraska. Roger S. Sharpe reviewed the chapter and gave me useful advice. Ralph Harrington provided information for the Lincoln area, Ruth C. Green for the Omaha area, Margaret Morton for central Nebraska, Roy J. and Maud Witschy for western Nebraska, and Doris Gates for northwestern Nebraska.

Nevada. As junior author, Donald H. Baepler revised the chapter written for the first edition by the late Jean M. Linsdale.

New Mexico. Dale A. Zimmerman authored the chapter, which replaces the chapter in the first edition by the late J. Stokley Ligon.

North Dakota. Robert E. Stewart assisted me substantially in not only revising the entire chapter but in providing information for bird-finding sites.

Oklahoma. George Miksch Sutton was of invaluable help in revising the chapter and putting me in touch with authorities on particular areas in the state. The *Bird Finding Guide*, prepared by the Tulsa Audubon Society, and Elizabeth Hayes were sources of important information for sites in eastern Oklahoma. John Shackford provided information on Oklahoma City.

Oregon. Robert M. Storm declined to revise his chapter in the first edition. Harry B. Nehls subsequently prepared, and made available for my use, a 54-page manuscript on bird finding in the state. Practically all area accounts, except for Crater Lake National Park, are based on his manuscript. Mr. Nehls reviewed the introduction to the chapter.

South Dakota. Nathaniel R. Whitney, Jr. assisted in revising and augmenting my accounts for western and central South Dakota, Nelda Holden for the north-central and northeastern parts of the state, and Gilbert W. Blankespoor for the southeastern part (Sioux Falls).

Texas. Of great reference value to me, as well as some sources of information, were *A Birder's Guide to the Rio Grande Valley of Texas* by James A. Lane, *A Birder's Guide to the Texas Coast* by James A. Lane and John L. Tveton, *A Bird Finding and*

Naturalist's Guide to the Austin, Texas, Area by Edward A. Kutac and S. Christopher Caran, and *Birds of Big Bend National Park* by Roland H. Wauer. The following persons helped me revise and/or provided much new information on certain areas: George A. Newman, Abilene and Guadalupe Mountains National Park; Kenneth Seyffert, Texas Panhandle; Warren Pulich, Dallas; Kevin L. Zimmer, El Paso; Frances Williams, Davis Mountains; Carrol Richardson, Fort Worth; Charles H. Bender, San Antonio and Uvalde; Kathleen S. Harrington, Wichita Falls.

Utah. William H. Behle completely revised his chapter that was published in the first edition.

Washington. A very useful reference was *A Guide to Bird Finding in Washington* by Terence R. Wahl and Dennis R. Paulson. Mr. Wahl reviewed the chapter. Wayne C. Weber also reviewed the chapter, helped revise the Anacortes account, and supplied the information for Coulee City and Victoria and Vancouver, British Columbia.

Wyoming. Kenneth L. Diem and Oliver K. Scott reviewed the entire chapter. Providing substantially new information were Helen Downing for the Big Horn Mountains and B. C. and M. Raynes for the Jackson area.

Gracing the pages of this edition are sixty drawings by George Miksch Sutton. Twenty-six of the drawings are new in this edition. Of these, twenty-three, done for the first edition but not used, are reproduced for the first time in this edition. One—of the Sandhill Cranes—is repeated from my *A Guide to Bird Finding East of the Mississippi* (New York, Oxford University Press, 1951, 1977). And two—of the Black-throated Gray Warbler and Harris' Sparrow—were first reproduced in *Fundamentals of Ornithology* by the late Josselyn Van Tyne and Andrew J. Berger (New York: John Wiley & Sons, 1959, 1976). For permission to use these two drawings, I thank the artist, junior author, and publisher.

I wish to express my appreciation to the Department of Biology, Virginia Polytechnic Institute and State University, for secretarial assistance with this edition during my tenure as Visiting Professor of Biology in 1978.

Finally, I wish to thank four good friends: Edward F. Dana and

Phillps B. Street for their sharp-eyed help in proofreading, Mary Holland Richards and Josephine Stott Dawson for sharing with me the tedious task of preparing the Index.

OLIN SEWALL PETTINGILL, JR.

Wayne, Maine
January 1981

Contents

The Plan of This Book

Area Covered. This book covers the twenty-two states lying wholly, or in part (Minnesota and Louisiana), west of the Mississippi River. In Washington, adjacent Canadian areas are included. Places for bird finding in each state are chosen to show (1) the widest variety of regular species; (2) seasonal concentrations of birds and migratory movements; (3) representative types of bird habitats from ocean beaches and deserts to mountaintops; and (4) the best representation of birdlife in the vicinities of metropolitan areas and leading vacation centers. Included whenever they offer productive bird finding are National Parks, National (Nature) Monuments, National Wildlife Refuges, many state and municipal parks, refuges, and preserves, and numerous public or privately owned sanctuaries and nature centers.

Birds Covered. Attention is given to all species residing or appearing regularly in the twenty-two states. Subspecies or races receive attention if their populations are readily identifiable in the field.

Terminology. The English vernacular names of birds follow the *A.B.A. Checklist: Birds of Continental United States and Canada* (American Birding Association, 1975), with these exceptions: Black-bellied Whistling Duck and Fulvous Whistling Duck are used for Black-bellied Tree Duck and Fulvous Tree Duck. Traill's Flycatcher is used for both Willow and Alder Flycatchers; Yellow-shafted, Red-shafted, and Gilded Flickers for the Common Flicker; Myrtle and Audubon's Warblers for the Yellow-rumped Warbler; Baltimore and Bullock's Orioles for the Northern Oriole; and Slate-colored, White-winged, and Oregon Juncos for the Northern Junco. In some instances, where a subspecies is recognizable by sight or sound, the English name of the subspecies precedes the name of the species in parentheses—for example, (Belding's) Savannah Sparrow.

Frequently the following terms are used to indicate groups of species:

Waterfowl: swans, geese, and ducks
Waterbirds: herons, egrets, bitterns, storks, ibises, spoonbills, cranes, and all swimming birds other than waterfowl
Wading birds: herons, egrets, bitterns, storks, ibises, spoonbills, and cranes
Shorebirds: oystercatchers, plovers, turnstones, woodcock, snipes, curlews, sandpipers, godwits, avocets, stilts, and phalaropes
Alcids: murres, guillemots, murrelets, auklets, and puffins
Pelagic birds: chiefly albatrosses, fulmars, shearwaters, petrels, tropicbirds; other waterbirds that habitually prefer the open sea to inshore waters; often phalaropes
Landbirds: vultures, hawks, eagles, falcons, gallinaceous birds, doves, cuckoos, owls, goatsuckers, swifts, hummingbirds, trogons, kingfishers, woodpeckers, and passerine birds
Parulids: all wood warblers
Fringillids: grosbeaks, finches, sparrows, and buntings

Names for physiographic features follow in most cases those used in *Physiography of Eastern United States* (1938) and *Physiography of Western United States* (1931) by Nevin M. Finneman (New York: McGraw-Hill) Names of trees follow quite closely those in *Trees of North America* by C. Frank Brockman (New York: Golden Press, 1968).

Organization of the Chapters. Each chapter consists of an introduction followed by a series of bird-finding places.

The introduction presents the birdlife of the state with relation to its physiographical regions and associated habitats, to migration and to the winter season. After the lead paragraph or lead paragraphs, which usually point up one or more ornithological features of the state, comes a sequence of paragraphs describing briefly the regions and their habitats, with mention of characteristic breeding birds. Except in the introductions to Arizona, California, Kansas, Nevada, New Mexico, Texas, and Utah, two lists of breeding birds are used that require explanation. One list contains birds characteristic of open country or 'farmlands'—a general category for fields (cultivated or fallow), pastures, ranch lands, wet meadows, brushy lands (including forest edges and hedgerows), orchards, and dooryards (lawns, ornamental shrubbery, shade trees). The other list contains birds characteristic of wooded areas—sometimes called woodlands, wooded tracts, or wooded bottomlands—mainly of deciduous growth and usually at the approximate elevations of adjacent open country. Both lists, however, are incomplete. Some species are omitted because they are commonplace, others because their habitat preferences are either too specialized or too broad; still others because their presence in the state during the breeding season is scarce or questionable. Below are species consistently left out of such lists for one or more of the above reasons:

Red-tailed Hawk
Golden Eagle
Bald Eagle
Osprey
Prairie Falcon
Peregrine Falcon
American Kestrel
Ring-necked Pheasant
Chukar
Gray Partridge
Wild Turkey
Killdeer
American Woodcock
Rock Dove
Barn Owl
Great Horned Owl
Long-eared Owl
Short-eared Owl
Common Nighthawk
All swifts
All hummingbirds
Belted Kingfisher
Lewis' Woodpecker
Traill's Flycatcher
Hammond's Flycatcher
Violet-green Swallow
Bank Swallow
Rough-winged Swallow
Cliff Swallow
Purple Martin
American Crow
North American Dipper
Sedge Wren
Canyon Wren
Rock Wren
American Robin
Veery
Cedar Waxwing
European Starling
Black-and-white Warbler
Golden-winged Warbler
Blue-winged Warbler
House Sparrow
Yellow-headed Blackbird
Red-winged Blackbird
Brewer's Blackbird
Brown-headed Cowbird
Lark Sparrow

Following the paragraphs concerned with regions and their habitats are at least two paragraphs. One deals with the peculiarities of migration, pointing out the principal migration lanes (if any) through the state, and concludes with a timetable giving the dates within which one may expect the peak flights of waterfowl, shorebirds, and landbirds. The second paragraph deals with winter birdlife, emphasizing regular visitants from out of state.

The introduction concludes with the names of authorities who were sources of information in this edition. If there are any available publications of recent date that will be helpful to bird finders visiting the state, they are cited after the authorities.

The accounts of bird-finding places after the state introductions are usually presented under the nearest cities or towns that are indexed in the *Rand McNally Road Atlas*. If there is a publication pertinent to the birds in a place described, it is cited at the end of the account.

Index. There is one index incorporating both (1) bird-finding places and the states, cities, and towns under which they are described, and (2) the bird species mentioned in connection with the bird-finding places. Species mentioned in the introductions to chapters are indexed in a few instances.

Suggestions for Bird Finders

When Wanting to Know Locations for a Bird Species. Look up the name of the species in the Index, then consult the pages given. On each page the bird is mentioned in connection with a place. Reading this information will determine how to reach the bird from the city or town under which the place is given, what to expect as to terrain, vegetation, and other birds.

When Visiting a State. Consult the chapter devoted to the state, reading first the introduction, which gives the principal features of the state's birdlife. Then read the account of bird-finding places. A road map of the state and its index will readily reveal the cities or towns from which the places may be reached. Usually helpful is the official highway map issued by the state, as it shows local routes and gives information about state parks, campgrounds, picnic areas, and other recreational facilities.

When Visiting a National Park, or a National Monument. Consult the Index as to where either is described with directions for reaching it. At the entrance, or at headquarters, obtain a map and general information including interpretive services. If there is a visitor center, stop for supplementary information and a checklist of birds if available.

When Visiting a National Wildlife Refuge. The National Wildlife Refuges, whose boundaries and directional signs bear the figure of a flying goose, are federally owned and supervised by the Fish and Wildlife Service of the United States Department of the Interior. Because they embrace some of the finest places for birds, a great many of the Refuges having resident personnel and permitting access to the public are described in this book. Consult the Index. Directions for reaching the headquarters of each Refuge are usually included, with route directions from the nearest city or town.

Never visit a National Wildlife Refuge without first going to headquarters and making known the purpose of the visit. This is important. While bird finders are always welcome, there will doubtless be regulations and restricted areas that they should know about. Also, there may be available maps, a checklist of birds of the Refuge, and advice on the best vantages for viewing birds. Many such Refuges have self-guiding car tours and foot trails laid out especially for bird finding.

When Visiting Other Parks and Refuges, Sanctuaries, Nature Centers, and Wildlife Areas. This book describes, and gives directions to, a great number of parks and refuges owned and operated by states, counties, and municipalities. In many cases they are patterned after those of the Federal Government, and many have excellent interpretive centers, museums, self-guiding nature trails, and naturalists in residence. Included with directions is a considerable number of sanctuaries and nature centers, publicly owned or privately owned, with a resident staff and interpretive services. Make a point of consulting headquarters *first* when visiting any of them, but especially in the case of refuges, sanctuaries, and nature centers since there may be restrictions as to access as well as opportunities to be gained in learning precisely where to see certain birds. Also included with directions in this book are numerous state-owned-and-supervised areas—variously called wildlife areas, wildlife or game management areas, and wildlife research areas —that are sufficiently accessible for good bird viewing although entry may be prohibited in certain seasons. Some areas have resident personnel and/or self-guiding car tours; others do not. Regardless of this, when visiting any of the areas, respect all posted regulations.

When in Quest of Particular Birds. When looking for shorebirds on the mud flats and sand flats in tidal estuaries and bays, consult the tide schedules in local newspapers. At low tide the shorebirds feed on the flats, widely scattered and often far from shore. As the tide rises and comes in, the birds move closer to shore—and closer for viewing.

For observing birds of the open sea, consider taking 'pelagic trips' by boat offshore from the coasts of California, Oregon, and Washington. Some trips are regularly scheduled by bird organizations; others may be undertaken by private charter. All such trips are almost certain to yield a good variety of sea birds, many of which are rarely seen inshore. Although the trips are productive in any season, especially off California, they are generally not recommended off Oregon and Washington in winter when weather conditions are often severe. For information on trips from Oregon, contact the Audubon Society of Corvallis, Box 148, Corvallis, OR 97330, or the Audubon Society of Portland, 5151 Northwest Cornell Road, Portland, OR 97210. For information about trips from California and Washington, see the introduction to their respective chapters in this book.

Searching for owls after sundown can be facilitated by the use of a portable cassette recorder. Thus, when visiting woods where the sought-after owls reside, come equipped with a cassette on which their particular vocalizations are pre-recorded, and play them. If the owls are present, chances are that they will respond and possibly draw closer to the source.

The cassette recorder is useful also during the breeding season, for bringing small landbirds into view. On hearing the recorded songs of their species, males mistake them for those of rival males and consequently approach, revealing themselves. Pre-recorded songs help to turn up males that are not vocalizing at the time of the visit. If the males are already singing, although out of sight, and a view of them is desired, one may record the songs and play them back, to achieve the same result. *But heed this urgent warning:* Repeated broadcasts of the songs may be disruptive, causing the males to desert the area or distracting attention from their nests. Males as well as females of some species play a significant role in the incubation of the eggs and care of the young.

Such birds as pelicans, cormorants, many wading species, avocets, and gulls and terns gather in colonies for nesting. Because their aggregations are impressive, the locations of a few of the more accessible ones are given in this book. Bird finders must bear in mind, however, that a walk into any colony can drive the adults from their nests and seriously expose the eggs or young, or drive older young from their nests and subject them to starvation or to attacks by neighboring adults or by predators. Therefore, *observe these precautions:* Circle the colony but never go into it. Do not stay near the colony longer than a few minutes. If the air temperature is exceptionally cool or the sun's rays are uncomfortably hot, do not approach the colony at all. Instead, postpone the visit to a time when the weather is more suitable.

Miscellaneous Advice and Precautions. Although this book gives directions, where possible, to all areas described, highways are sometimes relocated and renumbered, and privately owned areas are 'developed'—to the exclusion of the desired birds. Failing to find the areas or their birds, contact local people knowledgeable about birds, in cities and larger towns through the chamber of commerce, in smaller communities through the postmaster.

Before entering any area fenced off or posted, determine the ownership and ask permission. Violation of property rights can mean the subsequent exclusion of all bird finders from choice areas that are highly productive.

Finally, always be alert for hazards and even personal danger. In all large metropolitan areas—in parks, preserves, anywhere —beware of muggers. Walk with another person, preferably with a group of people, but in any case, *never* alone! When exploring forests without well-marked trails, carry a good map and a compass and consult both before entering. When exploring any wild area, be informed as to the prevalence of venomous snakes and poisonous plants and be able to recognize them. Wear clothing that will serve as a protection from noxious insects.

Aids to Bird Finding

Birdwatcher's Guide to Wildlife Sanctuaries. By Jessie Kitching. New York: Arco Publishing Company, 1976. Information on 295 bird sanctuaries and refuges in the United States and Canada.

Guide to the National Wildlife Refuges. By Laura and William Riley. Garden City, N.Y.: Doubleday, 1979. Covers all the National Wildlife Refuges with information on how to reach them, where to stay or camp, and what to expect by way of wildlife including birds.

A Guide to North American Bird Clubs. Compiled and edited by Jon E. Rickert, Sr., Elizabethtown, Kentucky: Avian Publications, 1978. Assists in contacting local people for information about bird-finding opportunities.

Wilderness Areas of North America. By Ann and Myron Sutton.

New York: Funk & Wagnalls, 1974. Descriptions of more than 500 parks, refuges, and primitive areas from Panama to the Arctic.

Aids to Identification of Birds, North America

Birds of North America: A Guide to Field Identification. By Chandler S. Robbins, Bertel Bruun, and Herbert S. Zim. New York: Golden Press, 1966.

Aids to Identification of Birds, East of Rocky Mountains

Audubon Bird Guide: Eastern [Small] Land Birds, and *Audubon Water Bird Guide: Water, Game, and Large Land Birds.* By Richard H. Pough, Garden City, N.Y.: Doubleday, 1946 and 1951.

The Audubon Society Field Guide to North American Birds: Eastern Region. By John Bull and John Farrand, Jr. New York: Alfred A. Knopf, 1977.

A Field Guide to the Birds. By Roger Tory Peterson. 4th ed. Boston: Houghton Mifflin, 1980.

Aids to Identification of Birds, Rocky Mountains West

Audubon Western Bird Guide: Land, Water, and Game Birds. By Richard H. Pough. Garden City, N.Y.: Doubleday, 1957.

The Audubon Society Field Guide to North American Birds; Western Region. By Miklos D. F. Udvardy. New York: Alfred A. Knopf, 1977.

A Field Guide to Western Birds. By Roger Tory Peterson. 2nd ed. Boston: Houghton Mifflin, 1961.

A Field Guide to Western Bird Songs. Boston: Houghton Mifflin, 1962. Either three cassettes or three 12-inch phonograph records to accompany page by page the *Field Guide to Western Birds* by Peterson (*see above*). Available from the Cornell Laboratory of Ornithology, 159 Sapsucker Woods Road, Ithaca, NY 14850.

Information about Birds and Their Study

Bird finders wishing to be informed about birds, or to be in touch with others who enjoy finding, watching, or studying birds, should subscribe to *American Birds* and join to American Birding Association.

American Birds is a bimonthly journal published by the National Audubon Society (950 Third Avenue, New York, NY 10022). One issue each year reports the Christmas bird counts taken in all fifty states and the ten Canadian provinces. The other five issues contain, besides seasonal reports from both the states and provinces, descriptions of sites for bird finding, articles on the techniques of bird watching, photography, and sound-recording, and reports on various studies of birds.

The American Birding Association (Box 4335, Austin, TX 78765) meets every other year in a different area of North America where the birdlife is particularly enticing, and publishes six times a year the magazine *Birding*, replete with numerous subjects pertinent to bird finding in North America and throughout the world.

Two other publications that will interest bird finders are *Birding News Survey* (quarterly; Avian Publications, Box 310, Elizabethtown, KY 42701) and *Bird Watcher's Digest* (bimonthly; Box 110, Marietta, OH 45750).

Bird finders eager to increase their knowledge of birdlife generally can begin at home by taking the *Seminars in Ornithology* offered by the Cornell Laboratory of Ornithology. This is a college-level home-study course, readable in style and profusely illustrated, dealing with the way birds are structured and fly, how they live and behave from the time they are hatched, their migrations and how they determine direction, where they nest and how they rear their young. For further details on the course and how to enroll, write the Laboratory of Ornithology at 159 Sapsucker Woods Road, Ithaca, NY 14850.

A GUIDE TO BIRD FINDING

WEST OF THE MISSISSIPPI

Second Edition

Arizona

BY GALE MONSON

CACTUS WREN

The saguaro cactus is one of the nation's most familiar symbols. To many Americans, it is the mark of the desert. To the people of Arizona, it is the state flower. To the bird finder, it is a distinguished host to a unique variety of birds. In its fluted, spiky columns the Gilded Flicker and the Gila Woodpecker excavate their nest holes, which are later re-used by nesting American Kestrels, Elf Owls, Common Screech Owls, Wied's Crested Flycatchers, and Purple Martins; other birds, notably White-winged Doves, find a repast in the bright red fruit that succeeds the crowns of waxy white blossoms. This giant among cacti is found in the United States almost exclusively in southern Arizona. It is king among other desert growths, of which the most common are creosote bush, mesquite, catclaw, brittle bush, bur sage, cholla, and prickly pear. These plants, which make up the typical desert floral landscape, are spread over the southern and western parts of Arizona, the portions of the state that are the true desert—home of such birds as the Verdin, Cactus Wren, various thrashers, Black-tailed Gnatcatcher, and Black-throated Sparrow.

Rising above this desert floor in southeastern Arizona are several small but very rugged and wonderfully beautiful mountain chains, among them the Santa Catalinas, the Santa Ritas, the Huachucas and the Chiricahuas, and the Pinalenos. These mountains attain elevations of from 9,000 feet to more than 10,000 feet above sea level. They are clothed with evergreen oak, manzanita, juniper, and pinyon pine on the lower slopes, and with ponderosa pine, Douglas-fir, and other large trees toward the summits. The lower parts of their canyons are lined with sycamores and other deciduous trees. These mountains are as wonderful ornithologically as they are scenically, for here are found such striking birds as the Montezuma Quail, the large Blue-throated and Magnificent Hummingbirds, the Elegant Trogon, the Bridled Titmouse, the Sulphur-bellied Flycatcher, the Red-faced Warbler, and the Painted Redstart. There are many birds in these mountains and adjacent valleys that are found nowhere else in the United States.

Separating the low southern and western parts of Arizona from the higher Colorado Plateau country of the north is the Mogollon Rim—an irregular, discontinuous escarpment crossing the state in a northwesterly direction. Climbing up its face, the traveler leaves the hot desert and Mexican birdlife and meets a new climate, a new flora, and a new fauna. He now crosses a forest of mile upon mile of juniper and pinyon pine; the largest stand of ponderosa pine in North America; and stretches of dense aspen, Douglas-fir, and spruce. Elevations range from about 6,000 feet to well over 11,000 feet on Baldy Peak in the White Mountains, and to almost 13,000 feet on the San Francisco Peaks. Despite the forest, the country remains true to the desert in that water is scarce: only rarely is there a permanent trout stream; the lakes are few and far apart; canyon floors are usually dry. Many birds typical of the mountainous west breed here, including the Red-shafted Flicker, Lewis' and Northern Three-toed Woodpeckers, Gray and Steller's

Jays, American Robin, Audubon's Warbler, Western Tanager, Black-headed and Evening Grosbeaks, and even the Pine Grosbeak and Green-tailed Towhee. On the lakes are nesting Eared Grebes, ducks, Soras, and American Coots, and in some secluded canyons there are breeding Common Mergansers, Ospreys, North American Dippers, and Lazuli Buntings.

This lofty forested belt drops on the north to the grand mesas and colored buttes of the Navajo-Hopi section of the Colorado Plateau. Here are great grassy expanses, pygmy forests of juniper and pinyon, and sagebrush-filled valleys. West of this Navajo-Hopi land lies the Grand Canyon, cut between the cold Kaibab Forest and the state's loftiest mountains, the San Francisco Peaks, whose highest, Humphreys Peak, reaches 12,670 feet, the highest point in the state. Typical Navajo-Hopi birds are the Horned Lark, Sage Thrasher, Mountain Bluebird, and Vesper, Sage, and Brewer's Sparrows.

Arizona lies almost completely within the drainage basin of the Colorado River. Along it and its chief tributaries within the state—the Gila, Salt, Verde, Santa Cruz, San Pedro, and Little Colorado Rivers—lie blocks of irrigated farmlands, luxuriantly green and exceedingly fertile, in sharp contrast to the brown surrounding desert. Cottonwoods and eucalyptus, orange and grapefruit, date palms and pecans, olives and oleanders vie with alfalfa, cotton, melons, lettuce, sorghum, and small grains. In these low valleys sprawl the cities of Phoenix, Tucson, and Yuma. Huge dams have created Roosevelt Lake on the Salt River, San Carlos Lake on the Gila River, and Lakes Havasu, Mohave, and Mead on the Colorado.

Such a diversity of topographical features results in so great a variety of vegetation and climate that bird distribution and migrations are complex in the extreme. In ascending the slopes of any Arizona mountain range, less than a vertical mile will transport one as far ecologically as would hundreds of horizontal miles in the central and eastern states. The low, hot southern desert changes almost imperceptibly from east to west across the state, a rainfall of about a foot per year at Douglas diminishing to less than four inches at Yuma. Everywhere in the state are small tracts that are 'ecological islands,' differing in major aspects from the surrounding territory. To illustrate, one would think of the saguaro and juniper as occupying very different natural niches; yet there are 'islands' in

Arizona where they grow side by side. Despite all these variations and transitions, it is possible to separate the state into major natural vegetative associations, which largely determine the distribution of the birds and other animals.

The first of these is the *southern desert.* The basic plant is the creosote bush; associated with it in varying degrees of abundance are the saguaro, paloverde, ironwood, catclaw, brittle bush, bur sage, ocotillo, cholla, and prickly pear. The creosote bush association covers most of southern Arizona, excluding the higher mountain ranges, from the San Pedro and upper Gila Valleys west to the Colorado River and northwest to the Lake Mead area. Characteristic breeding birds are:

Red-tailed Hawk
American Kestrel
Gambel's Quail
Greater Roadrunner
White-winged Dove
Common Screech Owl
Great Horned Owl
Elf Owl
Lesser Nighthawk
Costa's Hummingbird
Gilded Flicker
Gila Woodpecker
Ladder-backed Woodpecker
Wied's Crested Flycatcher
Ash-throated Flycatcher
Purple Martin
Northern Raven (*near mountains*)
Verdin
Cactus Wren
Northern Mockingbird
Curve-billed Thrasher
Black-tailed Gnatcatcher
Phainopepla
Loggerhead Shrike
House Finch
Brown Towhee
Rufous-winged Sparrow (*local*)
Black-throated Sparrow

Much of the southeastern part of the state is *high grassland,* occupying the territory between the farms and brushlands of the valley bottoms and the bases of the mountain ranges. This association covers the greater part of the Sulphur Springs, upper Santa Cruz, San Pedro, and San Simon Valleys. There are scattered small tracts northwest to beyond Wickenburg, and west to the far side of the Baboquivari Mountains on the Papago Indian Reservation. Dominant plants are annual and perennial grasses, chiefly grama and tobosa. Years of grazing by cattle and horses have resulted in the replacement of the grasses in many areas by mesquite, yucca, and snakeweed. Typical breeding birds are:

Swainson's Hawk
Scaled Quail
Mourning Dove
Burrowing Owl
Poor-will (*rocky sites*)
Black-chinned Hummingbird
Ladder-backed Woodpecker
Western Kingbird
Ash-throated Flycatcher
Say's Phoebe
Horned Lark
White-necked Raven
Northern Mockingbird
Bendire's Thrasher
Loggerhead Shrike
Eastern Meadowlark
Scott's Oriole
House Finch
Grasshopper Sparrow (*local*)
Lark Sparrow
Botteri's Sparrow (*local*)
Cassin's Sparrow
Black-throated Sparrow

Flourishing along the *major watercourses* in southern Arizona are cottonwoods and willows, which grow with high thick brush—mainly mesquite, batamote or seep willow, arrowweed, and tamarisk. In places, mesquite is the principal plant, forming veritable thickets. This riparian association has been mostly cleared to make way for irrigated farmlands and is limited to relatively narrow strips along the streams, some of which are now intermittent or dry. It is characterized by the breeding birds listed below. Asterisks denote those species not residing along the Colorado and lower Gila Rivers.

Green Heron
Cooper's Hawk
* Gray Hawk
* Black Hawk
Gambel's Quail
White-winged Dove
Mourning Dove
Common Ground Dove
Yellow-billed Cuckoo
Common Screech Owl
Great Horned Owl
Elf Owl
Black-chinned Hummingbird
Gila Woodpecker
Ladder-backed Woodpecker
* Rose-throated Becard (*local*)
Western Kingbird
Cassin's Kingbird
* Thick-billed Kingbird (*local*)
Wied's Crested Flycatcher
Black Phoebe
Vermilion Flycatcher

Northern Beardless Flycatcher
Rough-winged Swallow
* Bewick's Wren
Northern Mockingbird
* Curve-billed Thrasher
Crissal Thrasher
Bell's Vireo
Lucy's Warbler
* Yellow Warbler
Yellow-breasted Chat
Hooded Oriole
Bullock's Oriole
Brown-headed Cowbird
* Bronzed Cowbird
Summer Tanager
* Northern Cardinal
* Pyrrhuloxia
Blue Grosbeak
* Varied Bunting (*edges*)
Lesser Goldfinch
Abert's Towhee
Song Sparrow

To the bird finder, perhaps the most alluring plant association is the *oak-pine woodland* on the lower slopes of the southeastern mountains. The key plants are evergreen oaks, of which there are a number of species, and Chihuahua pines. On the driest slopes is 'brushy' chaparral, consisting for the most part of manzanita, buckthorn, mountain mahogany, scrub oak, sumac, and cliff rose; on other slopes, grasses and agaves may be the dominant plants. In canyon bottoms, usually dry and rocky courses that carry water only during cloudbursts, sycamores, ashes, Arizona cypresses, and Arizona walnuts grow with the oaks and pines. In such settings, typical breeding birds, which include some of the rarest United States species, are:

Zone-tailed Hawk
Montezuma Quail
Band-tailed Pigeon
Common Screech Owl
Whiskered Screech Owl
Elf Owl
Whip-poor-will
Poor-will
Magnificent Hummingbird
Blue-throated Hummingbird
Broad-billed Hummingbird
Elegant Trogon (*rather local*)
Acorn Woodpecker
Brown-backed Woodpecker
Cassin's Kingbird
Sulphur-bellied Flycatcher (*local*)
Wied's Crested Flycatcher
Olivaceous Flycatcher
Western Pewee
Gray-breasted Jay
Bridled Titmouse
Bushtit
White-breasted Nuthatch

Bewick's Wren
Canyon Wren
Eastern Bluebird (*local*)
Blue-gray Gnatcatcher
Hutton's Vireo
Black-throated Gray Warbler
Painted Redstart
Hooded Oriole
Scott's Oriole
Hepatic Tanager
Black-headed Grosbeak
Rufous-sided Towhee
Black-chinned Sparrow (*local*)

On the higher parts of the southeastern mountains, over great areas of the Mogollon Rim country, along the rims of the Grand Canyon, on the San Francisco Peaks, and atop the great mesas of the Navajo-Hopi country, there stands a magnificent *ponderosa pine forest* with admixtures of Gambel oak, New Mexico locust, aspen, and Douglas-fir. Breeding birds of this association include the species listed below. Those marked with an asterisk reside in the more southern mountains only.

Northern Goshawk
Golden Eagle
Wild Turkey
Band-tailed Pigeon
Flammulated Screech Owl
Northern Pygmy Owl
Whip-poor-will
White-throated Swift
Broad-tailed Hummingbird
Red-shafted Flicker
Acorn Woodpecker
Hairy Woodpecker
Western Flycatcher
* Buff-breasted Flycatcher (*very local*)
Western Pewee
* Coues' Flycatcher
Violet-green Swallow
Steller's Jay
* Mexican Chickadee (*Chiricahuas only*)
Mountain Chickadee
* Brown Creeper
White-breasted Nuthatch
Pygmy Nuthatch
American Robin
Western Bluebird
Mountain Bluebird
Solitary Vireo
Virginia's Warbler
* Olive Warbler
Audubon's Warbler
Grace's Warbler
Red-faced Warbler
Western Tanager
Hepatic Tanager
Black-headed Grosbeak
Evening Grosbeak
Pine Siskin
Red Crossbill
* Mexican Junco
Chipping Sparrow

On the highest peaks of the state is *fir-spruce forest*—Douglas-fir, white fir, Engelmann spruce, and aspen, with thickets of willow and alder along streams—that supports another association of birds. Here is the ultimate contrast to the low, hot desert; and the birds are as different as those of the Minnesota north woods are from the birds of the Texas plains. The following are characteristic breeding species:

Blue Grouse (*White and Chuska Mountains*)
Red-shafted Flicker
Yellow-bellied Sapsucker
Williamson's Sapsucker
Northern Three-toed Woodpecker
Dusky Flycatcher
Western Flycatcher
Olive-sided Flycatcher
Violet-green Swallow
Gray Jay (*White Mountains*)
Steller's Jay
Clark's Nutcracker
Mountain Chickadee
Red-breasted Nuthatch
Brown Creeper
House Wren
American Robin
Hermit Thrush
Townsend's Solitaire
Golden-crowned Kinglet
Ruby-crowned Kinglet
Warbling Vireo
MacGillivray's Warbler (*mainly White Mountains*)
Cassin's Finch
Pine Grosbeak (*mainly White Mountains*)
Pine Siskin
Red Crossbill
Green-tailed Towhee
Gray-headed Junco
White-crowned Sparrow (*local*)
Lincoln's Sparrow (*in the north*)

Above the limits of timber on the San Francisco Peaks, Water Pipits are summer residents.

All of the Navajo-Hopi country, except that occupied by ponderosa pine and fir-spruce forests, can be divided into areas characterized by three separate types of growth: (1) *open, weedy grassland*—chiefly grama and galleta grasses and snakeweed; (2) *stretches of sagebrush, greasewood, and saltbush;* and (3) the *'pygmy forest' of pinyon pine and juniper.* In the grassland, the Poor-will, Horned Lark, and Western Meadowlark are virtually the only breeding birds. Birdlife in the sagebrush-greasewood-saltbush type is also scarce, typified by such species as the Mourning

Dove, Common Nighthawk, Say's Phoebe, Northern Mockingbird, Bendire's Thrasher, Sage Thrasher, Loggerhead Shrike, House Finch, Vesper Sparrow, Black-throated Sparrow, Sage Sparrow, and Brewer's Sparrow. In the pygmy forest, the following are the typical breeding birds:

Red-tailed Hawk
American Kestrel
Mourning Dove
Great Horned Owl
Common Nighthawk
Cassin's Kingbird
Ash-throated Flycatcher
Gray Flycatcher
Scrub Jay
Pinyon Jay
Plain Titmouse
Bushtit
Bewick's Wren
Rock Wren (*cliffy situations*)
Mountain Bluebird
Blue-gray Gnatcatcher
Solitary Vireo
Black-throated Gray Warbler
Rufous-sided Towhee
Chipping Sparrow

In sharp contrast to the rest of the state are places along the Colorado River with extensive growths of *cattail and bulrush* that provide an aquatic habitat. Here typical breeding birds are:

Western Grebe (*Lake Havasu*)
Pied-billed Grebe
Double-crested Cormorant
Great Blue Heron
Green Heron
Great Egret
Snowy Egret
Black-crowned Night Heron
Least Bittern
Clapper Rail
Common Gallinule
American Coot
Marsh Wren
Common Yellowthroat
Yellow-headed Blackbird (*local*)
Red-winged Blackbird
Song Sparrow

The bird finder will surely be bewildered by bird migration in Arizona. There are a few well-defined migratory pathways, such as the Colorado River, which are extensively used by waterbirds and landbirds alike. Elsewhere a jumble of mountains, valleys, canyons, mesas, and plains serve to break up migration routes and upset timetables. Some birds migrate only through the mountains; others migrate only through the valleys and plains. Species that migrate through both mountain and valley often do so at different times, and frequently are of different subspecies as well. A few species nest in the lowlands, then migrate to higher elevations for the rest of the summer season; others nest in the highlands, then immediately fly to the lowlands to linger before traveling south. The Audubon's Warbler is found in the Navajo-Hopi country only as a transient; in the high country only as a summer resident; in the lower parts of the state only as a winter resident.

Transient birds often tarry unusually late in Arizona, and the tenderfoot ornithologist is likely to class lingering migrants as summer residents or winter residents. The migrations of some species, such as the Black Phoebe and Phainopepla, are little known. Some species occasionally leave their home territory to make unusual flights; among these are the Lewis' Woodpecker, Steller's Jay, Clark's Nutcracker, and Red-breasted Nuthatch. These are mountain birds that are seldom seen in the lowlands, even during the winter. Some birds are invaders—the Lawrence's Goldfinch from California and the Wood Stork from Mexico. Along the Colorado River and lower Salt and Gila Valleys, ocean birds, probably from the Gulf of California, occasionally put in their appearance.

Arizona's climate being relatively snow-free, many birds come to spend the winter, especially in the warmer parts of the state. Among these are Western Grebes, American White Pelicans, and Canada and Snow Geese; a variety of ducks; Rough-legged Hawks and Northern Harriers; Merlins; Sandhill Cranes, Soras, Common Snipes, Spotted and Least Sandpipers, and Greater Yellowlegs; Belted Kingfishers; Yellow-bellied Sapsuckers; and Cedar Waxwings. There are also large concentrations of Tree Swallows along the Colorado River; and often waves of Western and Mountain Bluebirds come to the lowlands, to feed on mistletoe berries and to hawk for insects over the fields. Flocks of Water Pipits come to the irrigated fields and pastures, and hordes of Yellow-headed Blackbirds, Red-winged Blackbirds, Brewer's Blackbirds, Great-tailed Grackles, and Brown-headed Cowbirds visit the stockyards and grainfields. There are large gatherings of Lark Buntings from the Great Plains; myriads of Horned Larks and McCown's and Chestnut-collared Longspurs go to the southeastern grasslands. And one must not overlook the cheerful flocks of juncos, and

Chipping, Brewer's, and White-crowned Sparrows that throng the low woodlands and brushlands during the winter months.

Authorities

H. C. Bryant, Harry L. Crockett, Steve Gallizioli, Fred Gibson, Joe T. Marshall, Allan R. Phillips, Shirley Spitler, Sally H. Spofford, Walter R. Spofford, Charles Wallmo, Janet Witzeman.

References

The Birds of Arizona. By Allan Phillips, Joe Marshall, and Gale Monson. Tucson: University of Arizona Press. 1964.

Birds of Maricopa County, Arizona. By Salome R. Demaree, Eleanor L. Radke, and Janet L. Witzeman. Phoenix: Maricopa Audubon Society. 1972.

Birds of Organ Pipe Cactus National Monument. By Richard A. Wilt. Globe, Arizona: Southwest Parks and Monuments Association. 1976.

Birds of the Grand Canyon Region. By Bryan T. Brown, Peter S. Bennett, Steven W. Carothers, Lois T. Haight, R. Roy Johnson, and Meribeth M. Riffey. Grand Canyon: Grand Canyon Natural History Association. 1978.

Birds in Southeastern Arizona. By William A. Davis and Stephen M. Russell. Tucson: Tucson Audubon Society. 1979.

COOLIDGE
Picacho Reservoir

This community southeast of Phoenix is best known to bird finders for **Picacho Reservoir,** a wetland about one square mile in extent, used as a holding area for excess runoff water by the San Carlos Irrigation District. To reach the Reservoir, drive about 8 miles south from Coolidge on State 87, then turn left (east) onto Selma Highway. In a short distance, turn right after crossing the railroad and follow a dirt road along a canal for about 1.0 mile. The Reservoir dike can now be seen. Approach it by one of several dirt tracks, rather rough but traversable, that lead to the top of the dike along the west side of the Reservoir. The roadway atop the dike is a single lane, so it is best not to drive on it, to avoid backing up long distances in case a District vehicle is met. Shallow shorebird areas are best seen by walking along the dike bordering the south side of the Reservoir.

Levels vary from year to year, but generally there is a considerable expanse of open water with bordering bulrushes and cattails, and shallow water areas especially good for shorebirds. Along the east side are extensive growths of rather impenetrable willow and tamarisk. Pied-billed Grebes, Black-crowned Night Herons, and Common Gallinules nest, and Great Blue Herons and Great Egrets are usually common. More unusual but to be expected in at least some years are the Roseate Spoonbill, Black-bellied Whistling Duck, and Clapper Rail. The Snowy Egret and Least Bittern are occasional.

A special feature is the tremendous numbers of wintering Yellow-headed Blackbirds that roost here from early November to March, sometimes between 500,000 and a million, mostly males. The daily flights in and out are a real spectacle. To witness it, be on hand well before sunrise or sunset on any particular day.

Among the common non-swimming, non-wading, and non-shorebird species are hordes of White-winged and Mourning Doves in summer, plus Common Ground Doves, Vermilion Flycatchers, Bendire's Thrashers, Great-tailed Grackles, Pyrrhuloxias, and Abert's Towhees. Sage Sparrows are common in winter.

DOUGLAS
Guadalupe Canyon | San Bernardino Ranch

Douglas is one of only two Arizona cities bordering Mexico, the other being Nogales. Two attractions for bird finders lie to the east of Douglas, **Guadalupe Canyon** and the **San Bernardino Ranch.** The former is privately owned, but generally accessible to bird finders, provided they observe the posted signs or obtain visiting permission from the owner in each case. Both sites are reached by turning east from US 80 in Douglas onto 15th Street, which becomes an improved county road, at first paved and soon gravel-surfaced. The two sites are about 32 and 18 miles from Douglas, respectively; signs indicate turnoffs.

Guadalupe Canyon is a normally dry watercourse in the extreme southeastern corner of Arizona and the extreme southwestern corner of New Mexico. The longer portion of it—about 3 miles—is in Arizona and flows into adjacent Mexico. The attraction for birds are large cottonwoods and sycamores. The Canyon is the only locality in the United States where a Buff-collared Nightjar may be found. Its presence, however, seems intermittent, so do not be sure of finding it. Other notable birds—and usually easily seen—include the Zone-tailed Hawk, Violet-crowned and Broad-billed Hummingbirds, Acorn Woodpecker, Thick-billed Kingbird, Olivaceous, Vermilion, and Northern Beardless Flycatchers, Bridled Titmouse, Crissal Thrasher, and Varied Bunting. Elegant Trogons and Rose-throated Becards, as well as Lucifer and Costa's Hummingbirds, have been seen.

For further bird finding in Guadalupe Canyon, *see under* **Lordsburg, New Mexico.**

San Bernardino Ranch is a historical property that also lies on the Mexican border. Ponds and trees at Ranch headquarters, fields and ditches, and trees and brush along Black Draw (also known as San Bernardino River) provide a variety of habitat. Among the unusual birds here are the Snowy Egret, White-faced Ibis, Crested Caracara, and Tropical Kingbird. Quite easily observed during the warmer months are White-winged and Common Ground Doves, Greater Roadrunner, Lesser Nighthawk, Wied's Crested and Vermilion Flycatchers, Barn Swallow, White-necked Raven, Bewick's Wren, Bell's Vireo, Lucy's and Yellow Warblers, Great-tailed Grackle, Bronzed Cowbird, Summer Tanager, Blue Grosbeak, and Lesser Goldfinch.

FLAGSTAFF

Oak Creek Canyon | Lower Lake Mary | Upper Lake Mary | Mormon Lake | Arizona Snow Bowl

This city in north-central Arizona is famed for the variety of its tourist attractions, among them the magnificent San Francisco Mountains towering 6,000 feet above it to the north, Walnut Canyon and Wupatki National Monuments (both Pueblo Indian ruins), and Sunset Crater National Monument.

At an elevation of 6,900 feet above sea level, Flagstaff is surrounded by extensive stands of ponderosa pine. In spring and summer, and to some extent in fall, the bird finder will have little difficulty seeing birds typical of the ponderosa pine association. The loud rattle of the Broad-tailed Hummingbird as it dashes from one flower patch to another soon becomes a familiar sound. Woodpeckers are represented by the Red-shafted Flicker and Acorn and Hairy Woodpeckers. Violet-green Swallows feed above the treetops. Big Steller's Jays troop through the trees, scolding the visitor raucously; flocks of Pinyon Jays are common. A spritely duo of bird species—the Mountain Chickadee and Pygmy Nuthatch—seems inseparable from the pine boughs. American Robins and Western Bluebirds are conspicuous residents. Each patch of low brush has its pair of Virginia's Warblers, with Grace's Warblers above them in the trees. The rich, rolling songs of Western Tanagers and Black-headed Grosbeaks are sometimes heard during the quiet summer days, the only bird voices to disturb the silence. During winter, the landscape is usually snowy, and comparatively few birds are about—mostly woodpeckers, jays, chickadees, nuthatches, and juncos.

South of Flagstaff, some 20 miles distant, is well-known **Oak Creek Canyon.** US 89A, which threads through the Canyon, drops sharply from an almost solid stand of ponderosa pine into a wooded wonderland of huge alders, willows, ashes, and sycamores. Through this rich growth rush the waters of rock-bedded Oak Creek; on either side rise high, rocky walls of marvelous red and yellow hues. Bright butterflies and wildflowers add their colors to the scene.

Oak Creek Canyon is remarkable ecologically, for here the flora and fauna of the ponderosa pine country and the desert regions meet and often intermix. At the north end of the Canyon, at Pine Flat, such birds as the Broad-tailed Hummingbird, Red-shafted Flicker, Hairy Woodpecker, Steller's Jay, Mountain Chickadee, Plain Titmouse, Pygmy Nuthatch, American Robin, Virginia's and Grace's Warblers, Painted Redstart, Western Tanager, and Black-headed Grosbeak are common. Only a few miles farther south, at Indian Gardens, are such desert birds as the Black-chinned Hummingbird, Wied's Crested Flycatcher, Black Phoebe, Yellow Warbler, Yellow-breasted Chat, Bullock's Oriole, Summer Tanager, Blue Grosbeak, and Song Sparrow—together with a goodly representation of the same species at Pine Flat. The number of ecological puzzles is well-nigh endless.

Band-tailed Pigeons are one of the more common birds of the Canyon, as are the White-throated Swifts and Violet-green Swallows coursing along its walls. The bird finder, by careful searching, will discover the Red-faced Warbler where there are pines and Douglas-firs. Two other birds that he should look especially for are the Lesser Black Hawk and North American Dipper.

Spring, summer, and fall are the best times to visit the Canyon, for then birds are most numerous and active. The winter bird population is relatively small, but it does include such birds as the Bridled Titmouse and North American Dipper.

Driving about 6 miles south and east of Flagstaff on State 209 brings the bird finder to **Lower Lake Mary,** and about 3 miles farther on, to 4-mile-long and narrow **Upper Lake Mary,** both set among pines. On these lakes, except in winter, are large concentrations of Eared and Pied-billed Grebes and ducks, including impressive numbers of Mallards, Common Pintails, Cinnamon Teal, Ring-necked Ducks, Canvasbacks, Lesser Scaups, Buffleheads, and Ruddy Ducks. In late summer and much of the fall, numerous shorebirds, gulls, and terns are present. American Crows can be expected. Another 5 miles farther south on State 209 comes large **Mormon Lake,** also rewarding for waterbirds, waterfowl, shorebirds, gulls, and terns. Bald Eagles are nearly always in view, except in summer. Under the rimrocks on the east side, look for Green-tailed Towhees. From the drive completely around the Lake, stop frequently to explore the surrounding ponderosa pines for such birds as the Grace's Warbler and Hepatic Tanager. During protracted droughts, all three lakes may recede substantially or even dry up completely.

Northwest of Flagstaff, leaving town on State 180 and turning off onto Forest Service Road 516 at Fort Valley, one continues on and up a winding road about 5 miles to the **Arizona Snow Bowl,** a resort open all year except in severe weather. The bird finder should take the mile-long ski lift up the west shoulder of Agassiz Peak, a high point at the timber line in the San Francisco Mountains. (Caution: This can be a cold ride in any season, so be prepared.) Here look for Clark's Nutcrackers, Evening Grosbeaks, and Cassin's Finches in any season, for hawks and Water Pipits in summer, and for Pine Grosbeaks and rosy finches as possibilities in winter.

Northeast of Flagstaff, along US 89 in winter, watch for Northern Shrikes sitting on fences that border 'parks' and fields.

FREDONIA
Pipe Spring National Monument

Fifteen miles west of this mile-high town near the Utah border, following State 389, one comes to **Pipe Spring National Monument,** a small enclave in the Kaibab Indian Reservation. Water from the spring has been used to create a green oasis of a few acres in a vast extent of desert. This draws birds in migration like a magnet draws iron filings, and there is no telling what vagrant species may appear. Acre for acre, Pipe Spring may produce more out-of-range birds than any other place in the state. Its special forte seems to be eastern warblers.

The oasis consists of silver and Lombardy poplars, ailanthus, willows, Siberian elms, cottonwoods, a plum-pear-peach orchard, gardens, and open pasture, with ponds and small canals. The best time for bird finding is in spring and fall when the Monument is crowded with transients.

For additional bird finding across the Utah border, *see under* **Kanab.**

GRAND CANYON
Grand Canyon National Park

Arizona's foremost tourist attraction, **Grand Canyon National Park,** has several kinds of birds commonly associated with it: the White-throated Swifts that dash and skitter along the Canyon's brink, the Violet-green Swallows that skim heedlessly into the abyss, and the Turkey Vultures and Northern Ravens that enjoy gliding on the air currents that rise from the depths. These the visitor will often see with the first glimpse into the mighty chasm.

Grand Canyon National Park is so big, covering over 800 square miles, and so varied in environment that it is virtually impossible for the bird finder to explore it with any degree of completeness. To visit the *South Rim* only, exit from I 40, 2 miles east of Williams, on State 64, which, after 59 miles north, ends directly on the South Rim at Grand Canyon village, Park headquarters, and visitor center. If the *North Rim* is the objective, turn south off US Alternate 89 onto State 67 at Jacob Lake, which is 51 miles west of the famous Navajo Bridge, or 30 miles east and south of Fredonia. It is then 45 miles, over the forested Kaibab Plateau, to Bright Angel Point on the North Rim. A trail connects Grand Canyon village and Bright Angel Point, crossing the Colorado River at the Canyon's bottom by means of a suspension bridge near Phantom Ranch. The visitor may hike down this trail to the bottom, or go by mule, in a day's time from the South Rim; and if he desires, he can make an overnight stop at Phantom Ranch. However, he must make reservations for the mule ride or the stay at Phantom Ranch generally far in advance.

The North Rim is roughly 1,000 feet higher than the South Rim at Grand Canyon village, from which it is separated by some 11 miles of space. At this elevation of over 8,000 feet the bird finder is in a forest chiefly of Engelmann spruce, white fir, Douglas-fir, and aspen, much like what he would encounter far to the north of Arizona. Ponderosa pine also stands in lower and drier spots. Birds characteristic of the fir-spruce forest association and common are the Wild Turkey, Williamson's Sapsucker, Olive-sided Flycatcher, Clark's Nutcracker, Red-breasted Nuthatch, Brown Creeper, Hermit Thrush, Mountain Bluebird, Townsend's Solitaire, Golden-crowned and Ruby-crowned Kinglets, Warbling Vireo, Virginia's and Audubon's Warblers, Western Tanager, Evening Grosbeak, Pine Siskin, Green-tailed Towhee, and Gray-headed Junco. Northern Ravens, Brewer's Blackbirds, and Cassin's Finches may be seen in the open 'parks.' The assiduous bird finder may observe here such rarities in Arizona as the Northern Goshawk, Blue Grouse, Downy and Northern Three-toed Woodpeckers, and in some years even the Pine Grosbeak.

The lower South Rim is forested by pinyon pine and juniper, with frequent patches of ponderosa pine. Among the common birds are the Band-tailed Pigeon, Broad-tailed Hummingbird, Red-shafted Flicker, Acorn and Hairy Woodpeckers, Gray Flycatcher, Western Pewee, Violet-green Swallow, Steller's and Scrub Jays, Mountain Chickadee, Bushtit, White-breasted and Pygmy Nuthatches, American Robin, Western Bluebird, Solitary Vireo, Black-throated Gray and Grace's Warblers, Western Tanager, Black-headed Grosbeak, Rufous-sided Towhee, and Chipping Sparrow. Less common are the Sharp-shinned Hawk, Say's Phoebe, Plain Titmouse, and Red Crossbill. The bird finder may catch a glimpse of that seldom-seen bird, the Flammulated Screech Owl.

A descent into the Canyon from either the North or South Rim is a marvelous experience for anyone. For the bird finder this is especially so, for he will be able to witness the change in birdlife as he goes from rim top to bottom, a classical example of zonation. A short distance below either rim, the ponderosa pine association gives way to pinyon pines and junipers—the pygmy forest of the Navajo-Hopi country—together with other trees and shrubs. Golden Eagles, Poor-wills, Scrub and Pinyon Jays, Bushtits, Blue-gray Gnatcatchers, House Finches, and Rufous-sided Towhees will be observed with the first drop in elevation. Pinyon Jays often become abundant. All of these birds are replaced, as the bottom is approached, by such species as the Black-chinned Hummingbird, Ash-throated Flycatcher, Say's Phoebe, Canyon and Rock Wrens, and Rufous-crowned, Black-throated, and Black-chinned Sparrows. In addition, along the Colorado River at the bottom, and along inflowing streams, there is a strange mixture of species: Belted Kingfisher, Black Phoebe, Traill's Flycatcher, North American Dipper, Bell's Vireo, Lucy's and Yellow Warblers, Yellow-breasted Chat, Hooded and Bullock's Orioles, Summer Tanager, Blue Grosbeak, Lazuli Bunting, Lesser Goldfinch, and Song Sparrow. The mesquite and willows at Phantom Ranch are in vivid contrast to the pines at the rims.

During winters, many of the birds that enliven the higher areas of the Canyon are gone. The hardy woodpeckers, jays, nutcrackers, chickadees, and nuthatches remain, however. Townsend's Solitaires and Evening Grosbeaks are common, and Oregon Juncos are abundant. In the Canyon's depths, winter birdlife includes species that frequent higher country in summer.

GREEN VALLEY
Madera Canyon | Santa Rita Range Reserve

At this community of retirement homes a little more than 20 miles south of downtown Tucson, proceed beyond the main intersection of I 19 to the next exit, known as Continental Exit but also signed 'Madera Canyon.' Turn east here, cross the usually dry Santa Cruz River with pecan groves on either side, pass through the hamlet of Continental, and follow a paved road and directional signs eastward and upward to **Madera Canyon,** the principal defile in the Santa Rita Mountains. The pine-oak woodland in Madera Canyon is succeeded on the higher slopes of the Santa Ritas by ponderosa pine and even aspen and Gambel oak.

Madera Canyon abounds with private homes and United States Forest Service picnic areas, but it also has large Arizona sycamores that attract many 'Mexican border' birds. From the picnic areas, the bird finder has opportunities to investigate the Canyon. His principal quarry is usually the Elegant Trogon and Sulphur-bellied Flycatcher; both, especially the latter, can be sighted without much difficulty from May to September at or near the picnic grounds at the end of the road. At the same time of year, by proceeding up trails beyond most of the picnic areas, the bird finder may see or hear a great variety of birds, including the Zone-tailed Hawk, Golden Eagle, Wild Turkey, Band-tailed Pigeon, Broad-tailed and Broad-billed Hummingbirds, Magnificent and Blue-throated Hummingbirds, Acorn and Brown-backed Woodpeckers, Cassin's Kingbird, Olivaceous Flycatcher, Steller's and Gray-breasted Jays, Bridled Titmouse, Brown Creeper, Hermit Thrush, Eastern Bluebird, Hutton's and Solitary Vireos, Black-throated Gray Warbler, Painted Redstart, Scott's Oriole, Hepatic Tanager, Black-headed Grosbeak, and Mexican Junco. Going up higher on the trails, the bird finder can add to his list the Hairy Woodpecker, Coues' Flycatcher, Grace's and Red-faced Warblers, and possibly the Red Crossbill. During evenings it is easy to hear, but not to see, Whiskered and Fammulated Screech Owls.

On the way to Madera Canyon, the road passes through the Forest Service's **Santa Rita Range Reserve,** which is a mixture, in varying degrees, of mesquite brushland and grassland. Here in the last two miles before the entrance to Madera Canyon, Cassin's Sparrows reside the year round, although they can be readily spotted only when they are singing—and this is almost entirely during the summer rainy season from late June to early September. Harder to spot in the same locality, when they are usually present in July and August, are singing Botteri's Sparrows. Rufous-winged Sparrows are also present. Among other species to watch for are the Gambel's Quail, White-winged Dove (summer), Ash-throated Flycatcher (not in winter), Verdin, Bewick's and Cactus Wrens, Bell's Vireo (summer), Lucy's Warbler (spring-summer), Northern Cardinal, Pyrrhuloxia, Blue Grosbeak (not in winter), Brown Towhee, and Black-throated Sparrow. In winter there are innumerable Lark Buntings and additional sparrows—the Vesper, Chipping, Brewer's, and White-crowned.

JOSEPH CITY
Cholla Lake Park | Little Colorado River

About 2.5 miles southeast of this town, off of I 40 in northeast-central Arizona, is **Cholla Lake Park.** Its principal feature is a sizable reservoir whose waters are used for a large, steam electricity-generating plant. A variety of waterfowl, including Common and Red-breasted Mergansers, is present the year round. In migration periods, look for Common Loons, American White Pelicans, and various gulls and terns. Great-tailed Grackles are summer residents in the area. In winter, expect to see one or more Bald Eagles and, probably, Northern Shrikes.

By crossing the railroad tracks that run along the south side of Cholla Lake, one can gain access to the nearby wooded **Little Colorado River.** Although normally dry or nearly so, at times it floods heavily. In and about the stunted cottonwoods with associated camel-thorn, rabbitbrush, tamarisk, sandbar willow, and

New Mexico forestiera, look for such breeding birds as Red-tailed Hawks, Hairy Woodpeckers, Western Kingbirds, Phainopeplas, Yellow Warblers, Bullock's Orioles, and Blue Grosbeaks, and for transient flycatchers, vireos, warblers, and sparrows.

KINGMAN

Davis Dam | Katherine Landing | Hualapai Mountain Park

From this northwestern Arizona community, at an elevation of 3,325 feet, drive northwest on US 93 for 5 miles, then west on State 68 for 26 miles over Union Pass to **Davis Dam,** a large barrier on the Colorado River forming Lake Mohave. In late fall and winter, the tail waters below the Dam attract a variety of birds, probably to feed on the fish and fish remains that come through the Dam. Among the ducks that appear, some are unusual for Arizona—e.g., Common Goldeneyes, Oldsquaws, Surf Scoters, Hooded Mergansers, and, in recent winters, at least one flock of Barrow's Goldeneyes. Among the gulls, chiefly California and Ring-billed, there are occasional species such as Bonaparte's, or even a Mew Gull. The best observation point is on the Nevada side of the Dam, reached by driving across it from the west, turning left on the first paved road, then left again into a recreational area for parking.

Four miles above Davis Dam via paved road and indicated by directional signs is **Katherine Landing,** a resort on the lower part of 50-mile-long Lake Mohave. By renting a boat at the resort, one can take a ride on the Lake, which can be rewarding for birds in season (caution: not on windy days). During early winter, expect to see both Common and Arctic Loons, thousands of Eared Grebes and a sprinkling of Horned and Western Grebes, and both Common and Red-breasted Mergansers. From midsummer to early fall, do not be surprised to see an occasional frigatebird, jaeger, or some other sea bird that has probably wandered upriver from the Gulf of California.

For nearby bird finding on the west side of the Colorado River, *see under* **Boulder City, Nevada.**

In the Hualapai Mountains, 14 miles southeast of Kingman by

paved road, is **Hualapai Mountain Park,** a pine-girt retreat from the surrounding desert. To reach it, exit south from I 40 in east Kingman. The road climbs rapidly southeastward to 6,200 feet in the Park.

The Hualapais are rugged and high—the only high mountains on the edge of the lower Colorado Valley—with stands of majestic ponderosa pine on the slopes, and some aspen, fir, and narrowleaf cottonwood in the deep, cold canyons. In the Park, which is in the northern and highest portion of the range, the characteristic trees are ponderosa pine and Gambel oak, with an understory of New Mexico locust, scrub oak, manzanita, and other low growth. Among the readily observed breeding birds are the Band-tailed Pigeon, Broad-tailed Hummingbird, Acorn and Hairy Woodpeckers, Western Pewee, Steller's Jay, White-breasted and Pygmy Nuthatches, House and Canyon Wrens, Solitary Vireo, Virginia's and Grace's Warblers, Painted Redstart, Hepatic Tanager, and Black-headed Grosbeak. If he is fortunate, the bird finder will see one or two Zone-tailed Hawks—but he must look sharply, for they are easily confused with the more numerous Turkey Vultures. Flammulated Screech Owls are common.

During May, and in late August and September, small flocks of migrating warblers are common. These include the Orange-crowned, Nashville, Audubon's, Black-throated Gray, Townsend's, Hermit, MacGillivray's, and Wilson's. Olive-sided and other flycatchers are common then, as are various transient hummingbirds, vireos, and Western Tanagers.

En route to and from the Park, the road passes through several miles of scrub-oak chaparral, nesting habitat for such birds as the Scrub Jay, Plain Titmouse, Bushtit, Bewick's Wren, Blue-gray Gnatcatcher, Gray Vireo, Scott's Oriole, Rufous-sided Towhee, and Black-chinned Sparrow. In October and later, the Fox Sparrow is a visitant.

MAMMOTH-WINKELMAN
San Pedro River Valley

These two towns lie in the **San Pedro River Valley** northeast of Tucson. Winkelman is actually at the confluence of the San Pedro with the Gila River. Mammoth is 21 miles upstream on the San Pedro. Perhaps the most important remaining stand of river-valley cottonwood in Arizona lines the first one to five miles of the San Pedro above Winkelman. The more common breeding birds here include the Red-tailed Hawk, Gambel's Quail, White-winged and Mourning Doves, Greater Roadrunner, Lesser Nighthawk, Black-chinned Hummingbird, Gila and Ladder-backed Woodpeckers, Cassin's Kingbird, Wied's Crested Flycatcher, Western Pewee, Bridled Titmouse, Bewick's Wren, Crissal Thrasher, Phainopepla, Bell's Vireo, Lucy's and Yellow Warblers, Yellow-breasted Chat, Hooded and Bullock's Orioles, Summer Tanager, Northern Cardinal, Abert's Towhee, and Song Sparrow.

Of special ornithological interest are nesting Mississippi Kites which first appeared here in 1970. To find them as well as other birds listed above: From State 77 at a point 15.5 miles north of Mammoth, or 5.5 miles southeast of Winkelman, turn off toward the San Pedro River, bearing to the right at first, then left, after passing a small church, to the River—a little more than 0.5 mile from the highway. The kites are present in the adjacent cottonwoods from April through September.

NOGALES

International Sewage Ponds | Kino Springs | Patagonia Lake State Park | Pena Blanca Lake

This border city, the gateway to wonderful bird-finding opportunities in Mexico, is justly acclaimed as one of the most rewarding localities for bird finding on the United States border. The outskirts of town, especially along Nogales Wash, which runs through the city from its sister city of the same name in Mexico, and among the hills that ensconce the town, afford looks at such birds as the Black Vulture, Gray Hawk, Montezuma Quail, Broad-billed Hummingbird, Cassin's Kingbird, Olivaceous Flycatcher, White-necked Raven, Bewick's Wren, Phainopepla, Lucy's Warbler, Hooded and Scott's Orioles, Great-tailed Grackle, Pyrrhuloxia, Blue Grosbeak, and Rufous-winged, Cassin's, and Botteri's Sparrows.

About 7 miles north of Nogales are the **International Sewage**

Ponds. To reach them, turn east at the Ruby (Pena Blanca Lake) Interchange onto I 19; then turn north onto the frontage road until a road is reached that turns right to cross a railroad. Take this road across the track to the entrance. Stop at the office for permission to enter, and be sure to observe the closing hours. The Ponds are almost invariably host to a wide variety of birds, particularly grebes, waterfowl, shorebirds, and gulls. Outside the fence, look for Vermilion Flycatchers and other birds.

About 5 miles northeast of Nogales the Santa Cruz River flows northwestward. At privately owned **Kino Springs,** a mile south on the west side of the River, a grove of elderberries, a spring-fed pond, and trees and weeds along the Santa Cruz are attractive to birds. Usually a long list may be obtained, including, according to season, the Olivaceous Cormorant, Black-crowned Night Heron, Black-bellied Whistling Duck, Green Kingfisher, Vermilion Flycatcher, Marsh Wren, Lucy's Warbler, Summer Tanager, Pyrrhuloxia, Lawrence's Goldfinch, Lark Bunting, and White-throated, Swamp, and Song Sparrows. Note: This is privately owned property and do not visit it, if doing so violates posted signs, or before securing the owner's permission.

By traveling northeast from Nogales on State 82 for 10 miles, one reaches the turnoff at left to **Patagonia Lake State Park** (4,000 acres). The Lake itself, an impoundment of Sonoita Creek, is visited by grebes, Double-crested and Olivaceous Cormorants, ducks including mergansers, Ospreys, phalaropes, and gulls. The most exciting areas for birds are along Sonoita Creek above and below the Lake. Reaching the area below the Lake involves a walk of 2 or 3 miles—including stone-hopping across a spillway—after leaving the car near a locked gate before the Park road makes its final descent to the lakeshore. The area above the Lake involves a somewhat shorter walk, along the south side of the Lake to what might be termed a delta, which necessitates some walking over rocky terrain. If the season is right, one may be rewarded by seeing numerous Arizona rarities—e.g., the Gray Hawk, Violet-crowned Hummingbird, Green Kingfisher, Rose-throated Becard, Thick-billed Kingbird, Northern Beardless Flycatcher, Rufous-backed Thrush, and Varied Bunting—plus other species such as the Yellow-billed Cuckoo, Broad-billed Hummingbird, Wied's Crested and Olivaceous Flycatchers, Black Phoebe, Vermilion Flycatcher, Gray-breasted Jay, Bridled Titmouse, Crissal Thrasher, Bell's Vireo, Lucy's Warbler, Yellow-breasted Chat, Hooded Oriole, Pyrrhuloxia, Northern Cardinal, Blue Grosbeak, and Brown Towhee.

Pena Blanca Lake is reached by exiting west from I 19, about 7 miles north of Nogales, and then continuing over a graveled road for about 8 miles. An artificial body of water in a canyon-like setting and within a unit of Coronado National Forest, Pena Blanca Lake often has herons and waterfowl. Surrounding trees and brush are worth searching for birds, particularly during migration periods.

ORGAN PIPE CACTUS NATIONAL MONUMENT

This unit of the National Park System, comprising 516 square miles, lies along the Mexican border 125 miles west of Tucson. If the visitor approaches from Tucson on State 86, he drives for miles through the center of the vast Papago Indian Reservation, where, in the vicinity of Sells, its capital, he may see Crested Caracaras along the highway. If he approaches from Gila Bend, his route south on State 85 will lead through the colorful town of Ajo with its enormous open-pit copper mine. Monument headquarters is on State 85 in the Monument, about 15 miles south of its junction with State 86 from Tucson.

The Monument is best known for its array of spectacular cacti, including particularly the long-armed organ pipe cactus, so named because of its fancied resemblance to rows of huge organ pipes. The saguaro is abundant. Along the International boundary grows the sinita, or whisker cactus. All the cacti provide an intriguing background for the birdlife of the area.

Coursing through the Monument are desert washes, lined by mesquite, ironwood, and paloverde. A good example is near headquarters. During the delightful winter months, as well as any other time of year, permanent-resident birds characteristic of the southern desert association can be expected here regularly. These include the Red-tailed Hawk, American Kestrel, Gambel's Quail,

PHAINOPEPLA

Mourning Dove, Greater Roadrunner, Common Screech and Great Horned Owls, Gilded Flicker, Gila and Ladder-backed Woodpeckers, Northern Raven, Verdin, Cactus Wren, Northern Mockingbird, Curve-billed and Crissal Thrashers, Black-tailed Gnatcatcher, Phainopepla, Loggerhead Shrike, House Finch, Brown Towhee, and Black-throated Sparrow. Among winter visitants to look for are the Say's Phoebe, House and Rock Wrens, Blue-gray Gnatcatcher, Ruby-crowned Kinglet, Audubon's Warbler, Green-tailed Towhee, Lark Bunting, Oregon Junco, and Brewer's and White-throated Sparrows. In spring and summer, the Turkey Vulture, White-winged Dove, Elf Owl, Poor-will, Lesser Nighthawk, Costa's Hummingbird, Wied's Crested and Ash-throated Flycatchers, Lucy's Warbler, Hooded Oriole, and Brown-headed Cowbird may be expected.

Recommended is a side trip to the locality in the southwestern part of the Monument that is known as Quitobaquito, reached by an unpaved road—perfectly passable except in wet weather—along the Mexican border for about 16 miles. The road leaves from the west side of State 85 shortly before it ends at Lukeville. Quitobaquito is an oasis with a spring-fed pond—a veritable oasis for birds, especially wading birds, waterbirds, waterfowl, and shorebirds, that drop in to feed and rest while on their migratory flights. One can practically depend on seeing Black Phoebes and Vermillion Flycatchers around the pond, and it would be unusual not to see one or more Arizona rarities. During this trip, always be alert for Black Vultures and Harris' Hawks.

PARKER

Colorado River Indian Reservation | Bill Williams Delta Unit of the Havasu National Wildlife Refuge

This town is at the northern end of the **Colorado River Indian Reservation,** a 40-mile-long mélange of irrigated fields, open desert, mesquite and tamarisk thickets, canals, small ponds, and the Arizona bank of the Colorado River with its associated marshes. To reach the Reservation, simply drive out of Parker to the southwest. During winter, along the road, the bird finder will be struck by the large number of Red-tailed Hawks, American Kestrels, Gambel's Quail, Mourning Doves, Greater Roadrunners, Say's Phoebes, Verdins, Phainopeplas, Western Meadowlarks, Red-winged and Brewer's Blackbirds, and sparrows—Savannah, Sage, White-crowned, and Lincoln's. Several hundred Sandhill Cranes are generally present, as well as large flocks of Killdeers and Water Pipits in moist fields. Often in winter there are flocks of American Robins, Western and Mountain Bluebirds, and American, Lesser, and Lawrence's Goldfinches. During the summer months, the bird finder can expect White-winged Doves, Lesser Nighthawks, Western Kingbirds, Yellow-breasted Chats, Bullock's Orioles, Blue Grosbeaks, and even Indigo Buntings.

About 15 miles up the Colorado River via State 95 from Parker is the **Bill Williams Delta Unit of the Havasu National Wildlife Refuge,** which is isolated from the main part of the Refuge (*see under* **Topock**) by some 25 miles. To reach it, turn off east on a dirt road that leaves State 95 shortly before it crosses the long bridge over the Bill Williams Arm of Lake Havasu. The area is recognizable by dense growths of large willows on the left along the Bill

Williams riverbed, with the best sections about 3 to 5 miles from State 95. Small streams and beaver ponds are in these willow groves. Possibly more species of birds appear here than in any similarly sized area elsewhere in Arizona. Among the possibilities, especially in migration periods and winter, are the Wood Stork, Wood Duck, Hooded Merganser, Red-shouldered and Zone-tailed Hawks, Merlin, Virginia Rail, Eastern Phoebe, Winter Wren, many warblers including eastern species, Indigo Bunting, and Swamp Sparrow.

PATAGONIA

Sonoita Creek Sanctuary | 'Roadside Rest Stop' | San Rafael Valley

The small town of Patagonia, situated scenically between the Santa Rita and Patagonia Mountains south of Tucson, is best known among bird finders for the **Sonoita Creek Sanctuary** of 312 acres, a property of The Nature Conservancy. To visit the Sanctuary, check first with the local custodian. The property is just south of the town, with road access on the west side of Sonoita Creek, which at Patagonia is dry but through the Sanctuary is a perennial stream. This water, plus the huge cottonwoods, are the principal attributes, although there is a spring-fed marsh together with productive areas of brush and tall grass.

Formerly a place where both Gray and Lesser Black Hawks nested, only the Gray now nests and its tenure may be limited because of heavy use of the Sanctuary by bird finders. However, many notable birds are permanent or summer residents, among them the Green Heron, Common Ground Dove, Yellow-billed Cuckoo, Broad-billed Hummingbird, Wied's Crested and Olivaceous Flycatchers, Black Phoebe, Vermilion and Northern Beardless Flycatchers, Bridled Titmouse, Eastern Bluebird, Phainopepla, Bell's Vireo, Lucy's Warbler, Yellow-breasted Chat, Summer Tanager, Blue Grosbeak, and Lesser Goldfinch. During migrations, warblers of several species are numerous; some eastern species among them have been identified. In winter, both the Gray and Dusky Flycatchers are frequent, as are the Gray-breasted Jay, Pyrrhuloxia, Lazuli Bunting, Green-tailed Towhee, and even the White-throated and Fox Sparrows. Sometimes in winter one can find the Green Kingfisher, Williamson's Sapsucker, Winter Wren, Townsend's Warbler, and Hepatic Tanager. The Sanctuary is one of those places where the unexpected is often seen.

About 4 miles southwest of Patagonia on State 82 is the '**Roadside Rest Stop,**' famed among local bird finders for the number of unusual birds that can be, and have been, seen in its vicinity. Although the creek bottom is closed to entry by the owners, in trees along the highway, or by observing from the edges of the highway, it is possible to spot the Rose-throated Becard, Thick-billed Kingbird, and Varied Bunting. This is the locality where the Yellow-green Vireo and Yellow Grosbeak have been identified.

Patagonia is the point of departure for the **San Rafael Valley.** Leave town on the unsurfaced Harshaw Canyon Road. This winds eastward up Harshaw Canyon through oak woodland. Along the way, be alert for the Zone-tailed Hawk, Montezuma Quail, Gray-breasted Jay, and Eastern Bluebird. After about 10 miles, the Road tops out, overlooking the grass-covered swells of the San Rafael Valley. Even though the Valley is largely in private ownership, many birds can be seen from the roads that lead across and down the Valley. Possibilities in summer are the Swainson's Hawk, Scaled Quail, Horned Lark, White-necked Raven, Eastern Meadowlark, and Grasshopper and Cassin's Sparrows; in winter, the Ferruginous and Rough-legged Hawks, Prairie Falcon, Short-eared Owl, Sprague's Pipit, Baird's Sparrow, and McCown's Longspur, besides large numbers of Lark Buntings, Savannah and Vesper Sparrows, and Chestnut-collared Longspurs.

PHOENIX

Encanto Park | Papago Park-Phoenix Zoo | Granite Reef Dam | Verde River Valley

A large, spreading city is generally a sterile place for bird watching, but Phoenix has a number of parks and ponds, with residential sections rich in trees and shrubs, and cultivated farmlands and old fields on its outskirts. The bird finder walking about town will see

White-winged and Inca Doves, Black-chinned Hummingbirds, Gila Woodpeckers, Western Kingbirds, Cactus Wrens, Northern Mockingbirds, Curve-billed Thrashers, Great-tailed Grackles, Northern Cardinals, and House Finches.

Of the Parks, **Encanto Park** (about 300 acres) at 15th Avenue and Encanto Boulevard is centrally located, well-planted, and expansive. Its ponds bring in several species of ducks. In winter it is possible to see Anna's Hummingbirds, Mountain Chickadees, all three nuthatch species, Brown Creepers, Golden-crowned Kinglets, and Red Crossbills, most of them attracted by Scotch pines and other conifers. Crossbills have even been known to breed here. **Papago Park–Phoenix Zoo,** between McDowell Road and Van Buren Avenue at 64th Street, has many attractive ponds where zoo-provided food is consumed by wild waterfowl, especially during winter.

The municipal sewage ponds and nearby Salt River ponds, between 27th and 35th Avenues, have for many years been Phoenix's number one bird-finding site, yielding a great variety of water species, from loons all the way through gulls and terns, including many rarities. Access is controlled by the city; inquiry as to entry should be made at the treatment plant's headquarters, Durango Street and 23rd Avenue. Another treatment plant with accompanying ponds is on 91st Avenue, 0.5 mile south on Broadway Road, along the Salt River several miles southwest of the city limits. This facility is generally open to the public, but a stop at headquarters office is advised to state the purpose of being on the premises. Along 91st Avenue, on its west side just north of Baseline Road, along the south side of an effluent channel, is a good place to search successfully for Crissal Thrashers.

Recently a considerable volume of treated sewage effluent has flowed down the old Salt River bed to beyond 115th Avenue, several miles southwest of the Phoenix city limits and at the point where the Salt and Gila Rivers join. The effluent has produced an extensive growth of willow, cottonwood, cane, cattail, and dock, which, with adjoining mesquite brushland and farmland, has created a varied habitat for birds. One can park his car at the point where 115th Avenue crosses the Gila River and follow the effluent stream eastward. The area is particularly good for warblers, especially from September through December. Other birds to see are Cinnamon Teal, many Common Gallinules, Yellow-billed Cuckoos, Crissal Thrashers, Marsh Wrens (winter), Common Yellowthroats, Yellow-breasted Chats, Blue Grosbeaks, Abert's Towhees, and a desert subspecies of Song Sparrow. Arizona being such a water-deficient state, it is likely that the effluent creating this linear oasis will be appropriated at some future time for irrigation or some other purpose.

Granite Reef Dam, and vicinity, is excellent for bird finding. To reach it, drive 9 miles east from Mesa on US 60-80, then turn off left onto the Bush Highway for 6 miles. The Dam is on the Salt River below its confluence with the Verde. Cattail habitat both above and below this irrigation diversion dam is good for herons—including Least Bitterns—as well as Virginia Rails, Soras, and Marsh Wrens. East from the Dam toward Saguaro Lake are picnic grounds along the Salt River that afford good bird finding, notably at Coon Bluff and Blue Point. Many unusual riparian birds of the desert country may be expected.

The **Verde River Valley** east of Phoenix, on the Fort McDowell Indian Reservation, offers good bird finding in cottonwood and mesquite habitats. It is most conveniently reached by taking State 87 (the Beeline Highway) from downtown Phoenix for about 25 miles. Instead of crossing the River, take the roads that lead off north and south parallel to the River and the side roads that go down to the River itself. Irrigated fields and brush here add to the overall bird habitat. Thus the list of birds one will see runs the gamut from Great Blue and Green Herons to Bald Eagles, Ladder-backed Woodpeckers, Vermilion Flycatchers, Bridled Titmice, five wren species, Phainopeplas, Lucy's Warblers, Hooded and Bullock's Orioles, to Summer Tanagers. In winter, this is a reliable area for Gray Flycatchers.

PORTAL

Cave Creek Canyon | Barfoot Park | Rustler Park

This hamlet on the east side of the Chiricahua Mountains, about 60 miles northeast of Douglas on the Mexican border, lies at the east

ELEGANT TROGON

entrance to **Cave Creek Canyon,** one of the principal objectives of bird finders in Arizona. Hummingbird feeders at the American Museum of Natural History's Research Station, up the Canyon 5 miles from Portal, are a main attraction. These feeders and temporary ones at campgrounds bring in rare hummers—e.g., Lucifer, Violet-crowned, Berylline, and White-eared—as well as the more usual Black-chinned, Broad-tailed, Rufous, Calliope, Magnificent, Blue-throated, and Broad-billed species.

All through the South Fork of Cave Creek Canyon, Elegant Trogons reside in summer. A favorite spot for seeing them is the vicinity of the United States Forest campground. But heed the Forest Service's request: Do not use tape recorders to attract the trogons, as overuse disrupts their normal behavior. Above the campground, one may sometimes find the Mexican Chickadee, in addition to the Montezuma Quail, Band-tailed Pigeon, Whiskered Screech and Elf Owls, Blue-throated Hummingbird, Acorn and Brown-backed Woodpeckers, Sulphur-bellied, Olivaceous and Coues' Flycatchers, Gray-breasted Jay, Bridled Titmouse, Bushtit, Brown Creeper, Townsend's Solitaire (not in summer), Hutton's Vireo, Black-throated Gray Warbler, Painted Redstart, Hepatic

Tanager, and Mexican Junco. Spotted Owls are more or less regular both in the South Fork and the main canyon below South Fork. Both Flammulated Screech Owls and Northern Pygmy Owls breed in the Canyon. Above the Research Station in the vicinities of John Hands and Herb Martyr 'Lakes' (actually small ponds), North American Dippers have been recorded.

Driving past the Research Station and keeping to the main road, the bird finder can reach the scenic main ridge of the Chiricahuas at Onion Saddle. South from Onion Saddle a 2-mile drive brings him to a T intersection where he may turn left to **Barfoot Park** or right to **Rustler Park.** Both places are rather small openings in the dark coniferous forest. Here he may locate such birds as the Wild Turkey, Band-tailed Pigeon, Northern Pygmy Owl, Broad-tailed Hummingbird, Red-shafted Flicker, Hairy Woodpecker, Western and Coues' Flycatchers, Violet-green Swallow, Steller's Jay, Mexican Chickadee, Red-breasted and Pygmy Nuthatches, Brown-throated Wren, American Robin, Hermit Thrush, Western Bluebird, Golden-crowned Kinglet, Virginia's, Audubon's, Grace's, and Red-faced Warblers, Western Tanager, Black-headed Grosbeak, and Mexican Junco. Evening Grosbeaks and Red Crossbills are possibilities. Townsend's and Hermit Warblers are common in migration.

During winter, in the open country along US 80 toward Douglas from Portal, the bird finder should be alert for Ferruginous and Rough-legged Hawks, Golden Eagles, Prairie Falcons, Lark Buntings, and Chestnut-collared Longspurs, in addition to the more usual birds such as Northern Harriers, American Kestrels, Greater Roadrunners, Horned Larks, Northern and White-necked Ravens, Loggerhead Shrikes, Eastern Meadowlarks, and sparrows—Savannah, Vesper, Lark, Brewer's, and White-crowned. In summer, watch for Burrowing Owls, Lesser Nighthawks, Bendire's Thrashers, and Cassin's Sparrows.

SAFFORD

Gila Valley | Pinaleno Mountains

Lying in the upper **Gila Valley** in eastern Arizona, Safford is the center of a verdant irrigated district, with the high **Pinaleno**

Mountains to the south and the lower Gila Mountains to the north. The Valley's many trees—cottonwood, ash, mulberry, umbrella, and others—offer abundant cover to White-winged, Mourning, and Inca Doves, Yellow-billed Cuckoos, Black-chinned Hummingbirds, Gila and Ladder-backed Woodpeckers, Western Kingbirds, Northern Mockingbirds, Phainopeplas, Yellow Warblers, Hooded and Bullock's Orioles, Great-tailed Grackles, Summer Tanagers, Blue Grosbeaks, House Finches, and Lesser Goldfinches. In the cotton and alfalfa fields, Western Meadowlarks and Red-winged Blackbirds abound. In and along thickets near the Gila River, and on the bordering desert, are Gambel's Quail, Greater Roadrunners, Vermilion Flycatchers, Verdins, Curve-billed Thrashers, Black-tailed Gnatcatchers, Lucy's Warblers, Northern Cardinals, and Brown Towhees. Limited to the vicinity of the River and to wet areas south of town are Yellow-breasted Chats, Abert's Towhees, and Song Sparrows. All the birds mentioned above are nesting species, many of them year-round residents. During winter, large flocks of blackbirds, including Yellow-headed Blackbirds, are a feature of birdlife, as are numerous White-crowned Sparrows and, in some years, large numbers of American Robins and Western and Mountain Bluebirds.

The Pinaleno Mountains rise rapidly from the Valley floor's 3,000-foot elevation to almost 11,000 feet. Their upper parts are accessible by car from Safford. Turn south onto US 666 in Safford; about 8 miles from town, turn right onto a paved road, known as the Swift Trail, that winds up the east face of the Mountains past a summer resort called Pinecrest, or Turkey Flat. The road soon 'tops out' in Ladybug Saddle and loops along the mountain top for several more miles. The vistas of the desert below from this road are among the finest in the state. To the bird finder, however, the most impressive aspects are the changes in vegetation and climate, with the corresponding changes in birdlife, from the foot of the Mountains to their summits. Just a few road miles transport him from the hot, cactus-dotted, shrubby desert up through the oak-juniper belt into the tall ponderosa pines, and finally to the cold top elevations where there is one of the most extensive stands of subalpine fir and Engelmann spruce in Arizona, with occasional wet grassy patches and aspen glades.

During this trip the bird finder will see or hear: *In the desert lowland* at the foot of the Swift Trail, the Gambel's Quail, Ash-throated Flycatcher, Verdin, Cactus Wren, Black-tailed Gnat-catcher, and Black-throated Sparrow. *In the oak-juniper belt* (with tall sycamores along the canyons), the Zone-tailed Hawk, Brown-backed Woodpecker, Cassin's Kingbird, Olivaceous Flycatcher, Scrub and Gray-breasted Jays, Bridled Titmouse, Bushtit, Bewick's Wren, Blue-gray Gnatcatcher, Scott's Oriole, and Rufous-crowned Sparrow. *Among the pines,* the Band-tailed Pigeon, Broad-tailed Hummingbird, Red-shafted Flicker, Hairy Woodpecker, Coues' Flycatcher, Violet-green Swallow, Steller's Jay, Mountain Chickadee, White-breasted and Pygmy Nuthatches, American Robin, Solitary Vireo, Olive, Grace's, and Red-faced Warblers, Painted Redstart, Western and Hepatic Tanagers, Black-headed Grosbeak, and Mexican Junco. *In the firs and spruces,* the Western Flycatcher, Red-breasted Nuthatch, Brown Creeper, Hermit Thrush, and Golden-crowned Kinglet.

SHOW LOW

Fool Hollow Lake | Show Low Lake | Rainbow Lake | White Mountain Lake | Silver Creek

This town in eastern Arizona at an elevation of 6,400 feet is on the north edge of the great ponderosa pine forest that extends along the Mogollon Rim. Northward from the town are broad juniper-pinyon pine tracts that give way after a few miles to a countryside characterized by rolling and for the most part overgrazed desert grasslands. There are many access points for the bird finder to investigate either the ponderosa pine or juniper-pinyon pine association.

The best bird-finding sites around Show Low are the several lakes, usually man-made. Just northwest of town off State 260 is **Fool Hollow Lake,** overused by visitors but nevertheless a possible site for unusual waterbirds and waterfowl. The same is true of **Show Low Lake,** between 3 and 4 miles south of town off State 260. After another 3 to 4 miles along State 260 is the town of Lakeside with its adjacent **Rainbow Lake.** Access to the Lake is limited to its north end. Waterfowl here are sometimes abundant, as are swallows. Lakeside itself is a 'sure-fire' site for Lewis' Woodpeckers; they are prominent.

Excellent for waterbirds, waterfowl, and landbirds is **White Mountain Lake,** reached by driving out of Show Low on State 77 for 7 miles, then turning right and going 3 miles. The Lake is accessible at various points, and generally has a wide assortment of birds, including Double-crested Cormorants, Snowy Egrets, ducks, shorebirds, and swallows, plus—in the bordering trees and shrubs—Common Nighthawks, Black-chinned Hummingbirds, Cassin's Kingbirds, Black Phoebes, Scrub and Pinyon Jays, Bendire's Thrashers, Mountain Bluebirds, Yellow-breasted Chats, Bullock's Orioles, Blue Grosbeaks, and Lark Sparrows. Southeast of the Lake, along **Silver Creek,** are nesting Cinnamon Teal, Ruddy Ducks, and Yellow-headed Blackbirds.

SIERRA VISTA

Huachuca Mountains | Ramsey Canyon Preserve | San Pedro River Valley

A recent addition to the towns of Arizona, Sierra Vista in the southeastern part of the state is the result of the greatly expanded activities, beginning in the 1950s, of the old border military post, Fort Huachuca. The grasslands surrounding Sierra Vista are some of the finest in the state. The bird finder will delight in looking for Scaled Quail, Horned Larks, White-necked Ravens, Eastern Meadowlarks, and Grasshopper, Cassin's, and Botteri's Sparrows—the last two sing freely, and therefore are easily sighted, after the summer rains commence in June.

Sierra Vista is the gateway to the **Huachuca Mountains,** a bird wonderland if there ever was one. Take State 92 south from town to reach a number of canyons—first, Ramsey Canyon, then Carr, Miller, and Ash. Another canyon for productive bird finding is Garden Canyon, but it is on the Fort Huachuca Military Reservation. Fort Huachuca is an 'open post,' except when one or more parts are closed because some military exercise is in progress. The other canyons mentioned above are in the Coronado National Forest and thus open to the public, excepting when they may be closed due to fire hazards and, in the case of Ramsey Canyon, where The Nature Conservancy controls access and at times restricts the number of visitors. All of these canyons are famous in ornithological history.

To see Mexican species, it is best to be on hand in late spring or early summer. The dominant vegetation along these canyons is evergreen oak and sycamore, with chaparral of varying composition on the slopes; the upper, more inaccessible portions of the canyons have Douglas-firs, pines, and maples. The list of birds that reside in the canyons is a long one that includes the Montezuma Quail, Whiskered Screech, Northern Pygmy, and Spotted Owls, (Stephens') Whip-poor-will, Magnificent and Blue-throated Hummingbirds, Elegant Trogon, Brown-backed Woodpecker, Sulphur-bellied, Olivaceous, and Buff-breasted Flycatchers, Gray-breasted Jay, Bridled Titmouse, Hutton's Vireo, Olive Warbler, Painted Redstart, Hepatic Tanager, and Mexican Junco.

Ramsey Canyon is the locale of the renowned 'Hummingbird Capital,' also known as Mile Hi Ranch, but now officially the 300-acre **Ramsey Canyon Preserve** of The Nature Conservancy. Here no less than fourteen species of hummingbirds have been recorded, among them the Lucifer, Berylline, and White-eared. Numerous feeders placed about the premises attract the hummers and other birds, especially Scott's Orioles (mainly in spring) and Black-headed Grosbeaks. The Yellow Grosbeak has been observed here.

Lying to the northeast, east, and southeast of Sierra Vista, accessible by State 82, 90, and 92, is the **San Pedro River Valley.** The land here is all privately owned and must not be entered without the individual owner's permission. Verdant farms occupy much of the Valley. Large cottonwoods and willows and good growths of mesquite stand in uncultivated sections and along the immediate banks of the River. Birds of the major watercourse association are plentiful, notably the White-winged and Common Ground Doves, Yellow-billed Cuckoo, Gila Woodpecker, Cassin's Kingbird, Vermilion Flycatcher, Bewick's Wren, Northern Mockingbird, Crissal Thrasher, Lucy's and Yellow Warblers, Yellow-breasted Chat, Eastern Meadowlark, Bullock's Oriole, Summer Tanager, Blue Grosbeak, Lesser Goldfinch, and Abert's Towhee.

SONOITA
Sonoita Plains | Canelo Hills Cienega | Parker Canyon Lake | Gardner Canyon

Little more than a crossroads, but likely to be 'developed,' Sonoita sits on the **Sonoita Plains** about 50 miles southeast of Tucson, at an elevation of 4,900 feet. The Plains are high grassland of about 150 square miles, an area unique in Arizona except for the nearby San Rafael Valley to the south (*see under* **Patagonia**). The grassland is relatively uninvaded by shrubs and weeds, probably due to a combination of favorable soil and moisture factors with good range management. As in the San Rafael Valley, the land is almost entirely in private hands but many birds can be seen from the roads that radiate from Sonoita. Should the bird finder want to explore away from the roads, he must check with the nearest ranch headquarters.

Being grassland and unique for Arizona, the Sonoita Plains are expectedly occupied by a noteworthy assortment of birds. Common breeding species include the Swainson's Hawk, Scaled Quail, Horned Lark, Eastern Meadowlark, and Grasshopper, Lark, and Cassin's Sparrows. Where there is brush, or cut banks along drainageways, look for the American Kestrel, Mourning Dove, Western and Cassin's Kingbirds, Say's Phoebe, Northern Mockingbird, Loggerhead Shrike, Blue Grosbeak, and Brown Towhee. Winter bird finding on the Plains is most rewarding, for then permanent residents such as the Eastern Meadowlark and Grasshopper Sparrow are joined by large numbers of Lark Buntings, Savannah, Vesper, and Brewer's Sparrows, and Chestnut-collared Longspurs, with fewer numbers of Sprague's Pipits, Baird's Sparrows, and McCown's Longspurs, and a sprinkling of Ferruginous and Rough-legged Hawks, Prairie Falcons, and Short-eared Owls.

By taking the Parker Canyon Lake Road south and east from Sonoita, the bird finder can visit the **Canelo Hills Cienega**, a 165-acre project of The Nature Conservancy. But he must telephone ahead to let the custodian know that he is coming, preferably giving him a day's notice, as well as to secure directions. The Cienega (a Mexican word for 'marsh') is a delightful spot, with a marsh-swamp of sedges, grasses, horsetails, and willows, and large cottonwoods and willows along the stream below the wet area. Here look for the Northern Goshawk, White-winged Dove, Yellow-billed Cuckoo, Common Nighthawk, Black-chinned Hummingbird, Acorn and Ladder-backed Woodpeckers, Cassin's Kingbird, Olivaceous Flycatcher, Black Phoebe, Western Pewee, Gray-breasted Jay, Bewick's Wren, Lucy's Warbler, Common Yellowthroat, Yellow-breasted Chat, Summer Tanager, Black-headed and Blue Grosbeaks, Lesser Goldfinch, and Rufous-crowned Sparrow.

Other good sites for birds in the Sonoita area are **Parker Canyon Lake,** farther along the same road to Canelo Hills Cienega and about 25 miles from Sonoita (explore the area below the dam, if time permits), and **Gardner Canyon** on State 83, 4 miles north of Sonoita—a good spot for Botteri's Sparrow in July and August.

SPRINGERVILLE-EAGAR

'Big Lake Country' | Baldy Peak | Nutrioso, Alpine, Greer | Lyman State Park

These neighboring towns in central-eastern Arizona near the New Mexico line are jumping-off places for numerous rewarding areas for birds. The towns lie about 7,000 feet above sea level, hence really good bird finding does not hold up in the winter months when many access roads are too deep in snow for travel.

About 25 miles to the southwest and 2,000 feet higher in the White Mountains is what is known as the **'Big Lake Country.'** It is open and rolling, with conifer- and- aspen-clad hills or buttes here and there and frequent lakes, ponds, and cienegas (marshes). Swainson's Hawks, Horned Larks, Mountain Bluebirds, Eastern Meadowlarks, Brewer's Blackbirds, and Savannah and Vesper Sparrows inhabit the grasslands. The pines, spruces, firs, and aspens may yield such 'un-Arizona' birds as the Wild Turkey, Saw-whet Owl, Northern Three-toed Woodpecker, Gray Jay, and Pine Grosbeak. Two foot trails at Sheep Crossing and Phelps Cabin lead up 11,590-foot **Baldy Peak** in the White Mountains from State 273. Along the trails, look for Northern Goshawks, Blue Grouse, Gray Jays, Golden-crowned Kinglets, Pine Grosbeaks, Cassin's Finches,

Red Crossbills, and, at the summit in summer, Water Pipits and White-crowned Sparrows.

The country south and west of Springerville and Eagar has several canyons draining into the Little Colorado River. All are scenic and good for bird finding; all are accessible by United States Forest Service roads. Typical birds include the Broad-tailed Hummingbird, Red-shafted Flicker, Williamson's Sapsucker, Hairy Woodpecker, Western Flycatcher, Western Pewee, Violet-green Swallow, Steller's Jay, Mountain Chickadee, Red-breasted and Pygmy Nuthatches, American Robin, Hermit Thrush, Western Bluebird, Townsend's Solitaire, Ruby-crowned Kinglet, Solitary and Warbling Vireos, Audubon's, Grace's, and Red-faced Warblers, Black-headed Grosbeak, and Gray-headed Junco. Included in the region are the towns of **Nutrioso, Alpine,** and **Greer.**

Some lakes and ponds in the region that are marshy and not too heavily frequented by livestock and fishermen have nesting Eared and Pied-billed Grebes, Blue-winged and Cinnamon Teal and other ducks, Bald Eagles, Virginia Rails and Soras, Spotted Sandpipers, and Yellow-headed Blackbirds. Included here are waters in the Big Lake Country (Big Lake, Crescent Lake, Mexican Hay Lake, Lee Valley Reservoir), Becker Lake at Springerville, Nelson Reservoir southeast of Springerville, and Luna Lake east of Alpine.

About 20 miles north of Springerville on the Little Colorado River is **Lyman Lake State Park** (6,000 acres). Cliff Swallows are numerous. At times during migrations there is a wide variety of herons, ducks, shorebirds, gulls, and terns. In brush below the dam, one may find Bendire's Thrashers, Phainopeplas, Virginia's Warblers, and Lazuli Buntings.

TOPOCK

Havasu National Wildlife Refuge | Lake Havasu

On the Colorado River above and below Topock lies **Havasu National Wildlife Refuge** (41,500 acres) with headquarters in nearby Needles, California. The Topock Swamp section of the Refuge is above Topock and can be reached by car at signed points on the Arizona side of the River. Grebes, Double-crested Cormorants, wading birds, and waterfowl are common in most seasons, their numbers lowest in summer. In cattails or on shore in mesquites, look for, in one habitat or the other, the Gambel's Quail, Clapper Rail, Common Gallinule, White-winged Dove, Greater Roadrunner, Lesser Nighthawk, Gila and Ladder-backed Woodpeckers, Vermilion Flycatcher, Verdin, Marsh Wren, Northern Mockingbird, Crissal Thrasher, Black-tailed Gnatcatcher, Phainopepla, Common Yellowthroat, Hooded Oriole, Blue Grosbeak, House Finch, Abert's Towhee, and Song Sparrow.

Downstream from Topock, on Havasu Refuge, the Colorado River runs through the scenic Topock Gorge section of the Refuge to the upper end of **Lake Havasu,** partly filled by river-transported sand on which large, tall cattails thrive. This area can be satisfactorily visited only from a boat. Rent one, either at Park Moabi on the California side above Topock, or at the Lake Havasu City marina, reached by driving east for 10 miles on I 40, then south for 21 miles on State 95. The interspersion of open fish-filled water with cattails has created ideal habitat for breeding Western Grebes, hundreds of which dot the water. Other conspicuous birds in season are the Common Loon, Pied-billed Grebe, American White Pelican, Double-crested Cormorant, Great Blue and Green Herons, Great and Snowy Egrets, Black-crowned Night Heron, Least Bittern, White-faced Ibis, Ruddy Duck, Common Gallinule, American Coot, Wilson's and Northern Phalaropes, California and Ring-billed Gulls, Forster's, Common, Caspian, and Black Terns, Marsh Wren, Common Yellowthroat, Yellow-headed and Red-winged Blackbirds, and Great-tailed Grackle. Always be alert for a Brown Pelican, Wood Stork, Bald Eagle, Osprey, Bonaparte's Gull, or some other rarity.

TUCSON

Gene C. Reid Park | Himmel Park | Evergreen Cemetery | Arizona-Sonora Desert Museum | Santa Catalina Mountains | Lower Sabino Canyon

This is a 'birdy' city. Its trees and shrubbery attract many birds in addition to the ubiquitous European Starling and House Sparrow.

White-winged Doves are prominent through spring and summer. Trim little Inca Doves flutter from the lawns and alleys up into oleanders and palms; their *coo-coo* is one of the most frequently heard bird voices. Great-tailed Grackles are a common sight, especially about parks and golf courses. Black-chinned Hummingbirds may be seen feeding at the many cultivated flowers that adorn homes; they build their nests in low-hanging branches of olive and mulberry. Gila Woodpeckers are noisy inhabitants of date palms, large cottonwoods, and eucalyptus. Purple Martins, which nest in saguaros outside the city, fly over at dusk en route to their roosts. Verdins scold from mesquite and catclaw remnants of the original desert vegetation. Cactus Wrens and loudly calling Curve-billed Thrashers skulk in every tract of thick shrubbery. Hooded Orioles nest in palms and cottonwoods. Northern Cardinals, brighter in Arizona than elsewhere, are seen almost in mid-city. House Finches bid fair to outnumber House Sparrows; their beautiful song is heard nearly year round. During winter, when hummingbirds, martins, and orioles are gone, White-crowned Sparrows populate the hedges and sing their dreamy songs on warm, sunny days.

Within the city itself there are productive places for bird finding in winters. **Gene C. Reid Park** (73 acres), at Broadway and Alvernon Way, has ponds where waterbirds and waterfowl, such as Pied-billed Grebes and Canvasbacks, show up, and has a small population of Ringed Turtle Doves the year round. **Himmel Park** (20 acres), at First Street and Tucson Boulevard, is good in all seasons but in summer for unusual birds, among them such possibilities as Mountain Chickadees, Solitary Vireos, Black-throated Gray Warblers, and northern juncos. In **Evergreen Cemetery** (180 acres), at Oracle Road and Miracle Mile, flocks of Cedar Waxwings and Red Crossbills appear in winter, as well as Anna's Hummingbirds, American Robins, Red-breasted Nuthatches, and Orange-crowned Warblers. Vagrant species are reported from time to time in all these cases.

For the visitor new to Arizona, a good introductuion to the birds and other animal life of the desert, as well as the desert flora, is provided at the extraordinarily fine **Arizona-Sonora Desert Mu-**

BLACK-CHINNED HUMMINGBIRD

seum, 12 miles west from Tucson, reached by exiting west from I 10 on West Speedway Boulevard and continuing on it and Gates Pass Road to Kinney Road, following signs. Not only are numerous live examples of desert birds on display, but wild birds throng the grounds. Here it is easy to make the acquaintance of Gambel's Quail, White-winged Doves, Gila Woodpeckers, Cactus Wrens, Curve-billed Thrashers, Hooded Orioles, House Finches, and Brown Towhees. From early February until April, it is usually possible to see Costa's Hummingbirds about the Museum's many flowering plants.

Tucson is the starting point for a delightful drive into the nearby **Santa Catalina Mountains** north of the city. From I 10 in Tucson, exit east on East Speedway Boulevard for 7 miles, then turn off left onto Wilmot Road and curve right into Tanque Verde Road. Continue for another 2 miles, then take a left turn onto a signed road which goes a few miles to **Lower Sabino Canyon** on the south side of the Mountains. This is a popular picnicking spot along a usually running stream where a large grove of willows behind a sand-filled dam, and sycamores and saguaros growing almost side by side, attract a variety of birds. These include the Broad-billed Hummingbird, Ladder-backed Woodpecker, Cassin's Kingbird, Wied's Crested Flycatcher, Canyon Wren, Black-tailed Gnatcatcher, Phainopepla, Bell's Vireo, Lucy's Warbler, Pyrrhuloxia, Lesser Goldfinch, and Abert's Towhee. In winter, add to the list of birds the Anna's Hummingbird, Black Phoebe, Gray-breasted Jay, Bridled Titmouse, Winter Wren, Hermit Thrush, Black-throated Gray Warbler, Rufous-sided Towhee, and Oregon Junco.

If the side trip to Lower Sabino Canyon is not made, continue on Tanque Verde Road across Tanque Verde Wash about 2 miles, veering to the left onto the Catalina Highway. Signs point the way. As one proceeds on the Highway, it is soon apparent that the car is climbing and that the roadway is winding sharply upward, along the saguaro-studded mountainside. Turkey Vultures float by at this level, and White-throated Swifts may dash alongside. Farther on, after more climbing and winding, the road enters *Molino Basin*, where there are many evergreen oaks, junipers, and sotols. This is the oak-juniper belt, habitat of Montezuma Quail, Poor-wills, Brown-backed Woodpeckers, Cassin's Kingbirds, Ash-throated Flycatchers, Gray-breasted Jays, Bridled Titmice, Bushtits, Bewick's Wrens, Crissal Thrashers, Blue-gray Gnatcatchers, Hutton's and possibly Gray Vireos, Black-throated Gray Warblers, Scott's Orioles, Rufous-sided Towhees, and Rufous-crowned Sparrows.

Continuing upward, the road enters a ponderosa pine forest with its own assemblage of birds. Roadside stops are conveniently situated, and there are picnic areas, the best of which for birds are in *Bear Canyon*, at *Rose Canyon Lake*, and at *Bear Wallow.* Here the bird finder will want to look especially for Zone-tailed Hawks, Band-tailed Pigeons, Whip-poor-wills, Broad-tailed Hummingbirds, Hairy Woodpeckers, Williamson's Sapsuckers (not in summer), Coues' and Western Flycatchers, Violet-green Swallows, Steller's Jays, Mountain Chickadees, Pygmy Nuthatches, Western Bluebirds, Solitary and Warbling Vireos, Virginia's Warblers, Painted Redstarts, Western and Hepatic Tanagers, Black-headed Grosbeaks, and Mexican Juncos.

The mountain road eventually reaches its end at the top of cool *Mt. Lemmon* (9,157 feet above sea level, 6,700 feet above Tucson). Often one cannot travel farther than Ski Valley, a few hundred feet below the top, due to deep snow or other road restrictions. At this elevation, pines have disappeared from the north-facing slopes and are replaced by Douglas-fir, Engelmann spruce, limber pine, and white and subalpine firs. Here the bird finder will see such fir-spruce forest species as the Red-breasted Nuthatch, Brown Creeper, Golden-crowned Kinglet, and Pine Siskin. In some years, Clark's Nutcrackers, Evening Grosbeaks, and Red Crossbills may be sighted.

Along this 40-mile drive—beginning among palms and cacti, Inca Doves and Cactus Wrens, and ending among spruces and firs, nuthatches and kinglets—four distinct associations of birdlife are experienced; and accomplished so easily in such a short time. The road is open all year, except briefly at times in winter. Although bird finding is worthwhile in any month, the best ones for seeing the most birds are April and May, August and September.

WILLCOX

Sulphur Springs Valley | Willcox Dry Lake | Chiricahua National Monument

This old 'cow town' lies in the center of **Sulphur Springs Valley** in southeastern Arizona, a vast grassland with considerable stretches of irrigated farms. Fields of alfalfa, small grains, corn, and cotton lie northwest of Willcox for more than 20 miles, as well as 10 to 20 miles southeast of town, in the Kansas Settlement area. Familiar breeding birds in the grassland include the Swainson's Hawk, Scaled Quail, Greater Roadrunner, Burrowing Owl, Horned Lark, White-necked Raven, Eastern Meadowlark, and Cassin's Sparrow. In winter come large numbers of Lark Buntings, Savannah and Vesper Sparrows, and Chestnut-collared Longspurs.

The farms usually have these same birds, plus, in winter, numbers of Ferruginous and Rough-legged Hawks and great flocks of Yellow-headed, Red-winged, and Brewer's Blackbirds, and Brown-headed Cowbirds. In recent years, since the Kansas Settlement farms were developed, as many as 4,000 Sandhill Cranes have been spending the winters, feeding in the harvested fields.

South of Willcox lies **Willcox Dry Lake**, the lowest point in the landlocked Sulphur Springs Valley. In years of good rainfall this mirage-prone playa fills with water and attracts waterbirds and waterfowl of all kinds. About 10 miles southeast of Willcox, just west of the Kansas Settlement Road, is the Willcox Playa Mexican Duck Nesting Area, managed by the Arizona Game and Fish Department. To visit it, check with the local Department office in Tucson.

Willcox is the main access point to **Chiricahua National Monument,** 36 miles to the southeast via State 186. The approach to the Monument is past the ghost town of Dos Cabezas and through miles of beautiful grassland. Lying on the northwest side of the Chiricahua Mountains, the 17 square miles of the Monument embrace the fantastic Wonderland of Rocks. The Monument is a restful place, with evergreen oak woodland and sycamores, along the main Bonita Canyon. Among the breeding birds easily observed are the Band-tailed Pigeon, White-winged Dove, Whip-poor-will and Poor-will, Blue-throated and Broad-tailed Hummingbirds, Acorn and Brown-backed Woodpeckers, Cassin's Kingbird, Olivaceous Flycatcher, Violet-green Swallow, Gray-breasted Jay, Mexican Chickadee, Bridled Titmouse, Bushtit, Brown Creeper, Bewick's and Canyon Wrens, American Robin, Hutton's and Solitary Vireos, Audubon's, Black-throated Gray, and Grace's Warblers, Painted Redstart, Scott's Oriole, Western and Hepatic Tanagers, Black-headed Grosbeak, Rufous-sided and Brown Towhees, and Mexican Junco.

YUMA
Imperial Dam | Mittry Lake | Imperial National Wildlife Refuge

The agricultural areas about Yuma on the lower Colorado River offer attractive opportunities for bird finding, particularly in winter, when the weather in southwestern Arizona is at its best. Extensive acreages of grapefruit, lemons, and oranges afford habitats for birds that dwell in trees, while fields of alfalfa, small grains, sorghum, lettuce, melons, and carrots lure birds that like open areas.

During winter, hundreds of Cattle, Great, and Snowy Egrets feed in flooded fields, often under large, swirling flocks of Ring-billed Gulls. In fall, the same fields attract Killdeers, 'peeps,' dowitchers, American Avocets, Black-necked Stilts, and Water Pipits, and occasionally flocks of Mountain Plovers, which are rare in Arizona. Greater Roadrunners and Gambel's Quail are frequent sights along the roads, as are large gatherings of Mourning Doves the year round. During late summer, hordes of White-winged Doves descend upon the grainfields. Common Ground Doves are often flushed from the citrus groves. Gila Woodpeckers are everywhere, and brilliant male Vermilion Flycatchers are frequent at all seasons.

Yuma is known for its tremendous spring and fall flocks of swallows—Tree, Rough-winged, Barn, and Cliff—which flash about over the fields and bead utility and fence wires by the thousands. The Cactus Wren, strangely enough, has adapted itself to this irrigated country. Northern Mockingbirds nest commonly. American Kestrels and Loggerhead Shrikes perch on posts or wires. Wintering flocks of Yellow-headed, Red-winged, and Brewer's Blackbirds and Brown-headed Cowbirds consort with feeding cattle. Songs of Western Meadowlarks are fluted from alfalfa fields nearly every month of the year. House Finches and White-crowned Sparrows gather during fall and winter along fencerows and hedges.

To reach some of the agricultural areas, the following routes from Yuma are recommended: (1) US 95 from I 8 west and south toward Somerton. (2) California State S24 north and west from I 8 through Bard, which is on the west side of the River. (3) US 95 from I 8 east to Gila Center, then north on Avenue 7E, which continues on to the Mittry Lake area. (4) US 95 from I 8 east to Gila River crossing, then east on the paved Dome Valley Road.

Some of the summering birds in and about lakes, ponds, and marshes along the Colorado River in the Yuma region are the Clapper Rail, Common Gallinule, White-winged Dove, Lesser Nighthawk, Black-chinned Hummingbird, Wied's Crested Flycatcher, Cliff Swallow, Verdin, Marsh Wren, Crissal Thrasher, Bell's Vireo, Lucy's Warbler, Common Yellowthroat, Yellow-headed and Red-winged Blackbirds, Hooded Oriole, Summer Tanager, Blue Grosbeak, Abert's Towhee, and Song Sparrow. In winter, look for a variety of waterfowl, Common Snipes, Ring-billed Gulls, and large flocks of Tree Swallows.

Desert areas away from the River do not produce the quantity of birdlife found along the River and in the fields; nevertheless one may see in the deserts, even during winter, the Costa's Hummingbird, Ash-throated Flycatcher, Rock Wren, Le Conte's Thrasher, Black-tailed Gnatcatcher, and Phainopepla. Winter visitants and transients include the Ruby-crowned Kinglet, Orange-crowned, MacGillivray's, and Wilson's Warblers, Sage Sparrow, and, in spring, the Sage Thrasher.

Both desert and River areas produce Red-shafted Flickers (in winter) and Ladder-backed Woodpeckers. Burrowing and Long-eared Owls favor fields and riverside, respectively, while the Great Horned Owl will appear almost anywhere. Cooper's and Sharp-shinned Hawks and Northern Harriers are common except in summer. Red-tailed Hawks reside the year round in the desert. Golden and Bald Eagles can sometimes be seen soaring over the hills along the River in winter.

The **Imperial Dam** complex is accessible by either US 95 or California State S24 (*see above*). The area immediately around the Dam provides diversity of habitat teeming with birdlife. Senator Wash, a holding reservoir northwest of the Dam, and adjacent Squaw Lake often yield loons and diving ducks in winter and grebes throughout the year. West Pond, just southwest of the California end of the Dam, is always a source of waterfowl, wading birds, rails (including Black Rails), shorebirds, Ospreys, Northern Harriers, gulls, and terns. The settling ponds below the Dam abound with waterfowl in winter.

On the Arizona side of the River above the inoperative Laguna Dam is **Mittry Lake,** a partly open body of water with dense growths of cattail, bulrush, and other aquatic plants. Here one may see in any season American White Pelicans, Double-crested Cormorants, Green Herons, Great and Snowy Egrets, Black-crowned Night Herons, American and Least Bitterns, waterfowl, and sometimes gulls and terns. Marsh Wrens, Common Yellowthroats, Abert's Towhees, and Song Sparrows are abundant. Clapper and Black Rails also reside here, but they are hard to detect. To reach Mittry Lake, take US 95 east from Yuma to Gila Center, then turn north on Avenue 7E and follow the main road, which becomes graveled at Laguna Dam.

Some 21 miles from Yuma on US 95 north, a paved road branches off left to the **Imperial National Wildlife Refuge.** Follow the signs to headquarters. During migrations especially, the Refuge—whose 25,765 acres include the main channel of the Colorado River and adjacent lakes, ponds, and sloughs—hosts great numbers of birds preferring water and wetlands. In winter there are usually large flocks of Canada Geese and, on occasion, Snow and Greater White-fronted Geese. At headquarters, inquire about access to the best vantage points for observing bird concentrations at the time of the visit.

Arkansas

BY EDITH AND HENRY HALBERG

MISSISSIPPI KITES

Arkansas offers the bird finder a wide variety of breeding species, a scattering of western immigrants, a good representation of northern migrants, and a fine influx of winter visitants.

The state's total area of 53,335 square miles is divided about equally into two regions, the lowlands of the Coastal Plain and the uplands. All rivers drain into the Mississippi, which forms the state's eastern boundary. The northern two-thirds of the state is drained by the Arkansas, White, and St. Francis Rivers, the southern third by the Ouachita and Red Rivers.

The lowlands lie east and south of a line drawn roughly from a point a few miles west of Corning in northeast Arkansas, through Little Rock, to Arkadelphia and west through DeQueen to the

Oklahoma state line. The eastern part of the lowlands constitutes the floodplains of the Mississippi, White, and Arkansas Rivers and is locally called the Delta. It lies east of a line from Corning to Little Rock, thence east of a line southeast to Monticello and south to the Louisiana state line. The Delta is interrupted by Crowley's Ridge, which ranges in width from 0.5 mile to 12 miles and extends about 150 miles from the Missouri state line near Piggott, through Jonesboro and Forrest City, to the Mississippi River at Helena. According to one theory, the Ridge represents an uneroded bank between former channels of the Mississippi that flowed west of the Ridge until it moved east to coalesce with the ancestral Ohio River.

Originally the Delta's 10 million acres contained vast swamps supporting large stands of bald cypress, tupelo, overcup oak, water locust, and other trees associated with southern swamps. The only unforested areas were the thousand or so square miles of natural prairies such as the Grand Prairie, which extends from Lonoke east to the White River and south to the confluence of the Arkansas and White Rivers. These natural prairies are treeless areas unique in a region of high precipitation. They once supported tall grasses and a luxuriance of wildflowers, whereas today they have been largely modified for the production of rice, soybeans, and cotton. Only a few hundred acres of natural prairies remain unplowed. The forests of the Delta, now reduced for agricultural purposes to less than 1.5 million acres, are confined mainly to Crowley's Ridge and streamsides and to federal wildlife refuges, state game-management areas, state parks, and privately owned tracts. These lowland woods, where extensive, are rich in summer-resident warblers, including the Black-and-white, Prothonotary, Swainson's, Northern Parula, Cerulean, and American Redstart. Mississippi Kites reside in lowland woods along the Arkansas River below Little Rock and along the Mississippi.

The lowlands west of the Delta are level to somewhat rolling. Except for bottomland hardwoods which border the rivers, much of the area is forested with loblolly and some shortleaf pines mixed with hardwoods such as black and post oaks. East of I 30 the forest is primarily monoculture pine, grown for wood products. Among the characteristic breeding birds of the pinelands are the Red-cockaded Woodpecker, Brown-headed Nuthatch, and Pine Warbler (*see under* **Crossett** *and* **El Dorado**).

The uplands, occupying the remainder of the state northwest of the lowlands, comprise mountainous country cut by the southeastward-flowing Arkansas River into the Ozarks north of the River and the Ouachita Mountains on the south. The Ozarks, a continuation of the dissected plateau of the same name in Missouri and Oklahoma, attain altitudes of more than 2,000 feet. The Ouachita Mountains are more rugged and have elevations higher than those in the Ozarks. Magazine Mountain, near Paris, reaches 2,753 feet, the highest point in the state. The Rufous-crowned Sparrow has nested among the rock ledges near the summit. The summit itself is a good vantage for watching the migration of hawks in fall as are other high points (for example, *see under* **Little Rock** *and* **Morrilton**). Both the Ozarks and Ouachitas are forest-covered. In the Ozarks, hardwoods predominate, principally oaks, gums, hickories, maples, and ashes. On the south and west slopes are shortleaf pines intermingled with the hardwoods. In the Ouachitas the situation is reversed, with shortleaf pines predominating and hardwoods intermingled or in pure stands on the better sites.

In the forests of both the Ozarks and Ouachitas there are relatively few bird species and numbers of individuals. Ornithologically, however, these highlands are of interest, for they constitute the southern limits of the regular breeding range of several species of northern affinities, among them the Whip-poor-will, Blue-winged Warbler (northern Ozarks), Ovenbird, Scarlet Tanager, and American Goldfinch.

One of the major man-created changes in the state has been the channeling and damming of the Arkansas River for navigation. At the twelve major locks and dams are recreation areas ('public use areas') that have many good sites for bird finding. The levees of the lower Arkansas and the Mississippi are topped by all-weather roads, which provide access to other good bird-finding sites.

The modification of the state's natural prairies and the reduction of forest acreage in the lowlands and in the valleys of the Ozarks and Ouachitas for agricultural purposes have encouraged the immigration of open-country bird species from the west. Greater Roadrunners nest in Little Rock and have been seen as far east as

Jonesboro. Scissor-tailed Flycatchers breed widely in the western half of Arkansas where it is not unusual to see them on utility wires; some individuals have appeared almost as far east as the Mississippi.

Breeding regularly throughout the state on farmlands (fallow fields, pastures, fencerows, brushy areas, woodland borders, orchards, and dooryards) and woodlands (both bottomland and upland) are the following:

FARMLANDS

Common Bobwhite
Mourning Dove
Red-headed Woodpecker
Eastern Kingbird
Eastern Phoebe
Barn Swallow
Carolina Wren
Northern Mockingbird
Gray Catbird
Brown Thrasher
Eastern Bluebird
Loggerhead Shrike
White-eyed Vireo
Bell's Vireo
Prairie Warbler
Common Yellowthroat
Yellow-breasted Chat
Eastern Meadowlark
Orchard Oriole
Common Grackle
Northern Cardinal
Blue Grosbeak
Indigo Bunting
Painted Bunting (*southern Arkansas*)
Dickcissel
Rufous-sided Towhee (*mainly northern Arkansas*)
Grasshopper Sparrow
Lark Sparrow
Chipping Sparrow
Field Sparrow

WOODLANDS

Turkey Vulture
Black Vulture
Cooper's Hawk
Red-shouldered Hawk
Broad-winged Hawk
Yellow-billed Cuckoo
Common Screech Owl
Chuck-will's-widow
Yellow-shafted Flicker
Pileated Woodpecker
Red-bellied Woodpecker
Hairy Woodpecker
Downy Woodpecker
Red-cockaded Woodpecker (*pine woods, southern Arkansas*)
Great Crested Flycatcher
Acadian Flycatcher

WOODLANDS (*Cont.*)

Eastern Pewee
Blue Jay
Carolina Chickadee
Tufted Titmouse
White-breasted Nuthatch
Brown-headed Nuthatch (*pine woods*)
Wood Thrush
Blue-gray Gnatcatcher
Yellow-throated Vireo
Red-eyed Vireo
Warbling Vireo
Prothonotary Warbler
Swainson's Warbler
Worm-eating Warbler (*upland woods*)
Northern Parula Warbler
Cerulean Warbler
Yellow-throated Warbler
Pine Warbler (*pine woods*)
Louisiana Waterthrush
Kentucky Warbler
Hooded Warbler
American Redstart
Summer Tanager
Baltimore Oriole
Bachman's Sparrow (*open pine woods*)

Spring and fall with migrations under way are exciting seasons for bird finding in Arkansas. Waterbirds, including American White Pelicans and Double-crested Cormorants migrate along the Arkansas and Mississippi Rivers as do many herons and egrets, gulls and terns. Vast numbers of waterfowl stop off to feed and rest on lakes, ponds, and reservoirs, sometimes forming huge concentrations—for example, in Big Lake and White River National Wildlife Refuges (*see under* **Blytheville** *and* **Stuttgart**). Many shorebirds migrate through the state, dropping down for interludes on mud flats and grassy wetlands. Flycatchers, kinglets, vireos, and nearly all the warbler species breeding north of Arkansas in the United States and Canada enliven the wooded slopes of the Ozarks and Ouachitas and the wooded bottomlands of the major watercourses. The following timetable gives the dates when one may expect the main migratory flights.

Waterfowl: 20 February–1 April; 20 October–15 December
Shorebirds: 15 April–20 May; 10 August–15 October
Landbirds: 25 March–10 May; 10 September–5 November

Remaining in winter rather than continuing southward are impressive numbers of ducks—especially Mallards, Gadwalls, Common Pintails, American Wigeons, Northern Shovelers, Ring-necked Ducks, Canvasbacks, Lesser Scaups, Buffleheads, and Ruddy Ducks along with vast numbers of American Coots, and small numbers of Common Loons and Horned, Eared, and Pied-billed Grebes. Red-tailed Hawks, a few Rough-legged Hawks, and many Northern Harriers appear every winter in open country, as do Bald Eagles along major watercourses and around lakes and reservoirs. Throughout the winter, millions of European Starlings and blackbirds—chiefly Red-winged, Common Grackles, and Brown-headed Cowbirds—swarm over the countryside and converge at sundown to their favorite roosts. The state is a wintering ground for Yellow-bellied Sapsuckers and a variety of passerines that breed in more northerly latitudes. These include the Red-breasted Nuthatch, Winter Wren, Hermit Thrush, Golden-crowned and Ruby-crowned Kinglets, Water Pipit and a few Sprague's Pipits, Cedar Waxwing, Orange-crowned and Myrtle Warblers, Purple Finch, Slate-colored Junco, and numerous sparrows—Savannah, Le Conte's, Vesper, American Tree (a few), White-crowned, White-throated, Fox, Lincoln's, Swamp, and Song. In some winters there are Evening Grosbeaks, Pine Siskins, and Red Crossbills. When snow is on the ground, big flocks of Lapland Longspurs sometimes appear on certain highway shoulders.

Authorities

Carl Amason, W. J. Baerg, B. W. Beall, Jimmie J. Brown, Ben B. Coffey, Jr., Earl L. Hanebrink, Douglas A. James, David M. Johnson, Charles Mills, Max Parker, H. H. Shugart.

Reference

Birds of Northeastern Arkansas. By Earl L. Hanebrink. 1980. Available from the author, Box 67, State University, AR 72467.

BLYTHEVILLE

Big Lake National Wildlife Refuge | Mallard Lake | Burdette Heronry

From Blytheville in northeastern Arkansas, leave I 55 at Exit 67 and drive west on State 18 for about 17 miles to the headquarters of **Big Lake National Wildlife Refuge,** just west of the dam on the Little River and about 2 miles east of Manila.

Big Lake lies in 'sunken lands' formed by the violent New Madrid earthquakes in 1811–12. The Refuge embraces about 11,000 acres of which 2,500 are open water, 7,000 are swamps, and 1,000 are hardwood forests subject to seasonal flooding. In suitable habitat, the Prothonotary Warbler is an exceedingly common breeding bird, the song being heard from mid-April through June. Hundreds of Wood Ducks nest, encouraged by nesting boxes provided for them, and a few Hooded Mergansers also nest. Vast numbers of ducks—especially Mallards, but also Common Pintails, Blue-winged Teal, American Wigeons, Ring-necked Ducks, and Lesser Scaups—stop off in migration; many remain in winter. At headquarters, obtain a Refuge checklist of birds and inquire about reaching the best sites for viewing them.

A particularly good area to see birds is in the vicinity of 300-acre **Mallard Lake** in the 12,160-acre Arkansas Game and Fish Commission's Big Lake Wildlife Management Area. When driving west from Blytheville on State 18 to Big Lake National Wildlife Refuge, turn north (right) onto a gravel road (marked by a Mallard Lake directional sign) 3.1 miles west from the intersection of State 181 with State 18. Upon reaching a parking area and small store, turn left and drive along the levee, then turn left again and drive for more than 5 miles through flooded bottomland woods. Stop frequently at boat-launching ramps to search for birds. At the end of the road, park and walk ahead on the grass-covered levee lined with trees—always rewarding for viewing passerine migrants. Breeding species include the Green Heron, Wood Duck, Mississippi Kite, Mourning Dove, Yellow-billed Cuckoo, Pileated, Red-bellied, and Red-headed Woodpeckers, American and Fish Crows, Brown Thrasher, Wood Thrush, Blue-gray Gnatcatcher, White-eyed, Yellow-throated, and Red-eyed Vireos, Prothonotary and Kentucky Warblers, Common Yellowthroat, Yellow-breasted Chat, Orchard Oriole, Summer Tanager (possibly the Scarlet), Blue Grosbeak, Indigo Bunting, and Rufous-sided Towhee.

To visit the **Burdette Heronry,** leave I 55 at Exit 63 south of Blytheville, proceed south 4.1 miles on US 61, then turn left (east) onto a gravel road, cross the railroad, and go 0.7 mile. The Heronry, owned by the Arkansas Game and Fish Commission, occupies a 4.6-acre woods in the middle of a large field. To reach the Heronry, park the car and walk north across the field about 0.5 mile. Nesting in the Heronry during late spring and early summer are Little Blue Herons, Cattle, Great, and Snowy Egrets, and Black-crowned Night Herons. In recent years, Louisiana Herons and Glossy Ibises have nested. Do not enter the Heronry, but walk around it for views of birds on their nests. Many of the adults will have been observed from the road as they pass overhead to and from the colony.

CONWAY
Beaver Fork Lake

From this central Arkansas city, 31 miles north of Little Rock, leave I 40 at Exit 124 and drive north on State 25 for about 3 miles to man-made **Beaver Fork Lake** (900 acres). Just 0.2 mile after crossing the west end of the Lake, turn right off the highway onto a paved road and go to its end and park. From here in fall the Lake may be scanned for migrating geese and ducks and in winter for such birds as Common Loons, Horned Grebes, Common Goldeneyes, Buffleheads, and Ruddy Ducks. When the water level is low in fall, ducks, shorebirds, gulls, and terns may be viewed along the north shore of the Lake west of the parking area.

Return to State 25 and go east for 0.7 mile to a point where the highway veers north away from Beaver Fork Lake. Continue straight ahead on another road for 0.1 mile to the Lake's spillway. Greater Roadrunners and Bewick's Wrens are seen occasionally in the vicinity. Park the car and walk east down a road along the north base of a ridge, passing the edges of soybean fields and through stretches of dense brush. In these areas, look for Rufous-sided Towhees and Fox and Swamp Sparrows in winter, Bell's Vireos, Yellow-breasted Chats, Eastern Meadowlarks, and Dickcissels in summer. If the soybean fields are left fallow, in winter Savannah and Le Conte's Sparrows may be present.

Reference

Birds of Faulkner County, Arkansas. By David M. Johnson. 1979. Available from Arkansas Valley Audubon Society, 53 Meadowbrook Drive, Conway, AR 72032.

CROSSETT
Lake Georgia-Pacific

For Red-cockaded Woodpeckers and Brown-headed Nuthatches, explore the large pine forests, owned by the Georgia-Pacific Paper Company, that surround **Lake Georgia-Pacific,** an artificial reservoir of some 1,700 acres in southeastern Arkansas. From Crossett, take US 82 west for 3.3 miles, then turn off right onto a blacktopped county road, marked by a Lake Georgia-Pacific sign, and follow subsequent signs to the Lake.

Upon reaching the Lake, park near the grocery store on the right. Look for the woodpeckers near the picnic shelter and the nuthatches anywhere in the pines. Red Crossbills appear here almost every winter. In early winter the Lake attracts many ducks, and all winter there are a few Bald Eagles.

In the 10 miles from Crossett to Lake Georgia-Pacific, numerous all-weather roads lead off through mature stands of pine, young second-growth woodlands, and some deciduous forested bottomlands—all productive for bird finding.

EL DORADO
Calion Lake | Mud Lake | El Dorado Sewage Oxidation Pond

The extensive forests in the vicinity of El Dorado, about 15 miles north of the Louisiana state line, have good habitat for the Red-cockaded Woodpecker. A favorite site is in the pine forests along US 167 about 28 miles north of the city. Look for mature live pines with resin running down from nesting cavities and peck holes. If possible, be present just after daybreak or before sunset when the birds are more likely to be near the cavities. Locate the birds by listening for their calls.

For good bird finding near El Dorado, visit the vicinities of **Calion Lake** and **Mud Lake** northeast of the city.

From El Dorado, take US 167 north for 11 miles to Calion; turn right off US 167, go straight through town and turn right on the street after the last house; continue a short distance, follow the road up onto the levee of Calion Lake, and drive for a mile to the end. Stop frequently on the levee to investigate the willows and shrubby thickets. In spring, be alert for migrating warblers—e.g., the Tennessee, Nashville, and Wilson's—and for the following species, most of which will remain to breed: American Woodcock, Common Snipe, Yellow-billed Cuckoo, Eastern Kingbird, Great Crested and Acadian Flycatchers, Eastern Pewee, Fish Crow, Carolina Wren, Eastern Bluebird, White-eyed and Red-eyed Vireos, thirteen warblers (Prothonotary, Swainson's, Worm-eating, Northern Parula, Cerulean, Yellow-throated, Prairie, Louisiana Waterthrush, Kentucky, Common Yellowthroat, Yellow-breasted Chat, Hooded, and American Redstart), Orchard and Baltimore Orioles, Blue Grosbeak, and Chipping Sparrow. In fall, watch for wrens—the House, Winter, Marsh, and Sedge. In winter, expect a variety of sparrows that should include the Savannah, Chipping, Field, White-crowned, White-throated, Lincoln's, Swamp, and Song. On reaching the end of the road on the levee, return to Calion on the lower road along the Ouachita River for another chance to see many of the birds named above.

From Calion, return to US 167 and drive north for about 2 miles, crossing the Ouachita River; then take the first gravel road on the left (about 0.5 mile north of the River) and follow it back to the River; go under the bridge and drive downstream for 0.25 mile to the only house, park the car and walk down the road on the left for about 0.4 mile through bottomland woods to Mud Lake. Along the way in spring and early summer, watch for warblers and for breeding White-eyed, Yellow-throated, and Red-eyed Vireos. After returning from Mud Lake, continue driving downstream through bottomland woods and along a cypress brake on the left where Red-headed Woodpeckers and Yellow-bellied Sapsuckers are plentiful in late fall and winter.

An **El Dorado Sewage Oxidation Pond** on the north side of the city is a rewarding site for bird finding. To reach it, drive 0.5 mile east from the intersection of North West Avenue (US 167B) and 19th Street, turn left onto North Smith Street, at the golf course, and continue north for 1.0 mile to a Y; bear right on Calion Road for 0.1 mile, then turn right on a gravel road, cross the railroad, and stop at the gate on the right. Walk to the Pond, which is in sight. Around it is appreciable open space where grass-loving birds reside in season. In spring the willows are often alive with migrating warblers; during the last three weeks of April it is ordinarily possible to identify more than twenty species. Although transient warblers such as the Golden-winged, Tennessee, Nashville, Yellow, Magnolia, Black-throated Green, and Chestnut-sided are a primary attraction, the year-round presence of woodpeckers in the many dead and dying trees runs a close second. All the Arkansas woodpeckers, except the Red-cockaded, are present; seeing or hearing a Pileated Woodpecker is almost a certainty. During late spring and summer, while woodpeckers are tending their nests, one may hear Pine Warblers singing in the pines and find such birds as the Prothonotary and Kentucky Warblers, Common Yellowthroats, and Hooded Warblers in suitable habitats. In fall the pond hosts ducks for brief periods and, in winter, Horned, Eared, and Pied-billed Grebes.

FAYETTEVILLE

Lake Fayetteville Park | Lake Wedington Area | West Fork White River

This university town in northwestern Arkansas is surrounded by a variety of good habitats for birds: lakes, upland and bottomland woods, and brushy fields.

On the north side of Fayetteville, east of US 62-71 (College Avenue), is city-owned **Lake Fayetteville Park.** Drive to the public boat dock at the northwest end of the Lake for viewing waterbirds and waterfowl. In all seasons, except summer, one may expect Pied-billed Grebes, Mallards, Gadwalls, Green-winged Teal, Ring-necked Ducks, Lesser Scaups, and American Coots. Others may be Common Pintails, Redheads, Canvasbacks, Common Goldeneyes, Buffleheads, and Common Mergansers. From the dock, continue driving along the north side of the Lake to a point where the road turns north at the site of the Nature Center and Preserve. Park the car to look for more waterbirds and waterfowl from the

DOWNY WOODPECKER

shore and to search the adjacent fields, brushy thickets, and woods for such birds as the Great Horned Owl, Red-headed Woodpecker, Carolina Chickadee, and Tufted Titmouse. In winter, investigate the thickets and overgrown fencerows for Harris' and White-crowned Sparrows and the possibility of American Tree Sparrows, and the dense grasses for Le Conte's Sparrows.

About 15 miles west of Fayetteville on State 16 is the **Lake Wedington Area** in the Ozark National Forest. Back from the 100-acre Lake is an extensive upland oak-hickory forest in which one may hear or see such breeding birds as the Chuck-will's-widow, Pileated, Red-bellied, Hairy, and Downy Woodpeckers, Blue Jay, White-breasted Nuthatch, Wood Thrush, and Summer Tanager. A bird speciality of the Area is the Blue-winged Warbler, found in the few thicket fields.

For a good bird-finding area near Fayetteville that includes river-bottomland woods, drive south on US 71 to a point about 2 miles south of its junction with US 71B where the highway crosses the **West Fork White River.** Park along the old highway crossing, or in a churchyard next to the old bridge. Follow the gravel road along the River north from the bend in US 71 just east of the River. In spring and early summer, keep an eye on the wires along the railroad and River for Eastern Phoebes and Painted and Indigo Buntings. Inspect the hedgerows south of the highway along the railroad for Common Yellowthroats, Yellow-breasted Chats, and Blue Grosbeaks. In the adjacent bottomland woods, look for a variety of birds including the Acadian Flycatcher, Carolina Wren, White-eyed Vireo, and Northern Parula, Yellow-throated, and Kentucky Warblers.

FORT SMITH
Lock and Dam Number 13

From I 540 in this city on the Oklahoma line, exit east on State 22 (Rogers Avenue) for 4.8 miles; just before reaching downtown Barling, turn north on the access road to **Lock and Dam Number 13** on the Arkansas River. The public use area of 385 acres, east of the Dam and most of it on the south side of the River, offers good bird finding except when crowded with people in the summer months.

Tree groves and shrubby thickets are inviting to transient small landbirds in spring and fall, especially the first week of May and in mid-October. Among the many breeding birds are the Green Heron, Common Bobwhite, Yellow-billed Cuckoo, Scissor-tailed and Acadian Flycatchers, Blue Jay, Gray Catbird, Bell's and Warbling Vireos, Prothonotary Warbler, Louisiana Waterthrush, Kentucky Warbler, and both Indigo and Painted Buntings. In winter, look for the Ring-billed Gulls and possibly other gulls along the River. If the road has been extended across the Dam, investigate the cultivated bottomlands for Horned Larks in any season, shorebirds in April–May and August, and Water Pipits and Lapland Longspurs in winter.

HOT SPRINGS
Lake Hamilton Fish Hatchery

From Hot Springs in southwest-central Arkansas, drive south on State 7 for 7 miles, turn left onto State 290 and continue through pleasant, rolling farmland, worth an occasional stop for viewing birds from the road. After about 2.5 miles, watch for a sign pointing the way to **Lake Hamilton Fish Hatchery** on the left.

Upon entering the Hatchery, park beside the road and walk along the grass-covered levees overlooking numerous ponds. If there are mud flats in April–May and late summer, look for shorebirds: Common Snipes, Spotted and Solitary Sandpipers, Greater and Lesser Yellowlegs, Pectoral Sandpipers, Short-billed and Long-billed Dowitchers, and Wilson's Phalaropes. No cars are permitted on the levees but one can drive to the shore of Lake Hamilton, visible on the left. Among the birds on the Lake—and on the ponds too—during late fall and winter are Common Loons and grebes, as well as many ducks: Mallards, Gadwalls, Green-winged Teal, American Wigeons, Northern Shovelers, Lesser Scaups, and Buffleheads. In all seasons, Turkey and Black Vultures wheel overhead and Bald Eagles are usually present in winter. In adjacent wooded areas are Carolina Wrens, Eastern Bluebirds, and Northern Cardinals and wintering Myrtle Warblers and many sparrows, chiefly White-throated.

JONESBORO

Municipal Airport | Crowley's Ridge State Park

For Grasshopper Sparrows in summer, for Sprague's Pipits and possibly a Smith's Longspur in winter, go to the **Municipal Airport** in this northeastern Arkansas city. From US 63 on the south side of Jonesboro, drive north on State 1, east on Nettleton Avenue to Airport Road, cross the railroad, and drive around the Airport, stopping to scan or search all grassy areas on its outskirts. In fall, be alert for Short-eared Owls, Bobolinks, and Savannah and Sharp-tailed Sparrows.

For the best year-round bird finding in this part of the state, visit **Crowley's Ridge State Park** north of Jonesboro, reached by taking State 141 from the city for 15 miles.

The Park embraces 271 acres of upland oak-hickory and climax pine forest, also a man-made lake, on elevated terrain overlooking farmlands for miles around. For many migrating passerines, the

Park is a stopping point; at the height of their spring passage, the Park often abounds with flycatchers, kinglets, vireos, and warblers. The lake usually attracts fair numbers of transient ducks and shorebirds. In winter there are Red-breasted Nuthatches, and in some winters there is an invasion of finches, particularly Evening Grosbeaks, Pine Siskins, and Red Crossbills.

LAKE VILLAGE
Lake Chicot State Park

Lake Chicot in the southeast corner of Arkansas is an oxbow (meander cutoff) of the Mississippi River, lined with cypresses. In fall its surface is sprinkled with Ruddy Ducks, while migrating swallows—Tree, Bank, Barn, and Cliff—swarm over it as they forage for insects. **Lake Chicot State Park,** occupying 132 acres on the north end of the Lake, provides good opportunities for finding birds in this fascinating area. Some of the species regularly breeding here and along the levee are the Brown Thrasher, Orchard Oriole, Summer Tanager, Blue Grosbeak, and Indigo and Painted Buntings.

The Park is reached from Lake Village by driving 8 miles north on State 144. For further exploration, after leaving the Park, turn right, drive through a farm gate across the road and follow the road to the Mississippi River levee on which it is possible to drive all the way to the Louisiana state line. Where side roads lead down to the River, follow some of them on foot. During summer months, always watch for Mississippi Kites, woodpeckers and flycatchers, Blue-gray Gnatcatchers, White-eyed Vireos, Cerulean Warblers, Louisiana Waterthrushes, and Yellow-breasted Chats. Cattle Egrets are common in late fall, as are Brown-headed Cowbirds.

LITTLE ROCK
Boyle Park | Burns Park | Faulkner Lake | Lake Maumelle | Pinnacle Mountain State Park | Rebsamen Golf Course

This capital city lies partly on the Coastal Plain and partly in the highlands. From an altitude of 300 feet downtown, one climbs abruptly to 600 feet in the highlands in the western part of the city and to 780 feet at Walton Heights. The Arkansas River bisects the uplands, providing bottomlands that diversify the habitat for birds.

Flocks of Cedar Waxwings gather in January and February to feast on the berries of ornamental hollies in the gardens of Little Rock. Hundreds of these birds work their way along in the hilly residential sections, stripping the trees. A 40-foot American holly draped with berries is picked clean in a day. American Robins trail along, eating berries that drop to the ground.

In Little Rock and vicinity there are several parks and other areas for especially good bird finding.

On the west side of the city, between 12th Street and Asher Avenue, lies **Boyle Park** (270 acres). From downtown Little Rock, the Park may be reached by driving west on I 630 (Wilbur Mills Expressway), taking Exit 5B south onto University Avenue, going right on 12th Street for one block, left on Cleveland Street, then right at the corner church and down the hill on Boyle Park Road to the entrance. Follow the Park's paved road and bear right at every intersection to a small pond on the right and a picnic pavilion on the left. Leave the car and walk ahead to a bicycle path that goes to the right through woods, or cross the dam at the foot of the pond and take any of the footpaths through the woods of tall pines and other trees draped with vines. In April and early May this part of the Park is unsurpassed in Little Rock for its host of migrating flycatchers, kinglets, vireos, and warblers. The narrow valley tends to funnel these birds into it. Among the birds nesting in the Park are Mississippi Kites, Broad-winged Hawks, Yellow-shafted Flickers, Red-bellied, Hairy, and Downy Woodpeckers, Carolina Chickadees, Tufted Titmice, White-breasted Nuthatches, Northern Mockingbirds, Wood Thrushes, White-eyed and Yellow-throated Vireos, Prothonotary, Yellow-throated, and Pine Warblers, Ovenbirds, Louisiana Waterthrushes, Kentucky Warblers, Yellow-breasted Chats, Hooded Warblers, Baltimore Orioles, Northern Cardinals, and Chipping Sparrows.

Burns Park (1,523 acres) in North Little Rock lies along the Arkansas River for 2 miles. Leave I 40 at the Burns Park Exit (Exit 150) and drive south in the Park across the golf course to the boat-launching area. Explore the banks of Shillcut Bayou and the River. Drive west along the River; watch for Sharp-shinned and Cooper's

Hawks and Ring-billed Gulls in winter and for American White Pelicans in migration. At the end of the paved road, continue ahead on a gravel road through the woods to a paved road; turn left onto it and search all side roads and around the camping area. At White Oak Bayou, walk east along the bank where there are usually Fox Sparrows in the bottoms below during winter. Follow the paved road as it leads to brushy roadsides and embankments, weedy fields, and both deciduous and pine woods. Inspect all these habitats suitable for such wintering species as the Hermit Thrush, Myrtle and Pine Warblers, Purple Finch, Pine Siskin, Slate-colored Junco, and numerous sparrows—Savannah, Field, White-throated, Lincoln's, Swamp, and Song. In any season, look over the golf course for the American Kestrel, Killdeer, Mourning Dove, Eastern Bluebird, and Loggerhead Shrike.

From I 40, leave at the Galloway Exit (Exit 161), go south, cross US 70, take State 391, and drive across Hill Lake and past cotton and soybean fields for a short distance to **Faulkner Lake.** Stop to look and listen for birds in the bordering trees. Continue past a wooden bridge for a wide-open view of a large marsh. Among the birds to expect here in summer are Least Bitterns, Wood Ducks, Purple Gallinules, and Red-winged Blackbirds. At the next intersection, turn left onto Faulkner Lake Road and left again, onto

LOGGERHEAD SHRIKE

Walker Corner Road, to arrive at US 70. Along the way in spring and summer, watch the utility wires for Dickcissels and stop to search the brushy roadsides for Painted Buntings. There should be Eastern Meadowlarks and Grasshopper Sparrows in the nearby fields.

One of Little Rock's water-supply reservoirs, **Lake Maumelle,** lies about 20 miles west of the city. From I 40, just west of North Little Rock, exit south on I 430 (Exit 147), cross the Arkansas River, and take the Cantrell Road Exit (Exit 9) west on State 10. (Or from downtown Little Rock, go west on State 10 to the same point.) Drive west on State 10 for 6 miles; turn right (north) onto State 300 (Roland Road) and go 2.8 miles; turn left onto a paved driveway that slopes uphill to the pine grove and the caretaker's house at Lake Maumelle Dam. Park here and, with a spotting scope, scan the Lake. In winter, look for Common Loons, grebes, diving ducks, Ring-billed and possibly Bonaparte's Gulls. Watch for Bald Eagles. Continue north on State 300, turn left onto State 113 and left again onto State 10 to encircle the Lake. For additional views of the Lake, take side roads to the left leading to the shore. The picnic area on State 10, just east of the causeway across the Lake, is a good vantage. Barn and Cliff Swallows nest in the bridge opening under the causeway.

An observation platform in **Pinnacle Mountain State Park** (1,354 acres) at an elevation of 560 feet affords a panoramic view of the Arkansas River Valley and, in fall, is an excellent vantage for watching migrating raptors—Sharp-shinned, Red-tailed, Red-shouldered, and Broad-winged Hawks, Northern Harriers, Ospreys, and American Kestrels—and for migrating American White Pelicans. To reach the Park from Lake Maumelle Dam, drive north on State 300 for 0.2 mile, turn right onto Pinnacle Valley Road, and follow signs to the Park's visitor center.

Among the several areas along the Arkansas River in Little Rock that are good for bird finding, **Rebsamen Golf Course** provides the easiest walking. From I 30 in the city, take Exit 141A west onto State 10 (becomes Cantrell Road) for about 4 miles, bear right on Rebsamen Road at the sign to Murray Lock and Dam. Stay on Rebsamen Road, cross the railroad, and drive to the Golf Course Clubhouse. Leave the car in the parking lot and walk west along the road under elms, cottonwoods, sycamores; a few cypresses are on the left, the Golf Course is on the right. Some of the birds to watch or listen for in late spring and summer are the Green Heron, Common Bobwhite, Eastern Phoebe, Eastern Pewee, Purple Martin (over the River), Blue-gray Gnatcatcher, Red-eyed and Warbling Vireos, several warblers including the Kentucky, Common Yellowthroat, and Hooded, Orchard and Baltimore Orioles, and Blue Grosbeak. Among the transients during migration time are Rose-breasted Grosbeaks, large flocks of American Goldfinches, and usually a few Philadelphia Vireos.

Continue following the road as it circles right close to the River and walk east along the bank where several jetties lead out to sandy islands. Take the footpath slanting down to the island covered with willows and walk with care out on the jetty—a good vantage for seeing American White Pelicans and Caspian Terns in spring and fall. Sometimes in fall, late in the day, Common Nighthawks pass overhead by the hundreds and many Chimney Swifts too.

Return to the bank and continue walking east; cross the Golf Course wherever convenient and return to the parking lot. On leaving the Golf Course road, turn right and proceed to Murray Recreation Area, where a bicycle trail and a network of roads run through woods attractive to many birds. In winter the entire area along the River shelters woodpeckers, Blue Jays, Carolina Chickadees, Tufted Titmice, House, Winter, and Carolina Wrens, Ruby-crowned Kinglets, finches, and sparrows.

LONOKE
Joe Hogan State Fish Hatchery

In the flat farmlands extending in all directions from Lonoke, about 25 miles east of Little Rock, are countless ponds and small lakes where shorebirds as well as ducks stop off during their spring and fall migrations. For a good sampling of variety and numbers, visit the **Joe Hogan State Fish Hatchery.**

From I 40 at Exit 175, go south on State 31 to the center of Lonoke, then west on US 70 for 1.7 miles to the Hatchery sign. Although the Hatchery is open to the public, permission must be obtained to drive on the levees. Scan all drained ponds for shorebirds, depending on the season. From mid-July on through September, the ponds swarm with transient plovers, sandpipers, yellowlegs, dowitchers, and phalaropes; more than twenty-eight species have been recorded on the Hatchery and nearby private ponds. Frequenting the Hatchery at one time or another are Little Blue Herons, Cattle and Great Egrets, Franklin's Gulls, and Forster's and Black Terns.

MARIANNA
St. Francis National Forest

In eastern Arkansas, **St. Francis National Forest** embraces the southern end of Crowley's Ridge for 18 miles. A late-April trip, as described below in this relatively undisturbed National Forest, is certain to yield a fine variety of birds.

From Marianna, drive southeast on State 44 into the National Forest to a point, less than 2 miles, where the road turns right. On the left is the District Ranger's office where maps and other information are available. Continue south for 5 miles to the south end of Bear Creek Lake Dam, then keep on south on Forest Route 1900 for 12 miles to Storm Creek Lake, stopping to explore side trails. From Storm Creek Lake, make a circuit by bearing left on FR 1909, then returning north in the bottomlands on FR 1901 (parts may be under water at times) with the ridge on the left.

During the trip, one should see Mississippi Kites at both Bear Creek and Storm Creek Lakes, Turkey Vultures, and Chimney Swifts above open areas almost anywhere. One should hear Common Bobwhites and hear or see both Yellow-billed and Black-billed Cuckoos, the Yellow-billed having arrived for the summer, the Black-billed passing through in migration. Some northbound Philadelphia Vireos may be lingering along with many northbound warblers including Palm Warblers, whose principal migration route in Arkansas seems to be along Crowley's Ridge. There may also be some late-migrating Scarlet Tanagers and Rose-breasted Grosbeaks. Among the summer residents that will have arrived are the Ruby-throated Hummingbird, Eastern Pewee, Gray Catbird, Brown Thrasher, Wood Thrush, Red-eyed and Warbling Vireos, Summer Tanager, Indigo Bunting, and the following warblers: Black-and-white, Prothonotary, Swainson's, Northern Parula, Cerulean, Prairie, Ovenbird, Kentucky, Common Yellowthroat, Yellow-breasted Chat, Hooded, and American Redstart. Some of the permanent residents to expect are the Common Screech and Barred Owls (both may respond to imitation), Red-bellied Woodpecker, Blue Jay, Carolina Chickadee, Tufted Titmouse, White-breasted Nuthatch, Carolina Wren, and Northern Cardinal.

MORRILTON
Petit Jean State Park

Some 60 miles northwest of Little Rock, flat-topped, horseshoe-shaped Petit Jean Mountain rises to an elevation of 1,200 feet. Lying on top of the Mountain is **Petit Jean State Park** (3,809 acres), which has numerous scenic attractions—a waterfall, strange rock formations, and an Indian cave—to which roads and trails lead through mixed hardwood-pine woods.

To reach the Park, leave I 40 at Morrilton (Exit 108), go south on State 9 for about 8 miles, through Morrilton and across the Arkansas River to Oppelo; then turn west onto State 154 and continue about 12 miles to the Park. From Oppelo, the first 6 miles to the foot of the Mountain traverse open farming and cattle country where Scissor-tailed Flycatchers, Eastern Meadowlarks, Dickcissels, and Lark Sparrows reside in summer.

After the steep climb up Petit Jean Mountain to the Park, stop at a sign explaining Petit Jean's grave and walk to the grave. Just beyond it the cliffs on the east end of the Mountain make a fine vantage from which to watch migrating hawks—chiefly Sharp-shinned and Broad-winged—during September and October on clear windy days after the arrival of a cold front. In the woods of the Park, during the first week of May, nearly all the species of thrushes and warblers that migrate north through Arkansas can be seen. Anytime in spring and early summer, listen for Bachman's Sparrows where there are open pine woods.

PINE BLUFF
Arkansas Levee System

Like the levee system on the Mississippi River, the **Arkansas Levee System** offers productive bird finding. The roads on the levee, good in all weather, are fringed with grasses and shrubby thickets, and during their course pass close to farmlands on one side, open water or bottomland woods on the other. The following trip will produce many rewards.

From Pine Bluff, drive southeast on US 65; 9 miles beyond its intersection with State 81, turn left at Linwood onto a paved road and go 3.6 miles to the Rising Star Public Use Area on the Arkansas, where there are woods worth exploring. From here, drive 4.1 miles downstream on the levee, turn left toward the River and drive 0.7 mile to Huff's Island Public Use Area at Lock and Dam Number 3. Habitats between the two Areas comprise grasses on the levee and in pastures, small oxbow lakes, farm ponds, croplands, and some stands of cottonwood, willow, elm, boxelder, sycamore, maple, and pecan. Return to US 65 via State 11 to Grady and continue southeast to Gould; turn left onto State 114, go north for 7 miles to the levee, and drive on it southeast for good views of the countryside.

From mid-April through May, woods and thickets throng with flycatchers, kinglets, vireos (watch carefully for the Philadelphia), and warblers migrating to their northern breeding grounds. Besides blackbirds, among the more common resident or summer-resident birds are the Yellow-crowned Night Heron, Mississippi Kite, Mourning Dove, Yellow-billed Cuckoo, Red-headed Woodpecker, Scissor-tailed and Acadian Flycatchers, Horned Lark, Barn Swallow, Eastern Bluebird, Prothonotary Warbler, Orchard and Baltimore Orioles, Northern Cardinal, Blue Grosbeak, Indigo Bunting, and Grasshopper Sparrow. In late summer and early fall, the woods are often alive with migrating passerines, this time southbound. All six species of eastern swallows swarm over the River. Watch for Sharp-shinned Hawks, American Kestrels, and Common Screech Owls. Herons and egrets are numerous along the shores. In fall, American White Pelicans pass down the River. In winter, Double-crested Cormorants, Ring-billed Gulls, and occasionally Herring Gulls are seen over the River. Wintering sparrows include the Savannah, Vesper, White-crowned, White-throated, Fox, Lincoln's, Swamp, and Song.

RUSSELLVILLE
Holla Bend National Wildlife Refuge

In west-central Arkansas, **Holla Bend National Wildlife Refuge** is a 6,367-acre 'island' between the old and new channels of the Arkansas River. Wooded areas, lakes, small ponds and sloughs, croplands, grassy fields, and numerous shrubby thickets provide diversity of bird habitats.

To reach the Refuge, leave I 40 at Russellville (Exit 81), drive south on State 7 across the Arkansas River for about 12 miles to Centerville, left (east) on State 154 for 4 miles, then left (north) on State 155 for 3.7 miles to the entrance. At the visitor information booth, or farther in along the road, obtain a checklist of Refuge birds. From the road to headquarters, walk down the slope into fields where there may be transient Bobolinks in May, and continue to tall thickets for summer-resident Bell's Vireos.

Winter is ideal for bird finding on the Refuge. Hundreds of Snow and some Canada Geese over-winter and Greater White-fronted Geese pause in their spring and fall migrations. Also over-wintering are thousands of ducks, chiefly Mallards but with Gadwalls, Common Pintails, Green-winged Teal, and American Wigeons well represented. Horned Larks and flocks of Lapland Longspurs are attracted to fields, and many sparrows to thickets and dense grassy sites. A Sharp-shinned Hawk or two, Red-tailed Hawks, and Northern Harriers are usually present as well as a few Bald Eagles often perched on isolated trees standing in ponds and fields.

STUTTGART
White River National Wildlife Refuge

In the vast, almost treeless flatland of farms in eastern Arkansas, **White River National Wildlife Refuge** is a veritable oasis of forests and waterways. Comprising 113,300 acres and ranging in width from 3 to 9 miles, the Refuge extends north for 54 miles along both sides of the White River from a point a few miles above its confluence with the Arkansas and Mississippi Rivers. During a wet winter or spring, many acres are flooded, but during low-water periods they constitute, in addition to the main river channel and innumerable bayous and meander cutoffs, 169 lakes and 98,000 acres of bottomland forests in which oak, gum, cypress, sycamore, willow, and hackberry are the predominating trees.

To reach the Refuge from Stuttgart, take State 130 east and south for 26 miles to DeWitt. At Refuge headquarters here (704 South Jefferson Street), ask for a checklist of Refuge birds and information on which roads are currently passable into the Refuge. In dry seasons, drive northeast from DeWitt on State 1 for 14 miles to St. Charles and continue to the River; but just before reaching it, take the last road on the right. This leads into the Refuge and follows the White River south, through bottomland forests, with intermittent views of open water, coming to a secondary headquarters after 5 miles. Park here and follow the road south on foot through more forests.

Wintering waterfowl are the Refuge's greatest ornithological attraction. Their appearance begins in early fall and by December they are present in enormous numbers, chiefly Mallards, although other species—Canada Geese, Gadwalls, Common Pintails, Green-winged Teal, American Wigeons, Northern Shovelers, Wood Ducks, Ring-necked Ducks, and Lesser Scaups—are well represented. Their presence invariably attracts Bald Eagles and sometimes one or more Golden Eagles. A few Snow Geese and occasionally Greater White-fronted Geese over-winter. In late summer and early fall there is an influx of herons and egrets.

For landbirds partial to bottomland forests and their brushy edges, the Refuge is excellent. Breeding warblers include the Prothonotary, Swainson's, Northern Parula, Cerulean, Louisiana Waterthrush, Common Yellowthroat, Yellow-breasted Chat, Hooded Warbler, and American Redstart. Some of the other breeding birds are the Common Screech and Barred Owls, Pileated, Red-bellied, and Red-headed Woodpeckers, Great Crested Flycatcher, Eastern Phoebe, Eastern Pewee, Carolina Chickadee, Carolina Wren, Wood Thrush, White-eyed and Red-eyed Vireos,

Orchard and Baltimore Orioles, Summer Tanager, and Indigo Bunting. With luck, Wild Turkeys may be seen. In winter, Hermit Thrushes are common and brush piles are havens for Winter Wrens, Rufous-sided Towhees, and White-throated Sparrows.

Not far from the Refuge, at Arkansas Post on the Arkansas River, John James Audubon first identified the Traill's Flycatcher.

TEXARKANA
Millwood Lake

Located in southwestern Arkansas, man-made **Millwood Lake** (29,500 acres) soon became a must for bird finders after its completion in 1966. The Christmas bird count in 1978 was 111 species, highest ever in the state. Every three years the water level is lowered after Labor Day to control the balance of the small game-fish population. The exposed flats then attract many shorebirds, among them Red Knots and Buff-breasted Sandpipers.

To reach Millwood Lake from the Texas state line at Texarkana, drive northeast on I 30 for 18 miles, take Exit 18 at Fulton and go north on State 355 for 12 miles to Saratoga, then turn left onto State 32 leading to the Lake and across the dam. Explore all side roads leading to the Lake. Stop at suitable vantages and, using a spotting scope, scan the Lake. In summer and early fall there should be, besides many herons and egrets, American Anhingas, White Ibises, and possibly a Wood Stork or two. In fall, Ospreys and Caspian Terns are often seen and good flights of Broad-winged Hawks have been recorded over the area. In winter, look for loons, grebes, geese, ducks, and—best of all—Bald Eagles perched on the many dead trees standing in the water.

WEST MEMPHIS
Wapanocca National Wildlife Refuge | Horseshoe Lake | Porter Lake

Two good areas for bird finding on the west side of the Mississippi River can be conveniently reached from this east Arkansas city.

From West Memphis, drive north on I 55 for 15 miles; take Exit 21 near Turrell and go east on State 42, cross State 77, pass under the railroad, and turn north toward Turrell. Enter the **Wapanocca National Wildlife Refuge** just after the highway turns.

The Refuge encompasses 5,500 acres of croplands, bottomland woods, swamps, and open water, including cypress-studded Wapanocca Lake. The road, after entering the Refuge, swings around headquarters (here obtain a checklist of Refuge birds), crosses a bridge, and goes onto the levee (sometimes closed during wet weather) for 2 miles. At 0.3 mile after leaving the levee, there is a turnoff at right to an observation platform at the Lake; at 1.3 miles the road passes through fine bottomland hardwoods for 1.0 mile.

Wood Ducks are common residents, readily seen on the Refuge tour. Other breeding species to be seen or heard include the Mississippi Kite, Red-shouldered Hawk, Common Bobwhite, Wild Turkey, Yellow-billed Cuckoo, Common Screech and Barred Owls, Pileated, Red-bellied, and Red-headed Woodpeckers, Chuck-will's-widow, Great Crested Flycatcher, Eastern Pewee, Carolina Chickadee, Carolina Wren, Gray Catbird, Wood Thrush, White-eyed and Red-eyed Vireos, Swainson's Warbler, Louisiana Waterthrush, Kentucky and Hooded Warblers, Orchard and Baltimore Orioles, Indigo Bunting, and Rufous-sided Towhee. In winter, on the Lake and easily viewed from the observation platform, are many ducks—Ruddy Ducks often being abundant—and American Coots.

From West Memphis, drive west on I 40 for 4 miles; at Exit 271, go south on State 147 for 15 miles, then turn left onto State 131. **Horseshoe Lake** is immediately ahead. (State 131 goes around the east and south sides of the Lake to rejoin State 147 on the west side.) After 1.0 mile on State 131, a pier just beyond an inlet provides opportunity to scan the Lake. At the south end of the Lake, beyond Seypel, the highway goes up on the Mississippi levee for less than a mile, then leaves it and skirts **Porter Lake** on the right. Several pullouts are available for viewing the Lake.

In winter, Horseshoe and Porter Lakes are host to vast numbers of ducks, Porter Lake hosting the larger number. A total of 10,000 Ruddy Ducks on both bodies of water has been estimated. Other ducks in large numbers are Mallards, Gadwalls, American

Wigeons, Ring-necked Ducks, and Buffleheads; others in small numbers are Common Pintails, Canvasbacks, Lesser Scaups, and Hooded Mergansers. American Coots are abundant, as many as 6,000 have been counted on Horseshoe Lake alone. A few Horned and Pied-billed Grebes are usually present.

WINSLOW
Devil's Den State Park

In the densely wooded, rugged terrain of the Ozarks in northwestern Arkansas is **Devil's Den State Park** (4,885 acres), noted for its rock fissures and limestone caves. To reach the Park from Winslow on US 71, about midway between Fayetteville and Fort Smith, turn west onto State 74 and go 13 miles to the entrance. Then follow a winding road down a steep slope to Park headquarters on Lee's Creek. From here foot trails and bridle paths lead off to wooded sections of the Park. Recommended is the trail that follows Lee's Creek upstream close to rocky slopes forested with hardwoods.

Among the summer-resident bird species are several of northern affinities—the Whip-poor-will, Blue-winged Warbler, and Scarlet Tanager—along with species of more southern affinities such as the Chuck-will's-widow and Summer Tanager. Some of the other breeding birds include the Black Vulture, Red-shouldered and Broad-winged Hawks, Great Horned and Barred Owls, Great Crested and Acadian Flycatchers, Eastern Pewee, Carolina Wren, Gray Catbird, Brown Thrasher, Wood Thrush, Blue-gray Gnatcatcher, three vireos (White-eyed, Yellow-throated, and Red-eyed), nine warblers (Black-and-white, Northern Parula, Yellow-throated, Prairie, Ovenbird, Louisiana Waterthrush, Kentucky, Hooded, and American Redstart), Orchard Oriole, Northern Cardinal, Indigo Bunting, Rufous-sided Towhee, and Chipping Sparrow.

California

CALIFORNIA CONDOR

California has been described in superlatives so often that there are few who have not heard that it possesses the biggest trees, the highest mountain in conterminous United States, and the lowest desert in North America, as well as one of the largest flying landbirds in the world—the California Condor. But even for those more interested in variety than in size there is, perhaps, no comparable area in North America. A bird finder may observe shearwaters before breakfast and before sunset see Clark's Nutcrackers and Cactus Wrens.

The topography of California is so diversified that a brief description is certain to be inadequate. From Oregon to the Mexico border the distance is 780 miles; west to east the width varies from 150 to 350 miles. The dominant topographical feature is the great range of the Sierra Nevada in the east, 385 miles long and averaging 80 miles in width. The Sierra rises as it trends southward to culminate in Mr. Whitney (*see under* **Bishop**), 14,494 feet above sea level, the third highest point in the United States after Mt. McKinley and Mt. Foraker in Alaska. Forty other peaks in the range exceed 10,000 feet. Mt. Shasta, a dormant volcano in northern California, is 14,162 feet, and Mt. Lassen (*see under* **Redding**), a volcano active early in this century, is 10,457 feet. The Coast Ranges, fringing the ocean for 500 miles and 3,000 feet in elevation, parallel the Sierra Nevada, with the Central Valley lying between them. The Tehachapi Range on the south and the Siskiyou Mountains on the north complete the encirclement of the Valley, which thus forms a vast inland bowl more than 400 miles long and averaging 50 miles in width. In the northeastern corner of the state is the Modoc Plateau and the western edge of the Great Basin, which impinges upon the eastern side of the Sierra Nevada southward. The southeastern third of the state is occupied by the Mojave and Colorado Deserts, together covering more than 24,000,000 acres. From the summit of Mt. Whitney it is only 60 miles southeast to Death Valley, the lowest spot in the United States. West of the Deserts rise the San Bernardinos, the San Gabriels, and the San Jacintos, with peaks over 10,000 feet.

The principal drainage system of the state is composed of the Sacramento and San Joaquin Rivers and their tributaries, which carry the runoff from the Sierra Nevada to San Francisco Bay. These rivers and those that drain seaward from the northern Coast Ranges, such as the Klamath, Eel, and Russian Rivers, flow the year round; but numerous small streams in the more arid sections of the state contain water only during winter and spring.

The climate of California is as diverse as its topography. The lowlands and coastal areas enjoy the mild winters of which Californians are justly proud, but winter in the mountains earned for the Sierra Nevada its Spanish name, which means 'snowy range.' Over most of the state the rains fall between October and April, summer rains being rare except in the southeastern deserts. Annual rainfall varies from 80 inches along the northwest coast to less than 5 inches in the deserts. Along the coast, summer fogs maintain a cool, moist climate, permitting the growth of the coast redwood and its associates.

It is apparent that such variations in climate and topography have proportional effects upon the plants and animals. The bird finder in California thus enjoys a much greater variety of habitats, and consequently of birds, than he will see in an inland area of low relief. More than 525 species of birds are recorded from the state, and of these more than 300 have bred.

The Central Valley of California was originally an area of grassland, with extensive marshes near the rivers. Parts of it were arid and not used for agriculture. Today almost the entire Valley is used for farming and pasture, its orchards, vineyards, and grainfields being in a large measure dependent on irrigation. Over 20 million acres in the state are under cultivation. In addition to the Sacramento and San Joaquin Valleys, the Santa Clara, Salinas, Napa, and Imperial Valleys are largely devoted to agriculture.

With irrigation and the planting of many types of trees and shrubs, the avian habitats have been both modified and multiplied. Many native, and a few introduced, species of birds are more abundant in the fields and farmyards, along roadsides, and in city parks than in any natural plant associations. Birds that reside in *man-made habitats* are the following:

American Kestrel
California Quail
Ring-necked Pheasant
Killdeer
Mourning Dove
Spotted Dove (*southwestern California*)
Ringed Turtle Dove (*southwestern California*)
Barn Owl
Red-shafted Flicker
Western Kingbird
Black Phoebe
Barn Swallow
Cliff Swallow
Scrub Jay
Yellow-billed Magpie (*west-central California*)
American Crow
Bushtit
Northern Mockingbird
American Robin
Loggerhead Shrike
Yellow Warbler
Western Meadowlark
Hooded Oriole (*southern California*)
Bullock's Oriole
Brewer's Blackbird
House Finch
American Goldfinch
Lesser Goldfinch
Rufous-sided Towhee
Brown Towhee
White-crowned Sparrow
Song Sparrow

Grassland occupies approximately 10 per cent of the state's area; small patches exist widely in the lowlands, but the most extensive areas are in the southern and western portions of the Central Valley and in a belt around the Valley at the base of the surrounding hills. The grassy areas of the lowlands are green from November through May; those of northern California and the montane meadows of the Sierra Nevada are green throughout the summer. Typical birds of the *low altitude grassland* include:

Prairie Falcon
American Kestrel
Killdeer
Mourning Dove
Burrowing Owl
Western Kingbird
Say's Phoebe
Horned Lark
Northern Raven
American Crow
Loggerhead Shrike
Western Meadowlark
Brewer's Blackbird
Brown-headed Cowbird
Lark Sparrow

The foothills encircling the Central Valley support a park-like woodland whose principal components are the Digger pine and blue oak. The valley oak often grows on the more level areas. In the northwestern part of the state, areas of oak woodland are usually dominated by the Oregon oak; in the central Coast Ranges and in southern California, by coast live oak. The following birds are associated in the breeding season with *oaks or with intervening open areas* filled by grasses, other herbaceous plants, or low shrubs. Many of the same species are also found in riparian situations:

Cooper's Hawk
Red-tailed Hawk
Golden Eagle
American Kestrel
California Quail
Mourning Dove
Common Screech Owl
Black-chinned Hummingbird
(*southern California*)
Anna's Hummingbird
(*southwestern California*)
Red-shafted Flicker
Acorn Woodpecker
Nuttall's Woodpecker
Ash-throated Flycatcher
Western Pewee
Violet-green Swallow
Plain Titmouse

Bushtit
White-breasted Nuthatch
House Wren
Western Bluebird
Blue-gray Gnatcatcher
Hutton's Vireo
Black-throated Gray Warbler
Bullock's Oriole
Brewer's Blackbird
Brown-headed Cowbird
Black-headed Grosbeak
House Finch
Lesser Goldfinch
Brown Towhee
Lark Sparrow
Chipping Sparrow

Few species in streamside thickets and trees are restricted to that habitat. Most of them are also in either the chaparral or the oak-woodland association. Riparian trees in the arid regions of the state are largely cottonwoods, willows, and sycamores. The humid coastal region and the mountain canyons of southern California are characterized by alders, bigleaf maple, coast live oak, and California-laurel, with a dense ground cover of thimbleberry, wild rose, snowberry, blackberry, coffeeberry, and bracken fern. The following birds breed in *riparian regions* of the state:

Red-shouldered Hawk
California Quail
Yellow-billed Cuckoo (*rare*)
Common Screech Owl
Great Horned Owl
Long-eared Owl
Black-chinned Hummingbird (*southern California*)
Red-shafted Flicker
Downy Woodpecker
Black Phoebe
Traill's Flycatcher
Western Flycatcher
Western Pewee
Bushtit
House Wren
Bewick's Wren
Swainson's Thrush
Bell's Vireo (*rare*)
Warbling Vireo
Yellow Warbler
Common Yellowthroat
Yellow-breasted Chat
Wilson's Warbler
Bullock's Oriole
Brown-headed Cowbird
Black-headed Grosbeak
Blue Grosbeak
Lazuli Bunting
American Goldfinch
Lesser Goldfinch
Rufous-sided Towhee
Brown Towhee
Lincoln's Sparrow (*mountains*)
White-crowned Sparrow (*mountains*)
Song Sparrow

Before agricultural development there were extensive marshes in the river bottoms of the Sacramento and San Joaquin Valleys. Today there are only remnants. Some marshy areas have been set aside as wildlife refuges for breeding and wintering waterfowl. Small patches of fresh-water marsh will be found in the shallows of most lakes and ponds and in the sloughs and backwaters of the larger rivers. Cattails, bulrushes, sedges, and willows are the usual dominant emergent plants. Birds that breed in *fresh-water marshes* or along the shores of lakes or streams are the following:

Eared Grebe
Western Grebe
Pied-billed Grebe
Double-crested Cormorant
Great Blue Heron
Green Heron
Great Egret
Snowy Egret
Black-crowned Night Heron
Least Bittern
American Bittern
Mallard
Gadwall
Common Pintail
Green-winged Teal
Blue-winged Teal (*northeastern California*)
Cinnamon Teal
Northern Shoveler (*Central Valley and northeast*)
Wood Duck
Redhead (*Central Valley and northeast*)
Ruddy Duck
Common Merganser (*northern California*)
Northern Harrier
Virginia Rail
Sora
Common Gallinule (*southern California*)
American Coot
Killdeer
Spotted Sandpiper
Willet (*northeastern California*)
American Avocet
Black-necked Stilt
Wilson's Phalarope
Forster's Tern
Caspian Tern
Black Tern (*local*)
Belted Kingfisher
Black Phoebe
Marsh Wren
Common Yellowthroat
Yellow-headed Blackbird
Red-winged Blackbird
Tricolored Blackbird
Song Sparrow

The North American Dipper is a resident along turbulent permanent streams in the mountains and near the coast.

Coniferous forests cover more than one-fifth of the state: most of northern California above the Valleys, except for the northeast corner, which is invaded by the Great Basin sagebrush and juniper; the Sierra Nevada; the San Bernardinos, San Gabriels, San Jacintos, Lagunas, and other mountains in southern California; the inner Coast Ranges south to San Franicsco Bay; and the Santa Cruz and Santa Lucia Mountains which border the ocean from San Francisco Bay south to San Luis Obispo. From Monterey northward, the coastal coniferous forest is dominated by the towering coast redwood, which with its associates—Douglas-fir, madrone, tanbark oak, black oak, golden chinquapin, and California-laurel—forms a belt 450 miles long and from one to 150 miles wide. Patches of closed-cone pine—Monterey, Bishop, knobcone, and Torrey—stand at intervals along the coast.

The montane coniferous forest of the Sierra Nevada may, for ornithological convenience, be roughly divided at the 6,000-foot level. Below this point the dominant trees are the ponderosa pine, sugar pine, incense cedar, Douglas-fir, white fir, and black oak. Above 6,000 feet the lodgepole pine, Jeffrey pine, red fir, and western white pine are characteristic; near the timber line, whitebark pine, foxtail pine, limber pine, and mountain hemlock form an open subalpine forest. In the list below, the symbols H (high, mainly above 6,000 feet), L (low, mainly below 6,000 feet), and C (coastal) indicate the distribution of the more common breeding birds of the *coniferous forests*. Species residing in all three divisions are unmarked. Most species in the Sierra Nevada also range into the inner Coast Ranges of northern California, but in the following list these species are not considered 'coastal,' since they do not usually reside in the redwood or coastal pine associations:

Sharp-shinned Hawk (L, C)
Blue Grouse (H, L, *north coast*)
Ruffed Grouse (*north coast, local*)
California Quail (C)
Mountain Quail
Band-tailed Pigeon (L, C)
Common Screech Owl (L, C)
Flammulated Screech Owl (L, C)
Great Horned Owl
Northern Pygmy Owl (L, C)
Spotted Owl
Saw-whet Owl
Common Nighthawk (H, L)
Vaux's Swift (C)

Allen's Hummingbird (C)
Calliope Hummingbird (H, L)
Red-shafted Flicker
Pileated Woodpecker
Acorn Woodpecker (L, C, *in oaks*)
(Red-breasted) Yellow-bellied Sapsucker (H, L, *north coast*)
Williamson's Sapsucker (H)
Hairy Woodpecker
White-headed Woodpecker (H, L)
Black-backed Three-toed Woodpecker (H)
Hammond's Flycatcher (H)
Western Flycatcher (C)
Western Pewee
Olive-sided Flycatcher
Violet-green Swallow
Tree Swallow
Gray Jay (C, *north*)
Steller's Jay
Clark's Nutcracker (H)
Mountain Chickadee (H, L)
Chestnut-backed Chickadee (C)
White-breasted Nuthatch
Red-breasted Nuthatch (H, *local coastally*)
Pygmy Nuthatch (*pines only*)
Brown Creeper
Winter Wren (*mainly* C, *local* L)
American Robin
Varied Thrush (C, *north*)
Hermit Thrush
Swainson's Thrush (L, C)
Western Bluebird (L, C)
Mountain Bluebird (H)
Townsend's Solitaire (H)
Golden-crowned Kinglet
Ruby-crowned Kinglet (H, L)
Solitary Vireo
Nashville Warbler (L, *in deciduous growth*)
Audubon's Warbler
Hermit Warbler (H, L, *local coastally*)
MacGillivray's Warbler
Western Tanager
Black-headed Grosbeak
Evening Grosbeak (H)
Purple Finch (L, C)
Cassin's Finch (H)
Pine Grosbeak (H)
Pine Siskin
Red Crossbill (H, *local coastally*)
Oregon Junco
White-crowned Sparrow
Fox Sparrow
Lincoln's Sparrow (H, *mountain meadows*)

The dense stands of woody shrubs that occupy large tracts of the arid foothills in the state are composed of plants which the early Spanish settlers called 'la chaparra.' The area where they grow, in accordance with the Spanish word structure, is therefore 'chaparral.' Chaparral is characteristic of the dry inner Coast Ranges along the west side of the Sacramento and San Joaquin Valleys, and of the western slopes of the Sierra Nevada, and reaches its greatest extent in southwestern California. It occupies nearly 10 per cent of the surface of the state, or approximately 10 million acres. The component plants vary greatly but are always stiff, heavily branched, woody perennials. The manzanitas, scrub oak, chamise, mountain lilac or buckthorn (several species), poison oak, baccharis, and certain sages are plants of the arid chaparral. Along the coast and in riparian situations in the Sierra there exists a moister phase that is often designated as 'soft chaparral.' Although the component plants are usually different, the growth form is similar, and the bird associates are nearly the same. MacGillivray's Warbler breeds in this type, both coastally and in the Sierra. At middle altitudes (4,000 to 8,000 feet) in the Sierra is a chaparral association of mountain mahogany, snowbush, huckleberry oak, and manzanita. The Dusky Flycatcher, Green-tailed Towhee, and Fox Sparrow are birds breeding in this montane chaparral, with the Nashville Warbler where scattered trees are admixed. The following are characteristic breeding birds of *low-altitude chaparral:*

Poor-will
Costa's Hummingbird (*southern California*)
Ash-throated Flycatcher
Scrub Jay
Bushtit
Wrentit
Bewick's Wren
California Thrasher
Blue-gray Gnatcatcher
Black-tailed Gnatcatcher (*southern California, local*)
Orange-crowned Warbler
Lazuli Bunting
Lesser Goldfinch
Rufous-sided Towhee
Brown Towhee
Rufous-crowned Sparrow
Sage Sparrow
Black-chinned Sparrow
White-crowned Sparrow (*coastal*)

Where the Great Basin impinges upon the eastern side of the Sierra Nevada, pinyon pine, juniper, and sagebrush are the dominant plants. This semidesert association exists in the northeastern corner of the state (eastern Modoc and Lassen Counties), in the central-eastern portion (Mono and Inyo Counties), in western Kern County, on the higher ranges in the Mojave Desert, and along the lower slopes of the southern California mountains. The breeding birds of this association include:

American Kestrel
Sage Grouse (*northeast and east of Sierras*)
Mourning Dove
Poor-will
Ash-throated Flycatcher
Gray Flycatcher (*northeast and east of Sierras*)
Horned Lark
Scrub Jay
Black-billed Magpie (*northeast and east of Sierras*)
Northern Raven
Pinyon Jay
Plain Titmouse
Bushtit
Sage Thrasher
Blue-gray Gnatcatcher
Loggerhead Shrike
Gray Vireo
Vesper Sparrow
Black-throated Sparrow (*east of Sierras*)
Sage Sparrow
Brewer's Sparrow

The Mojave Desert extends northward to Death Valley and westward to the San Gabriel and Telachapi Mountains. It varies in altitude from 2,000 to 5,000 feet, with peaks rising to over 7,000 feet in some of the desert ranges. The lower Colorado Desert extends eastward from the Salton Sea Basin to the Colorado River. The bird finder driving across these deserts will pass mile after mile of dull-green creosote bush. In the washes are mesquite, paloverde, ironwood, and smoke tree; the few permanent streams are bordered with cottonwood and willow. The angular Joshua tree is a prominent feature in the higher Mojave; the spiny, red-flowered ocotillo is characteristic of the Colorado Desert. There are over twenty species of cacti, at least half of which grow in both Deserts. The giant saguaro stands in a few scattered localities near the Colorado River. Among the birds regularly breeding in both the *Mojave and Colorado Deserts* are the following:

Prairie Falcon
American Kestrel
Gambel's Quail
Mourning Dove
Greater Roadrunner
Great Horned Owl
Lesser Nighthawk
Poor-will
Costa's Hummingbird
Ladder-backed Woodpecker

Say's Phoebe
Northern Raven
Verdin
Cactus Wren
Rock Wren
Northern Mockingbird
Le Conte's Thrasher (*local*)
Black-tailed Gnatcatcher
Phainopepla
Loggerhead Shrike
Scott's Oriole
House Finch
Black-throated Sparrow

The following species breed regularly only in the *Colorado Desert:*

White-winged Dove
Gila Woodpecker
Vermilion Flycatcher (*local*)
Crissal Thrasher
Hooded Oriole
Abert's Towhee

California has numerous coastal salt marshes, usually dominated by pickleweed (*Salicornia*) and to a less extent by spartina grass and low bushes (*Grindelia*). Few bird species are entirely restricted to salt marshes, but many wintering shorebirds retire to rest in the marshes when high tides cover their feeding areas on the nearby mud flats. Wintering ducks find refuge on tidal sloughs and ponds; herons and egrets visit the marshes to feed. Breeding birds include the Northern Harrier, Clapper Rail, Short-eared Owl, and Savannah and Song Sparrows. During winter these residents are joined by American Coots, Virginia Rails, Soras, Marsh Wrens, Water Pipits, and Common Yellowthroats. When the highest tides of the year flood the salt marshes of Tomales and San Francisco Bays, it is possible to see the secretive Black Rails which at other times remain well hidden in the pickleweed tangles. They may also be seen in early spring in the fresh-water marshes along the lower Colorado River (*see under* **Yuma, Arizona**)

California's coastline, approximately 1,500 miles in actual length, is extremely varied, with rocky cliffs and headlands providing shelter for small bays and sandy beaches. The mild climate along the coast permits numerous transient waterbirds, waterfowl, and shorebirds to spend the winter. The bays of Humboldt, Tomales, Bolinas, San Francisco, Monterey, Morro, Newport, and others provide shelter from ocean storms. Rocky headlands and offshore islets have nesting sites for Brandt's Cormorants, Pelagic Cormorants, Western Gulls, Thin-billed Murres, and Pigeon Guillemots. The following lists include regular residents, transients, and winter visitants. An asterisk indicates that the species nests along the California coast. Species in the first list may be observed on all types of *coast or on coastal waters.*

Common Loon
Arctic Loon
Red-throated Loon
Horned Grebe
Eared Grebe
Western Grebe
* Brown Pelican
* Double-crested Cormorant
White-winged Scoter
Surf Scoter
Red-breasted Merganser
Ruddy Turnstone
Whimbrel
Willet
Red Phalarope
Wilson's Phalarope
Northern Phalarope
Glaucous-winged Gull
* Western Gull
Herring Gull
Thayer's Gull
California Gull
Ring-billed Gull
Mew Gull
Bonaparte's Gull
Heermann's Gull
* Forster's Tern
Royal Tern
Elegant Tern
Caspian Tern

Rocky shores provides nest sites or forage areas for the following:

* Brandt's Cormorant
* Pelagic Cormorant
* Black Oystercatcher
Surfbird
Black Turnstone
Spotted Sandpiper
Wandering Tattler
* Thin-billed Murre
* Pigeon Guillemot

Sandy beaches and adjacent dunes are the favored habitat for the following:

* Snowy Plover
Marbled Godwit
Sanderling
* Little Tern
Water Pipit
* Savannah Sparrow

On the more *sheltered bays and estuaries*, and on the tidal mud flats, the bird finder may see the following transients and/or winter visitants:

Pied-billed Grebe
American White Pelican
Great Blue Heron
Great Egret
Snowy Egret
Black Brant
Mallard
Common Pintail
American Wigeon
Canvasback
Lesser Scaup
Common Goldeneye
Bufflehead
Ruddy Duck
American Coot
Semipalmated Plover
Killdeer
Black-bellied Plover
Long-billed Curlew
Greater Yellowlegs
Lesser Yellowlegs
Red Knot
Least Sandpiper
Dunlin
Short-billed Dowitcher
Long-billed Dowitcher
Western Sandpiper
American Avocet
Black-necked Stilt
Belted Kingfisher

Certain rocky islands and islets off the California coast are nesting sites for the Brown Pelican, Double-crested Cormorant, and four alcids: Xanthus' Murrelet, Cassin's Auklet, Rhinoceros Auklet, and Tufted Puffin. The Marbled Murrelet nests high in trees inshore from the northern California coast. At least three species of storm petrels—the Fork-tailed, Leach's, and Ashy—breed on offshore islands but they seldom come close to the coast except when forced by strong onshore winds or by violent storms.

Groups of pelagic birds may sometimes be viewed from headlands, breakwaters, or other vantages jutting oceanward. From April through November, but primarily in summer, Sooty Shearwaters by the thousands mill about over the sea or pass in long lines. Less common among them are Pink-footed Shearwaters. Both species are visitants from the Southern Hemisphere. A frequent winter visitant, occasionally common, is the Northern Fulmar, which breeds far to the north.

Pelagic birds are best viewed off the coast on boat trips, scheduled by bird clubs and by private charter, leaving such favored ports as Mission Bay (San Diego), Newport Bay, San Pedro, and Oxnard-Ventura in southern California, Morro Bay, Monterey, Berkeley (San Francisco Bay), and Bodega Bay in central California, and Eureka in northern California. *In spring* (particularly late April and early May), a trip from any port is rewarding for the aforementioned shearwaters and, additionally, for northbound Red and Northern Phalaropes, Pomarine and Parasitic Jaegers, Sabine's Gulls, and Arctic Terns. *In midsummer*, from any port for good numbers of shearwaters; from southern ports for Black Storm Petrels as well as for Xanthus' Murrelets near the southern California islands where they nest; from Monterey and ports north to the Oregon border, for Black-footed Albatrosses, Thin-billed Murres, Pigeon Guillemots, and Cassin's Auklets. *In late summer and fall* (particularly late September and early October), from southern California toward San Clemente Island or the Osborne Bank, for the common pelagic species plus Least Storm Petrels, Craveri's Murrelets, and occasionally Red-billed Tropicbirds; from Monterey to the rim of the deep submarine canyon in Monterey Bay for storm petrels, shearwaters, jaegers, alcids, and gulls in great abundance and the strong possibility of a Flesh-footed Shearwater (usually among the swarms of Sooty and Pink-footed), Buller's Shearwater, South Polar Skua, and other rarities in California waters. *In winter*, from any port for Black-footed and Laysan Albatrosses, Northern Fulmars, Short-tailed Shearwaters, Fork-tailed Storm Petrels, Black-legged Kittiwakes, and numerous alcids including Rhinoceros Auklets and Marbled and Ancient Murrelets.

In most of California there are no sharply distinct migration periods as in the eastern United States; multitudes of transients pass through the state over such extended seasons that there is never a time of year when none is on the move. Northbound landbird transients sweep into the state from the southeast, passing across the deserts and adjacent foothills with little regard for their normal habitats. One main line leads up from Imperial Valley along the Little San Bernardino Mountains (*see under* **Twentynine Palms**) and then splits into two: (1) A major route following the southern foothills of the San Bernardino and San Gabriel Mountains to the Tehachapi Range and the Sierra Nevada, or the Central Valley. (2) A route along the north bases of the San Bernardinos and San Gabriels, thence across the western Mojave Desert and up the east flank of the Sierra Nevada. Swainson's and other hawks, Vaux's Swift, hummingbirds, flycatchers, swallows, thrushes, vireos, warblers, orioles, and tanagers move in waves along both of these routes; and those that use the second may switch to the first one at the Tehachapis. Parallel spring routes in the low mountains of the San Diego area converge with these two routes above Los Angeles, while many swallows continue northward along the coast or valleys near the coast. In the San Francisco Bay and Central Valley regions, however, there seem to be less spectacular movements, the birds gradually arriving and departing, with few or no waves. Only a few species, such as the Vaux's Swift, Rufous Hummingbird, Townsend's Warbler, and Western Tanager, pass along the north Coast Ranges in numbers. Swainson's Thrushes are very widespread as transients and may be heard overhead at night along many of these routes.

At the beginning of the southward migration, from late June into September, most small landbirds use the high-altitude route, the foothills being hot and dry at this season. In July the flower-carpeted meadows of the high Sierra are in the noisy possession of hordes of Rufous Hummingbirds already southward bound; and by late August an amazing array of transients and up-mountain, post-breeding wanderers spreads to and above the timber line. Participating in this altitudinal movement, an important aspect of western migration, are flycatchers, most of the vireos and warblers, Swainson's Hawk, American Robin, Hermit Thrush, Swainson's Thrush, Western Tanager, Black-headed Grosbeak, Purple and Cassin's Finches, and even the familiar Brewer's Blackbird. The southbound transients find spring just past at these altitudes and flowers and insects plentiful, although night temperatures may be near freezing and winter not far off.

Many species that nest in the mountains carry out a complementary down-mountain migration as the cold weather sets in, the Mountain Quail making the trip of 25 to 40 miles down the west slope of the Sierra on foot. Meanwhile the Blue Grouse moves up into the snow to feast on the fir needles all winter. By October the foothills are teeming with transients, particularly the later parulids and fringillids.

Waterbirds, waterfowl, and shorebirds in spring use the same routes along the southern California mountains in passing from the Imperial Valley to the coastal areas, or into the Central Valley with its series of marshes. Flocks of American White Pelicans, geese, and gulls are especially frequent high over the mountains, the California and Ring-billed Gulls following an almost east-west route from the coast across the central part of the state to reach breeding grounds in the Great Basin. Similarly, in the north, the main movement of waterfowl is northeastward from the Sacramento Valley wintering grounds to the elevated Modoc Plateau and the Great Basin region. A strictly coastwise route is used by the Black Brant, many gulls and terns, some shorebirds, and by loons and scoters, which may be counted as they pass headlands.

Herewith is a general timetable of migration periods:

American White Pelican	October–April
Waterfowl and most waterbirds	(August), September–December; February–April
Hawks	February–early April; September–November
Most shorebirds	April–early May; late July–October
Phalaropes, terns	May; (July), August–September
Vaux's Swift	late April–early May; August–September
Hummingbirds	February–April (May); (June), July–September
Flycatchers	April–May; July–September (October)
Swallows	March–April; late August–September
Swainson's Thrush	late April–May; late August–September

Vireos and warblers	April–early May; August–September (October)
Orioles, Black-headed Grosbeak	late March–early May; August–early September
Western Tanager	late April–May; August–September
Sparrows	March–April; late September–November

For the bird finder, winter in California holds special attractions. There are many species of birds at the lower altitudes from October to April, as during the rest of the year; and many of the species are present in far greater numbers than are any of the summer birds.

Beginning as early as early August with the arrival of Common Pintails in flocks, the winter waterfowl population of the Sacramento Valley is estimated at many millions from November through February. In the moist stubble fields one can then see, often from a paved road, flocks of several thousand Canada Geese and Greater White-fronted Geese and scores of thousands of Mallards, Common Pintails, Green-winged Teal, and American Wigeons. Snow Geese and Northern Shovelers are also abundant, and there are usually small numbers of Ross' Geese, but they keep somewhat closer to the ponds and marshes of the refuges and gun clubs, as do the diving ducks, such as the Canvasback, Lesser Scaup, Common Goldeneye, Bufflehead, and Ruddy Duck. Whistling Swans and Sandhill Cranes are common in several areas in December and January. Throughout the state, except in the northeastern corner where lakes freeze over, ducks winter abundantly on all bodies of water where food and protection are provided; American Coots are ubiquitous and exceedingly abundant in many areas; large numbers of Eared and Western Grebes frequent the larger lakes; and both Great and Snowy Egrets winter commonly in central and southern California.

Along the coast the common winter birds include three kinds of loons, five grebes, three cormorants, eight gulls, and a variety of shorebirds and waterfowl, as well as alcids and other pelagic species that frequent rocky headlands. From Santa Barbara southward, Western Grebes congregate in flocks of thousands to fish beyond the breakers. The larger bays all along the coast support great rafts of Black Brant and diving ducks. Many thousands of the smaller shorebirds winter in the San Francisco Bay area and southward; the larger species are most common along the southern coast, where in a single midwinter group there may be hundreds of Willets, both dowitchers, Marbled Godwits, American Avocets, and dozens of Long-billed Curlews.

Throughout the state at the lower altitudes, in city and country, the following winter visitants are normally common from October to March in almost any habitat that has food and cover:

Sharp-shinned Hawk
Cooper's Hawk
American Robin
Hermit Thrush
Ruby-crowned Kinglet
Cedar Waxwing
Audubon's Warbler
Oregon Junco
White-crowned Sparrow
Golden-crowned Sparrow
Fox Sparrow

Besides the above, several species are common winter visitants to certain habitats or regions: Band-tailed Pigeon, in or near oak woodland; (Red-breasted) Yellow-bellied Sapsucker, in foothill orchards and in pepper and oak trees; Myrtle Warbler, in the northwest coast region and, less commonly, to the south in riparian and nearby habitats; Townsend's Warbler, in live oaks and riparian woodland, usually near the coast from central California south; Varied Thrush, in the northwest coast region and erratically throughout the state in canyon and oak woodlands; Lewis' Woodpecker (erratic), in oaks; and Golden-crowned Kinglet (erratic), in conifers and other evergreens.

In the open fields of the valleys, flocks of California and Ring-billed Gulls follow the plows; Ferruginous and Rough-legged Hawks join the resident hawk species, the latter chiefly in the north. Many Water Pipits, Red-winged and Tricolored Blackbirds, and Savannah Sparrows roam about the grasslands and fields, the

Red-winged and Tricolored Blackbirds often in huge massed flocks with which may be associated European Starlings, Brewer's Blackbirds, and a few Brown-headed Cowbirds. Vesper Sparrows are widespread but only locally common. Say's Phoebes and Mountain Bluebirds favor fields with little or no vegetation, and flocks of Mountain Plovers are frequent in similar situations (*see under* **Brawley**).

The deserts also receive winter influxes of several species that nest in other habitats to the north—for example, the Bewick's Wren, Sage Thrasher, Blue-gray Gnatcatcher, and four sparrows, the Sage, Chipping, Brewer's, and White-crowned. The cold northeastern corner of the state (*see under* **Tulelake**) receives northern visitants, not found in other parts of the state, such as the Northern Shrike, Gray-crowned Rosy Finch, Common Redpoll, American Tree Sparrow, Lapland Longspur, and an occasional Snow Bunting, but there are fewer resident species here. The greatest variety of wintering landbirds is in the foothills of the Sierra Nevada and the southern Coast Ranges, and along the coast where the temperature is seldom below freezing.

Authorities

E. Turner Biddle, Laurence C. Binford, Jean Brandt, Theodore A. Chandik, Howard L. Cogswell, Elizabeth Copper, Richard A. Erickson, Stanley W. Harris, James A. Lane, Jeri Langham, Paul Lehman, Susanne A. Luther, William Reese, Charles G. Sibley, Arnold Small, S. K. Stocking.

References

The Birds of California. By Arnold Small. New York: Winchester Press. 1974.

Water Birds of California. By Howard L. Cogswell. Berkeley: University of California Press. 1977.

Annotated Field List of Birds of South California. By Robert L. Pyle; revised by Arnold Small. Los Angeles Audubon Society, 7377 Santa Monica Boulevard, Los Angeles, CA 90046. 1961.

Birds of Northern California: An Annotated Field List. By Guy McCaskie and others. 2d. ed. Golden Gate Audubon Society, 2718 Telegraph Avenue, Suite 206, Berkeley, CA 94705. 1979.

A Birder's Guide to Southern California. By James A. Lane. Distributed by L & P Press, Box 21604, Denver, CO 80221. 1979.

BAKERSFIELD
Mt. Pinos | Edmundson Pumping Station | Dough Flats | Sespe Wildlife Area

Almost anywhere over the foothill grazing country south of Bakersfield there is always a slight chance of sighting a soaring California Condor. Opportunities for seeing the great birds are best, however, in the rugged mountains south of Bakersfield and east of Santa Barbara.

One good site is from the summit of **Mt. Pinos** (8,831 feet), accessible by car. To reach it, take the trip (*see below*) in summer, preferably in August, and plan to be at the summit before mid-day in order to watch for the condors in the afternoon when they are most likely to be riding the air currents. Even if the condors are missed, the trip will be rewarding for other birds en route and in the montane coniferous forest near the summit.

From I 5, about 42 miles south of Bakersfield and 60 miles north of Los Angeles, exit west through the town of Frazier Park (elevation 4,750 feet) and continue west to the town of Lake of the Woods and on through Cuddy Valley, largely sagebrush country, worth investigating for Lazuli Buntings and such sparrows as the Lark, Sage, Brewer's, and Black-chinned.

From the road in Cuddy Valley, at the intersection of Mill Potrero Road, continue left on the 19-mile paved road up Mt. Pinos. Be alert for Mountain Quail all the way. At the McGill Campground, the first of two at 7,500 feet, there are Band-tailed Pigeons, (Red-breasted) Yellow-bellied Sapsuckers, White-headed Woodpeckers, Dusky Flycatchers, Steller's Jays, Mountain Chickadees, all three nuthatches, Brown Creepers, both kinglets, Audubon's Warblers, Western Tanagers, Cassin's Finches, sometimes Red Crossbills and possibly Northern Pygmy Owls.

Farther up the main road, numerous named logging roads—all dirt and dead-ending—intersect and lead through coniferous-forest habitats. One is Fir Ridge Road on the right, especially worth a walk to the north end for Williamson's Sapsuckers and Townsend's Solitaires. The pavement of the main road ends in a parking area near Iris Meadow. The bordering buckthorn thickets here should be investigated for Calliope Hummingbirds, Green-tailed Towhees, and Fox Sparrows. Flammulated Screech Owls and Saw-whet Owls may be in the nearby woods.

The condor lookout is at the west end of the summit, where it commands a 360-degree panoramic view. To reach the lookout from the parking area, go up a dirt road (rough, closed in winter) for about 2.5 miles. When coming out on the rocky area near the summit, watch for Rock Wrens, both Western and Mountain Bluebirds, and Lawrence's Goldfinches. Clark's Nutcrackers are common in the area.

Condor watching at the lookout means patience and eyestrain—constantly watching the sky for a big gliding bird—and sometimes momentary hope when a Golden Eagle, always a possibility, is first sighted. Among the raptors more usually sighted are the Cooper's and Red-tailed Hawks, Prairie Falcon, and American Kestrel.

If no condor appears, there are two other promising lookouts for later afternoons, preferably in winter.

One is the observation station at the **Edmundson Pumping Station** adjacent to an aqueduct. From I 5, 10.5 miles north of the exit to Frazier Park, take the Grapevine Exit and proceed east 1.6 miles, following signs to the Station. During the winter months, this is good country for Rough-legged and Ferruginous Hawks, Prairie Falcons, and Merlins, as well as Lewis' Woodpeckers in the large oaks.

The other is an overlook from **Dough Flats.** From I 5 (34 miles south of the exit to Frazier Park and 34 miles north of Los Angeles), exit west on State 126 for 19 miles to Fillmore. Here turn off right to A Street (becomes Goodenough Road) and go north for 3 miles, then turn right onto a dirt road (treacherous in wet weather), following signs to Dough Flats—15 miles from Fillmore. Dough Flats overlooks part of the 53,000-acre **Sespe Wildlife Area,** established primarily as a sanctuary for the California Condor and closed to the public.

BISHOP
Owens Valley | Mt. Whitney | Mammoth Pass

In eastern-central California this bustling supply center for many high-altitude Sierra Nevada resorts, and outfitting center for packers and hikers, is at the northern end of the long trough of **Owens Valley,** at the point where numerous streams converge to form the Owens River. In the fields and brushlands, from the vicinity of Bishop southward, California Quail, Greater Roadrunners, Burrowing Owls, Black-billed Magpies, Northern Ravens, Bushtits, Lesser Goldfinches, and Savannah Sparrows reside all year. In summer, Common Nighthawks forage daily over the Valley from nesting grounds on higher mountains to the west, and Lesser Nighthawks nest in limited numbers on the floor of the Valley, this area being one of the few places where both may be seen. The Bell's Vireo, Yellow-breasted Chat, Yellow-headed Blackbird, and Blue Grosbeak nest along streams or irrigation ditches; many other riparian-woodland species and the Lewis' Woodpecker inhabit the cottonwoods lining the Owens River and its tributaries. Winter visitants in the open Valley include the Northern Harrier, Water Pipit, Audubon's Warbler, and the Vesper, Sage, White-crowned, and Lincoln's Sparrows.

From Bishop southward for 119 miles, US 395 parallels one of the greatest fault scarps in the world, marked by the east face of the Sierra Nevada, which rises abruptly to 10,000 feet above the floor of Owens Valley, affording spectacular scenery along the way. Side roads lead from the main highway westward up many of the rugged canyons. For the bird finder interested in nesting Pine Grosbeaks, White-crowned Sparrows, and especially in Gray-crowned Rosy Finches, the following two roads leading to the high elevations are recommended: (1) From Lone Pine on US 395, 59 miles south of Bishop, the road up Lone Pine Canyon. This ends at Whitney Portal, 8,371 feet elevation, beyond which is a 7-mile trail up **Mt. Whitney** to Outpost Camp, at 10,350 feet, where rosy finches are occasional, and thence to Consultation Lake, where they are common. (2) From Casá Diablo Hot Springs on US 395, 42 miles northwest of Bishop, the road to Mammoth Lakes and beyond, ending at about 9,000 feet, just short of **Mammoth Pass.** Rosy finches are here in summer and probably nest in the vicinity. All the roads and trails leading to the high altitudes mentioned are open, of course, only in summer, and it is usually advisable to check with the Inyo National Forest ranger station in Bishop before attempting them.

BRAWLEY

Salton Sea and Environs | Ramer Lake | Finney Lake | Mouth of New River | Salton Sea National Wildlife Refuge | Red Hill Marina | Wister Waterfowl Management Area

The **Salton Sea and Environs,** on the floor of the Imperial Valley in the Colorado Desert in southern California, constitute one of the most productive areas for year-round bird finding in the state. The Salton Sea itself, of man's creation in 1905 when waters from the Colorado River were accidentally diverted into the Salton Basin, is about 35 miles long and 16 miles in greatest width, its surface between 235 and 240 feet below sea level. Although subject to considerable evaporation, the Sea is continually replenished by waters from the Whitewater River at the north end, the Alamo and New Rivers at the south end, and small irrigation canals. The Sea is nonetheless highly saline; yet it supports an abundance of plankton and other small organisms which in turn support numerous fish species for waterfowl and waterbirds. Silt, carried by the inflow of waters, has formed extensive mud flats that are rich in aquatic invertebrates for shorebirds. The once natural, outlying desert is largely supplanted by irrigated fields with tamarisk, mesquite, catclaw, and paloverde along rivers and canals and around the few marshy, fresh-water lakes—a situation that now attracts a wide variety of landbirds, besides desert species, and certain waterfowl and waterbirds from the Sea to feed on insects, grains, and succulent plants.

Among the birds that may be expected at the Salton Sea in suitable habitats the year round are the Pied-billed Grebe, American White Pelican (except in summer), Double-crested Cormorant, Great Blue and Green Herons, Cattle, Great, and Snowy Egrets, Black-crowned Night Heron, Least Bittern (scarce in winter), White-faced Ibis, Cinnamon Teal, Ruddy Duck (scarce in early winter), Gambel's Quail, Clapper and Virginia Rails, Sora, Common Gallinule, American Coot, Snowy Plover, Killdeer, American Avocet, Black-necked Stilt, Common Ground Dove, Greater Roadrunner, Burrowing Owl, Ladder-backed Woodpecker, Horned Lark, Verdin, Cactus and Rock Wrens, Crissal Thrasher,

Loggerhead Shrike, Common Yellowthroat, Western Meadowlark, Yellow-headed and Red-winged Blackbirds, Abert's Towhee, and Song Sparrow.

Winter is the ideal time to visit the Salton Sea; temperatures are comfortable—seldom above 70 degrees F.—and birds are in greatest numbers. The Sea itself teems with Eared Grebes and Ruddy Ducks and somewhat fewer numbers of Western Grebes, Redheads, Canvasbacks, Lesser Scaups, Common Goldeneyes, and Buffleheads. Shallows and lakes are alive with surface-feeding ducks, particularly Common Pintails, Green-winged Teal, American Wigeons, and Northern Shovelers. At sundown, thousands of geese pass in skeins over grainfields. The majority are Snow Geese; others are mainly Canada Geese, small family groups of Ross' Geese, and rarely a few Greater White-fronted Geese that have not gone to wintering grounds farther south.

The Sea and all waterways and fields frequently abound with wintering gulls, chiefly Ring-billed but with Herring Gulls well represented. On dirt fields recently plowed there are often wintering flocks of Mountain Plovers and Mountain Bluebirds. Trees and shrubby thickets harbor wintering passerines such as Orange-crowned and Audubon's Warblers.

Spring and fall are rewarding seasons. From late March through April, transient swifts, swallows, vireos, parulids, tanagers, and fringillids sweep up the Imperial Valley. The numbers of Tree Swallows in passage are impressive, especially during the first week of April. In fall, the movements of all these small landbirds tend to be more scattered and leisurely, thus less noticeable. What is remarkable about both spring and fall migrations is the appearance of birds that one usually expects along the coast. In spring migration, for instance, the Whimbrel is sometimes abundant and the Black Brant, Red Knot, and Sanderling are common. In early fall, the Parasitic Jaeger shows up regularly as does the Magnificent Frigatebird.

Summer, despite the intense heat, can also be rewarding. By then the resident variety of breeding birds has been increased with the arrival of Fulvous Whistling Ducks, Gull-billed Terns, and Black Skimmers for nesting and hundreds of Black Terns which stay through the summer but do not nest. Late in the summer, from mid-July until October, Brown Pelicans, Wood Storks, Western Gulls (yellow-legged form), and Laughing Gulls are regular visitants; and, occasionally, boobies and other sea birds are also visitants, probably coming north from the Gulf of California.

The following trip from Brawley, south of the Salton Sea, takes in several sites northward that will yield a representative variety of bird species and bird aggregations in one season or another. Caution: Most roads leading to the sites from the state highway are dirt and consequently treacherous in wet weather. In summer, be prepared for extremely high temperatures, said to be the hottest in the United States.

Drive north from Brawley on State 111, watching all irrigated fields. In winter, scan any dirt fields immediately north of Brawley for Sandhill Cranes, Mountain Plovers, and longspurs. In any season there may be Common Ground Doves near farmhouses and in ranchyards.

Before reaching Calipatria, turn off east onto Albright Road for 0.4 mile, south on Kershaw Road for 1.5 miles, then west onto Quay Road across a railroad grade to **Ramer Lake.** Nesting in its marshes are Yellow-headed Blackbirds among other species. From Ramer Lake, drive south on Perimeter Road about 1.0 mile to **Finney Lake.** Least Bitterns as well as Yellow-headed Blackbirds and a few Great-tailed Grackles nest in its marshes. In spring and summer, look for Fulvous Whistling Ducks. Inspect the adjacent tamarisks and brushy growth for breeding, transient, or wintering birds. Cactus Wrens and Abert's Towhees are year-round residents as are a pair or two of Crissal Thrashers, which can be found by diligent searching.

Return to State 111 and continue north to Calipatria; turn west onto County S30 to Gentry Road, south 0.5 mile to Bowles Road, then west onto Bowles Road (becomes dirt) to the end. Here at the **Mouth of New River** are reeds and tall grasses for Least Bitterns and Clapper Rails, and mud flats for many transient shorebirds in spring and for Laughing Gulls and Gull-billed Terns in summer.

Return to State 111 and continue north for 3.8 miles, then turn off west onto Sinclair Road (paved) and go 5.6 miles to headquarters of the **Salton Sea National Wildlife Refuge.** Along the way, watch for Greater Roadrunners and Burrowing Owls on the banks of the irrigation ditches. At headquarters, obtain a bird checklist and inquire about road conditions and good sites for birds at the time of the visit.

The 36,527 acres of the Refuge, bordering the southern end of the Salton Sea, comprise croplands and marshes for hordes of geese and other waterfowl during the winter months and for many other birds in season. Rock Hill, reached from headquarters by driving west and then north on a dirt road is sometimes an excellent vantage for viewing great aggregations of Snow Geese and many wintering gulls on the lagoons extending east and south.

Another good vantage lies north of the Refuge at **Red Hill Marina,** reached from headquarters by driving east on Sinclair Road for 1.0 mile, then north on Garst Road to the entrance. From the nearby terraces in summer, look for Wood Storks, Laughing Gulls, Great-tailed Grackles, and such wanderers as the Magnificent Frigatebird and Roseate Spoonbill.

Return to State 111 and continue north. Five miles north of Niland, turn off left to the **Wister Waterfowl Management Area.** This state-owned property embraces extensive fields which, in May, bring many shorebirds including Long-billed Dowitchers, Western Sandpipers, and sometimes Stilt Sandpipers. Thousands of geese are here in winter.

CAPISTRANO BEACH
San Juan Capistrano Mission

When driving between Los Angeles and San Diego on I 5, the bird finder may wish to exit west on State 74 (3 miles north of the Capistrano Beach Exit) and drive one block to see the **San Juan Capistrano Mission** with its nationally famous Cliff Swallows. The experienced bird finder will not, of course, expect the swallows to conform as faithfully in arrival and departure dates—19 March and 23 October—as the legend would have them, and he must not be disappointed if White-throated Swifts are the only prominent aerial foragers here on appointed days.

DEATH VALLEY JUNCTION
Death Valley National Monument

In the eroded, semidesert country of eastern-central California and southwestern Nevada, the 2,891 square miles of **Death Valley National Monument** includes the 165-mile-long Death Valley and bordering precipitous mountains. Of the Monument's total area, 550 square miles are below sea level, reaching at Bad Water (−282 feet) the lowest point in North America. Telescope Peak, 15 miles away in the Panamint Mountains, attains an elevation of 11,049 feet.

Headquarters, a visitor center, and museum are within the Monument on State 190, 35 miles north of Death Valley Junction, which is outside the Monument near the Nevada line. Besides the entrance from Death Valley Junction, there are other entrances on paved roads leading into the Monument and interconnecting. At any one of the ranger stations or the visitor center, the bird finder should obtain a map showing the road system and the particular localities of birds that will be mentioned below.

From May to September, mid-day temperatures at sea level and below range from 90 to 110 degrees F., sometimes higher; bird finding is consequently recommended only in the early morning or late afternoon. During other months the temperatures are usually milder throughout the day.

In the areas below sea level, most covered with salt deposits and sand dunes, there are a few birds—e.g., Say's Phoebes, Rock Wrens, and Northern Ravens; but along the streambeds entering the Valley, growths of mesquite and arrowweeds harbor such nesting birds as Gambel's Quail, Costa's Hummingbirds, Verdins, Lesser Goldfinches, and others. In or near the several permanent watering places around the edges of the Valley and on the floor, many transients congregate, especially in April and early May, October and early November. Among the species are the Great Blue Heron, Great and Snowy Egrets, American Bittern, at least seven species of surface-feeding ducks, Cooper's Hawk, American Coot, Mourning Dove, Rough-winged Swallow (may nest locally), Water Pipit, Audubon's Warbler, Yellow-headed Blackbird, and the Savannah, Vesper, and Chipping Sparrows. Species breeding in the vicinity of the larger springs include the Yellow Warbler, Yellow-breasted Chat, Red-winged Blackbird, Bullock's Oriole, and House Finch. Regular wintering landbirds, some of which stay into April, include the Sage Thrasher, American Robin, Blue-gray Gnatcatcher, and the Sage, Brewer's, White-crowned, and Song Sparrows.

Ideal for seeing many of these birds is the golf course at Furnace Creek Ranch on US 190 south of headquarters. At the northwest corner of the golf course is a mesquite-bordered pond; elsewhere there are more open ponds and a group of sewage-disposal ponds at the northwest corner of the airstrip. Besides all these ponds, the bird finder should pay special attention to the rows of tamarisk trees for transient warblers, the mesquite edging the course for transient Indigo Buntings and nesting Lucy's Warblers, and the grassy areas and lawns for spring-transient Bobolinks.

Another fine oasis for birds is Scotty's Castle at the northern extremity of the Monument. Cottonwoods shading the picnic area and the mesquite southwest of it, also the jungle-like growths northeast of the Castle and easily reached from the parking lot, yield a good variety of birds in season. Chukars are permanent residents in the area.

The height of the nesting season is mid-April, when the shrub-dotted slopes at medium altitude in the Monument have nesting Greater Roadrunners, Burrowing Owls, Le Conte's Thrashers, and Black-throated Sparrows. The Prairie Falcon, Great Horned Owl, White-throated Swift, Ash-throated Flycatcher, and Rock Wren nest where cliffs or rock jumbles provide suitable crevices; and the Canyon Wren nests in the deeper canyons. Although the lowermost steep slopes of the mountains around the Valley are practically devoid of vegetation, there is an extensive pinyon pine-juniper woodland at higher altitudes and small areas of bristlecone pine and limber pine near the highest peaks.

Birds typical of the pinyon pine-juniper woodland may be expected along the Mahogany Flat Road which runs eastward from the Emigrant-Wildrose Road within the Monument's western boundary. From the end of the Road a 7-mile trail leads to the summit of Telescope Peak, passing through pines along the way. The Clark's Nutcracker, Mountain Chickadee, White-breasted Nuthatch, Mountain Bluebird, Oregon Junco, and even the Evening Grosbeak, Red Crossbill, and occasionally other montane species reside here. Spectacular views of Death Valley, 2 miles below, will be enjoyed on this climb.

EUREKA
Humboldt Bay | Trinidad State Beach

A visit to **Humboldt Bay** on the northern California coast is worthwhile for viewing birds any time of the year, but the best times are in winter, spring, and fall when waterbirds, waterfowl, and shorebirds congregate in vast numbers.

Humboldt Bay comprises South Bay (about 4 square miles) and North Bay (25 square miles) connected by a 7-mile shipping channel—the whole separated from the ocean by long, narrow sand spits. Entrance to Humboldt Bay is near the north end of South Bay, between the rock-and-concrete North and South Jetties. South Bay is deeper and thus better for waterbirds and waterfowl; North Bay is much shallower and more attractive to shorebirds. US 101 passes close to the east side of Humboldt Bay.

Among the several access points to South Bay, the very best for birds is reached by driving south from Eureka on US 101 for about 6.5 miles, exiting west on Hookton Road to Table Bluff (a headland), driving down its northwest extremity to the base of the sand spit, and then continuing north for 3 miles on the sand spit to its end at the South Jetty. Stop frequently along the sand spit to scan the offshore waters; later, walk out on the Jetty, waves permitting.

From mid-February to late April, expect to see thousands of Black Brant in the Bay. Some of the other birds to expect: *In any season:* Double-crested and Pelagic Cormorants, Snowy Plover (nests among dunes back from the beach), and Western Gull. *In summer and early fall:* Brown Pelican, Brandt's Cormorant, Osprey, and Heermann's Gull. *In winter and/or migration:* Common, Arctic, and Red-throated Loons, rarely the Yellow-billed; all five species of grebes; many ducks, including the Common Pintail, American Wigeon, Redhead, Greater Scaup, Bufflehead, all three scoters, and Red-breasted Merganser; gulls, including the Glaucous-winged, California, Ring-billed, Mew, and Bonaparte's, and Black-legged Kittiwake (on the Jetty, usually after storms); shorebirds, especially the Semipalmated and Black-bellied Plovers, Willet, Least and Western Sandpipers, Dunlin, Short-billed Dowitcher, and Marbled Godwit on the sand-mud flats in the Bay, the Whimbrel and Sanderling on the ocean beach, and the Black Turnstone, Surfbird, Wandering Tattler, and sometimes the Rock Sandpiper on the Jetty.

Among the access points to North Bay, two are highly recommended.

1. From US 101, exit north on State 255. This soon crosses the Eureka-Somoa Bridge over a narrow portion of North Bay, including two islands, the second of which—Indian Island—has a nesting colony of Great Blue Herons, Cattle, Great, and Snowy Egrets, and Black-crowned Night Herons in view from the highway. After reaching the sand spit, turn left off State 255 and drive south to the base of the North Jetty. Patches of trees and shrubs along the last 0.5 mile of the road should be investigated in winter for numerous sparrows—e.g., White-crowned, Golden-crowned, Fox, and Song—and warblers, among them both Myrtle and Audubon's, the former predominating. The Jetty, property of the United States Coast Guard and open to the public in good weather, is similar to the South Jetty in avian attractions. Often seen from the North Jetty in late summer and early fall are Common Terns being harassed by Parasitic Jaegers, and in winter numerous alcids such as Thin-billed Murres and Marbled Murrelets.

2. From US 101 in Arcata, 7 miles north of Eureka, exit west on Somoa Boulevard (State 255); one block after the second stop light, turn left onto I Street and drive to its end in a parking lot. On two sides of the lot are mud flats at low tide; on the other two sides, fresh-water marshes owned and managed as wildlife habitat by the city. The mud flats at near high tide are ideal for shorebirds in winter and/or migration. Besides species appearing also on the sand-mud flats in South Bay (*see above*), there are Killdeers, Ruddy Turnstones, a few Long-billed Curlews, Spotted Sandpipers, Greater Yellowlegs, Lesser Yellowlegs (fall only), Red Knots, Pectoral Sandpipers (fall only), Baird's Sandpipers (occasional in fall), Long-billed Dowitchers, Red Phalaropes (in fall

after storms), Northern Phalaropes (spring and fall), and many American Avocets. Peregrine and Prairie Falcons and Merlins show up regularly in winter, attracted by the abundance of shorebirds.

A visit to the rugged coast north of Humboldt Bay is rewarding for waterbirds and shorebirds frequenting the rocky ocean shores. From Arcata, drive north on US 101 for about 12 miles; within 2 miles of Trinidad, take the Westhaven Exit, turn left (west) under the highway and then right (north) onto Scenic Drive. In the next 2 miles to Trinidad the Drive winds along cliffs overlooking the sea. Stop now and then to look for birds, scanning the rocky shoreline and islets and the waters beyond the crashing surf.

Continue north from Trinidad to the Elk Head-College Cove section of **Trinidad State Beach** (159 acres), reached by leaving Trinidad on Stagecoach Road (begins at the northeast corner of the elementary school) for about a mile to the intersection on the right with Anderson Road; continue on Stagecoach Road for about 200 yards, then turn off west onto a dirt road, park the car, and follow the trail system north and west for about a mile to the northwest extremity of Elk Head. This overlooks two islands. On the nearer island, from mid-April to late June, Tufted Puffins may be seen resting at the entrances to their nesting burrows. Other birds nesting on the island are Double-crested, Brandt's, and Pelagic Cormorants, Western Gulls, Thin-billed Murres, and Pigeon Guillemots. Watch for Black Oystercatchers nesting on offshore rocks above high tide and foraging on the intertidal rocks at low tide. Also look carefully for Marbled Murrelets and Rhinoceros Auklets on the inshore waters just outside the breaking surf where they do much of their fishing.

LOS ANGELES AREA BY JEAN BRANDT

Griffith Park | South Coast Botanic Garden | Ballona Creek | Big Sycamore Canyon | McGrath State Beach | Placerita Canyon State Park | San Gabriel Mountains | Los Angeles State and County Arboretum | Santa Anita Canyon

The Los Angeles Area includes Los Angeles County of about 4,000 square miles and a few nearby localities, all within a 1.5-hour drive from downtown Los Angeles. Despite the fact that the County itself has a population of over seven million people, there are nevertheless within the Area numerous parks, recreation areas, arboretums, and other sites where birds may be observed in great variety, over four hundred species having been recorded. Given below is a selection of sites along the coast and in the mountains and deserts that will, in the aggregate, yield the widest variety. The bird finder *must* obtain a detailed road map of the Area and bear in mind in planning his trips that all sites are usually crowded with people on weekends and holidays, though less so in early morning when birds tend to be more active and thus more visible.

The best park for bird finding in Los Angeles is **Griffith Park,** centrally located and by far the largest, covering over 5 square miles of chaparral-covered mountain slopes and live oak-sycamore woodlands, as well as riparian habitats, large golf courses and other lawn areas, picnic grounds, the world-famous Griffith Observatory and Planetarium, and the equally renowned Los Angeles Zoo. There are several entrances to the Park, one from Franklin Avenue, five from Los Feliz Boulevard, three from Golden State Freeway (I 5), two from Ventura Freeway (State 134), and one from Forest Lawn Drive. The Park is partially serviced by the RTD, but one should telephone for bus schedules. Otherwise a car is necessary. There are many hiking trails throughout the Park.

The several entrances into the Park from Los Feliz Boulevard lead into the southwest part of Park. One of them, Vermont Avenue, goes to the Planetarium, to the east of which, in Vermont Canyon, is a tiny 'bird sanctuary' where California Quail, California Thrashers, Scrub Jays, Wrentits, both Rufous-sided and Brown Towhees, and other woodland-border species reside. The adjacent chaparral slopes are good for Fox and Golden-crowned Sparrows in winter. Above the Planetarium, winding roads and trails give access to the slopes of Mt. Hollywood and join other roads from the north end of the Park.

The greatest variety of habitats is near the Zoo on the northeast side of the Park. Here are great live oaks and sycamores with resident Anna's Hummingbirds, Downy and Nuttall's Woodpeckers, Scrub Jays, Plain Titmice, Bushtits, and wintering Purple Finches. Common as summer residents in the canyon above the Zoo are the

Black-chinned Hummingbird, Western Flycatcher, Western Pewee, House Wren, Orange-crowned Warbler, and Black-headed Grosbeak. The Costa's Hummingbird and Phainopepla are on the chaparral slope to the north. At any time of year, a flock of White-throated Swifts may be seen flying about Bee Rock, which juts from the mountainside high above the Zoo. In winter, Cedar Waxwings are common and Chipping Sparrows are visitants to the lawn areas. (Red-breasted) Yellow-bellied Sapsuckers, though scarce, are sometimes in the pepper trees lining some of the roads and paths.

Nearly all the oak-woodland and chaparral species are somewhere in the Park. Common transients include Allen's and Rufous Hummingbirds, all the swallows except martins, Warbling Vireos, and many species of warblers, which reach peak numbers in April, especially the Black-throated Gray, Townsend's, Hermit, and MacGillivray's (scarce), as well as Swainson's Thrushes and Western Tanagers, which remain common into May. Particularly good sites during migration are the Mineral Wells Picnic Grounds and vicinity, on the northeast side of the Park, and the areas on the south side from the Western Avenue or the Vermont Avenue entrances.

South of downtown Los Angeles, on the Palo Verdes Peninsula, is the **South Coast Botanic Garden.** The main entrance to the parking lot and headquarters is off Crenshaw Boulevard, reached by turning south from Pacific Coast Highway (State 1).

The Botanic Garden is excellent for hummingbirds. Anna's, Costa's, and Allen's are resident; the Rufous is a spring transient and the Calliope is sometimes present from mid-April to early May. On a peninsula, the Garden is a natural 'trap' for migrants and vagrants. A walk from the entrance passes a meadowy area where California Quail, Ring-necked Pheasants, and Western Meadowlarks nest. Farther in among the gardens the walk comes to a small lake which has ducks in season and gulls the year round. A willow-lined stream from the lake is the best site for viewing warblers in migration.

In Playa del Ray, the south side of **Ballona Creek,** as it enters Santa Monica Bay west of downtown Los Angeles, offers highly productive bird finding in late fall and winter. From Culver City, drive west on Culver Boulevard, then north (right) on Pacific Avenue to its end and park at the foot bridge across the Creek.

Walk over the bridge; go left to the end of the rocky jetty into the Bay. From the jetty in winter, look for Double-crested, Brandt's, and Pelagic Cormorants, Surfbirds, both turnstones, and Wandering Tattlers. At the mouth of Ballona Creek, look for Red-throated Loons, Eared, Horned, and Western Grebes, Greater and Lesser Scaups, Surf Scoters, and Red-breasted Mergansers. Common and Arctic Loons may be seen in migration and White-winged Scoters in some winters. Scan the outlying breakwater for Brown Pelicans, Great Blue Herons, numerous gulls such as Heermann's, and possibly a few Black Oystercatchers.

Walk back across the jetty and recross the bridge, then go east (left) along the upper channel of Ballona Creek, watching for Burrowing Owls on the banks. An impressive assemblage of transient and wintering shorebirds favors the mud flats on the right (south of the channel). The adjacent pickleweed marsh is a good area for the White-tailed Kite, Short-eared Owl, Belted Kingfisher, Say's Phoebe (in winter), Loggerhead Shrike, Western Meadowlark, and (Belding's) Savannah Sparrow.

In winter the fresh-water pond just east of Pacific Avenue is worth checking for both scaups, Buffleheads, and Ruddy Ducks. During spring and fall migrations, the willow thickets behind (east of) the apartment buildings on Vista del Mar host goodly numbers of small landbirds including *Empidonax* flycatchers and vagrant warblers.

Big Sycamore Canyon in Point Mugu State Park (13,358 acres), a one-hour drive from downtown Los Angeles, has birds typical of oak woodland and chaparral plus a varied riparian community. To reach the Park, drive northwest on Pacific Coast Highway (State 1) to the entrance which is 5 miles past the Ventura County line.

In the Canyon, Western Flycatchers, Hooded and Bullock's Orioles, and Black-headed Grosbeaks reside in summer, California Quail, Downy and Nuttall's Woodpeckers, Wrentits, and California Thrashers all year. Along the cliffs, Canyon and Rock Wrens and Rufous-crowned Sparrows are common resident birds. Sage Sparrows inhabit the upper reaches of the Canyon the year round and Black-chinned Sparrows in summer.

Big Sycamore Canyon, owing to its favorable position on the coast, is principally notable among bird finders for migrants and vagrants. In three years an impressive total of twenty-five species of warblers and fourteen species of sparrows were recorded, plus rare species, all in the fall and many 'firsts' for California. In addition to the usual transients, the willows along the dry streambed attract such migrating species as the Rufous and Allen's Hummingbird and Traill's Flycatcher.

McGrath State Beach (250 acres), embracing the estuary of the Santa Clara River and a 1.5-hour drive north from downtown Los Angeles, offers the widest variety of waterbirds, waterfowl, and shorebirds on the southern California coast. To reach the State Beach from Los Angeles, drive west on US 101 to Oxnard, exit at Victoria, and then proceed west along Olivas Parkway to Harbor Boulevard, turning left to the entrance.

The principal ornithological attraction of McGrath State Beach is the extensive area of mud flats which, at various times, permit easy viewing of such representative shorebirds as the Semipalmated, Snowy, and Black-bellied Plovers, Ruddy and Black Turnstones, Whimbrel, Greater and Lesser Yellowlegs, Least Sandpiper, Dunlin, Short-billed and Long-billed Dowitchers, Western Sandpiper, American Avocet, and Black-necked Stilt. Late summer brings thousands of Wilson's and Northern Phalaropes, while fall produces small flocks of such usually scarce transients as Pectoral and Baird's Sandpipers, plus Solitary Sandpipers and a few Red Knots. This is the best place in southern California for variety of shorebirds; many rarities have been recorded here.

The ponds at the north end of the fenced-in sewage plants are good for Black-crowned Night Herons, phalaropes, and wintering ducks. Some of the Cinnamon Teal and Ruddy Ducks are present all year and nest. The Yacht Harbor to the north of the estuary is equally good for wintering Common and Red-throated Loons and migrating Arctics, and for wintering Horned, Eared, and Western Grebes and scoters—occasionally all three species. Among the terns, Forster's, Common, Little, Royal, Elegant, and Caspian are present at appropriate times of the year. Common gulls include the Western, California, Ring-billed, and Heermann's. Herring, Mew, and Bonaparte's Gulls appear in winter as do Black-legged

Kittiwakes (in some years) and a few Glaucous-winged Gulls and the scarce Thayer's. Black Skimmers occasionally show up in late summer.

Belted Kingfishers frequent the lagoon in winter and migration, and Brown Pelicans are common offshore in any season. Also offshore during spring and summer, one may see large concentrations of Sooty Shearwaters, and in fall, Parasitic Jaegers harassing terns.

The coastal pickleweed marsh south of the estuary is habitat for the (Belding's) Savannah Sparrow. At the upper end of the lagoon, the reeds have Soras and Virginia Rails and Marsh Wrens in winter, Green Herons mostly in summer, and Great Blue Herons, Common Yellowthroats, and Song Sparrows the year round.

Placerita Canyon State Park (351 acres), a 45-minute drive northwest from downtown Los Angeles, is situated at the western edge of the San Gabriel Mountains at an elevation of 1,500 feet. Here there is a good representation of birds characteristic of oak woodland and chaparral together with a few montane and riparian species. To reach the Park, drive northwest on I 5, then north on State 14 to the Placerita Canyon offramp; turn right and follow signs to the entrance.

Oak woodland dominates the Canyon floor and supports such familiar resident birds as the Cooper's and Red-shouldered Hawks, Anna's Hummingbird, Acorn and Nuttall's Woodpeckers, Scrub Jay, Plain Titmouse, Bushtit, House Wren (scarce in winter), Western Bluebird, and Hutton's Vireo. These are joined in winter by the Hermit Thrush and Purple Finch and, in some years, by the Varied Thrush. Summer residents of the oaks include the Western Flycatcher, Western Pewee, Bullock's Oriole, Black-headed Grosbeak, and Lawrence's Goldfinch. The montane component is represented by the Mountain Quail, Steller's Jay, and Oregon Junco. Lark Sparrows frequent the grassy areas among the oaks, most commonly in summer.

The sparse riparian area along the stream has Black Phoebes and summering Black-chinned Hummingbirds. Canyon Wrens reside on the adjacent rocky slopes.

The hillsides bordering the valley are covered with 'hard' chaparral, habitat for such typical resident birds as the Wrentit, Bewick's Wren, California Thrasher, Lesser Goldfinch, and Rufous-sided and Brown Towhees. Winter brings Golden-crowned Sparrows, and summer adds such breeding birds as the Poor-will, Costa's Hummingbird, Ash-throated Flycatcher, and Orange-crowned Warbler. Resident White-throated Swifts may be seen overhead, plus Violet-green Swallows in spring and summer.

The Canyon attracts a variety of spring and fall migrants. Among them in the oaks and willows are the Hammond's Flycatcher, Solitary Vireo, Black-throated Gray, Townsend's and Hermit Warblers as well as the Swainson's Thrush and Western Tanager. Transient MacGillivray's Warblers may be observed in patches of brush, Lazuli Buntings in the chaparral, and Vaux's Swifts overhead.

A trip north from Los Angeles into the **San Gabriel Mountains,** above the noise and smog of the great city, is invariably rewarding for bird finders. Since the San Gabriels are part of the Angeles National Forest, there are numerous areas for picnicking and/or camping. From Foothill Freeway (I 210) at La Cañada, exit north onto the Angeles Crest Highway (State 2), which winds through the east-west trending range from La Cañada, attaining an elevation of 7,900 feet at Dawson's Saddle. Northern Ravens may be seen soaring from most anywhere along the Crest Highway. Described below are some of the sites especially worth exploring.

Switzer Picnic Area (elevation 3,000 feet), 12 miles from La Cañada. Here along the streambed of the Arroyo Seco, lined with sycamores and oaks, with chaparral on the adjacent slopes, is a fine variety of riparian and lower montane birds. Breeding species to look or listen for in suitable situations are the Mountain Quail, Band-tailed Pigeon, Great Horned and Spotted Owls, White-throated Swift, Anna's Hummingbird, Red-shafted Flicker, Acorn, Hairy, and Nuttall's Woodpeckers, Ash-throated and Western Flycatchers, Violet-green Swallow, Steller's and Scrub Jays, Plain Titmouse, White-breasted Nuthatch, Wrentit, House, Bewick's, Canyon, and Rock Wrens, California Thrasher, Hutton's, Solitary (scarce), and Warbling Vireos, Yellow (in migration) and Wilson's Warblers, Bullock's Oriole, Black-headed Grosbeak, Lazuli Bunting, Purple Finch, Lawrence's Goldfinch, and Black-chinned Sparrow. Poor-wills also breed. Always be alert for a Cooper's Hawk or a Golden Eagle.

Charlton Flats Area (5,000 feet) and *Chilao Campground* (5,300

NORTHERN PYGMY OWL

feet), 23 and 25 miles from La Cañada. Here in the forests of black oak and ponderosa pine such birds as the White-headed Woodpecker, Mountain Chickadee, Pygmy Nuthatch, and Brown Creeper are resident. In summer, look for the Olive-sided Flycatcher.

Buckhorn Campground (6,300 feet), 31 miles from La Cañada. From the road and down into the Campground, through stands of sugar pine and white fir, search or listen for the Northern Pygmy Owl, (Red-breasted) Yellow-bellied Sapsucker, Dusky Flycatcher, Townsend's Solitaire, Hermit and MacGillivray's Warblers (uncommon but regular), Western Tanager, Purple and Cassin's Finches, Pine Siskin, Green-tailed Towhee, and Fox Sparrow—all breeding species.

Dawson's Saddle, 38 miles from La Cañada. On reaching this highest point on the highway, walk behind the maintenance buildings on the north side of the highway and up the trail for Clark's Nutcrackers and Red-breasted Nuthatches. Williamson's Sapsuckers have been found along the trail.

Just south of the San Gabriel Mountains is the **Los Angeles State and County Arboretum** (127 acres), reached from Foothill Freeway (I 210) by exiting south for 0.5 mile on Baldwin Avenue in Arcadia. Comprising various habitats—lagoons with reedy borders, a bit of riparian woodland with brushy thickets, live oak woodland, and stands of exotic trees—the Arboretum is particularly good for bird finding in winter. At the entrance gate, obtain a checklist of Arboretum birds and a map of the trails, then walk to the following sites:

South African Section for hummingbirds in season or during migration—Black-chinned, Costa's, Anna's (resident), Rufous, and Allen's (scarce)—attracted to the clumps of bottlebush. *Lasca Lagoon,* for wintering Wood, Ring-necked, and Ruddy Ducks on the open water, Soras and Common Gallinules in the reeds. Pied-billed Grebes and Green Herons are residents. *Upper Lagoon,* in spring and fall for many migrating warblers and, in spring, for Hammond's, Western, and Olive-sided Flycatchers in the willows at the west end. Look particularly for the Olive-sideds in the tops of the taller trees. *Meadowbrook,* for resident Killdeers and for migrating Western Kingbirds. *Tallac Knoll,* covered with oaks, for Hutton's Vireos in any season and for warblers and other passerines in migration.

Throughout the Arboretum in suitable shrubby situations, check all wintering sparrows for such species as the Lincoln's, White-crowned, Golden-crowned, and the possibility of a Clay-colored, Harris', or White-throated. Spotted Doves may be seen anywhere in the Arboretum.

Of the many canyons on the southern face of the San Gabriel Mountains, **Santa Anita Canyon** is one of the most accessible. With its perennial stream, lush riparian growth, steep hillsides, and surrounding chaparral, it is also one of the most scenic and rewarding for the bird finder. From Foothill Freeway (I 210), exit at Santa Anita Avenue in Arcadia and proceed north to the end of the road that winds up through the foothills to the Chantry Flats ranger station. From here a steep road, a mile long and closed to vehicles, descends to the Canyon's floor; thence comes a mile hike upstream to Sturdevant Falls.

North American Dippers are year-round residents along the stream, nesting in the riprap below the many checkdams. House, Bewick's, and Canyon Wrens are resident, the Canyon

Wrens nesting under the porches of homes. Black Swifts, which nest under the Falls, are best seen at either dawn or dusk; they are summer residents as are Olive-sided Flycatchers. Townsend's Solitaires are winter visitants.

The 4-mile round-trip hike required to visit beautiful Santa Anita Canyon can be strenuous for one unaccustomed to steep trails and unrelenting sun. A sensible procedure is to plan the trip so as to hike back up to the car in the cool of the late afternoon or evening.

Reference

Birding Locations in and around Los Angeles. Compiled by Jean Brandt. Los Angeles: Los Angeles Audubon Society, 1976. Available from the Los Angeles Audubon Society, 7377 Santa Monica Boulevard, Los Angeles, CA 90046.

LOS BANOS
San Luis National Wildlife Refuge

For wintering concentrations of waterfowl in the San Joaquin Valley of central California, the 7,340-acre **San Luis National Wildlife Refuge** is excellent. Bordered by Salt Slough on the east and the San Joaquin River on the west, the area contains a maze of ponds and marshes (constituting about a third of the total acreage), meandering wooded sloughs, and lush grasslands. To reach the Refuge from Los Banos, drive north on County Road J-14 (North Mercy Springs Road) for 8 miles, then northeast 2 miles on Wolfsen Road to the entrance. Here obtain information on a self-guided auto tour.

Geese and ducks start appearing in October and reach peak numbers by December. Snow Geese, Mallards, Gadwalls, Common Pintails, Green-winged and Cinnamon Teal, Northern Shovelers, and Ruddy Ducks are the waterfowl in greatest abundance. Canada Geese, Greater White-fronted Geese, and Ross' Geese are much less numerous. A few small flocks of Whistling Swans usually show up by December. Besides waterfowl, many Sandhill Cranes arrive in October and November and White-faced Ibises in December.

Birds breeding on the Refuge, in addition to representatives of all the duck species mentioned above, include Pied-billed Grebes, Great Blue Herons, Great and Snowy Egrets, Black-crowned Night Herons, American Bitterns, Killdeers, American Avocets, Black-necked Stilts, Marsh Wrens, and Red-winged and Tricolored Blackbirds.

MERCED
Yosemite National Park

Within the 1,189 square miles of **Yosemite National Park** in east-central California where elevations attain 13,000 feet, the bird finder can observe all the avian associations of the Sierra Nevada amidst unrivaled scenic grandeur. Indeed, he can be torn between looking for birds and taking in more of Nature's handiwork.

Yosemite National Park is reached from Merced by driving northeastward on State 140 for 84 miles to the Arch Rock Entrance. Here, or at anyone of the other three entrances, if arriving by another highway, obtain a map showing the 200 miles of roads and some 700 miles of trails as well as places and geographical features. Use this map for reaching the bird-finding sites in summer that are described below.

The 'heart' of the Park is 7-mile-long Yosemite Valley with its famed scenic wonders—Bridalveil Fall, Half Dome, Yosemite Falls, El Capitan, and Cathedral Rocks, to name a few. Although a visit is a must for the scenery alone, it is also excellent for bird finding despite the crowds of people.

The floor of the Valley, at an elevation of 4,000 feet and forested largely with pine and fir, abounds with birds. Seemingly every picnic table is a feeding station for Steller's Jays, American Robins, Western Tanagers, and Oregon Juncos. Usually in sight overhead are White-throated Swifts and Violet-green Swallows, occasionally a Golden Eagle or Black Swift. Six species of woodpeckers are present. Look for the Pileated Woodpecker in large pines above Mirror Lake, and for Red-shafted Flickers, Acorn Woodpeckers, Hairy Woodpeckers, Downy Woodpeckers, and White-headed Woodpeckers in the oaks and cottonwoods. (Red-breasted) Yellow-bellied Sapsuckers are sometimes present.

Along the Merced River and inflowing streams there are Spotted Sandpipers, Belted Kingfishers, and North American Dippers. Streamsides with brushy deciduous growth attract Swainson's Thrushes, Warbling Vireos, Yellow, MacGillivray's, and Wilson's Warblers, Black-headed Grosbeaks, and Song Sparrows. A search of the brushy talus slopes along the south side of the Valley may be rewarded with such species as the Mountain Quail, Dusky Flycatcher, Nashville Warbler, Black-throated Gray Warbler, and Fox Sparrow.

Other Valley species include the Band-tailed Pigeon, four owls, the Great Horned, Northern Pygmy, Spotted, and Saw-whet, Western Flycatcher, Western Pewee, Olive-sided Flycatcher, Mountain Chickadee, White-breasted and Red-breasted Nuthatches, Brown Creeper, Canyon Wren, Golden-crowned Kinglet, Solitary Vireo, Audubon's Warbler, Brewer's Blackbird, Purple Finch, and Chipping Sparrow. Hermit Thrushes may be found along Tenaya Creek above Mirror Lake.

For specific bird species residing in the higher country of Yosemite Park, the following two trips are recommended:

1. Take the Glacier Point Road (closed in winter) which climbs from Chinquapin at 6,039 feet to Glacier Point at 7,214 feet elevation, passing forests of red fir and lodgepole pine that are often interrupted by lush meadows. Watch for Townsend's Solitaire nests—marked by trailing 'aprons' of pine needles—in the cut banks of the highway. Stop at Bridalveil Campground, 6 miles from Chinquapin, from which trails lead to three meadows. First, try the trail to Westfall Meadow. Through the coniferous woods, be alert for Blue Grouse, Williamson's Sapsuckers, and Black-backed Three-toed Woodpeckers. Ruby-crowned Kinglets, Cassin's Finches, Pine Grosbeaks, and Red Crossbills should be observed. Keep looking up for a Northern Goshawk. Once at the Meadow, watch for Calliope Hummingbirds visiting flowers and scan the wooded edges of the Meadow for Great Gray Owls. In early morning or late afternoon, the owls may be perched in full view, awaiting sight of their small furry prey in the Meadow. If not all of the above bird species have been observed, then try the trails to Peregoy and McGurk Meadows.

2. Take the Tioga Road (closed in winter) that winds up and

across the Park from Big Oak Flat Road on the west to the Tioga Pass Entrance on the east, reaching elevations well above 9,000 feet. In the vicinity of Tuolumne Meadows Campground (elevation 8,600 feet), 12 miles west from the Tioga Pass Entrance, there are stands of lodgepole pine and open tundra. Clark's Nutcrackers are conspicuous as well as noisy. Hammond's Flycatchers, Mountain Bluebirds, and Red Crossbills may be expected. White-crowned and Lincoln's Sparrows nest in shrubby areas lining streams. Early in summer, Gray-crowned Rosy Finches sometimes forage for weeds and frozen insects on the margins of snow that has not yet melted. Later in the season the rosy finches can be found only by hiking on trails to higher tundras, to talus slopes as on Mt. Dana and Mt. Lyell, and along the margins of Saddlebag Lake.

MONTEREY

Monterey Peninsula | Carmel River State Beach | Point Lobos Reserve State Park | Pfeiffer Big Sur State Park | Elkhorn Slough

The **Monterey Peninsula** offers a great diversity of bird habitats: rocky shores, sandy beaches, bay and ocean, woods.

From State 1, as it passes through Monterey, exit west on Del Monte Avenue and proceed first to the *Municipal Wharf* by turning off right onto Figueroa Street. Then return to Del Monte Avenue and continue through the tunnel, staying on the right side and following signs toward Cannery Row (a coastal roadway) and soon turning right off Cannery Row to the *Coast Guard Wharf and Breakwater.* Both the Municipal Wharf and Breakwater, jutting into Monterey Bay, are excellent vantages in winter for seeing numerous loons, grebes, ducks, gulls, and alcids. At the end of the Breakwater there will be cormorants (mostly Brandt's) and Brown Pelicans. Look for California sea lions on the rocks at the end of the Breakwater and scan the waters of the Bay for sea otters.

Return to Cannery Row and continue northwestward along the Bay, eventually on Ocean View Boulevard to Pacific Grove. Upon reaching a golf course with a small pond on the left, walk out to rocky *Point Pinos* on the right—by far the best vantage for viewing pelagic birds on the Monterey Peninsula. In summer, especially in the early morning or evening, line after line of Sooty Shearwaters, including some Pink-footeds, pass by. During early mornings in winter, there are spectacular offshore foraging activities of Rhinoceros Auklets and other alcids. Except in summer, one can expect to see large concentrations of gulls—Western, California, Mew, and Heermann's among others.

From Point Pinos, proceed south on Ocean View Drive, which becomes Sunset Drive, to the beginning of *Asilomar State Beach*, where there is one house only on the oceanside. Just south of it, the rocky shore at high tide is excellent for Black Oystercatchers the year round, for Surfbirds, Black Turnstones, and an occasional Rock Sandpiper in winter, and for Wandering Tattlers in spring and fall, a few in winter.

Continue south on Sunset Drive, following it inland, then turn off right through the north entrance to Seventeen Mile Drive (toll). This passes through extensive groves of Monterey pines and along several miles of rocky oceanfront. Stop at the following sites, indicated by directional signs: *Point Joe*, second only to Point Pinos for pelagic birds. *Bird Island*, in view offshore where Brandt's Cormorants nest. *Fan Shell Beach*, for shorebirds. *Cypress Point*, for Monterey cypresses, grotesquely shaped from constant buffeting by onshore winds.

After exiting from Seventeen Mile Drive through the south (Carmel) gate, go south, staying on drives along Carmel Bay, to **Carmel River State Beach** (106 acres). The beach and mud flats here at the River's mouth are frequented by many transient shorebirds, particularly in May and August. Black-bellied and Snowy Plovers, Whimbrels, Willets, Baird's Sandpipers, Marbled Godwits, and Sanderlings are among the species to be expected.

Leave the State Beach by going north on Carmelo Street, right on 15th Avenue, right on Dolores Street which becomes Lasuen Drive past Carmel Mission, and then right to Rio Road to State 1. Here turn right (south) and proceed for 3 miles to **Point Lobos Reserve State Park** (1,276 acres). Ask for a map at the entrance and drive to the end of the road near the south boundary. In view offshore is Bird Island. Brandt's Cormorants nest on its flat surface, Pelagic Cormorants on its cliffs, and Western Gulls all over.

Pigeon Guillemots breed on rocky islets and cliffs in the vicinity. On the nearby rocky shores, look for the same shorebird species mentioned above for Point Pinos.

The State Park inland has a natural grove of Monterey cypresses, large tracks of Monterey pines, and mixed stands of pine and oak, all accessible by trails and worth investigating. In pines, listen for the Northern Pygmy Owl and be alert for Pygmy Nuthatches and Brown Creepers. In mixed stands primarily, expect to see Band-tailed Pigeons, Acorn Woodpeckers, Steller's Jays, Chestnut-backed Chickadees, Hutton's Vireos, and Oregon Juncos. In summer there will be Western and Olive-sided Flycatchers; in winter, an abundance of Audubon's Warblers and usually a few Townsend's and Hermit Warblers, and roving groups of Red-breasted Nuthatches, Evening Grosbeaks, and (possibly) Red Crossbills. A pair of White-tailed Kites is often seen at Carmelo Meadow.

From Point Lobos State Reserve, continue south on State 1. In the next 24 miles to Pfeiffer Big Sur State Park the highway follows the coastline, often high above the waves breaking on rocky cliffs. One site of special interest is Hurricane Point, 1.2 miles south of Bixby Creek (watch for the sign) and 13.5 miles from the Point Lobos Park entrance. From here one can look down at the large rock offshore where there is a colony of several hundred Thin-billed Murres during the breeding season.

The kelp beds in view offshore from this 24-mile-stretch of highway should be scanned for sea otters. A spotting scope will be helpful. Some of the kelp beds for otters are just south of Mal Paso Creek, 2.7 miles south of Point Lobos; opposite the intersection of the Palo Colorado Road, 9.3 miles south of Point Lobos; and from Hurricane Point, between the mainland and the rock with the murre colony.

Pfeiffer Big Sur State Park (821 acres) has stands of the coast redwood. Numerous trails give access to a variety of habitats including riparian, oak woodland, and chaparral. Watch for North American Dippers along the Big Sur River. Among other breeding species in suitable habitats are the Cooper's and Sharp-shinned Hawks, Saw-whet Owl, Nuttall's Woodpecker, Ash-throated Flycatcher, Scrub Jay, Winter Wren, Swainson's Thrush, Solitary

Vireo, Audubon's and Black-throated Gray Warblers, Bullock's Oriole, Western Tanager, Lazuli Bunting, and Pine Siskin. Steller's Jays will come to the picnic table and often approach to within a few inches of a person's hand—if the hand is offering food.

Off State 1, 17 miles north of Monterey, is the town of Moss Landing at the mouth of **Elkhorn Slough.** This area has ponds, salt marshes, tidal mud flats, and sandy beaches attractive in spring and fall to transient shorebirds, in summer to breeding American Avocets, Black-necked Stilts, and Forster's Terns, and in winter to many ducks. For access to parts of the area best for birds, turn east off State 1, about 0.5 mile north of the bridge crossing Elkhorn Slough, onto Jetty Road and follow it over a causeway across the lagoon. Stop on the causeway and scan the tidal flats north and south of it. Continue on to the end of the road at the mouth of Elkhorn Slough. Park and walk up the sandy beach, which, in summer, is a breeding site for a few Snowy Plovers.

MORRO BAY
Morro Bay State Park | Morro Rock

The only tidal estuary of any size between San Francisco and Santa Barbara is triangular Morro Bay. Most of its northeast shore and adjacent marshy areas, as well as the long sand-dune peninsula opposite it, are included in **Morro Bay State Park** (1,483 acres). This is reached by exiting south from State 1 in the business district of Morro Bay.

The mile-wide pickleweed marsh in the Park has several winding tidal channels along which large numbers of herons and egrets, waterfowl, and shorebirds congregate during most of the year. Long-billed Curlews are sometimes here by the hundreds in August and September, and scores of Great Blue Herons stand in the marsh during December. The open flats of the Bay beyond hold thousands of probing shorebirds at low tide, and the sandier areas have Baird's Sandpipers (late July to September), Red Knots (fall), and all the western plovers. Hundreds of Black Brant feed in the Bay in winter but retire across the sand dunes to the open ocean when disturbed. In winter, loons, grebes, pelicans, cormorants, diving ducks, and all the gulls that commonly frequent the coast may be seen in the Bay, especially in the U-shaped harbor and channel portion west of the town.

Morro Rock, a tall, dome-shaped island just offshore from the mouth of the Bay, can be reached by walking out to it and to the breakwater beyond it—when the waves are not too high. The opportunity is here for seeing, in season, Surfbirds, turnstones, and other shorebirds of the rocky-coast habitat. Brown Pelicans, all three species of cormorants, and various gulls roost on the Rock. All the western species of terns are to be expected in and about the mouth of the Bay during migration. High on the Rock a pair of Peregrine Falcons nests every year.

OCEANSIDE
Palomar Mountain

A trip from Oceanside, 25 miles north of San Diego, by exiting east from I 5 onto State 76 for 34 miles, then turning north onto State Secondary 6 for about 5 miles, brings the bird finder up **Palomar Mountain.** Atop this, at an altitude of 5,202 feet, is the Palomar Observatory.

The steep winding road up the Mountain passes through a forested belt of oaks and conifers with birdlife much the same as on Cuyamaca Peak in Cuyamaca Rancho State Park (*see under* **San Diego Area**). However, the summit area of Palomar Mountain is covered with dense oak and mixed chaparral—habitat for nesting California Thrashers and Black-chinned Sparrows.

The wooded slopes of Palomar are noted for owls: Northern Pygmy, Spotted, Saw-whet, and in summer the Flammulated Screech. If they cannot be heard at night along the roads, try the Fry Lake Campground before reaching the Observatory, or the campground in Palomar Mountain State Park (1,879 acres), reached, after turning onto S 6 from State 76, by turning left onto S 7 for 3 miles.

ORANGE
Tucker Wildlife Sanctuary

In the Santa Ana Mountains southeast of Los Angeles is a most unusual sanctuary, the 9-acre Tucker Sanctuary, owned and operated by California State University, Fullerton. Here hummingbirds of six species are attracted to a battery of feeders along a shelf outside the screen of a large porch on which one may sit and watch at a range of 2 or 3 feet.

The Anna's Hummingbird is a resident, the Black-chinned and Costa's arrive for the summer in late March or early April; the Rufous, Allen's, and Calliope are transients, chiefly in early April, February and March, and late April, respectively. In the fall migration, large numbers stay into October in this favorable spot, long after the main migration at higher altitudes.

To reach the Sanctuary from Orange, which is on the Newport Freeway (State 55), take Chapman Avenue east to the Santiago Canyon Road, thence south on the Modjeska Canyon Road to its location. For days and hours when visitors are welcome, contact the Sanctuary at its address: Star Route, Box 858, Orange, CA 92667.

ORICK
Redwood National Park | Prairie Creek Redwoods State Park

US 101, in its 200-mile course between Crescent City and Leggett in the northern coast region of California, passes near many of the magnificent fern-carpeted groves of giant coast redwoods—'the world's tallest trees'—that have been preserved for posterity. **Redwood National Park,** extending mainly south from Crescent City (Park headquarters), incorporates about 91 square miles of redwood forest and ocean shore, including three state parks. One of them, **Prairie Creek Redwoods State Park** (12,240 acres) in the southern portion of the National Park, lies 6 miles north of Orick, about midway between Crescent City and Eureka.

Birdlife in the National Park–Prairie Creek State Park complex is typical of what one may find in most coast redwood forests. At the National Park information center in Orick on US 101, or at the State Park visitor and trail center farther north, obtain a map of roads and trails. Cal-Barrel Road, leading east off US 101, provides a scenic tour through stands of splendid old-growth redwoods. Davison Road, going down to the ocean from US 101, gives access to Fern Canyon and Gold Bluffs Beach, 4 miles of which are reserved for hiking only and wonderfully unspoiled. Cars with trailers are prohibited on both Cal-Barrel and Davison Roads.

For bird finding in the redwoods, take almost any one of the several trails through old-growth stands, along creeks and forest edges, and into cutover areas. Spring, preferably the last of April or the first of May, is the best time, when many of the breeding birds mentioned below are more vociferous and hence easier to spot in towering trees and their rich undergrowth.

Always listen for the Varied Thrush's moaning, whistled notes that carry a considerable distance, and for the Hermit Warbler's lisping song high in the canopy. Toward evening, listen for the three-to-four successive *hoots* of the Spotted Owl and keep an eye on the upper canopy for a Marbled Murrelet in flight to an old-growth redwood where it may nest. Some of the other birds to be alert for are the Ruffed Grouse, Vaux's Swift, Western Flycatcher, Gray Jay as well as Steller's Jay, Chestnut-backed Chickadee, Brown Creeper, Winter Wren, Swainson's Thrush, and Golden-crowned Kinglet.

Elsewhere, *in riparian situations,* look for the Common Merganser and North American Dipper along rivers and the Orange-crowned and Wilson's Warblers in the brushy edges of rivers and streams; *in cutover areas and forest edges,* the California Quail, Northern Pygmy Owl, Allen's Hummingbird, (Red-breasted) Yellow-bellied Sapsucker, Violet-green and Tree Swallows, Purple Martin, Cedar Waxwing, Hutton's Vireo, Black-headed Grosbeak, Pine Siskin, Purple Finch, American and Lesser Goldfinches, and Song Sparrow. Ospreys have aeries in some of the higher trees and Wrentits are resident in brushy openings.

Prairie Creek Redwoods Park is a prime area for Roosevelt elk. They are more frequently seen in cutover forests and forest edges along Gold Bluffs Beach.

PALM SPRINGS
Coachella Valley | Mt. San Jacinto | Chino, Tahquitz, Andreas, and Palm Canyons | Covington Park | Seven Level Hill | Pinyon Flat | Hemet Valley | Mount San Jacinto State Park

At the head of **Coachella Valley** and close to the eastern desert face of **Mt. San Jacinto,** this resort town 120 miles southeast of Los Angeles is centered conveniently for winter and spring bird finding in desert and mountain country.

Desert birds are so widespread that giving directions to particular sites for a good representation is difficult. Many of the roads leading out from Palm Springs and from communities on State 111 south from Palm Springs in the lower Coachella Valley traverse much desert country in which the predominant plant growth consists of creosote bush and cactus on the gentle upland slopes, mesquite, catclaw, paloverde, and smoke trees along low sandy washes. Breeding birds are similar to those in Joshua Tree National Monument (*see under* **Twentynine Palms**).

Well worth visiting are one or more of the canyons at the base of Mt. San Jacinto—for example, **Chino Canyon** (reached by a 2-mile road west from State 111, 1.5 miles north of Palm Springs) and **Tahquitz, Andreas,** and **Palm Canyons** (reached by well-marked roads west from State 111 at the south end of Palm Springs). The native Washingtonian palms, amid gigantic boulders, are a great scenic attraction, especially in Palm Canyon. In these canyons the willows and sycamores along the streambeds, as well as the rocky slopes and the proximity of desert plant associations, bring together a variety of birds that includes the Gambel's Quail, Ladder-backed Woodpecker, Canyon and Rock Wrens, Phainopepla, and Black-throated Sparrow.

Summer residents arrive in the Coachella Valley as early as February. Among them is the Costa's Hummingbird which does not usually get to the Pacific slope until late March or April. The Phainopepla nests in February and March near Palm Springs, long before it begins nesting on the Pacific slope. The height of the general nesting season, however, is in April. The northwesternmost known nesting site of the Vermilion Flycatcher is in **Covington Park** bordering the village of Morongo Valley, reached from Palm Springs by driving north on Indian Avenue, crossing I 10, and continuing north for 9 miles.

To reach the higher altitudes of the San Jacintos, one may take the Palm Springs Aerial Tramway, which leaves Valley Station in Chino Canyon (elevation 2,643 feet) and goes up to Mountain Station (8,516 feet) on Mt. San Jacinto. After enjoying the grand view of the entire Coachella Valley, the Salton Sea Basin southward, and the desert ranges eastward, one should start looking for Clark's Nutcrackers in the area.

For more productive bird finding at higher altitudes, one may drive up the Palms to Pines Highway (State 74). This leaves State 111, 13 miles southeast of Palm Springs, and climbs along a broad alluvial fan where there are typical desert plants and birds. It then leads up the mountain by the switchbacks of **Seven Level Hill** where ocotillo and agave blooms add to the colorful scene in spring. Rock Wrens and chuckwalla lizards live in the crevices of rocky prominences, and Prairie Falcons nest nearby. Higher up the Highway passes **Pinyon Flat,** a belt of pinyon pines and junipers, where one should see Pinyon Jays. Beyond Pinyon Flat, which is 15 miles from State 111, the Highway soon reaches the Jeffrey pines of **Hemet Valley,** leaving the desert behind. Although Hemet Valley and the reservoir in it are 2,300 feet or so lower in elevation than Bear Valley in the San Bernardinos (*see under* **San Bernardino**), bird-finding opportunities are somewhat comparable. Some of the species to look for especially are the Band-tailed Pigeon, White-headed Woodpecker, Pygmy Nuthatch, and Pine Siskin.

Thirty-five miles from State 111, State 243 leads north for 4.5 miles to Idyllwild at 5,300 feet, near **Mount San Jacinto State Park** (13,515 acres) set amid pines and cedars. From here a trail to the 17 square miles of primitive high country starts from the end of Fern Valley Road in Idyllwild and eventually reaches the summit of 10,831-foot Mt. San Jacinto from which views are unsurpassed. Birdlife of this region is quite similar to that in the San Bernardinos.

REDDING
Lassen Volcanic National Park

In northern California, 47 miles east of this city on State 44, is the northwest entrance to **Lassen Volcanic National Park,** established to preserve the spectacular volcanic area surrounding Mt. Lassen. Within the 167 square miles of the Park are numerous volcanic cones, hot springs, lava flows, and other evidences of volcanism. Mt. Lassen (10,457 feet) erupted violently in 1914 and 1915 and did not become quiescent until 1921.

From the Park's northwest entrance, the 30-mile Park Road winds and twists halfway around Mt. Lassen to the southwest entrance. During its course the Road attains elevation above 7,000 feet and consequently is closed because of snow from the end of October until early June. Park headquarters is at Mineral, reached by leaving the southwest entrance on State 89, then turning right onto State 36 and proceeding 5 miles.

The forests of the Park are primarily montane coniferous with such breeding-bird associates as the Hammond's and Olive-sided Flycatchers, Steller's Jay, Clark's Nutcracker, Mountain Chickadee, Red-breasted Nuthatch, Brown Creeper, Townsend's Solitaire, Golden-crowned Kinglet, Audubon's Warbler, Evening Grosbeak, Cassin's Finch, Pine Siskin, and Red Crossbill. North American Dippers reside along streams.

Recommended for the best variety of birds in the Park is the immediate vicinity of Manzanita Lake—a beautiful body of water with Mt. Lassen towering beyond—just within the northwest entrance and accessible all year. Here at the relatively low elevation of 5,800 feet there are not only stands of pine and fir, but also willows and other deciduous growth rimming the Lake and bordering Manzanita Creek. Thus one may find, in addition to some of the montane-forest birds already mentioned, numerous other species including the White-headed Woodpecker, Dusky Flycatcher, Western Pewee, Warbling Vireo, Nashville, MacGillivray's, and Wilson's Warblers, Western Tanager, Purple Finch, Oregon Junco, and Fox Sparrow.

SACRAMENTO

Upper Sunrise | Scott Road | Folsom Lake State Recreation Area | Yolo Bypass and Vicinity

For bird finding near this capital city at the confluence of the Sacramento and American Rivers, four areas are recommended.

Upper Sunrise. From Sacramento, drive east on I 50 about 10 miles; take the Sunrise Boulevard (Fair Oaks) Exit and go north on Sunrise Boulevard for 1.5 miles, turn right onto South Bridge Street, which is just before the bridge across the American River and parallels a portion of the American River Parkway known as Upper Sunrise, and park the car. Walk up along the River on the American River Bicycle Trail, between Bridge Street and the River, for 3 miles, passing the Nimbus Fish Hatchery. The Trail passes through vegetation which ranges from riparian growth rich in cottonwoods and willows to dense oak forests to grass fields. Among the birds to expect in any season by investigating all habitats are the California Quail, California Gull, Anna's Hummingbird, Nuttall's and Acorn Woodpeckers, Black Phoebe, Scrub Jay, Yellow-billed Magpie, Plain Titmouse, Bushtit, Wrentit, California Thrasher, Western Bluebird, House Finch, Lesser Goldfinch, and Brown Towhee. In winter, look for the Barrow's Goldeneye (a few regularly), Varied Thrush, Oregon Junco, and Golden-crowned

WRENTIT

Sparrow; in summer, several hummingbird species, Ash-throated Flycatcher, Western Pewee, Bullock's Oriole, and possibly Vaux's and White-throated Swifts, Violet-green Swallow, and Phainopepla. Passing through in migration are such birds as Western and other flycatchers, Hutton's Vireo, Townsend's, Hermit, Black-throated Gray, and MacGillivray's Warblers, Western Tanager, and Black-headed Grosbeak.

Scott Road. Return to I 50 and continue east about 7 miles and take the Prairie City Road Exit; drive 2 miles south on Prairie City Road to a T intersection (unmarked) with White Rock Road; here turn left, proceed 0.5 mile, and turn right onto Scott Road, which winds for 8 miles through open pastures with occasional large, scattered, or sometimes smaller, densely growing oaks. A few small streams add variety. Scott Road—a 'country road' and poorly paved—is recommended in winter when one may spot a White-tailed Kite, Ferruginous Hawk, or Prairie Falcon. Look for the Lewis' Woodpecker (most likely in large oaks beside Scott Road after driving on it about 2 miles), Mountain Bluebird, Say's Phoebe, Lawrence's Goldfinch, and Golden-crowned Sparrow.

Folsom Lake State Recreation Area. Drive northeast from Sacramento on I 80 for about 25 miles, take the Douglas Boulevard East Exit, and follow the Boulevard east for about 5 miles to the Granite Bay entrance to the Folsom Lake State Recreation Area (17,545 acres). Although there are various side roads leading to picnic areas or boat ramps, the main road winds through fields, oak woodland, and some Digger pine forest to dead-end about 4 miles from the entrance. From here a bridle path leading north may produce, in summer, Ash-throated Flycatchers and Lazuli Buntings. In any season there should be Wrentits, Canyon and Rock Wrens, California Thrashers, and Lawrence's Goldfinches, especially up the path into the chaparral. However, the ornithological highlight of Granite Bay is to be found by returning 0.8 mile toward the entrance, turning left, and proceeding 0.2 mile to a small parking lot on the left. In April and May, the hill beyond the parking lot has many clumps of bright orange monkey flowers which attract hummingbirds—Black-chinned, Anna's, Rufous, Allen's, and rarely Costa's and Calliope. A search of the area will produce Rufous-crowned Sparrows and perhaps a Phainopepla.

Yolo Bypass and Vicinity. This area is a 2- to 4-mile-wide and 30-mile-long floodplain, west of the city and the Sacramento River, which may be inundated during periods of heavy rain in winter but is normally farmland. Portions of it offer excellent bird finding. Drive north from Sacramento on I 5 (State 99). After passing the Sacramento Metropolitan Airport Exit and crossing the Sacramento River, take the first possible exit (Elkhorn), proceed to a T intersection, and turn left toward Woodland. Less than 0.5 mile up the road, an elevated railroad appears on the right. The next 1.5 miles is part of the Yolo Bypass. When flooded in winter the fields beyond the railroad may have Whistling Swans, Greater White-fronted Geese, and even a few Ross' Geese among hundreds of Snow Geese. Thousands of wintering ducks may be expected. Cinnamon Teal, American Avocets, and Black-necked Stilts nest in the ricefields farther down the road. Continue to the intersection with Country Road E-8 (Farm Road 102), turn north (right) and proceed for about 0.5 mile to FR 21. Between here and FR 20 ahead, the ponds on the left may have nesting Eared Grebes and Cinnamon Teal. Return to CR E-8 and turn left. After passing over I 5, continue about 1.5 miles to FR 25. Between here and FR 28 ahead, and all side roads east of CR E-8, check all plowed fields for wintering Mountain Plovers. Turn left onto FR 28 toward the Yolo County Central Landfill and go about 1.5 miles to the Hawk and Owl Preserve, managed by the Davis Audubon Society. Look here for the White-tailed Kite, Burrowing Owl, and other raptors in any season and for Long-billed Curlews in winter. Continue east on FR 28 past the Landfill to the sewage treatment ponds for Wilson's Phalaropes in migration. Backtrack and turn right onto FR 104 (Mace Boulevard, gravel) for a better view of the Hawk and Owl Preserve and another chance to see Mountain Plovers.

SAN BERNARDINO
San Bernardino Mountains

From this southern California city 65 miles inland from Los Angeles, State 18 leads up into and over the **San Bernardino Mountains** to the north. Although steep and rugged for the most part, the San

Bernardinos are unique among the southern California ranges in having an extensive area of gentle topography at a fairly high altitude, with several large lakes and adjoining meadows. Birdlife is consequently present in much greater variety.

From downtown San Bernardino, drive 5 miles north on Sierra Way to where State 18 begins a 10-mile climb through the chaparral belt and across the heads of several small wooded canyons. Stops for chaparral birds are easily made at the several large parking areas along the way. Costa's Hummingbirds are common in summer and Rufous-crowned Sparrows all year in the chaparral-covered foothills. From April to July, Black-chinned Sparrows may be heard singing from the end of the switchback, 2.5 miles above the upper end of the paralleling Waterman Canyon Road.

For the next 30 miles State 18 is known as the Rim of the World Drive, since it skirts for several miles the upper edge of the steep chaparral slopes from which there are excellent views of the valley and the distant San Jacinto and Santa Ana Mountains. Soon, however, State 18 enters a forest—ponderosa pine, incense cedar, and black oak—through which several roads lead to *Lake Arrowhead* (elevation 5,106), 2 miles to the north. Many coniferous-forest birds are here; but the trees come right down to the shore of the Lake and there are fewer waterfowl and shorebirds than at 7-mile-long Big Bear Lake (6,750 feet) in Bear Valley, 21 miles farther along the Rim of the World Drive.

Five miles after the last road leads off to Lake Arrowhead, a side road goes north from State 18, for about 10 miles, passing Green Valley Lake and Village to the *Green Valley Campground* (elevation 7,000 feet), on the north-facing slope of the San Bernardinos. Williamson's Sapsuckers nest in the area as do Dusky Flycatchers, Hermit and MacGillivray's Warblers, and Lincoln's Sparrows. Red Crossbills appear irregularly.

At *Big Bear Lake*, State 38 leads off State 18 to go around the north side of the Lake through Fawnskin; State 18 continues around the south side and crosses State 38 at the east end of the Lake. Thus one can encircle the Lake by car. On the south shore the best vantages for waterbirds, waterfowl, and shorebirds are from two roads that go out onto points in the Lake, the first a mile west of the town of Big Bear Lake, the second a mile east. On the north shore the best area for waterfowl is Grout Bay at Fawnskin. North from Fawnskin an unpaved road goes about 3 miles to the Hanna Flat Campground, near which, with diligent searching, one may find Saw-whet Owls.

When the water level of Big Bear Lake is normal, the concentrations of American Coots in fall and winter are enormous. Transient shorebirds include the Common Snipe, Willet, Greater and Lesser Yellowlegs, Least Sandpiper, Short-billed and Long-billed Dowitchers, Western Sandpiper, American Avocet, Black-necked Stilt, and Wilson's and Northern Phalaropes. Ring-billed Gulls are commonest through winter, with California, Herring, and Bonaparte's Gulls appearing chiefly as transients, as well as Forster's, Caspian, and Black Terns. Late April to May and September to October are the best times for these birds; but thousands of ducks are at their peak in November. Canvasbacks and hundreds of wintering Common Mergansers arrive later and are most numerous in midwinter. Bald Eagles winter regularly and Ospreys are transients. In late May and June many Pied-billed Grebes, American Coots, and a few waterfowl nest in the weedy area at the east end of the Lake, Cinnamon Teal being the most common, with fewer Mallards, Common Pintails, Redheads, and Ruddy Ducks. Killdeers, Spotted Sandpipers, and Horned Larks nest along the open-pastured shore in May.

The meadows back from the Lake—e.g., near the mouth of Rathbone Creek east of Eagle Point or in Bear Valley east of the Lake—are foraging areas for Ferruginous Hawks (winter), American Kestrels, Western Kingbirds (summer), Say's Phoebes, Western and Mountain Bluebirds, Western Meadowlarks, and Savannah and Vesper Sparrows. Violet-green and Cliff Swallows also forage over the meadows, the Cliff Swallows regularly building their mud nests on the tall boles of ponderosa pines along the south side of the Valley.

Baldwin Lake, in the east end of Bear Valley, 2 miles beyond Big Bear Lake, is without an outlet and may be intermittently dry for periods of years. When it has water and marshy borders, it supports a nesting colony of Eared Grebes, another of Yellow-headed Blackbirds, and possibly nesting Wilson's Phalaropes. The sewage ponds on the south side of the Lake attract shorebirds, sometimes sizable numbers of Baird's Sandpipers in August. The outlying dry flats and grasslands often bring flocks of Horned Larks and, occasionally, Lapland Longspurs in fall and early winter. Common Nighthawks can be seen overhead in late spring and summer. Pinyon Jays can be found on the adjacent pine-clad slopes the year round. From State 18, north of the Lake, an unpaved road leads around the east side, passing through a broad slope of sagebrush in which Green-tailed Towhees and Brewer's Sparrows nest in May and June.

SAN DIEGO AREA

Balboa Park | Silverwood Sanctuary | Cuyamaca Rancho State Park | Old Mission Dam Historical Site | Point Loma | Silver Strand | Buena Vista Lagoon | Tijuana River Valley

The San Diego Area in the southwestern corner of California offers a diversity of bird-finding sites from the Pacific Coast to inland mountains, canyons, and streamsides.

In the city, **Balboa Park** contains 1,158 acres of wooded canyons, brushy borders, ponds, and groves of eucalyptus, which not only attract the more common breeding birds of man-made habitats, but vast numbers of wintering warblers, orioles, and tanagers. When flowering, the eucalyptus groves bring an abundance of Costa's and Anna's Hummingbirds as well as other hummers including occasional Broad-tailed and Rufous. The San Diego Zoo, east of Park Boulevard as it extends north through the Park from downtown, displays one of the finest worldwide collections of tame and caged birds in the United States. Wild birds are encouraged in the Zoo grounds; many ducks come and go daily, and Black-crowned Night Herons roost and nest in tall trees over the animal pits. The Natural History Museum, close to the Zoo but operated separately by the San Diego Society of Natural History, gives helpful suggestions for bird-finding localities in the San Diego region.

Silverwood Sanctuary (405 acres), in Wildcat Canyon northeast of the city's center and operated by the San Diego Audubon Society, is one of the better places near San Diego for chaparral birds and, during migration, for hosts of western warblers. Among the regular breeding birds are the California Quail, Ash-throated Flycatcher, Scrub Jay, Wrentit, Bewick's Wren, California Thrasher, and Brown Towhee. The Botany Trail provides ready access to these birds and many others. To reach the Sanctuary from I 8 east of the city, exit north on State 67, turn off east onto Mapleview Avenue for one block, then left onto Ashwood Street (becomes Wildcat Canyon Road) and proceed for 5.5 miles. Watch for the entrance, consisting of a gate and stone pillars on the right.

Further northeast of San Diego, **Cuyamaca Rancho State Park** (24,677 acres) includes many square miles of black oak-Coulter pine forest, live oak woodland, and chaparral. Headquarters is reached by exiting north from I 8 on State 79 for 6.5 miles. Cuyamaca Peak, on the west side of the Park and accessible by road, attains 6,515 feet in altitude and has a forest of ponderosa pine on its higher slopes. Several of the coniferous-forest birds—e.g., the White-headed Woodpecker, Mountain Chickadee, and Pygmy Nuthatch—reach the southern limit of their distribution in the state here or on Laguna Mountain to the southeast. Other birds on the higher slopes include the Mountain Quail, Band-tailed Pigeon, Williamson's Sapsucker (in winter), and Olive-sided Flycatcher. Green-tailed Towhees nest in the chaparral along the road to Cuyamaca Peak.

North of the city, the **Old Mission Dam Historical Site** provides good riparian habitat for productive bird finding in any season. Nuttall's Woodpeckers, Western Flycatchers, Bell's and Warbling Vireos, Yellow-breasted Chats, Hooded and Bullock's Orioles, and American and Lesser Goldfinches nest. On the rocky hillsides just downstream from the Dam, Black-chinned Sparrows reside in spring and summer and Rufous-crowned Sparrows the year round. To reach the Site, exit north from I 8 on Mission Gorge Road for 7 miles, then turn left onto Father Junipero Serra Trail and proceed 1.5 miles to the parking lot on the left.

Point Loma, site of Cabrillo National Monument and overlooking San Diego Bay and the ocean, is an excellent vantage for viewing coastal birds the year round. In winter its rocky shore below the steep cliffs attracts Double-crested, Brandt's, and Pelagic Cormorants and such shorebirds as Surfbirds, Black Turnstones, and Wandering Tattlers. During fall the Point, projecting southward as it does, serves to funnel hawks in their southbound migration. From mid-December to mid-February the point is a popular vantage for watching gray whales swimming south from Arctic waters to Baja California. To reach Point Loma, exit from I 5 west of the city center on State 209 and proceed south for about 10 miles.

The **Silver Strand,** much of which is included in Silver Strand State Beach (24,677 acres), extends between Coronado and Imperial Beach with the ocean on one side and the west side of San Diego Bay on the other. To reach the Strand from I 5, south of city center, exit west on State 75 (Palm Avenue) which eventually bears north on the Strand (and called Silver Strand Boulevard). On the beach in winter, look for numerous shorebirds—Snowy Plovers, Long-billed Curlews, Whimbrels, Red Knots (often large concentrations), Dunlins, Marbled Godwits, and other species—and gulls, especially Glaucous-winged, Western, California, Mew, Heermann's, and an occasional Thayer's. American Avocets, Black-necked Stilts, Royal Terns, and Black Skimmers are present all year, particularly on the bayside near the southern end in the South Bay Marine Biological Study Area. Little and Elegant Terns nest here and there is always a good chance in winter of seeing one or more Louisiana Herons. Any time in winter, one should scan the south half of the Bay for a flock of Black Brant and rafts of scaups and scoters.

Several coastal lagoons north of San Diego attract many waterbirds and waterfowl as well as shorebirds. **Buena Vista Lagoon** is the largest and probably the most productive. I 5, between Oceanside and Carlsbad, crosses the Lagoon. To explore it, exit east on State 78, which passes the north side, then turn off sharply right onto Jefferson Street, which passes the south side. Least Bitterns and White-faced Ibises may nest at the Lagoon's east end. Redheads and Ruddy Ducks are present all year and nest. The best time to visit the Lagoon is in winter when there are vast numbers of herons, egrets, ducks, shorebirds, gulls, and terns.

In the extreme southern corner of the state, the **Tijuana River Valley** has a diversity of habitats for birds in any season and often yields species from across the Mexican border that are unusual in the United States. Exit west from I 8 on Coronado Avenue, then turn left onto 19th Street and right onto Sunset Road, which passes down the Valley. From side roads in winter, look over dirt fields recently plowed for Mountain Plovers and weedy fields for Short-eared Owls. Any of the chaparral-covered hillsides have resident Black-tailed Gnatcatchers and Sage Sparrows and wintering Golden-crowned Sparrows. Stands of tamarisk and other trees are sometimes alive with migrating passerines in spring and fall. The marsh at the mouth of the River, reached from the west end of Sunset Road, has resident Clapper Rails and attracts many transient and wintering shorebirds.

SAN FRANCISCO BAY AREA, EAST

Lake Merritt | Joaquin Miller Park | Redwood Regional Park | Bay Farm Island | Bay Bridge Toll Plaza | Charles Lee Tilden Regional Park | Berkeley Aquatic Park | Mount Diablo State Park

Famous as a refuge for wintering waterfowl is **Lake Merritt** in Oakland on the east side of San Francisco Bay. Although in the heart of the city, this 155-acre body of salt water is visited annually by thousands of ducks as well as by many grebes, cormorants, gulls, and terns. A feeding and banding station is on the northeast arm of the Lake at the intersection of Bellevue Avenue and Perkins Street. During winter the ducks are fed daily in the morning and at 3:30 p.m. The concentration of birds is largest between November and April, when the commoner species include the Horned, Eared, and Pied-billed Grebes, Double-crested Cormorant, Mallard, Common Pintail, Canvasback, Lesser Scaup, Common Goldeneye, Ruddy Duck, various gulls, and Forster's Tern. Barrow's Goldeneyes regularly winter in small numbers. Swans and geese, released as cripples and some of them breeding, are very tame. A small amount of food will bring dozens of wild ducks to a person's feet. Photographic opportunities are consequently excellent.

Favorable areas for landbirds are numerous in the vicinity of Oakland. One of the best is **Joaquin Miller Park** (500 acres), entered from Joaquin Miller Road or Skyline Road. Its fine stands of second-growth coast redwood are native, but the Monterey pines and cypresses have been planted. Adjacent to Joaquin Miller Park is **Redwood Regional Park** (2,489 acres), entered from Redwood Road, where grassland, coastal scrub, evergreen broadleaf forest, and riparian habitats are also accessible. Among the breeding birds are the Hairy Woodpecker, Olive-sided Flycatcher, Steller's Jay, Chestnut-backed Chickadee, Red-breasted Nuthatch, Brown Creeper, Wrentit, Hutton's Vireo, Purple Finch, and Oregon Junco. The Saw-whet Owl breeds in the redwoods along Redwood Creek in Redwood Park.

Bay Farm Island is no longer the mecca for shorebird enthusiasts that it once was. The best mud flats fell victim to landfill operations, but there is still good shorebird finding in the general vicinity. Migration periods in spring and fall produce the greatest numbers; mid-April, when the northbound transients are in nuptial plumage, is especially rewarding. Wintering flocks are present from October to March. The ideal time for a visit is the first hour after high tide. One can obtain tide tables at sporting-goods stores and some service stations.

To reach Bay Farm Island, take High Street west to Otis Drive in Alameda; turn left onto it and continue across the bridge to the Bay Farm Island side; then turn left almost immediately onto Doolittle Drive and continue along the Municipal Golf Course. Watch the ponds and the open water of San Leandro Bay on the left for waterbirds and waterfowl.

Another productive area is the south shore of Alameda. From High Street, turn right onto Otis Drive and proceed to Broadway; turn left onto it and continue to Shoreline Drive. Extensive mud flats are in view here at low tide. Among the more abundant birds to be expected in suitable situations are four grebes including the Western, Brown Pelican, Double-crested Cormorant, Great and Snowy Egrets, several species of surface-feeding and diving ducks, Clapper Rail, American Avocet, Black-bellied Plover, Marbled Godwit, Long-billed Curlew, Willet, Short-billed Dowitcher, Red Knot, Sanderling, Least Sandpiper, Dunlin, Western Sandpiper, several species of gulls, Forster's Tern, and Caspian Tern.

The **Bay Bridge Toll Plaza** is also worth a visit. Driving west on I 80, take the Oakland Army Base Exit just before reaching the toll gate for the San Francisco-Oakland Bay Bridge to the small road leading toward several radio transmitting stations north of the toll gate. Besides most of the waterbirds and waterfowl that one may see in the Bay Farm Island area, there are usually Elegant Terns present in large numbers from August through October.

From the intersection of Grizzly Peak Boulevard and Wildcat Canyon Road in the northeastern part of Berkeley, take Canyon Drive to the bottom of the hill and turn left to the nature area of **Charles Lee Tilden Regional Park** (2,065 acres). A wide variety of habitats here hold many landbirds of the central Coast Range, including the California Quail, Poor-will, and Allen's Hummingbird (spring and summer), Anna's Hummingbird, Wrentit, California Thrasher, and Hutton's Vireo. An extensive trail system allows thorough coverage of the area. The slopes to the west are heavily wooded with coast live oak, California-laurel, and madrone. Lush riparian growth follows Wildcat Creek northwest from the parking lot to Jewel Lake. To the east are extensive eucalyptus groves, and north of the Lake are open grassy hills. Birds are plentiful throughout the year, with April and May the favored months.

Berkeley Aquatic Park is reached by driving west on University Avenue, turning left (south) onto 6th Street and going one block, then right (west) onto Addison Street for four blocks. A paved drive encircles the area. From November through March, concentrations of ducks and gulls are the primary avian attractions. The foot of University Avenue leads to Berkeley Municipal Pier and Berkeley Yacht Harbor. Both provide vantages for viewing wintering waterbirds and waterfowl, as does the frontage road to the south.

Mount Diablo State Park (7,919 acres), southeast of Berkeley, offers optimal conditions for observing birds of the chaparral and blue oak-Digger pine associations. From Berkeley take State 24 to Walnut Creek, from which there are two possible entrances to the Park. (1) Turn left (north) onto I 680 at its junction with State 24 in Walnut Creek; then take the first exit and proceed east on Ygnacio Valley Road to Walnut Avenue; turn right onto it and continue, following signs to the north entrance. (2) Turn right (south) onto I 680 and go 6 miles, take the Diablo Road Exit in Danville, turn left onto Diablo Road and follow it to the south entrance. Both routes lead to the summit of Mt. Diablo (elevation 3,849 feet), from which the view on a clear day is remarkable. Look for Rufous-crowned, Sage, and Black-chinned Sparrows in the chaparral, especially along the road below the south entrance. In the oak-pine woodland is an assemblage of bird species markedly different from those on the Berkeley side of the hills. Some of the breeding birds likely to be found are Acorn and Nuttall's Woodpeckers, Ash-throated Flycatchers, White-breasted Nuthatches, Blue-gray Gnatcatchers, and Lark and Chipping Sparrows.

SAN FRANCISCO BAY AREA, WEST

Golden Gate Park | Cliff House | Lake Merced | Baylands Marsh Preserve | Bolinas Lagoon | Audubon Canyon Ranch | Point Reyes Bird Observatory | Tomales Bay | Tomales Bay State Park | Point Reyes | Dillon Beach | Muir Woods National Monument | Point Diablo

The visitor to **Golden Gate Park** in the city of San Francisco will see little evidence that the entire area now covered by woods, lawns, flower beds, and lakes has been developed from barren sand dunes since 1872. The long, narrow Park covers 1,013 acres between Stanyan Street and the ocean. From late September through March, the lakes attract waterfowl that may be approached and photographed at close range. The Mallard, American Wigeon, Northern Shoveler, Wood Duck, Ring-necked Duck, Canvasback, Lesser Scaup, and Ruddy Duck are regular in winter, and at least one male Eurasian Wigeon is present during most winters. Resident landbirds include the California Quail, Band-tailed Pigeon, Anna's Hummingbird, Allen's Hummingbird (February–August), Violet-green Swallow (March–August), Scrub Jay, Chestnut-backed Chickadee, Bushtit, Pygmy Nuthatch, Hutton's Vireo, Red Crossbill, Brown Towhee, and Oregon Junco. In most winters, Varied Thrushes and Golden-crowned Sparrows are common.

To the epicure the **Cliff House,** on the Great Highway, is known for its cuisine, but bird watchers of central California have long used its observation deck as a convenient vantage from which to see birds on and around the offshore Seal Rocks. From late July to mid-April, Black Oystercatchers, Surfbirds, Black Turnstones, and Wandering Tattlers may be viewed on the offshore and mainland rocks. Brandt's Cormorants and Western Gulls nest on the Seal Rocks; Pelagic Cormorants and Thin-billed Murres are visitants throughout the year. In summer and fall, Brown Pelicans, Pigeon Guillemots, Heermann's Gulls, and Elegant Terns are regular; in late summer, huge numbers of shearwaters, mostly Sooty, are sometimes seen offshore.

From the Cliff House, **Lake Merced** may be reached by driving south along the Great Highway to its junction with Skyline Boulevard. Formerly, Laguna de la Merced was a single body of water connected with the ocean in winter, but today it is divided into two parts, both permanently cut off from the sea. Certain estuarine invertebrates, however, still inhabit its now fresh waters. From late fall to early spring, Eared, Horned, Western, and Pied-billed Grebes may be observed here, as well as numerous wintering ducks and gulls. In any season one may expect Marsh Wrens and Common Yellowthroats in the cattails and rushes bordering the shore. Two of the best vantages for observation are the concrete bridge at the southeastern end of the southern lake and near the boathouse on the narrow strip of land that separates the two lakes.

On San Francisco Bay in Palo Alto, near its southern extremity, is the **Baylands Marsh Preserve** (120 acres), about 1.25 miles east of US 101 on Embarcadero Road. The city has preserved this remnant of natural salt marsh and built a nature interpretive center, with a boardwalk crossing the marsh to the Bay's edge. A high tide of more than 6.6 feet (Golden Gate level) forces the more secretive marsh birds and mammals out from cover, providing an opportunity to observe the endangered California race of the Clapper Rail, which resides in the cord grass. In winter, Virginia Rails and Soras reveal themselves on such tides, and there is always a chance of seeing the elusive Black Rail, which resides in the pickleweed. During the migration seasons and winter from late August to late April, the tidal mud flats are alive with thousands of shorebirds representing over fourteen species. For the closest views, try to be at the end of the city's boardwalk, or at the northern end of the P. G. and E. boardwalk, as the tide is beginning to ebb—one to two hours after maximum high. All winter the Preserve supports large numbers of ducks (thirteen species), gulls (eight species), and herons (four species). Every winter there are usually one or two Eurasian Wigeons and Glaucous Gulls. At the north end of the marsh is a winter roost of Black-crowned Night Herons, as many as 350 individuals in some years.

The bird finder interested in sea birds nesting in central California may observe them at Point Reyes, approximately 50 miles by road north of San Francisco. Take US 101 across the Golden Gate Bridge and exit on State 1. A stop just north of the town of Stinson Beach, where State 1 passes along the east shore of **Bolinas Lagoon,** will be productive for viewing waterbirds and, at low tide, for shorebirds in season. Three miles past the town on State 1 is **Audubon Canyon Ranch** (open only on weekends from 1 March to 4 July), where nesting Great Blue Herons and Great Egrets may be viewed closely.

Some of the common coastal landbirds may be seen in the hand at the Palomarin field station of the **Point Reyes Bird Observatory,** where birds are netted for banding. The area immediately around the station offers good landbird finding. To reach the station, leave State 1 at the north end of Bolinas Lagoon on a paved road going south (left) toward the town of Bolinas; after about 1.5 miles from State 1, turn west (right) onto Mesa Road (the first stop sign encountered) and follow it about 4 miles to the south entrance to Point Reyes National Seashore; continue on an unpaved road a few hundred yards to the station on the left.

Return to State 1, continue north through Olema, and, just before reaching the hamlet of Point Reyes Station, turn off west onto Sir Francis Drake Boulevard toward Inverness. Follow this road for 0.5 mile to Olema Marsh (worth a stop if flooded) and thence around the west side of **Tomales Bay** to Inverness. Stop whenever there is a turnout and look for waterbirds, waterfowl, and shorebirds on Tomales Bay, which is best in winter. About 2 miles past Inverness, take the Pierce Point Road at right for 2 miles to the entrance to **Tomales Bay State Park** (1,018 acres), which has a variety of forest birds, including the Pygmy Nuthatch. Northern Pygmy, Saw-whet, and Spotted Owls are heard here but rarely seen.

Return to Sir Francis Drake Boulevard and proceed south toward **Point Reyes.** Side roads to the left lead to Drake's Bay at Drake's Beach and the 'Fish Docks,' where wintering loons, grebes, diving ducks, and gulls are in abundance; Red-necked Grebes and Black Scoters are regular. Watch above the grasslands for Ferruginous Hawks and other wintering and/or migrating raptors. At the end of the main road is the Point Reyes Light Station. Park in the designated lot and walk along the gated road, past the restrooms, and down the steps. Search the rocks and water below for Brandt's and Pelagic Cormorants, Thin-billed Murres, and Pigeon Guillemots. Their nesting season is from May to July, but a visit any time of the year is rewarding. Keep an eye and ear open for Northern Ravens, Rock Wrens, and, in winter only, a rare Peregrine Falcon.

On the east side of Tomales Bay are numerous vantages from which the bird finder can, with a spotting scope, scan the Bay for wintering waterfowl. At the mouth of the Bay is **Dillon Beach,** which from mid-October through March is a vantage for sighting wintering loons (three species), grebes, cormorants, Black Brant, ducks, and shorebirds. April is an ideal time for migrating shorebirds. To reach Dillon Beach, continue north on State 1 to the village of Tomales. A turnoff left, indicated by a sign, leads to the Dillon Beach resort area, 4 miles distant. Turn left at Lawson's Resort Store and proceed for one block, then turn left again to the toll booth. Continue about 0.5 mile to Lawson's Landing, then walk southwest along the Beach, past Sand Point to the mouth of the Bay; also walk southeast to a good marsh for ducks and shorebirds.

Of the extensive areas of coastal redwood forests, the nearest to San Francisco is **Muir Woods National Monument** (550 acres). A visit here may be combined with a long-day trip to Point Reyes. Thus, after crossing Golden Gate Bridge on US 101 and exiting on State 1, proceed for about 3 miles only toward Stinson Beach and then turn off right on the Panoramic Highway, following the signs to the Monument about 3 miles distant. The variety of birdlife in Muir Woods is poor, but the trees alone are worth the visit. Spotted Owls are resident; search for them along the Alice Eastwood

Trail within 0.25 mile of the Bootjack Trail. Other residents are Steller's Jays, Chestnut-backed Chickadees, Brown Creepers, Winter Wrens, and Hermit Thrushes.

An excellent lookout for raptors in migration is **Point Diablo** in the headlands overlooking Golden Gate Bridge from the north. Eighteen species of raptors have been recorded, and passing rates may reach two birds per minute on exceptional days. The best period is 10 September to 31 October, with the peak about 1 October; the best time of day is from 10:00 a.m. to 1:00 p.m. after the morning fog has dispersed. Mid-March to early April is far less productive. Flocks of Band-tailed Pigeons, three species of swifts, and swallows also pass the lookout on their way south. To reach the lookout, take US 101 north across Golden Gate Bridge; then immediately take the Alexander Avenue Exit toward Sausalito for about 50 yards; turn left under US 101 and then right into the Golden Gate National Recreation Area; proceed 1.8 miles (ignoring the side road to the right), park at the crest of the road, and walk up the service road to the peak at the extreme end of the hill.

SAN JOSE

Alum Rock Park | Mt. Hamilton

Seven miles east of San Jose and about 55 miles south of San Francisco is **Alum Rock Park** (776 acres), a deeply cut, wooded canyon in the inner Coast Range on the east side of the Santa Clara Valley. From downtown San Jose, drive east on Santa Clara Street, which becomes Alum Rock Avenue east of US 101.

There are many trails in the Park. Along permanent Penitencia Creek the canyon bottom is heavily wooded with coast live oak, California buckeye, willow, western sycamore, white alder, and bigleaf maple. The walls of the canyon are clothed with patches of sagebrush, chaparral, grass, and pine-oak woodland. Rufous-crowned Sparrows are common in the sagebrush. In the chaparral, Wrentits and California Thrashers reside the year round and Allen's Hummingbirds in spring, when they arrive to court and nest. Golden Eagles often soar overhead. North American Dippers nest along the Creek, and in the bordering trees, Nuttall's Woodpeckers excavate their nest holes, which are sometimes used by the resident Northern Pygmy Owls. From the cliffs above, where White-throated Swifts nest, one may hear the tumbling song of Canyon Wrens. Although a visit at any time of year is rewarding, a morning in April, May, or June is best.

Mt. Hamilton is the site of the Lick Observatory of the University of California. The 18 miles of twisting paved road to the summit (elevation 4,209 feet) pass through pine-oak woodland. Yellow-billed Magpies are common in the open areas. In the oaks and pines can be found the Plain Titmouse, Western Bluebird, Lesser Goldfinch, and, with luck, the Phainopepla and Lawrence's Goldfinch. April and May are the months when the birds are most active and the grasslands may be covered with an impressive array of wildflowers. To reach the road up Mt. Hamilton, follow the directions to Alum Rock Park, but take the right turn (watch for the sign) approximately 5 miles east of San Jose. On a clear day, the view from the top of the mountain is well worth the trip.

SANTA BARBARA

Mission Canyon | Santa Barbara Botanic Gardens | La Cumbre Peak | Lake Cachuma | Cachuma Campground | Figueroa Mountain | Nojoqui Falls County Park | Andree Clark Refuge | Santa Barbara Cemetery | Harbor Area | Goleta Area

Situated picturesquely on a 3-mile-wide gentle slope between steep mountains and the sea about 95 miles northwest of Los Angeles, Santa Barbara has long been a center for bird watching.

In the northwestern part of the city, **Mission Canyon** between the Old Mission (founded in 1786) and the Santa Barbara Museum of Natural History is alive with birds all year. A mixture of native live oak and sycamore woodlands, with shrub and tree plantings around residences, on the grounds of the Museum, and in Rocky Nook Park just above the Museum, brings in all the widespread wintering species. Townsend's Warblers are common every winter. Allen's Hummingbirds and Hutton's Vireos begin nesting in March or before, but the height of the nesting season is in April and May, when Red-shouldered Hawks, Band-tailed Pigeons, Purple Finches, and Oregon Juncos have eggs or young. At the same time, many transient warblers, orioles, tanagers, and grosbeaks are passing through.

To reach Mission Canyon from US 101 in downtown Santa Barbara, exit east on Mission Street, turn left onto State Street for 2 blocks, then right onto Los Olivos Street, passing the Old Mission and entering Mission Canyon Road. Turn off left at the sign to the Museum and park for exploring the area. Some of the birds to look for in any season are the California Quail, Acorn Woodpecker, Scrub Jay, Plain Titmouse, Bushtit, Wrentit, California Thrasher, and Brown Towhee.

Return to Mission Canyon Road and continue on it 0.3 mile; turn right onto Foothill Road and go 2 blocks; then turn left and continue again on Mission Canyon Road to the **Santa Barbara Botanic Gardens.** The tree and shrub plantations here, below the open sagebrush and chaparral slopes, attract a wide variety of birds, breeding and transient. Small numbers of Phainopeplas are usually present in summer and fall.

Backtrack to Foothill Road; turn right onto it and go 3 miles; turn right onto State 154, the San Marcos Pass Drive, which ascends through chaparral and live oak woodland and crosses the crest of the Santa Ynez Mountains at an elevation of 2,182 feet, only 13.5 miles from downtown Santa Barbara. Stop here to admire the splendid view of the city and offshore islands and to scan the sky for soaring raptors.

At the top of the San Marcos Pass, turn east onto East Camino Cielo Road which proceeds along the crest of the Santa Ynez Mountains and is good for chaparral birds. The area around **La Cumbre Peak,** 9.4 miles from San Marcos Pass, is particularly rewarding in spring and early summer for the Mountain Quail, Greater Roadrunner, Canyon and Rock Wrens, and Rufous-crowned, Sage, and Black-chinned Sparrows.

Return to San Marcos Pass and continue northwestward. After about 5 miles, the east end of **Lake Cachuma** will appear on the right. Park on the opposite side of the highway. During winter the Lake supports a large and varied assortment of waterfowl and a few Bald Eagles and Ospreys. In view from the Bradbury Dam Observation Point at the west end of the Lake is usually a huge number of wintering Western Grebes. Anytime in winter, look for the Lewis' Woodpecker in the more open oak-woodland around the Lake.

Three miles west of Bradbury Dam on State 154, turn off right (north) onto the Armour Ranch Road, then right again onto Happy Canyon Road for about 10 miles through oak-savanna woodland—habitat for Yellow-billed Magpies—to **Cachuma Campground.** The area is well worth exploring for oak-woodland and riparian birds, including Northern Pygmy and Spotted Owls the year around and the Poor-will in spring and summer. Thirteen miles farther, Happy Canyon Road leads to the summit of **Figueroa Mountain** (elevation 4,530 feet), but do not attempt the drive up immediately after a rain. The area about the summit yields coniferous-forest birds such as the Steller's Jay, Mountain Chickadee, Pygmy Nuthatch, and Brown Creeper, and—additionally in winter—the Williamson's Sapsucker, White-headed Woodpecker, and Townsend's Solitaire.

Backtrack to State 154 and continue northwest through the Santa Ynez Valley and turn off left onto State 246 to Solvang. If Yellow-billed Magpies have not already been seen, they should have been as it is in the Santa Ynez Valley where they reach the southernmost limit of their breeding range. On coming into Solvang, turn off left near the Santa Ines Mission onto Alisal Street. **Nojoqui Falls County Park,** a small area 6.5 miles from State 246, has an abundance of Yellow-billed Magpies which are quite tame. Purple Martins nest here. By diligently searching the trees that stand along the upper part of the trail to the Falls, one will occasionally find a Spotted Owl.

At the southeast entrance to Santa Barbara, between US 101 and East Cabrillo Boulevard, is the **Andree Clark Refuge**—a park lake of fresh water that attracts grebes, cormorants, ducks, and many gulls. Usuaully among the gulls are Western, California, Ring-billed, and Heermann's in any season, Glaucous-winged, Mew, Bonaparte's, and occasionally Herring and Thayer's in winter. Most of the gulls get used to receiving food from people and hence become quite tame.

The Refuge may be reached from US 101 by turning off south onto Hot Springs Road or from the Boulevard by turning off north onto Los Patos Way. Across the Boulevard southeast of the Refuge is the **Santa Barbara Cemetery** with stands of conifers and oaks which sometimes swarm with migrating landbirds and hold many as winter residents.

The **Harbor Area** along East Cabrillo Boulevard and its continuation, West Cabrillo Boulevard, is good for the usual sandy-coast birds. The breakwater, reached from the far end of West Cabrillo Boulevard, is a fine vantage for viewing offshore birds. Elegant Terns are common in summer and fall; in late summer and fall they are frequently harassed by Parasitic Jaegers which are then regularly present. Besides loons, grebes, and diving ducks in winter, one can usually count on seeing Snowy Plovers on the sand spit at the end of the breakwater.

Farther up the coast from Santa Barbara is the **Goleta Area,** famed for its excellent sites for viewing birds the year round. Located on the campus of the University of California, Santa Barbara, it is reached by following US 101 north from the city for 6 miles to the well-marked University exit. At the entrance kiosk to the campus, ask for directions to Goleta Point, sometimes known as Campus Point.

The Point is an excellent vantage for watching sea birds, particularly during strong onshore winds that drive them close to shore, but also for various migrating waterbirds and waterfowl. From March to late May, thousands of loons (predominantly Arctic), cormorants, Black Brant, scoters (largely Surf), gulls, and terns head up the coast past the Point. In summer, Sooty Shearwaters pass by in lines or sometimes raft not far offshore. The rocks at the Point are frequented by Black Turnstones from fall to spring and by Surfbirds and Wandering Tattlers in spring only. Adjacent to the Point is the Campus Lagoon which has many ducks in season.

STOCKTON

New Hogan Reservoir | Victoria Island

In the Sierran foothills, northeast of this city, is the **New Hogan Reservoir.** From I 5 in Stockton, exit east on State 26 and proceed for about 33 miles; about 1.0 mile before reaching Valley Springs, turn off right onto a paved road with a sign indicating the way to New Hogan Dam, which is 3 miles distant. Park near the Dam. The rolling country above the Reservoir is wooded with Digger pines and blue oaks interspersed with chaparral areas of chamise, manzanita, and buckthorn, all worth exploring for the birds associated with these vegetation types. The Reservoir is a winter refuge for Canada Geese and other waterfowl. Ferruginous Hawks and Bald Eagles appear during winter in its vicinity, as does an occasional Golden Eagle.

The delta of the great Sacramento-San Joaquin drainage system west of Stockton attracts many transient and wintering waterfowl. Although the marshes have been diked and drained, the birds continue to come, frequenting the cornfields and pastures that were once marshland. Numerous sloughs and canals provide the necessary habitat for marsh dwellers. **Victoria Island** is a representative area for birds. From I 5 in Stockton, exit west on State 4 for approximately 13 miles to a directional sign. Between November and March there should be Whistling Swans, flocks of Greater White-fronted and Snow Geese, and great numbers of Sandhill Cranes. At the same time most of the species of fresh-water ducks that normally winter in California will be on the ponds and sloughs. In any season, look for White-tailed Kites either hovering over fields or perched in dead or leafless trees.

TULELAKE

Lower Klamath, Clear Lake, and Tule Lake National Wildlife Refuges

In this semidesert country of northeastern California, where the vegetation is largely sagebrush and juniper, there are nonetheless extensive lakes, marshes, and wetlands which attract enormous numbers of nesting geese and ducks. Probably more breed here than in any other part of the state. Some of the finest of their nesting areas are embraced by the **Lower Klamath, Clear Lake,** and **Tule Lake National Wildlife Refuges** near the Oregon line.

Nesting waterfowl on all three Refuges include the Canada Goose, Mallard, Gadwall, Common Pintail, Cinnamon Teal,

Northern Shoveler, Redhead, and Ruddy Duck. During migration the Refuges play host to great flights of waterfowl, many of which breed in the Arctic and spend the winter in marshes of the Central Valley south. The largest concentrations are in September and October, when, in addition to the breeding species, there are Whistling Swans, Greater White-fronted and Snow Geese, Green-winged Teal, American Wigeons, Ring-necked Ducks, Canvasbacks, Lesser Scaups, Common Goldeneyes, Buffleheads, and a few Ross' Geese. Waterfowl are not the only avian attractions. Among the other breeding birds dependent on the water and/or marshes are grebes, Great Blue Herons, Great and Snowy Egrets, Black-crowned Night Herons, Northern Harriers, American Coots, Willets, American Avocets, Black-necked Stilts, Wilson's Phalaropes, Forster's and Black Terns, Marsh Wrens, and blackbirds—Yellow-headed, Red-winged, Tricolored, and Brewer's.

Each of the Refuges has its distinctions. Lower Klamath Refuge (47,583 acres) has the widest diversity and greatest numbers of nesting waterfowl. Clear Lake Refuge (33,440 acres) has numerous small islands with nesting colonies of Double-crested Cormorants, California and Ring-billed Gulls, and Caspian Terns. On its dry, sagebrush-covered uplands are Sage Grouse, which may be observed in early spring (late March-April) on their strutting grounds, and such nesting passerine birds as the Sage Thrasher, and the Vesper, Lark, Sage, and Brewer's Sparrows. Tule Lake Refuge (38,811 acres) has hundreds of nesting Eared, Western, and Pied-billed Grebes.

Most of the migratory waterbirds and waterfowl leave the Refuges about November when the lakes freeze over, but a few Whistling Swans, with an occasional Trumpeter among them, remain on the few warm open ponds. During winter, bird finders in the area will see good numbers of Rough-legged Hawks, Bald Eagles, Prairie Falcons, Short-eared Owls, and Northern Shrikes. In addition, they will see large flocks of White-crowned Sparrows usually containing a few American Tree Sparrows and Harris' Sparrows, also flocks of Lapland Longspurs, sometimes with a Snow Bunting or two in their midst, and Gray-crowned Rosy Finches.

All three Refuges are administered from headquarters at Tule Lake Refuge, reached from Tulelake on State 139 by driving west on East-West Road for 5 miles, then south (left) on Hill Road for 0.5 mile to headquarters on the right. Inquire here about tour routes through the Refuges.

Lower Klamath headquarters may be reached by driving 4 miles north from Tulelake on State 139, then turning west onto State 161 (State Line Highway) and proceeding 9.5 miles. Clear Lake headquarters may be reached by driving south from Tulelake on State 139 for 18 miles, then turning off east at the railroad overpass onto the Forest Service Road to Double Head Mountain and continuing for 9 miles.

TWENTYNINE PALMS
Joshua Tree National Monument

This resort town on the southern Mojave Desert, reached from I 10 between Banning and Indio by exiting north on State 62 (Twentynine Palms Highway), is headquarters of the **Joshua Tree National Monument** whose area of 870 square miles may be entered immediately south. Since there are many roads throughout the Monument, it is advisable for the bird finder to stop at headquarters visitor center for a map of the road system. It is also recommended, before leaving for the Monument, that he take the self-guided nature trail through the adjacent Twentynine Palms Oasis, thus obtaining a good introduction to the vegetation in the Monument.

Spring is by far the best time to visit the desert foothills and mountains of the Monument, for, if there have been any appreciable rains during the winter, the ground between the scattered shrubs becomes a many-hued carpet of blossoming annuals. The grotesque arms of the Joshua trees are capped with mammoth columns of creamy white flowers in late March and April. Besides the unending variety of desert scenery provided by the 'woodlands' of Joshua tree and of pinyon pine and desert juniper, there are several oases where willows and cottonwoods and available water attract birds from a considerable surrounding area. There are also imposing rock outcrops, from one of which—Keys View at an elevation of 5,185 feet—is a magnificent view of the low-lying Colorado Desert and the Salton Basin stretching southward.

At Twentynine Palms (elevation 2,000 feet), Phainopeplas nest in March when most of the winter visitants are still present; other species of the desert nest chiefly in April and early May. Most of the oases within the Monument are near the boundary between the pinyon pine-juniper woodland of the higher slopes and the creosote bush and cactus desert of the lower altitudes, with the Joshua trees overlapping both habitats. The Gambel's Quail, Ladder-backed Woodpecker, Costa's Hummingbird (summer resident), Verdin, Cactus Wren, Rock Wren, Scott's Oriole (summer resident), and Black-throated Sparrow are readily observed about these oases. Mountain Quail are sometimes in the same areas with the Gambel's Quail, although the former usually nest at higher altitudes. Gray Vireos are local summer residents in the low, sparse shrubs on the not-too-rocky slopes. Golden Eagles and Prairie Falcons are present throughout the year. Pinyon Jays nest at the higher altitudes but may be found almost anywhere in the Monument at other times.

Common winter species include those widespread over the state and, in addition, the Anna's Hummingbird, Gray Flycatcher, Canyon Wren, Sage Thrasher, Western Bluebird, Townsend's Solitaire, and Chipping and Brewer's Sparrows. In April and October many warblers pass through and may be seen both around the oases and flitting from bush to bush across the desert. The Black-throated Gray Warbler is especially common, but the Orange-crowned, MacGillivray's, and Wilson's are also numerous. Other transients are the Swainson's Hawk, Vaux's Swift, House Wren, Hermit and Swainson's Thrushes, Lawrence's Goldfinch, Green-tailed Towhee, and the swallows, particularly the Violet-green.

Near the southern border of the Monument is Cottonwood Spring, reached by a paved road south through the Pinto Basin, a distance of about 45 miles, or by exiting north from I 10, 25 miles east of Indio, and proceeding north on a paved road into the Monument to the Spring, a distance of about 9 miles. At the Spring there is a small grove of cottonwoods and a few palms about a pool with willows and bulrushes—all on the main line of migration, which follows along the Little San Bernardino Mountains and into coastal southern California to the west.

UPPER NEWPORT BAY

With its salt marshes and mud or tidal flats, **Upper Newport Bay** southeast of Los Angeles is unquestionably one of the richest areas for waterbirds, waterfowl, and shorebirds in southern California. It may be reached by taking the Jamboree Road Exit southwest from I 405, then turning right onto Back Bay Road and following it along the length of the south side of the Bay.

Practically all the shorebirds listed for the mud flats in the introduction to this chapter can be seen almost any time, except summer, on the flats and sand bars in the Bay. The shorebirds include all three phalaropes, with the Northern predominating. Great Blue and Little Blue Herons are usually present all year, as are Clapper Rails, American Coots, Ring-billed Gulls, Forster's Terns, Marsh Wrens, and (Belding's) Savannah Sparrows. From fall to spring, Common Loons, Horned, Eared, and Pied-billed Grebes, Great and Snowy Egrets, various surface-feeding and diving ducks, Herring, California, and Bonaparte's Gulls, and Common, Royal, and Caspian Terns may be expected. American White Pelicans are sometimes present in winter, contrasting sharply in plumage and habits with the resident Brown Pelicans.

VISALIA
Sequoia and Kings Canyon National Parks

Access to the western flank of the Sierra Nevada in east-central California is provided by contiguous **Sequoia and Kings Canyon National Parks.** Together embracing 1,324 square miles and extending north and south for 65 miles, they reach from the foothills in the San Joaquin Valley at 1,500 feet to the 2-mile-high crest of the Sierra Nevada that forms their eastern boundary. Their principal natural feature is, of course, the groves of giant sequoias—'the world's largest trees'—which are generally at elevations between 6,000 and 7,000 feet. But the giant sequoias are not alone among the magnificent trees that make up the forests at these lower elevations.

Headquarters for both Sequoia and Kings Canyon is at the Ash Mountain entrance to Sequoia, reached from Visalia by driving east on State 198 for about 36 miles. Upon arriving here, obtain a map and information about interpretive services. From the entrance, the Generals Highway, open all year and connecting Sequoia with Kings Canyon farther north, winds for 46 miles through the sequoia belt and at Big Baldy Saddle attains an elevation of 7,643 feet. Along the way, bird walks are offered during summer in Giant Forest (elevation 6,409 feet) at *Crescent Meadow,* amid giant sequoias, firs, quaking aspens, willows, and dogwoods, and at *Wolverton,* firs, lodgepole and ponderosa pines, and willows; in Grant Grove (6,589 feet) at *Big Stump,* giant sequoias, firs, incense cedars, and willows; and in Cedar Grove (4,635 feet) at *Zumwalt Meadows,* ponderosa pines, incense cedars, black oaks, cottonwoods, willows, and alders.

Among the breeding birds at one or another of the above sites are the Blue Grouse, Mountain Quail, Calliope Hummingbird, Pileated Woodpecker, (Red-breasted) Yellow-bellied Sapsucker, White-headed Woodpecker, Olive-sided Flycatcher, Mountain Chickadee, Red-breasted Nuthatch, Hermit Thrush, Townsend's Solitaire, Golden-crowned Kinglet, Warbling Vireo, warblers (including the Orange-crowned, Nashville, Audubon's, Hermit, MacGillivray's, and Wilson's), Western Tanager, Evening Grosbeak, Purple and Cassin's Finches, and Fox and Lincoln's Sparrows. Birdlife in the canyons and at elevations above the sequoia belt is comparable to that in Yosemite National Park (*see under* **Merced**).

WILLOWS

Sacramento National Wildlife Refuge

Bird finders motoring through the Sacramento Valley on I 5 can readily visit the **Sacramento National Wildlife Refuge** by exiting east, 8 miles south of Willows, on Norman Road (goes to Princeton), then immediately turning north, and proceeding 1.5 miles to headquarters, which is at right on the Refuge. Approximately half of the Refuge's 10,766 acres are marshy with numerous ponds accessible by car. Inquire at headquarters about a self-guided tour route.

The Refuge is best visited in December and Janauary when wintering geese are in greatest numbers, with Snow Geese the most abundant. Others are Canada Geese, Greater White-fronted Geese, and usually a good representation of Ross' Geese. Already present are Whistling Swans together with thousands of ducks, some of which began arriving in September. Chief among them are Mallards, Common Pintails, Green-winged and Cinnamon Teal, American Wigeons (occasionally one or more Eurasian Wigeons in their company), Northern Shovelers, and Hooded and Common Mergansers. Numerous in the area at the same time are hawks—Red-tailed, Rough-legged, and Ferruginous—as well as Bald Eagles and Prairie Falcons.

Birds nesting on the Refuge are relatively few but include Pied-billed Grebes, Great and Snowy Egrets, Black-crowned Night Herons, Mallards, Green-winged and Cinnamon Teal, and Killdeers.

Colorado

WHITE-TAILED PTARMIGAN

In Colorado 1,500 peaks attain 10,000 or more feet of elevation; 300 exceed 13,000 feet; and 52 rise above 14,000 feet. Between the lowest point (3,350), near the Kansas line, and the highest point (14,433 feet), on the summit of Mt. Elbert, there is an altitudinal range of nearly two miles; moreover, Colorado as a whole has a mean altitude of 6,800 feet—higher than that of any other of the 48 contiguous states. Quite naturally visitors in Colorado become elevation conscious, and this is especially true of those in search of birds, for finding different species is as much a matter of changing from one altitude to another as covering miles of territory.

Not all of Colorado is mountainous; the towering mountains lie only across the central part of the state, from the northern to the southern boundary, and in the southwest. East of the mountains are extensive plains, and west of them is a high plateau. From east to west, the plains, the mountains, and the plateau constitute three natural regions, each occupying roughly a third of the state.

The eastern plains, a segment of the Great Plains, are a vast expanse of prairie extending for some 200 miles in a gradual upward slope from the eastern boundary to about 5,500 feet at the base of the mountain foothills. The only natural interruptions of an otherwise slightly undulating terrain are occasional sandhills and mesas and the shallow river valleys, notably those of the South Platte and Arkansas Rivers and their tributaries, which drain most of the area. Short grasses, chiefly grama and buffalo, yuccas, and cacti are the prominent native vegetation, except for scrub oaks on the sandhills, stretches of sagebrush, and (along streams or in moist draws) small stands of trees—cottonwoods, boxelders, and willows, together with clumps of wild plum, wolfberry, chokecherry, and other shrubs.

Though relatively dry, the plains are used for farming and cattle raising; hence their monotony is relieved in part by cultivated fields, by lines of fences, by widely scattered farm buildings, ranch houses, and small communities, most of which are sheltered by tree plantations. Birdlife is characteristically western, yet numerous species of eastern affinities—e.g., the Blue Jay, Eastern Bluebird, and American Redstart—now breed in the river valleys.

Throughout the plains the species listed below nest regularly in open country (prairie grasslands, sagelands, fallow fields, wet lowlands, brushy places, woodland borders, orchards, and dooryards) or in wooded tracts (tree plantations and deciduous woods along streams). Species marked with an asterisk also breed in suitable places in the valleys, foothills, and even higher slopes of the western mountains.

OPEN COUNTRY

Turkey Vulture
Swainson's Hawk
Ferruginous Hawk
Northern Harrier
Mountain Plover
* Mourning Dove
Burrowing Owl
Red-headed Woodpecker
Eastern Kingbird
Western Kingbird

OPEN COUNTRY (*Cont.*)

* Say's Phoebe
* Horned Lark
* Barn Swallow
* Black-billed Magpie
* House Wren
Northern Mockingbird
* Gray Catbird
Brown Thrasher
Eastern Bluebird
* Western Bluebird
Loggerhead Shrike
* Yellow Warbler
Common Yellowthroat
Western Meadowlark
Common Grackle
Blue Grosbeak
Lazuli Bunting
American Goldfinch
Brown Towhee (*southeastern Colorado only*)
Lark Bunting
* Savannah Sparrow
* Vesper Sparrow
Cassin's Sparrow
* Chipping Sparrow
* Song Sparrow
McCown's Longspur (*northeastern Colorado only*)
Chestnut-collared Longspur (*northeastern Colorado only*)

WOODED TRACTS

* Sharp-shinned Hawk
* Cooper's Hawk
* Common Screech Owl
Red-shafted Flicker
* Hairy Woodpecker
* Downy Woodpecker
Blue Jay
* Black-capped Chickadee
* White-breasted Nuthatch
* Red-eyed Vireo
* Warbling Vireo
American Redstart
Orchard Oriole
Bullock's Oriole
* Black-headed Grosbeak

The mountains are part of the Rocky Mountains system or 'Rockies,' which form, in conterminous United States, a barrier between the Great Plains on the east and the several plateaus on the west. In Colorado they are high, massive granite uplifts grouped into ranges; all trend north and south and are bordered by sedimentary foothills. The easternmost of the major ranges are the Front Range (north) and the Sangre de Cristo Range (south); those of greatest height and grandeur to the west are the Park Range (north), the Sawatch and Elk Ranges (central), and the San Juan Mountains (south). Winding from Wyoming to New Mexico across the Colorado Rockies, separating them into eastern and western slopes, is the Continental Divide. This follows the crest of the Park Range, swings eastward, proceeds southward along the Front Range, and then meanders southwestward along the Sawatch Range and through the San Juan Mountains.

Behind the eastern ranges is a chain of four wide and elevated valleys locally referred to as 'parks.' From the Wyoming line south these are the North, Middle, and South Parks—all west of the Front Range—and the San Luis Valley—west of the Sangre de Cristo Range. Originally grasslands, they are now for the most part cut up into farms and ranches. No bird species are peculiar to the parks but such species as the Western Meadowlark, Savannah Sparrow, and Vesper Sparrow are common as in other agricultural areas of the state. In North Park (*see under* **Walden**) and in the San Luis Valley (*see under* **Monte Vista** *and* **Saguache**) are numerous marshy ponds and lakes that attract a good variety of nesting waterbirds, ducks, and other birds attracted to marshes.

Like all high mountains, the Rockies in Colorado show from base to summit a vertical succession, or belts, of plant associations. Their elevations depend on such factors as latitude, exposure, soil, and moisture, and their width depends on steepness of the slopes.

East of the Continental Divide the mountain slopes have, in general, an abrupt incline; therefore their plant associations are relatively narrow and well defined.

In northern Colorado the foothills, rising from the edge of the plains at 5,500 feet elevation and reaching up to 7,800 feet, support an open forest of ponderosa pine, lodgepole pine, and Douglas-fir—nesting habitat for such birds as the Band-tailed Pigeon, Williamson's Sapsucker, Pygmy Nuthatch, Solitary Vireo, Ovenbird, and Western Tanager. Along watercourses, where cottonwoods, aspens, willows, chokecherries, alders, and other deciduous trees and shrubs thrive, the Broad-tailed Hummingbird, Dusky Flycatcher, Warbling Vireo, Yellow Warbler, and many other species found also on wooded stream bottoms on the plains are summer residents.

In southern Colorado the foothills rising from the edge of the plains and reaching up to 7,000 feet, or higher, support a scrub or 'pygmy' forest, consisting of juniper, pinyon pine, and scrub oak, as well as mountain mahogany and other shrubs. Low ridges and rims of canyons have the same growth. In this type of environment, which is warm and dry, the Poor-will, Scrub Jay, Pinyon Jay, Plain Titmouse, Bushtit, Bewick's Wren, Blue-gray Gnatcatcher, and Black-throated Gray Warbler are regular breeding residents. Above the pygmy forest, from 7,000 feet to 8,500 feet is the open pine-fir forest which is much the same in composition as in northern Colorado but begins at the edge of the plains since there is no true pygmy forest.

Above the forests of the foothills is a heavy mountain forest at elevations averaging from 7,800 to 10,200 feet in northern Colorado and from 8,500 to 10,800 feet in the south. At lower levels the tree growth is a mixture of lodgepole pine and quaking aspen, with the addition of white fir in southern Colorado. Blue spruce, alder, and willow commonly fringe the bogs and cool streams. At the higher levels the timber is an almost pure stand of Engelmann spruce, except on the cooler slopes in shaded ravines, and near bogs and streams, where it is intermixed with subalpine fir. Breeding more frequently in this forest than elsewhere in the mountains are the Northern Goshawk, Blue Grouse, Hammond's and Olive-sided Flycatchers, Gray Jay, Red-breasted Nuthatch, Brown Creeper, Hermit and Swainson's Thrushes, Townsend's Solitaire, Golden-crowned and Ruby-crowned Kinglets, Wilson's Warbler, Cassin's Finch, Pine Grosbeak, Red Crossbill, and (in willow thickets) Lincoln's Sparrow.

Between the mountain forest and the timber line, reached in northern Colorado at an average elevation of 11,200 feet and in southern Colorado at 11,800 feet, is the subalpine forest, in which spruce and fir are stunted and, on the Front Range and others, gradually replaced by limber pine, subalpine fir, and bristlecone pine, the characteristic trees of the timber line. Above timber line are alpine meadows or tundras where mats of grasses and perennial flowering plants are interspersed with thickets of dwarf willows. Loose boulders, cliffs, and talus slopes, cold streams, and lakes derived from melting snow, and, even in midsummer, low temperatures and winds of high velocity characterize this bleak region.

In the subalpine forest few birds species are represented that cannot also be found in the mountain forest below. From mountain forest to timber line, changes in birdlife involve mostly a steady reduction in numbers of individuals. At the timber line, however, where widely spaced dwarf conifers meet alpine willows, White-crowned Sparrows nest regularly. Above timber line the Horned Lark is a summer resident of the larger meadows, and the Northern Raven is occasionally seen patrolling a precipice where it probably nests; but the three species breeding regularly are the White-tailed Ptarmigan, Water Pipit, and Brown-capped Rosy Finch.

A number of species nest widely over the forested slopes of the Colorado Rockies, their habitat preferences being less restricted than those already mentioned. Among those that nest regularly are the following:

Northern Three-toed Woodpecker (*more common on higher forested slopes*)
Violet-green Swallow
Tree Swallow
Steller's Jay (*most common on lower forested slopes*)
Clark's Nutcracker
Mountain Chickadee
Audubon's Warbler
MacGillivray's Warbler (*deciduous thickets*)
Pine Siskin
Gray-headed Junco

Other species in the Rockies are confined during the breeding season to the following situations: Prairie Falcon, to cliffs in the foothills; White-throated Swift, to cliffs at all elevations save those above timber line; Lewis' Woodpecker, to open areas and canyons near forests in the foothills; Yellow-bellied Sapsucker, to aspens at all elevations; Traill's Flycatcher, to shrubs on stream borders at lower elevations; Western Flycatcher, to well-shaded spots near streams or buildings below 9,000 feet; Western Pewee, to forests containing either ponderosa pine, or cottonwood and aspen; North American Dipper, to fast-flowing mountain streams; Canyon Wren to cliffs below 9,000 feet; Rock Wren, to rough, rocky slopes at all elevations; House Finch to the vicinity of human habitations, both in the foothills and on the adjoining edge of the plains.

West of the Continental Divide, the mountain slopes are comparatively gradual, with the result that the plant associations not only occupy belts of considerable width but tend to overlap extensively; nevertheless the sequence and elevations of the plant associations are essentially the same as on the eastern slopes.

Most of the high plateau west of the mountains in Colorado is part of the Colorado Plateau, the greater part of which lies in eastern Utah, northern Arizona, and northwestern New Mexico. Lofty ridges and occasional broad mesas between deep meandering canyons and valleys, each with similar tributaries, characterize the terrain.

Quite appropriately, the section of the Colorado Plateau lying in Colorado and eastern Utah is referred to as the canyon lands. The floors of the canyons and valleys, with elevations from 4,500 to 6,000 feet, are typically arid or desert-like; year-round streams are few, and the vegetation consists to a large extent of sagebrush and cacti. An exception may be in the large river valleys—e.g., the Colorado River Valley—where the floors are several miles wide and have moist floodplains with soils suitable for farming. On the canyon and valley slopes, from 6,000 to 8,500 feet, grows a pygmy forest of pinyon pine and juniper, together with antelope bush, serviceberry, mountain mahogany, and other shrubs. Where slopes extend from 8,500 to 10,500 feet, as on the big mesas—e.g., Grand Mesa (*see under* **Grand Junction**)—there is a succession of foothills and mountain forests with associated birdlife comparable to that of the Rockies to the east. In the extreme north, the high plateau of western Colorado is an area of sagebrush plains at an elevation of about 6,000 feet, flanked by rough country in which the principal plant cover is a sparse growth of junipers. Trees and shrubs such as cottonwood, willow, and buffaloberry are confined to the banks of streams.

On the high plateau of western Colorado, the following birds are among the breeding species considered characteristic of the sagebrush country (the sagebrush-studded floors of canyons and valleys and the sagebrush plains) and of the pygmy forests on canyon and valley slopes:

SAGEBRUSH COUNTRY

Gambel's Quail (*southern plateau*)
Burrowing Owl
Horned Lark
Sage Thrasher
Loggerhead Shrike
Western Meadowlark
Vesper Sparrow
Lark Sparrow
Black-throated Sparrow
Sage Sparrow
Chipping Sparrow
Brewer's Sparrow

PYGMY FOREST

Poor-will
Gray Flycatcher
Scrub Jay
Pinyon Jay
Plain Titmouse
Bushtit
Bewick's Wren
Blue-gray Gnatcatcher
Gray Vireo
Black-throated Gray Warbler
Rufous-sided Towhee

The streams, small lakes, and reservoirs on both sides of the mountains and in the mountain parks attract modest numbers of transient waterbirds, waterfowl, and shorebirds during spring and fall; at the same time, the mountain valleys that trend north and south are paths for many transient landbirds. Periods during which peak flights may be expected are as follows:

Waterfowl: 1 March–10 April, 15 October–10 December
Shorebirds: 25 April–25 May, 5 August–10 October
Landbirds: 10 April–10 May, 5 September–1 November

There is no place in Colorado, except in the eastern foothills, where one may expect to observe huge aggregations or waves of migrating birds; but the mountains nevertheless provide a fertile field for the study of vertical migration. This is especially true of the Front Range, where different altitudinal environments are readily accessible by highways (*see under* **Boulder, Denver,** *and* **Loveland**). Certain species, such as the Brown-capped Rosy Finch, move down the mountains in the fall. In the spring a reverse movement takes place. Migrations from one level to another are said to proceed with the same regularity as horizontal migrations from one latitude to another.

In late summer and early fall, Northern Harriers, Prairie Falcons, American Kestrels, Mourning Doves, Eastern Kingbirds, Pinyon Jays, Western Meadowlarks, and other species that breed in the open country, foothills, and mountain parks appear frequently at high altitudes, even on alpine meadows. This phenomenon is explained by some authorities as being the result of an up-mountain wandering prior to a down-mountain return and subsequent southward migration. A few authorities advance the explanation that it represents a lingering during an early, though normal, north-to-south migration via high elevations.

Winter birdlife in the valleys and foothills of the Rockies contains an impressive mixture of species from north and northwest of the state, species that breed during the summer on the adjacent high slopes, and species resident in the immediate vicinity.

Authorities

Alfred M. Bailey, Paul H. Baldwin, Allegra E. Collister, William A. Davis, William Ferguson, Harold R. Holt, Nancy Hurley, James A. Lane, Robert J. Niedrach, Van Remsen, Ronald A. Ryder.

References

Pictorial Checklist of Colorado Birds (with brief notes on the status of each species in neighboring states of Nebraska, Kansas, New Mexico, Utah, and Wyoming). By Alfred M. Bailey and Robert J. Niedrach. Denver: Denver Museum of Natural History, 1967.

A Birder's Guide to Eastern Colorado (east of the Continental Divide). By James A. Lane and Harold R. Holt. Distributed by L & P Press, Box 21604, Denver, CO 80219. 1975.

BOULDER
Boulder Canyon | Brainard Lake | Mt. Audubon

The birdlife of the Front Range immediately west of Boulder illustrates particularly well the altitudinal changes in species from foothill canyons and forests to timber line and alpine meadows. To see this altitudinal succession to advantage, a pleasant trip may be made—preferably in July—from Boulder, at an elevation of 5,530 feet, through **Boulder Canyon,** up to **Brainard Lake,** and on up to the alpine meadows of **Mt. Audubon,** whose summit reaches 13,223 feet. This involves a climb of approximately 35 miles by car plus a little over 2 miles on foot. An indication of the birds that may be expected in the early summer is given in the paragraphs below.

Drive west from Boulder to Nederland on State 199, which at once enters Boulder Canyon. Its stream, Boulder Creek, is lined near the mouth with cottonwoods, willows, and various shrubs that provide habitats for such birds as MacGillivray's Warblers, Black-headed Grosbeaks, Lazuli Buntings, and Rufous-sided Towhees. On the south-facing slopes of the Canyon are scattered stands of ponderosa pine; on the north-facing slopes are close stands of Douglas-fir. At a point 9 miles up the Canyon, park the car and take a foot trail on the right to Boulder Falls, 75 yards distant, for a likely glimpse of North American Dippers. At this same spot, look for Steller's Jays, if they have not already been seen.

At Nederland, 18 miles west of Boulder and just beyond the large Barker Reservoir, turn north onto State 72 and proceed 12 miles north to Ward. Once above 8,000 feet in altitude, the highway passes through a forest of lodgepole pine and aspen. Birds to be expected along the way include the Olive-sided Flycatcher, Violet-green Swallow, Mountain Bluebird, Cassin's Finch, Pine Siskin, Green-tailed Towhee, and, possibly, the Band-tailed Pigeon and Red Crossbill.

At Ward (elevation 9,253 feet) take a road to the left for 4 miles to Brainard Lake (elevation 10,300 feet), a glacial body of water below the timber line. Walking around the Lake, one may chance to see or hear a number of species, among them the Spotted Sandpiper, Northern Three-toed Woodpecker, Hammond's Flycatcher, Gray Jay, Clark's Nutcracker, Mountain Chickadee, Ruby-crowned Kinglet, Audubon's and Wilson's Warblers, Pine Grosbeak, Gray-headed Junco, and White-crowned and Lincoln's Sparrows.

From Brainard Lake, take the Buchanan Pass Trail up Mt. Audubon, reaching the timber line east of the peak, about 2 miles' walking distance. The climb, which goes as high as 1,000 feet above Brainard Lake, requires about two hours. Possibly a Blue Grouse may be seen from the Trail. At the rocky area of the timber line, begin looking for White-tailed Ptarmigan. Protectively colored, they are difficult to see; but, if discovered, they may be approached closely before they become unduly alarmed.

For other birds of the alpine region, follow the rock cairns—the

Trail itself is less evident above the timber line—for another half-mile above and west of the big permanent snowbank on the east side of the mountain. Water Pipits are numerous in the alpine meadows not far from the snowbank and may be observed giving their aerial displays. A few Horned Larks may also be here. In addition to seeing White-tailed Ptarmigan and Water Pipits, perhaps the greatest reward for the climb is to watch Brown-capped Rosy Finches, sometimes as many as 200 at a time, as they search on the snowbank for insects that have been blown onto its surface and become immobilized by the cold temperature.

COLORADO SPRINGS

Garden of the Gods | Black Forest

On the northwestern outskirts of this plains city (elevation 5,900 feet) in central Colorado, almost in the shadow of Pikes Peak, is a point of great scenic attraction—the **Garden of the Gods.** To reach it, drive west from the city on US 24 (Midland Expressway), exit north on 6th Street in Manitou Springs, and continue north to the entrance. In this city park, great masses of red rock, some curiously ridged and pinnacled, loom above an area that is partly grassland and partly covered by thickets of scrub oak and stands of pinyon pine and juniper. Among the birds nesting in the precipices of the higher rock formations are—besides vast numbers of feral Rock Doves—White-throated Swifts, Violet-green and Cliff Swallows, and Rock Wrens. In the thickets below, look for Scrub Jays and, during evenings of early summer, listen for Poor-wills.

Of considerable ecological interest in central Colorado is the **Black Forest** (elevation 7,000–7,500 feet), a rolling area of 150,000 acres, situated between the Great Plains and the slopes of the Rockies and timbered with a nearly pure stand of ponderosa pine. Only along the creeks is the coniferous growth interrupted—by willows and aspens, associated with adjoining patches of alder, chokecherry, mountain mahogany, wild rose, and other shrubby growth.

Typical, uncleared sections of the Black Forest may be reached from Colorado Springs by driving north for 18 miles on I 25, exiting east on State 105 for 6 miles or more, and then taking side roads south for productive sections of the Forest. Be on the alert for Ferruginous Hawks and Golden Eagles. Some of the bird species breeding regularly in the pine and creek environments are the Mourning Dove, Broad-tailed Hummingbird, Hairy Woodpecker, Western Pewee, Black-capped Chickadee, Pygmy Nuthatch, House Wren, Western Bluebird, Solitary and Warbling Vireos, Yellow and Audubon's Warblers, Pine Siskin, Lesser Goldfinch, Green-tailed Towhee, and Gray-headed Junco.

CORTEZ
Mesa Verde National Park

In extreme southwestern Colorado, 52,074 acres of the canyon lands comprise **Mesa Verde National Park,** established in 1906 by the Federal government for conserving and protecting the remains of prehistoric Indian villages, scores of which are on mesa tops and in canyon caves. Their remarkably good state of preservation and romantic settings, coupled with their archeological importance, are sufficient to excite the interest of the most apathetic visitors. In fact, the ancient ruins are so much the focal point of interest to visitors that the Park's fauna and flora are usually overlooked.

Mesa Verde has a biota that may be considered typical of the high, semiarid country of southwestern United States. Plants and animals are sharply restricted in variety and numbers. The bird finder will find it fascinating to observe the kinds of birds that tolerate the rather severe conditions of this environment.

Geologically, the area embraced by the Park consists of one tableland or plateau—Mesa Verde—with a general elevation of 8,000 feet. On the north it has a 2,000-foot escarpment facing out over a rather dry plateau and, beyond that, the broad Montezuma Valley; on the south it is skirted by the canyon of the Mancos River, or Mancos Canyon, which has a depth of nearly 2,000 feet. Except for the north rim and its escarpment, the Mesa is strangely cut by many deep, steep-walled, ramifying canyons, all trending southward and converging on Mancos. On a topographic map this complex canyon system resembles the branching of a dense shrub, with Mancos Canyon as the main stem and the Mesa's canyons the branches emerging from one side.

A relatively thick growth of pinyon pine and juniper stands on the uncut surfaces of the southern half of the Mesa; a cover of scrub-oak thickets, interspersed with open areas in which grasses, sagebrush, and rabbitbrush are common, is distributed along the north rim. In certain spots, such as the heads of canyons, there are small stands of Douglas-fir and ponderosa pine; elsewhere in the pinyon pine and juniper forest and on the upper canyon walls are mountain mahogany, serviceberry, and other shrubs.

The Park entrance, or checking station, is on U.S 160 about midway between Cortez and Mancos and 35 miles west of Durango. A map and a schedule of guided trips to the ruins are available here. From the entrance it is 20 miles south to headquarters, which is in the Park on the west rim of Spruce Tree Canyon. Here, or in the immediate vicinity, are various services and accommodations for visitors, and a fine archeological museum. Although the Park is open all year, most services and accommodations are available only during the Park season, which extends from 15 May to 15 October.

One of the best places for birds on Mesa Verde is on the north rim, which the entrance road soon follows after ascending the north escarpment at its eastern end. From here, while taking in, on the right, the magnificent view of Montezuma Valley and, on the left, the sweeping panorama of the Mesa's canyon-gashed surface sloping downward from the road, the bird finder may see a few Turkey Vultures, White-throated Swifts, and Northern Ravens, and perhaps one or two Red-tailed Hawks and Golden Eagles. It is also possible to hear Canyon Wrens singing below the rim.

A side road to the right, 10 miles from the entrance, leads to Park Point at 8,572 feet, the highest elevation in the Park. Not far west of this side road, the entrance road passes three canyon heads on the left and then bears south to headquarters, passing additional canyon heads on the way.

At or near headquarters, as well as on the north rim and in some of the canyon heads passed by the entrance road, there are habitats suitable for the following birds that breed regularly: Blue

Grouse, Mourning Dove, Poor-will, Common Nighthawk, Broad-tailed Hummingbird, Lewis' Woodpecker, Say's Phoebe, Violet-green Swallow, Steller's and Scrub Jays, Black-billed Magpie, Pinyon Jay, Plain Titmouse, Bushtit, Bewick's Wren, Mountain Bluebird, Blue-gray Gnatcatcher, Solitary Vireo, Virginia's and Black-throated Gray Warblers, Western Tanager, and Green-tailed Towhee.

DENVER

Mt. Evans | Red Rocks Park | Genesee Park | Echo Lake Park | Barr Lake Drainage Area

On mile-high plains, with short-grass prairie on the east and the foothills and Front Range of the Rockies on the west, Denver has in its vicinity numerous prairie lakes and several 'Mountain Parks' where birds may be observed in great variety. The Mountain Parks are recreational areas owned and maintained by the city and are reached by highways leading south and west from downtown thoroughfares. All have such facilities as tables, fireplaces, and water supplies.

Of the many trips for birds one may take in the Denver area, the one that terminates with a drive to the summit of **Mt. Evans** (elevation 14,264 feet) over the highest paved road in continental United States is most exciting, particularly for those unfamiliar with birdlife of the high country west of the Great Plains. Directions for this trip, including stops in three of the Mountain Parks, follow.

Drive west from Denver on I 70; about 10 miles from the city limits, exit on State 26 south for 1.5 miles and turn right into **Red Rocks Park** (639 acres). Here ponderosa pines and junipers stand in striking contrast to the brilliant hues of the oddly upturned and eroded rock strata. Elsewhere in the Park are grassy areas and brushlands, all accessible by roads and trails. Breeding birds associated with the rock formations include White-throated Swifts, Say's Phoebes, Violet-green, Barn, and Cliff Swallows, Canyon and Rock Wrens, and hundreds of Rock Doves which here have reverted to their ancestral cliff-dwelling habits. Among breeding birds associated with other habitats are the Poor-will, Scrub Jay, Gray Catbird, Virginia's Warbler, Yellow-breasted Chat, Black-headed Grosbeak, Lazuli Bunting, House Finch, Pine Siskin, American and Lesser Goldfinches, Green-tailed and Rufous-sided Towhees, and Chipping and Song Sparrows. In winter, Townsend's Solitaires, Gray-crowned, Black, and Brown-capped Rosy Finches, and four juncos—White-winged, Slate-colored, Oregon, and Gray-headed—may be found in the Park Area.

Return to I 70; continue west and exit north on Stapleton Drive to **Genesee Park** (2,403 acres), situated higher in the foothills. Much of the Park is forested. Open stands of ponderosa pine and Douglas-fir comprise the principal growth, but here and there are aspen groves and, in the ravines, a variety of deciduous trees. By carefully searching different wooded areas, one is likely to turn up such nesting birds as the Broad-tailed Hummingbird, Yellow-bellied and Williamson's Sapsuckers, Western Pewee, Steller's Jay, Black-capped and Mountain Chickadees, Pygmy, Red-breasted, and White-breasted Nuthatches, Western and Mountain Bluebirds, Solitary and Warbling Vireos, Audubon's Warbler, Western Tanager, and Gray-headed Junco.

STELLER'S JAY

Again return to I 70; continue west and exit at Idaho Springs; then turn south onto State 103, which ascends gradually to Echo Lake at an elevation of 10,605 feet in **Echo Lake Park** (about 50 acres). In boggy habitat around the Lake, there is little difficulty in finding, during the height of the singing period in June, both Wilson's Warblers and Lincoln's Sparrows. Back from the Lake, on the slopes that are densely wooded with Englemann spruce and subalpine fir, more arduous searching will yield some of the high-mountain breeding birds, including the Northern Three-toed Woodpecker, Hammond's Flycatcher, Clark's Nutcracker, Gray Jay, Brown Creeper, Golden-crowned and Ruby-crowned Kinglets, Hermit Thrush, and Pine Grosbeak.

Beyond Echo Lake, State 103 becomes State 5 and winds upward toward the summit of Mt. Evans. At the timber line (11,500 feet), where stunted and twisted trees meet rock slides and alpine meadows, are the breeding habitats of Cassin's Finches and White-crowned Sparrows. Along the road from the timber line to the summit—the last few yards must be covered on foot—are alpine meadows with dense clusters of wildflowers in June and early July. Here the breeding birds most likely to be seen are Water Pipits, but other birds should be looked for: White-tailed Ptarmigan, protectively colored, slow moving, and consequently hard to spot; Horned Larks on the wider alpine meadows; one or more Northern Ravens flying about near precipitous cliffs; and Brown-capped Rosy Finches foraging for insects and seeds, frequently in the vicinity of snow patches or around the rocky shores of Summit Lake. In early fall the meadows just above the timber line are often places for Blue Grouse, which wander up to them from the forests below.

For an entirely different bird-finding trip in the vicinity of Denver, explore the **Barr Lake Drainage Area** northeast of the city. This is characterized by many prairie lakes, reservoirs, and drainage pools, shallow and often bordered by marshes and mud flats. The best way to cover the Area is to drive north out of Denver on I 25, then turn off right to I 76 for about 12 miles and exit right to Barr Lake, a reservoir, for initial bird finding. Return to I 76 and continue, taking any of the next exits to roads going east past various areas of water worth stops for more bird finding.

Birds usually breeding in good marshes include Eared and Pied-billed Grebes, American Bitterns, Mallards, Common Pintails, Blue-winged and Cinnamon Teal, Ruddy Ducks, Virginia Rails, Soras, American Coots, Common Snipes, American Avocets, Wilson's Phalaropes, Common Yellowthroats, Yellow-headed, Red-winged, and Brewer's Blackbirds, and occasionally Savannah Sparrows on their periphery. Frequenting many of the lakes and marshes from their few nesting colonies in the Drainage Area are Double-crested Cormorants, Great Blue Herons, Snowy Egrets, and Black-crowned Night Herons. Some of the water areas with extensive mud flats are excellent in late April and May, again in late summer, for many transient shorebirds, among the more abundant being Baird's Sandpipers.

Upland grassy areas such as fields and pastures are breeding sites for Horned Larks, Western Meadowlarks, Lark Buntings, and Grasshopper Sparrows; groves of trees for Swainson's and Ferruginous Hawks, American Kestrels, Eastern and Western Kingbirds, Loggerhead Shrikes, Bullock's Orioles, House Finches, and American Goldfinches; shrubs bordering creeks or fencerows for Blue Grosbeaks and Song Sparrows.

FORT COLLINS
Pawnee National Grassland

Beginning about 25 miles northeast of this city and stretching eastward over 50 miles is the **Pawnee National Grassland,** embracing some 775,000 acres of native prairie, primarily short-grass, administered by Roosevelt National Forest. About 27 per cent of the acreage is government-owned; the rest, consisting mainly of ranches, is in private hands.

For a sampling of breeding birdlife in this vast area, the following short trip is suggested. Although bird finding can be productive all year, the time recommended is spring, preferably in May or June, when wildflowers are in bloom and most birds are nesting.

From I 25, running north-south east of Fort Collins, exit east on State 14 for 37 miles to Briggsdale. Here turn off north onto the road to Hereford, passing Crow Valley Park, a Forest Service campground, just outside of town. After about 3 miles, turn off left through a cattle guard and proceed west on a road for 10 miles. Warning: Do not attempt driving on this road in a conventional car after a heavy rain as its dirt surface becomes muddy and in places practically impassable.

The road traverses prairie that lacks uniformity in vegetational composition. In the first mile or so there is short-grass prairie on both sides of the road; then comes prairie with taller grasses mixed with other herbaceous plants; and finally there is more short-grass prairie. As certain bird species prefer one vegetational type to another, the birdlife along the way also lacks uniformity. Moreover, as the bird finder will discover, man's use of the area—in building bridges, setting up fencerows, planting trees around dwellings, and so on—has further blurred uniformity by attracting bird species that were never native to the prairie.

The bird finder should drive slowly along the road, stopping frequently to look and listen. Where there is short-grass prairie, especially when grazed, be alert for Mountain Plovers and McCown's Longspurs—and the possibility of a pair or two of Long-billed Curlews. In tall-grass prairie, expect Grasshopper Sparrows

GRASSHOPPER SPARROW

and, by chance, a few Cassin's Sparrows and Chestnut-collared Longspurs. Where there are low, brushy patches, inspect them for Brewer's Sparrows.

Killdeers, Common Nighthawks, Mourning Doves, Horned Larks, Western Meadowlarks, Lark Buntings, and Vesper and Lark Sparrows, all common and less restricted to type of prairie, are likely to show up almost anywhere from the road. Other birds to watch for: Say's Phoebes and Barn and Cliff Swallows near bridges under which they nest or around abandoned dwellings where they also nest; Western Kingbirds, sometimes Eastern Kingbirds, and Loggerhead Shrikes on trees and fences around deserted home sites; Swainson's and Ferruginous Hawks and Northern Harriers in flight or perched on fences.

GRAND JUNCTION

Grand Mesa | Colorado National Monument

In west-central Colorado, where an angle is formed by the confluence of the Colorado and Gunnison Rivers, rises 'the grandfather of all mesas,' **Grand Mesa.** Actually a fragment of a great lava cap, it is sometimes referred to as the largest flat-topped mountain in the United States. A drive over the level top of this great plateau at an elevation of 10,500 feet from the starting point at Grand Junction at 4,500 feet cannot fail to impress even the most sophisticated traveler. Besides unrivaled scenery, the trip has, for the bird finder, abundant rewards.

Early summer is the time for observing the greatest variety of birds. Of the several routes over the Grand Mesa, the following two are suggested:

1. From Grand Junction, drive southeast on US 50. At Whitewater, turn left (east) onto an unnumbered road which becomes Lands End Road and climbs steeply and tortuously to the rim of the Mesa. Follow Lands End Road across the Mesa to State 65; turn right onto it and follow it down from the Mesa to State 92; turn right onto State 50 at Delta, and return to Grand Junction. The entire trip, about 130 miles, can be made in a day with ample time for bird finding. (Should one wish to avoid the steep climb up

unpaved Lands End Road, he can take the second suggested trip. Although a longer trip, the roads are of more gradual incline and are paved all the way.)

Sharply contrasting environments are passed through in the journey to the top of the Mesa. First is the valley floor of the Gunnison River, where desert-like wastes alternate with irrigated and well-cultivated farmlands and ranches. Next is the steep slope of the Mesa with its pygmy forest of scrub oak, pinyon pine, and juniper. Finally, on top of the Mesa, are meadows bright with wildflowers, small blue lakes, and extensive stands of Engelmann spruce and other conifers. (Grand Mesa does not rise above a timber line and thus there are no alpine meadows.) Almost anywhere along the way one is likely to see an American Kestrel, Violet-green Swallow, or Mountain Bluebird—species with which the bird finder is already familiar if he has been in Colorado very long; but most of the different kinds of birds dwell in particular places and hence must be searched for with knowledge of their habitat preferences.

Where the Lands End Road begins in the Gunnison Valley, Ring-necked Pheasants and Gambel's Quail may be observed, especially in the early morning and late afternoon, on the farmlands and brushy deserts. Horned Larks are common. Other breeding birds regularly residing include Red-tailed and Swainson's Hawks, Mourning Doves, Western Kingbirds, Say's Phoebes, Sage Thrashers, Loggerhead Shrikes, Western Meadowlarks, and Brewer's Sparrows. Barn and Cliff Swallows frequently fly low over the cultivated fields to catch insects.

As Lands End Road climbs to meet the main slope of the Mesa and farms become smaller and fewer, one should begin looking for Northern Ravens. The oak thickets along Kannah Creek, about 10 miles from Whitewater, have nesting Black-billed Magpies. Above the Creek, in growths of aspen, scrub oak, pinyon pine, and juniper, are Ash-throated Flycatchers, Scrub and Pinyon Jays, Plain Titmice, Bushtits, Western Tanagers, and Rufous-sided Towhees.

At the rim of the Mesa, not far from the Road, is a rugged promontory known as Lands End; on it is a glass-walled observatory. Overlooking a tremendous stretch of western Colorado, Lands End makes a fine vantage from which to watch soaring Turkey Vultures and Golden Eagles and the circling-sailing-flickering flights of White-throated Swifts. Back from the edge of the dropoff are spruces, whose seed-bearing cones often attract Clark's Nutcrackers.

From Lands End across the top of the Mesa to State 65, the Road traverses meadows alternating with spruce groves, whose edges or interiors should be investigated for Blue Grouse, Red-shafted Flickers, Williamson's Sapsuckers, Northern Three-toed Woodpeckers, Hammond's and Olive-sided Flycatchers, Steller's Jays, Mountain Chickadees, Pygmy Nuthatches, Ruby-crowned Kinglets, Cassin's Finches, Pine Grosbeaks, Pine Siskins, Gray-headed Juncos, White-crowned Sparrows, and other birds that nest in the coniferous forests of Colorado's high country.

2. From Grand Junction, drive east on I 70 for 17 miles; exit right on State 65 through the town of Mesa to Skyway, a resort; continue on State 65 up the Mesa and across the top. When Lands End Road appears on the right, turn off onto it to Lands End, described in the first trip. Bird-finding opportunities are quite similar to those of the first trip from Grand Junction to the top of the Mesa, except that the succession of habitats up the Mesa is not as sharply discernible.

When in Grand Junction during summer, no bird finder should miss **Colorado National Monument** (18,311 acres), which embraces a section of the great escarpment on the south side of the Colorado River Valley. Here, produced by the prolonged effects of erosion, are immense canyons with precipitous walls of red sandstone, gigantic, curiously carved monoliths towering as high as 500 feet, and other formations of herculean proportions. This being arid country, the woody vegetation of the Monument is sparse, consisting principally of scattered pinyon pine, juniper, and low shrubs including sagebrush. Breeding birdlife is thinly diffused.

The Monument may be reached from Grand Junction by driving southeast on State 340. Soon after it crosses the Colorado River, turn off left, following signs to the Monument. At the picnic area, 0.25 mile inside, park the car and explore the immediate vicinity for birds as well as the pinyon pine-juniper and sagebrush habitats along a path down to the streambed.

Scrub Jays frequent the picnic tables for food. Among the breeding birds to look for in appropriate habitats are the Gambel's Quail, Black-chinned Hummingbird, Ash-throated and Gray Flycatchers, Plain Titmouse, Bewick's Wren, Sage Thrasher, Blue-gray Gnatcatcher, Gray Vireo, Rufous-sided Towhee, and Lark and Black-throated Sparrows.

From the picnic area, follow the main road up to Cold Shivers Point and take the famous Rimrock Drive, the construction of which was a remarkable engineering feat—the road was laid along meandering edges of great chasms. Some of the birds to watch for are Turkey Vultures, Red-tailed Hawks, American Kestrels, White-throated Swifts, Say's Phoebes, Horned Larks, Black-billed Magpies, Northern Ravens, Pinyon Jays, Canyon and Rock Wrens, and Loggerhead Shrikes.

The Rimrock Drive eventually reaches Park headquarters and visitor center at the west end of the Monument. From here one may return to Grand Junction by taking the Monument road north across the Colorado River to Fruita and turning east onto I 70.

LOVELAND
Rocky Mountain National Park

In the Front Range of the Rockies in north-central Colorado, there is hardly an easier or more exciting way of seeing birds than to drive from Loveland to the village of Estes Park, thence into **Rocky Mountain National Park** and up to an elevation exceeding 12,000 feet. By making stops at different elevations and vegetational associations, it is possible to identify an astonishing variety of birds. Although Rocky Mountain Park is open all year, its main highway, the Trail Ridge Road (US 34), is closed above 9,620 feet at Many Parks Curve from about mid-October to Memorial Day weekend because of snow. Bird finding is profitable anytime but best in mid-June, when most species are in full song.

Rocky Mountain Park embraces 412 square miles of mountain terrain, which has no elevation lower than 7,800 feet; 65 peaks exceed 10,000 feet; the highest, Longs Peak, reaches 14,255 feet and is a perennial favorite among mountain climbers. The Continental Divide runs southward across the Park from the northwest corner, separating the area almost centrally into eastern and western portions.

Forests are extensive and are characterized at lower elevations by open stands of ponderosa pine, with blue spruce in the sheltered gorges on the eastern side of the Park and willow and aspen along the streams; at elevations of 8,000 to 9,500 feet, by dense stands of lodgepole pine and aspen; and at elevations above 9,500 feet, by Engelmann spruce and subalpine fir joined toward the timber line (11,000 feet) by limber pine, which becomes stunted and twisted, forming with dwarfed spruce and fir the so-called 'wind timber.'

Although the splendid forests are distinctive features, the Park's great charm may be attributed less to them than to a great variety of scenic attractions—the spectacular peaks and the wide intervening valleys, the gulches and gorges flanked by precipitous cliffs and talus slopes, the alpine meadows carpeted with a profusion of wildflowers, the small glaciers wedged in the heads of gorges, the mirrored pools and lakes, and the cascading streams—and the abundance of wildlife, which includes, in addition to birds, elk, mule deer, bighorn sheep, beaver, and many smaller mammals.

Directions for a June trip from Loveland (elevation 4,980 feet) to the highest point (12,183 feet) on the Trail Ridge Road in Rocky Mountain Park are given below. Stops for bird finding are selected chiefly for showing the variety of species and vertical succession.

Drive west from Loveland on US 34 to Estes Park (elevation about 7,500 feet). Continue through the village and west for 5 miles on US 34 to the Fall River entrance to Rocky Mountain National Park. Obtain here a map of the Park, together with a schedule of interpretive services. (If one wishes to visit Park headquarters, he should go to the Beaver Meadows entrance, reached from US 34 as it nears Estes Park, by turning off left onto US 36, following it west into the village, and then proceeding 2.7 miles southwest.)

One mile from the entrance, along the Fall River, is Horseshoe Park, a large grassy meadow with thickets and several beaver ponds at the western end. The Spotted Sandpiper, Belted Kingfisher, Savannah and Vesper Sparrows, and, possibly, the Traill's Flycatcher can be observed here.

A little farther on, at the upper end of Horseshoe Park, turn right off US 34 and drive to Roaring River. The aspen growth south of the road is worth investigating for Downy Woodpeckers, Yellow-bellied and Williamson's Sapsuckers, Western Pewees, Warbling Vireos, MacGillivray's Warblers, and, perhaps, a Blue Grouse. In the large aspens near the River, look for holes in which Violet-green and Tree Swallows nest. South of the aspens, along Fall River, is an area of shrubby willows where *Empidonax* flycatchers (probably Dusky among others), Yellow and Wilson's Warblers, and Lincoln's Sparrows are summer residents. Always watch for North American Dippers along the River itself.

Return to US 34, turn right on it to Deer Ridge Junction. From here to its highest point in the Park, US 34 is commonly known as the Trail Ridge Road, a highway unrivaled for its magnificent scenery.

After passing through Hidden Valley (9,240 feet elevation), a ski area, the highway swings sharply to the left and begins a steep grade. About 1.0 mile farther on, it switches back to the right. After taking this turn—called Many Parks Curve—stop the car in a parking place and look or listen for birds in the adjacent tree growth, which consists mainly of lodgepole pine mixed with Engelmann spruce. The Northern Goshawk, Blue Grouse, Olive-sided Flycatcher, Gray and Steller's Jays, Clark's Nutcracker, Mountain Chickadee, Brown Creeper, Hermit Thrush, Audubon's Warbler, Cassin's Finch, and Gray-headed Junco are some of the birds that may be expected here.

From Many Parks Curve (elevation 9,620 feet) the highway continues to climb. After a mile or two it passes through a virgin forest of towering spruces and firs. In this tree association, the bird population is not high, but if the area is worked intensively, preferably above the road, it should yield, in addition to species below Many Parks Curve, the Ruby-crowned Kinglet, Evening and Pine Grosbeaks, and, with luck, the Northern Three-toed Woodpecker.

Stop at Rainbow Curve (elevation 10,829 feet), the next big parking space. Owing to its popularity among tourists as a lunching spot, freeloading Clark's Nutcrackers abound as well as a few Gray Jays, Cassin's Finches, Gray-headed Juncos, a host of chipmunks and ground squirrels, and sometimes a pika or two. The outer edge of the parking space is a good vantage from which to hear the songs of the Townsend's Solitaire and to scan the sky, horizon, and valley below for Northern Ravens and such raptors as the Red-tailed Hawk, Golden Eagle, and even a Prairie Falcon.

Not far above Rainbow Curve the highway climbs through the timber line (11,500 feet elevation) with its stunted limber pine, spruce, and fir. This is a favorite habitat for White-crowned Sparrows. Just above the timber line, leave the car at one of several parking spaces and explore the area. Among the rocks along the ridge to the south, Rock Wrens are summer residents. Above the alpine meadows on the other side of the ridge, Horned Larks and Water Pipits may be giving their flight-songs. The same meadows may have White-tailed Ptarmigan, but they are hard to find. If they are not here, try some of the other alpine meadows and rock piles near the highway as it continues its ascent.

Next, stop at Lava Cliffs at 12,080 feet elevation; walk to the edge of the cirque on the left and peer over its edge to view the steep walls. Brown-capped Rosy Finches are likely to be perched on the ledge shelves or moving to and from their nests in various crevices.

A short distance up, above Lava Cliffs, the Trail Ridge Road reaches its highest point and then descends to the Alpine visitor center at 11,796 feet. The rosy finches may be here too. The meadows in the vicinity are a good place to search for White-tailed Ptarmigan and to observe Horned Larks and Water Pipits, in case they have been missed at previous stops.

Reference

Birds of Rocky Mountain National Park. By Allegra E. Collister. Denver: Museum of Natural History. 1970.

MONTE VISTA
Monte Vista National Wildlife Refuge

In the San Luis Valley of south-central Colorado, where the Sangre de Cristo Range rises on the east and the Sawatch Range on the west, lies the **Monte Vista National Wildlife Refuge** containing 13,547 acres. These include over 200 ponds, artificially developed with food and marsh cover for nesting birds, and over 400 acres of farmlands on which food is produced for wintering waterfowl.

Chief among the breeding birds associated with the ponds and their marshy borders are the Pied-billed Grebe, Snowy Egret, Black-crowned Night Heron, American Bittern, White-faced Ibis, Canada Goose, Mallard, Gadwall, Common Pintail, all three teal—Blue-winged, Green-winged, and Cinnamon—Northern Shoveler, Redhead, Ruddy Duck, Northern Harrier, Virginia Rail, Sora, American Coot, Common Snipe, Spotted Sandpiper, American Avocet, Wilson's Phalarope, Marsh Wren, Common Yellowthroat, Yellow-headed, Red-winged, and Brewer's Blackbirds, and Savannah Sparrow.

One of the ornithological features on the Refuge is the congregation of migrating Sandhill Cranes. The big birds begin arriving in early September and are gone by late November. In spring they are fewer in number, starting to show up in February and disappearing by late March.

Ducks, largely Mallards, are in greatest number on the Refuge in December.

To reach the Refuge from Monte Vista, drive south on State 15 for 6 miles to the entrance. At headquarters, obtain a checklist of Refuge birds and inquire about the 6-mile, self-guided auto route.

MONTROSE

Black Canyon of the Gunnison National Monument

The weirdly spectacular Black Canyon of the Gunnison River in western Colorado is famous for its 50 miles of sheer walls consisting of schists, predominantly bluish black, with white, gray, pink, and red granitic intrusions. The **Black Canyon of the Gunnison National Monument** embraces 14,464 acres, including the most formidable 10-mile section of the chasm. In places the rims are as close as 1,300 feet, yet the depth ranges from 1,730 to 2,425 feet.

Because the Canyon itself is so exceedingly rugged, it supports little vegetation; nevertheless, there are draws leading down from the rims that contain pinyon pines, junipers, scrub oaks, and some sagebrush. Among the birds to look for in these draws during the summer are the Broad-tailed Hummingbird, Scrub and Pinyon Jays, Mountain Chickadee, Mountain Bluebird, Plain Titmouse, Virginia's Warbler, Black-headed Grosbeak, Green-tailed Towhee, and Gray-headed Junco. From vantage points overlooking the gorge, watch especially for Black Swifts in flight. Other birds that should be seen meanwhile are Turkey Vultures, one or two Red-tailed Hawks and Golden Eagles, American Kestrels, White-throated Swifts, and Violet-green Swallows.

Both the north and south rims of the Canyon are accessible by car during the late spring and summer. To reach the south rim from Montrose, drive 8 miles east on US 50, then 9 miles northward over State 347. To reach the north rim, drive 21 miles northwest from Montrose on State 50 to Delta, then east and south for 39 miles on State 92 to a point south of Crawford, where a marked, graded road, on the right, leads to the rim, 14 miles distant.

SAGUACHE
Russell Lakes

In the San Luis Valley, 9 miles south of Saguache along the east side of US 285, are the **Russell Lakes,** shallow bodies of water fringed by marshes in which bulrushes are the predominant growth. Here, in a mountain park at an elevation of 7,580 feet, may be found nesting, beginning in May, a remarkable variety of birds including the Eared, Western, and Pied-billed Grebes, Snowy Egret, Black-crowned Night Heron, American Bittern, White-faced Ibis, Mallard, Gadwall, Common Pintail, Blue-winged and Cinnamon Teal, Redhead, Ruddy Duck, Virginia Rail, Sora, American Coot, American Avocet, Wilson's Phalarope, Marsh Wren, Common Yellowthroat, and Yellow-headed, Red-winged, and Brewer's Blackbirds.

WALDEN
Arapaho National Wildlife Refuge

For a good opportunity to see Sage Grouse, as well as a fine variety of waterbirds and waterfowl, visit the **Arapaho National Wildlife Refuge** (8,019 acres) in extreme north-central Colorado. Headquarters is in Walden. To reach sub-headquarters on the Refuge, drive 8 miles south from Walden on State 125, then east a short distance on County 32.

The Refuge is situated in a high mountain basin or valley called North Park, virtually isolated by a surrounding rim of lofty peaks. The valley bottomland within the Refuge has irrigated meadows, marshes, and ponds, their waters coming from the Illinois River which drains northward through the valley. Immediately back from the bottomland rise low, dry foothills on which sagebrush thrives.

Sage Grouse are year-round residents in the foothills. Inquire at sub-headquarters as to the best vantages for viewing them.

The nesting season for birds of the wetlands begins in May. Among the regular breeding species are the Canada Goose (recently introduced), Mallard, Common Pintail, Gadwall, Blue-winged and Cinnamon Teal, American Wigeon, Northern Shoveler, Redhead, Canvasback, Lesser Scaup, Ruddy Duck, Virginia Rail, Sora, Common Snipe, Willet, American Avocet, Wilson's Phalarope, Marsh Wren, and Yellow-headed, Red-winged, and Brewer's Blackbirds. Other birds one is likely to see in the same season, either on the Refuge or elsewhere in North Park, are the Great Blue and Black-crowned Night Herons, Swainson's Hawk, Northern Harrier, American Kestrel, possibly a Prairie Falcon, and Common Nighthawk.

Migrating waterfowl reach peak numbers in late April and early May, and again in late September and early October.

WALSENBURG
Cucharas River Valley

From this city (elevation 6,200 feet) among the foothills of southeast-central Colorado, the highway to Cucharas Pass (elevation 9,941 feet) in the Sangre de Cristo Range follows the **Cucharas River Valley.** In late spring and summer, bird finding is very good in many spots along the highway.

From Walsenburg, take US 160 southwest for 10 miles, then turn left onto State 12 and proceed to La Veta. Between Walsenburg and La Veta, the Valley is broad and has fertile soils used extensively for crop production. Peaks in the Sangre de Cristo Range rise in the west, but the most eye-catching scenic features are the massive twin mountains, East Spanish Peak (12,683 feet) and West Spanish Peak (13,610 feet), which stand apart from the Range on the south. Red-shafted Flickers, Western Kingbirds, Black-billed Magpies, Mountain Bluebirds, Western Meadowlarks, Bullock's Orioles, and House Finches are common in this part of the Valley, as are Lewis' Woodpeckers which should be watched for on utility poles.

From La Veta, continue on State 12, which soon ascends through the gradually narrowing Cucharas Valley to Cucharas Pass. Here the Valley floor has meadowland dotted with ponds and with marshes fringed by cattails and willow clumps; the slopes of the Valley have pinyon pine and scrub oak giving way at higher elevations to pine, fir, and aspen. Among the birds to be expected, if stops are made during the climb through the Valley, are Band-tailed Pigeons, Violet-green Swallows, Steller's Jays, Northern Ravens, MacGillivray's Warblers, Western Tanagers, and Green-tailed Towhees.

Idaho

CLARK'S NUTCRACKER

Idahoans will be the first to admit that their state lacks geographical unity. No great mountain chain, valley, or arid plain belongs exclusively to Idaho; those that exist here are shared with one or more of the six bordering states. The same may be said of Idaho's biota. Despite this peculiarity, 'the parts of other states' that comprise Idaho have some of the choicest scenery and, for the visiting bird finder, some of the most attractive settings for birds that he will come upon anywhere in the United States.

In southern Idaho, the widest part of the state, the principal physiographic feature is the Snake River Plain, which extends about 375 miles in length, from the eastern to the western boundary, looping southward during its course, and averaging between 75 and 100 miles in breadth. Mountain ranges clearly mark most of the Plain's boundaries. Some form abrupt rims; others encroach upon the Plain in parallel formation, leaving long peninsulas of the Plain between them. The Snake River, as it cuts through the Plain from Wyoming to the Oregon line, drops from 6,000 to 3,000 feet; on the north and south the Plain itself rises gradually from the River, like the sides of a shallow trough, to altitudes approaching 6,000 to 7,000 feet at the base of the mountain ranges.

The Snake River Plain was once desert-like, its rolling surface covered with sagebrush, greasewood, and clumps of grasses, in some places as far as the eye could see. The only major interruptions were localities—for example, the Craters of the Moon National Monument (*see under* **Carey-Arco**)— where lava beds, lava flows, and other rock features of volcanic origin discouraged vegetation, and along watercourses and in coulees where there were trees (cottonwoods, aspens, willows) and various shrubs. Today much of the Plain that is without rock has become a productive agricultural area. Under extensive irrigation, the lands adjacent to the Snake River yield Idaho's famous potatoes, as well as sugar beets, peas, alfalfa, and fruits. Upon the higher lands near the mountains, away from the River and beyond reach of irrigation, acres upon acres of wheat come from soils once thought useless for cultivation.

In southwestern Idaho, southwest of the Snake River, is the Owyhee Country, a vast highland roughly embraced by Owyhee County. Arid and bleak in aspect, broken by canyons and ridges, and by occasional mountains rising to 7,000 feet, it is an inhospitable land that has only sparse human population and few roads. Sagebrush and associated plants find sufficient soil in scattered spots; alder, cottonwood, aspen, wild currant, and wild rose grow along the small number of watercourses; and meadows and marshes exist in the few watered valley bottoms. Elsewhere the country is desert. In southeastern Idaho—the rest of the state south of the Snake River Plain—the country is decidedly different. Here are many small mountain ranges, all running north and south, with peaks in some cases attaining heights of 10,000 feet. Their slopes frequently have stands of juniper and mountain mahogany. (For an indication of some of the birds that may be expected in these mountain areas, *see under* **Burley**.)

Alternating with the mountain ranges are valleys of varying width which have elevations between 4,000 and 5,000 feet. In nearly all of the valleys there are streams with borders of deciduous trees and shrubs and broad stretches of sageland. Several have fine lakes and marshes. Through irrigation, many of the sagelands have been converted into areas for crop production.

Throughout the Snake River Plain and the mountain valleys to the south, the following breeding birds may be expected in the open country (sagelands, agricultural lands, wet meadows, brushy places, and dooryards) and in the wooded stream bottoms.

OPEN COUNTRY

Turkey Vulture
Swainson's Hawk
Ferruginous Hawk
Northern Harrier
Sharp-tailed Grouse
Sage Grouse
Long-billed Curlew
Mourning Dove
Burrowing Owl
Eastern Kingbird
Western Kingbird
Say's Phoebe
Horned Lark
Barn Swallow
Black-billed Magpie
Bushtit
House Wren
Sage Thrasher
Mountain Bluebird
Loggerhead Shrike
Yellow Warbler
Common Yellowthroat
Yellow-breasted Chat
Western Meadowlark
Lazuli Bunting
House Finch
American Goldfinch
Savannah Sparrow
Vesper Sparrow
Lark Sparrow
Sage Sparrow
Chipping Sparrow
Brewer's Sparrow
Song Sparrow

WOODED STREAM BOTTOMS

Common Screech Owl
Red-shafted Flicker
Hairy Woodpecker
Downy Woodpecker
Dusky Flycatcher
Western Pewee
Black-capped Chickadee
Red-eyed Vireo
Warbling Vireo
MacGillivray's Warbler
Bullock's Oriole
Black-headed Grosbeak

The development of agriculture in southern Idaho has had profound effects on birdlife. In the reclamation of the sagelands, the numbers of Sage Grouse, Sage Thrashers, and other birds preferring that type of environment have been markedly reduced; conversely, some birds such as American Robins, Yellow Warblers, Bullock's Orioles, Black-headed Grosbeaks, and House Finches have undoubtedly increased because of the growth of orchards and tree and shrub plantations about homes. The creation of irrigation reservoirs in this region of few natural lakes and marshes has greatly extended drinking and resting places for waterbirds, waterfowl, and shorebirds; where water levels have been sufficiently stabilized to permit growth of aquatic plants, feeding and breeding grounds for such birds have also been augmented. As a result of these artificial basins, water-loving birds have shown a notable increase in southern Idaho. There is no better evidence of this fact than in the National Wildlife Refugees (*see under* **Nampa, Rupert,** *and* **Soda Springs**) and Wildlife Management Areas (*see under* **Idaho Falls** *and* **Mountain Home**) where storage reservoirs or impoundments with controlled water supply attract impressive congregations of grebes, cormorants, herons, geese, ducks, shorebirds, gulls, and terns for resting, feeding, and nesting purposes.

Central Idaho, that part of the state lying between the Snake River Plain and the Panhandle (here defined as the remainder of the state north of Lewis and Idaho Counties), embraces a vast section of the northern Rocky Mountains. A great many peaks exceed 10,000 feet; Mt. Borah, 12,662 feet, is the highest point in the state. The general topography, some of the roughest in conterminous United States, is a veritable maze of mountain ranges and intermittent valleys, with a remarkable assortment of deep, sunless canyons, steep bare ridges, and jagged peaks. Dense forests blanket the wider valleys and sometimes extend high up the adjacent slopes. Almost entirely coniferous, save for scattered aspen groves, the forests consist principally of Douglas-fir, grand fir, and western red cedar, and occasionally of ponderosa pine, at elevations below 5,000 feet; Engelmann spruce, subalpine fir, and lodgepole pine between 5,000 and 7,000 feet; and whiteback pine from 7,000 feet to the timber line at about 7,500 feet. There are a few somewhat level surfaces above the timber line that are typical alpine meadows. Hidden away in nearly all parts of this great wilderness are lovely lakes, frequently forest-bordered.

Contributing in no small measure to the grandeur of central Idaho is the Salmon River, which gouges a deep path through seemingly impervious terrain from its source in the Sawtooth Mountains near Galena Summit. For 100 miles it takes a northeastward course, gathering volume from its tributaries and passing through the Upper Salmon River Gorge. Near the Montana line the River, already a turbulent stream, bends abruptly westward and enters the forbidding, 200-mile-long canyon bearing its name. By the time the Salmon River has descended 2,000 feet, receiving additional tributaries on the way, and emerged from its high-walled chasm at Riggins near the Oregon line, it is a roaring torrent. On seeing its final miles through the canyon, one readily understands how it earns another name, 'The River of No Return.' To navigate upstream through the Salmon River Canyon in anything other than a motorized craft is unthinkable.

Much of north-central Idaho is wilderness, largely undisturbed. Roads, in most cases secondary or unimproved, connect a few parts; towns, resorts, mines, and ranches are few in number and far between. Lying north and south of the Salmon River, about midway between US 93 on the east and US 95 on the west, are several million acres of wilderness designated as reserves or 'primitive areas' in which road construction and other human modifications are restricted, if not prohibited.

Although the mountain fastnesses of central Idaho are generally accessible only to hardy bird finders willing to organize pack trips by trail, or even by unmarked route, nevertheless there are many high country spots reached by car (for example, *see under* **Salmon**) that can be rewarding ornithologically. Listed below are the bird species breeding regularly in one or another of the coniferous forests, canyons, rocky slopes, timber lines, and alpine meadows.

Northern Goshawk
Sharp-shinned Hawk
Blue Grouse
Spruce Grouse
Ruffed Grouse

Northern Pygmy Owl
Yellow-bellied Sapsucker
Williamson's Sapsucker
Black-backed Three-toed
Woodpecker

Olive-sided Flycatcher
Gray Jay
Steller's Jay
Clark's Nutcracker
Mountain Chickadee
Red-breasted Nuthatch
Pygmy Nuthatch
Brown Creeper
Hermit Thrush
Swainson's Thrush
Townsend's Solitaire
Ruby-crowned Kinglet
Solitary Vireo
Orange-crowned Warbler
Audubon's Warbler
Western Tanager
Evening Grosbeak
Cassin's Finch
Black Rosy Finch
Pine Siskin
Red Crossbill
Oregon Junco
White-crowned Sparrow
Fox Sparrow
Lincoln's Sparrow

The many precipitous slopes and deep canyons that characterize so much of the rough country in both central and southern Idaho are often favorite breeding localities of such species as the following:

Turkey Vulture
Red-tailed Hawk
Golden Eagle
Prairie Falcon
American Kestrel
Poor-will
White-throated Swift
Violet-green Swallow
Cliff Swallow
Northern Raven
Canyon Wren
Rock Wren

The Idaho Panhandle is a mixture of lesser mountain ranges, beautiful lakes, and coniferous forests. Resorts, farmlands, many towns, and one large city (Coeur d'Alene) interrupt the natural landscape, which enjoys a fairly heavy rainfall and is thus more verdant than southern Idaho. Bird-finding possibilities in the Panhandle are rich. (e.g., *see under* **St. Maries**). Among the peculiarities of birdlife are the Western Bluebird and several breeding species of eastern affinities—the Gray Catbird, Cedar Waxwing, Northern Waterthrush, American Redstart, Bobolink, and Grasshopper Sparrow—which nest only locally, if at all, elsewhere in the state.

During spring and fall migrations, concentrations of waterbirds, waterfowl, and shorebirds are usually impressive in certain of the National Wildlife Refuges. For example, in the Minidoka Refuge (*see under* **Rupert**), several hundred Whistling Swans appear in spring and many thousands of geese and ducks in fall; in the Deer Flat Refuge (*see under* **Nampa**), transient shorebirds gather in early summer on mud flats; and in the Grays Lake Refuge (*see under* **Soda Springs**), Grays Lake is a staging area in mid-September for as many as 2,000 Sandhill Cranes. Small migratory landbirds breeding north of Idaho move north-south through the state in a relatively dispersed manner as there are no river or mountain valleys running continuously north-south in the state through which they may pass. The main migratory flights in southern Idaho take place within the following dates:

Waterfowl: 10 March–20 April; 10 October–25 November
Shorebirds: 15 April–1 June; 15 July–1 October
Landbirds: 15 March–20 May; 1 August–20 October

Owing to Idaho's generally high elevation, winters are cold and not very productive ornithologically. Open water in southern Idaho sometimes holds large wintering populations of waterfowl. West-central Idaho and the southern Panhandle, where the elevations are lowest in the state, have the mildest winter conditions. Here species breeding north of Idaho, such as the Northern Shrike, Bohemian Waxwing, Common Redpoll, Slate-colored Junco, and American Tree Sparrow, may be found in winter along with species such as the Mourning Dove, Cedar Waxwing, Ruby-crowned Kinglet, Audubon's Warbler, Western Meadowlark, Red-winged Blackbird, Brewer's Blackbird, Evening Grosbeak, American Goldfinch, Oregon Junco, and White-crowned Sparrow, which might migrate south were the winter conditions severe.

Authorities

Liven A. Peterson, Jr., Hadley B. Roberts, Charles H. Trost.

Reference

Birds of Idaho. By Thomas D. Burleigh. Caldwell, Idaho: Caxton Printers. 1972.

BOISE
Boise National Forest | Snake River Birds of Prey Natural Area

Idaho's capital city, Boise, lies on the Boise River in the Upper Boise Valley of the Snake River Plain. At an elevation of only 2,740 feet and sheltered on the north by the Boise Range, it is favored by mild winters.

For a profitable summer bird-finding trip, drive southeastward and then northward on State 21 to Idaho City—almost a ghost town—in the Boise Basin, a distance of about 45 miles. The last 25 miles of the trip is through **Boise National Forest,** notable for its fine stands of pine and spruce. As the road approaches Idaho City it passes numerous ponds, created by placer mining operations and now choked to some extent with vegetation. Nearly all of these water areas attract a few nesting waterbirds, ducks, and shorebirds. Blue Grouse, Ruffed Grouse, and many other landbirds breed in adjacent suitable habitats.

The bird finder may wish to continue on State 21 past Lowman (82 miles from Boise) to its terminus at Stanley (127 miles from Boise) on US 93. Like the trip on US 93 (*see under* **Salmon**), this one is through wild, rough, mountainous country.

Along the Snake River south of Boise, a 33-mile portion of the Snake River Canyon is set aside by the Bureau of Land Management as **Snake River Birds of Prey Natural Area** for the preservation of nesting habitat for raptors. Access is limited but the following directions will lead to good vantages. From Boise, drive west on I 80N for about 10 miles and exit south on State 69 toward Kuna, 9 miles distant. Just before Kuna there are two sharp right-angle turns. At the second on the outskirts of Kuna, turn left off State 69 onto a paved road—though poorly marked, it is the only paved road which crosses a creek and then railroad tracks. Follow this road which, after 3 miles, becomes a well-maintained gravel road, for about 16 miles to the east. Near its end the road skirts huge irrigated fields and in about 2 miles comes to a Canyon overlook before going down a steep grade to Swan Falls Dam. Any of the small roads heading off before the overlook lead to the Canyon's rim for good views.

The basaltic cliffs here, up to 500 feet in height, afford many

GOLDEN EAGLE

nesting sites for Northern Ravens and raptors. Red-tailed Hawks, Golden Eagles, and Prairie Falcons are the most common; Ferruginous Hawks and American Kestrels less so. Nothern Harriers are often present in the area. Eventually there may be Peregrine Falcons, if the attempt at cross-fostering them with Prairie Falcons succeeds.

The key to the nesting success of so many raptors is the abundance of Townsend's ground squirrels on the sagebrush desert back from the Canyon's rim. The time to visit the Area for raptors is from late February and early March to mid-June. By late June when the ground squirrels begin aestivation, most of the birds relying on them for food disappear, with the exception of a few ravens and eagles, and will not show up again until the following year.

BONNERS FERRY
Kootenai National Wildlife Refuge

In the Kootenai Valley of Idaho, near the northern tip of the Panhandle and only 18 miles south of the Canadian border, the **Koo-**

tenai National Wildlife Refuge occupies 2,762 acres of lowlands with large ponds maintained from the Kootenai River and tributary creeks primarily for waterfowl. Impressive numbers of migrating Whistling Swans stop off from late March to late April. In the breeding season, Wood Ducks and Common Goldeneyes are common as are Mallards, Blue-winged and Cinnamon Teal, and Lesser Scaups.

To reach the Refuge from US 95, drive 5 miles west from Bonners Ferry on the Dike Road, then turn north onto the West Side Road and continue 0.5 mile to headquarters. Here obtain directions to the best car routes from which to scan the ponds for ducks.

The West Side Road continues north from headquarters through conifer-timbered areas with trails leading east to Myrtle Creek bordered by deciduous thickets. Stopping frequently along the Road and following the trails, one may see or hear such breeding birds as the Ruffed Grouse, Rufous Hummingbird, Pileated and Lewis' Woodpeckers, Traill's Flycatcher, Western Pewee, Red-breasted Nuthatch, Varied and Swainson's Thrushes, Cedar Waxwing, Solitary Vireo, Orange-crowned and MacGillivray's Warblers, American Redstart, Pine Siskin, Oregon Junco, and Lincoln's Sparrow. One is certain to observe North American Dippers along Myrtle Creek, and, if he takes the time to explore all accessible parts of the Refuge, he will see all six species of swallows regularly found in Idaho.

BURLEY
Goose Creek | City of Rocks | Mt. Harrison

Two rewarding sites for bird finding in extreme southern Idaho, off the beaten path, are **Goose Creek** and the **City of Rocks.** The best time for viewing breeding birds is in late May and June.

To reach Goose Creek, drive south on State 27 for 22 miles to Oakley, then continue south on an unnumbered road (paved partway) to a point within 2 miles of the Idaho-Utah line. Here, in the vicinity of Goose Greek, are low ridges, covered with pinyon pine, juniper, and mountain mahogany where such birds nest as the Ash-throated and Gray Flycatchers, Scrub and Pinyon Jays, Plain

Titmouse, Blue-gray Gnatcatcher, and Black-throated Gray Warbler.

To reach the City of Rocks, drive south from Burley on State 27 to Oakley; here turn east and proceed for about a mile on an unnumbered road going to Elba, then turn off south onto an unnumbered road and go 14 miles. This is an arid, 25-square-mile area marked by sheer cliffs and by pinnacles of rock towering hundreds of feet above the floor of the valley. Pinyon pine, juniper, and mountain mahogany, though sparse, are the predominating vegetation; aspen and whitebark pine are prevalent. Some of the breeding birds here are the following: Turkey Vulture, Golden Eagle, Prairie Falcon, Burrowing Owl, Poor-will, White-throated Swift, Black-chinned Hummingbird, Yellow-bellied Sapsucker, Say's Phoebe, House and Rock Wrens, Hermit Thrush, Mountain Bluebird, Virginia's Warbler (unusual in the state), Audubon's Warbler, Green-tailed Towhee, Lark Sparrow, Gray-headed Junco, Brewer's Sparrow (prefers open ridges here), and Fox Sparrow.

Another good site off the beaten path is **Mt. Harrison,** reached from Burley as follows: Drive east on State 81 for 9 miles to Delco, turn south onto State 77 and go 9 miles to Albion and then continue south on State 77 which, after 10 miles, ascends Mt. Harrison via Howell Canyon. The highway goes almost to the top. Near the top, in a lovely setting at the head of Howell Canyon and accessible by an easy trail from the highway, is Lake Cleveland. The mountain slopes (elevations 4,500–8,000 feet) differ in their vegetation according to exposure; those on the north are covered with pine and fir, those on the south with sagebrush. During the trip up Mt. Harrison in June, look or listen for the Blue Grouse, Clark's Nutcracker, North American Dipper, Hermit Thrush, Townsend's Solitaire, Ruby-crowned Kinglet, Audubon's and MacGillivray's Warblers, Cassin's Finch, Pine Siskin, Green-tailed Towhee, and Oregon Junco.

CAREY-ARCO
Craters of the Moon National Monument

For weird, unworldly landscape, **Craters of the Moon National Monument** in southeastern Idaho is practically unrivaled. Here at an elevation of 5,900 feet have been set aside 53,545 acres of cinder cones, craters, fissure eruptions, lava flows, caves, and tunnels—some of the most striking volcanic phenomena that prevail over wide areas of the Snake River Plain.

On first seeing this vast panorama of black and gray desolation, the bird finder may conclude that it is lacking in birdlife except possibly for Poor-wills, Horned Larks, Violet-green Swallows, Northern Ravens, and Rock Wrens, but on closer inspection he will soon discover patches of juniper, limber pine, aspen, and various shrubs that hold small numbers of such breeding birds as the Lewis' Woodpecker, Western Flycatcher, Clark's Nutcracker, American Robin, Mountain Bluebird, Loggerhead Shrike, Yellow Warbler, Western Tanager, Green-tailed and Rufous-sided Towhees, and Fox Sparrow.

The Monument is traversed by US 20–26 midway between Carey and Arco and is open all year. Stop at the visitor center, on the highway about halfway through the Monument, for information on roads, trails, and interpretive services. Near the west entrance from Carey is a marsh well worth looking over for birds.

IDAHO FALLS

Market Lake Wildlife Management Area | Mud Lake Wildlife Management Area

North of this city on the Snake River in eastern Idaho are two state-owned tracts that incorporate excellent marshes for birds.

One tract is the **Market Lake Wildlife Management Area** (4,900 acres of which 2,450 are marshes), reached from Idaho Falls by driving north on I 15 for 17 miles, exiting east to Roberts, going north from the center of town on a paved road which, 1.2 miles out of town, angles right and then continues north, after 3 miles passing Area headquarters on the right and soon crossing the Area's huge marsh on a dike—a fine vantage for observations. Nesting in the marsh, besides some ten species of waterfowl, are such birds as the Eared and Western Grebes, Snowy Egret, Black-crowned Night Heron, American Bittern, White-faced Ibis, Virginia Rail, Franklin's Gull (in a colony), and Forster's and Black Terns.

The other tract is the **Mud Lake Wildlife Management Area** (8,900 acres, of which 200 are marshes, around an irrigation reservoir), reached by returning to, and continuing north on, I 15 for 8 miles, exiting west on State 88 for 12 miles to Terreton, then turning off north and proceeding 1.0 mile to headquarters. Here inquire about the best vantage points for viewing birds. Among the many species nesting in the Area are the Double-crested Cormorant, California and Ring-billed Gulls, Long-billed Curlew, and Willet. In late summer, when the reservoir is drained heavily, there are extensive mud flats attractive to transient shorebirds. In spring (March and April), when the reservoir is usually at full capacity, transient ducks are in great abundance.

MOSCOW

This university city on the western edge of the Panhandle is surrounded by rolling farmlands, actually a continuation of the Palouse Country of eastern Washington. Mountains and heavily forested areas are not far distant. Common breeding birds beyond the city's outskirts—in deciduous woods, pastures, fields, and shrubby places—include the Northern Harrier, Long-eared Owl, Calliope Hummingbird, Red-shafted Flicker, Traill's and Dusky Flycatchers, Western Pewee, Violet-green Swallow, Black-capped Chickadee, House Wren, Gray Catbird, Veery, Mountain Bluebird, Warbling Vireo, MacGillivray's Warbler, Yellow-breasted Chat, American Redstart, Bullock's Oriole, Black-headed Grosbeak, Lazuli Bunting, House Finch, American Goldfinch, and Savannah, Grasshopper, Fox, and Song Sparrows.

MOUNTAIN HOME

Bruneau Dunes State Park | C. J. Strike Wildlife Management Area | Bruneau Canyon

From I 80N in southwestern Idaho, exit south through the city of Mountain Home and thence drive farther south on State 51 for 16 miles, crossing the Snake River; then turn east onto State 78 and go 3 miles to **Bruneau Dunes State Park** (2,520 acres). Here, amid spectacular dunes, are a few small lakes with marshy borders that attract such transient or summer-resident birds as the Horned and Western Grebes, Great Blue Heron, Green-winged and Cinnamon Teal, Ruddy Duck, and American Coot.

For more birds frequenting open water or marshes, return to State 51 and go south on it for 5 miles through the town of Bruneau and across the Bruneau River; then turn west onto State 78 and drive 2 miles to the **C. J. Strike Wildlife Management Area,** which embraces 7,200 acres of reservoir shoreline and adjoining uplands. Since the reservoir is used for producing hydroelectric power, its water level stays almost constant the year round and thus supports thick stands of bulrushes and cattails in its shallower parts. Whistling Swans stop off commonly during their spring and fall migrations as do many other waterfowl. Among the birds one is likely to see in summer are American White Pelicans, Double-crested Cormorants, Black-crowned Night Herons, American Bitterns, Virginia Rails, Soras, Common Snipes, Willets, and American Avocets.

For upland birds in a scenic situation, return to Bruneau and drive south for 16 miles on an unnumbered road along the east side of the Bruneau River to **Bruneau Canyon,** a deep, narrow cleft between basaltic cliffs. After reaching Bruneau Hot Springs, 8 miles from Bruneau, the road ascends a plateau, through which the Canyon cuts, to an overlook. Some of the breeding birds regularly present in the Canyon area are the Turkey Vulture, Red-tailed Hawk, Golden Eagle, Prairie Falcon, Northern Raven, Cliff Swallow, Black-billed Magpie, and Canyon Wren.

NAMPA
Deer Flat National Wildlife Refuge

Just southwest of Nampa in southwestern Idaho, at the extreme western end of the Snake River Plain, is the **Deer Flat National Wildlife Refuge,** comprised of the Lake Lowell and the Snake River Sectors.

The Lake Lowell Sector (11,585 acres) consists primarily of Lake Lowell, an irrigation reservoir formed by diverting water from the Boise River, a tributary of the Snake. Nine miles long and 2.5 miles wide, the Lake has considerable shallow water, with bulrushes, smartweeds, and other emergent plants. Below the upper dam, west of headquarters, is a marsh of 20 acres, with cattails the predominating vegetation. Cottonwoods, black locusts, willows, and various shrubs stand along the shore of the Lake and the several radiating canals; sagebrush and native grasses constitute the principal cover on the adjacent higher ground.

Birds nesting regularly in suitable situations around Lake Lowell are the Eared and Western Grebes, Canada Goose, Mallard, Gadwall, Common Pintail, Green-winged, Blue-winged, and Cinnamon Teal, American Wigeon, Northern Shoveler, Redhead, Northern Harrier, American Coot, Killdeer, Common Snipe, Marsh Wren, and Yellow-headed, Red-winged, and Brewer's Blackbirds. Some of the birds breeding regularly in the general vicinity are the Mourning Dove, Red-shafted Flicker, Eastern and Western Kingbirds, Violet-green, Barn, and Cliff Swallows, Black-billed Magpie, Yellow Warbler, Bullock's Oriole, Lazuli Bunting, and Song Sparrow. The nesting season for most species is under way by the first week of May.

By late summer, when the water of Lake Lowell has been drawn down for irrigation, migrating shorebirds—Greater Yellowlegs, Long-billed Dowitchers, Western Sandpipers, and others—gather on the exposed mud flats.

From the third week of August to the end of April, transient waterfowl are present, but the populations are heaviest in November and December. Most abundant are Canada Geese, Mallards, and Common Pintails; less abundant, though impressive in numbers, are Green-winged Teal, American Wigeons, and Common Mergansers. A great many stay through the winter.

Attracted by the wintering concentrations of waterfowl, Bald Eagles commonly appear to scavenge on dead, maimed, or weakened birds. Other raptors appearing in winter include Red-tailed and Rough-legged Hawks and Golden Eagles.

Headquarters of the Deer Flat Refuge, situated in the Lake

Lowell Sector, may be reached from Nampa by driving south on 12th Avenue (State 45) and turning off west onto Lake Lowell Avenue.

The Snake River Sector consists of 86 islands in the Snake River from Walter's Ferry Bridge in Idaho downstream for 110 miles to Farewell Bend, Oregon. Forty of them, each with less than an acre of surface, are low-lying and covered with a dense growth of willows; the others, larger and higher, have willows on their borders, but their interiors support a desert type of vegetation. The smaller islands particularly are nesting grounds for many Canada Geese and Mallards and some are occupied by colonies of Great Blue and Black-crowned Night Herons and California and Ring-billed Gulls.

POCATELLO
Mink Creek Area

South of this large city in southeastern Idaho is the **Mink Creek Area,** mountainous country rich in birdlife. Herewith are directions for finding some of the birds in the breeding season.

From I 15, just after it passes through the gap about 5 miles south of the city, take the Portneuf Exit right and head south across railroad tracks and the Portneuf River, then bear right at the first crossroad which, after 2 miles, dead-ends at Mink Creek Road. Here turn left on Mink Creek Road and follow it up through a canyon to Crystal Summit at about 7,000 feet elevation. As the Road ascends, habitats change along the way from open juniper-sage woodland, where there are often Scrub and Pinyon Jays, to aspen-fir forests near the Summit.

Just after Mink Creek Road enters Caribou National Forest, about 7 miles from Pocatello, Kinney Creek comes down from the left. Walk up along the Creek through the canyon. In the course of about 3 miles, as juniper and sage are gradually replaced by aspen, search for such birds as the Long-eared Owl, Poor-will, Calliope Hummingbird, Plain Titmouse, Bushtit, Virginia's and Black-throated Gray Warblers, and Green-tailed and Rufous-sided Towhees.

Return to Mink Creek Road. About a mile farther up on the

right is Cherry Springs, a Forest Service campground (closed), adjacent to brushy bottomland through which Mink Creek meanders, interrupted by beaver dams. Paved paths up and down the Creek from the campground parking area are worth following for a variety of birds including the Cooper's Hawk, Traill's Flycatcher, Black-capped Chickadee, House Wren, Gray Catbird, Mountain Bluebird, MacGillivray's Warbler, Yellow-breasted Chat, and Fox and Song Sparrows.

Continue another mile up Mink Creek Road and take a paved road left up the East Fork of Mink Creek. Open from June until November, this road leads to Justice Park, a Forest Service campground at the base of Scout Mountain, where the elevation is about 8,000 feet. Among the birds occupying the fir forests in the vicinity are the Northern Goshawk, Blue and Ruffed Grouse, Saw-whet Owl, Hairy Woodpecker, Dusky Flycatcher, Steller's Jay, Clark's Nutcracker, Mountain Chickadee, Red-breasted Nuthatch, Brown Creeper, Hermit and Swainson's Thrushes, Townsend's Solitaire, Golden-crowned and Ruby-crowned Kinglets, Audubon's Warbler, Western Tanager, Cassin's Finch, Pine Grosbeak, Red Crossbill, Gray-headed Junco, and Chipping Sparrow.

Return to Mink Creek Road. From here one may continue up past Crystal Summit, investigating more side roads and creeks, then go down into Arbon Valley, turn right, through Fort Hall Indian Reservation, to I 15W and right again for 10 miles back to Pocatello.

RUPERT
Minidoka National Wildlife Refuge

The Snake River Plain, in the vicinity of Rupert in south-central Idaho, was once entirely desert-like except for scattered wet-weather marshes and the willow-bordered banks of the Snake River and its tributaries. But around the first of this century the Minidoka Dam was built across the Snake River northeast of Rupert, forming Lake Walcott, a huge reservoir for irrigation. Very soon it began attracting many waterfowl and induced President Theodore Roosevelt in 1909 to establish it as part of the Minidoka

Reservation. In the 1930s, after the name of the Reservation had been changed to the **Minidoka National Wildlife Refuge**, some of the shallow bays were cut off by dikes in order to assure a more stabilized water supply for aquatic plants when the level of the Lake drops. Such units attracted still more waterfowl.

Some 12,000 acres of the Refuge's 25,630 acres are open water and marsh. Lake Walcott is generally deep, with abrupt shorelines, but there are areas near shore, in the dammed-off upper reaches of the bays, and around low islands, where the water is shallow enough for cattails, bulrushes, and other marsh vegetation. The variety and numbers of birds that breed regularly are impressive. These include not only many pairs of Canada Geese, Mallards, Gadwalls, Common Pintails, American Wigeons, Redheads, Ruddy Ducks, and other waterfowl, but the following, which occupy particularly some of the islands—e.g., Bird and Tule Islands—or their marshy borders: Western Grebes, Double-crested Cormorants, Great Blue Herons, Snowy Egrets, Black-crowned Night Herons, White-faced Ibises (uncommon but regular), Northern Harriers, American Coots, Marsh Wrens, and Yellow-headed and Red-winged Blackbirds. The nesting season for all these birds extends from April through June.

Impressive as the Refuge is for nesting birds, it is even more impressive for migrating waterfowl. Several hundred Whistling Swans appear in the spring. At times in the fall, geese and ducks number as many as 250,000.

The Refuge may be reached from Rupert by driving northeast 6 miles on State 24 through Acequia, then east for 6 miles on County 400-North. Headquarters is in a lava-rock building on the point north of the Minidoka Dam. Normally, the south side of the Refuge, which is accessible by car, is the most advantageous for bird observations.

ST. MARIES

St. Maries–St. Joe River Bottomlands | Heyburn State Park | St. Joe National Forest

The community of St. Maries (elevation 2,145 feet), southeast of Coeur d'Alene Lake in the Panhandle, lies on low hills; but in its immediate vicinity, where the St. Maries River merges with the St. Joe River, are the wide **St. Maries–St. Joe River Bottomlands,** the rich soils of which are used for growing fruits and crops. Some of the bird species regularly breeding in fields, wet meadows, pastures, orchards, and shrubby places, in the remaining deciduous woods along the two watercourses, and about farm buildings are the following: American Kestrel, Killdeer, Mourning Dove, Common Nighthawk, Black-chinned, Rufous, and Calliope Hummingbirds, Red-shafted Flicker, Downy Woodpecker, Eastern Kingbird, Western Pewee, Violet-green, Tree, Rough-winged, Barn, and Cliff Swallows, Black-billed Magpie, Black-capped Chickadee, House Wren, Gray Catbird, Mountain Bluebird, Cedar Waxwing, Red-eyed and Warbling Vireos, Yellow Warbler, American Redstart, Bobolink, Western Meadowlark, Bullock's Oriole, Brewer's Blackbird, Black-headed Grosbeak, Lazuli Bunting, American Goldfinch, Rufous-sided Towhee, and Savannah, Chipping, and Song Sparrows.

West of St. Maries the St. Joe River (diked) continues on its course to enter Coeur d'Alene Lake, flanked, before reaching Coeur d'Alene Lake, by farmlands and by Benewah and Chatcolet Lakes, which have open water surrounded by cattail marshes, cottonwood and willow swamps, and mud flats. State 5, west from St. Maries, makes this productive area for birds readily accessible. Included among the nesting birds here in late May, June, and early July are the Pied-billed Grebe, American Bittern, Mallard, Cinnamon Teal, Wood Duck, Virginia Rail, Sora, American Coot, Common Snipe, Black Tern, Veery, Northern Waterthrush, Common Yellowthroat, and Yellow-headed and Red-winged Blackbirds. The Lakes are attractive to numerous waterfowl, particularly Whistling Swans in small flocks, Canada Geese, Common Pintails, Green-winged and Blue-winged Teal, American Wigeons, Northern Shovelers, Redheads, Ring-necked Ducks, Canvasbacks, Lesser Scaups, and Hooded Mergansers. Also attracted in spring—and during winter, as long as there is open water—are Common Goldeneyes and Common Mergansers. Ducks appear commonly in fall (October and November), but the area is open to hunting, which makes observations unsatisfactory. Shorebirds, such as the Solitary Sandpiper, Long-billed Dowitcher, Western Sandpiper, and Northern Phalarope, frequent the mud flats in late August and September, more rarely in May.

Farther west on State 5 is the entrance on the right to **Heyburn State Park,** a recreational and camping area on the south side of Chatcolet Lake and the southeastern extremity of Coeur d'Alene Lake. Of the Park's 7,838 acres, 5,500 are uplands, covered largely by timber, primarily coniferous except along watercourses, and to a less extent by shrubs. Among the birds that should be found in the Park are the Ruffed Grouse, Spotted Sandpiper, Great Horned and Northern Pygmy Owls, Belted Kingfisher, Lewis' Woodpecker (in open timber), Hairy Woodpecker, Dusky Flycatcher, Steller's Jay, Chestnut-backed Chickadee, Pygmy Nuthatch, Swainson's Thrush, Ruby-crowned Kinglet, Solitary and Red-eyed Vireos, Audubon's and MacGillivray's Warblers, American Redstart, Western Tanager, Evening Grosbeak, Pine Siskin, Red Crossbill, Oregon Junco, and Fox Sparrow.

To the east of St. Maries lies the 'upper division' of **St. Joe National Forest,** embracing rough, mountainous terrain with altitudes ranging from 2,400 feet on the St. Joe River, which cuts through the division from the east, to 7,000 feet on the higher peaks near the Idaho-Montana border. Most of the area, except ridge tops, burned-over spots, and the few alpine meadows, is forested with western red cedar and ponderosa pine at lower elevations; western white pine, Douglas-fir, and western larch at intermediate elevations; and Engelmann spruce, lodgepole pine, subalpine fir, and whitebark pine at higher elevations to timber line.

From St. Maries, National Forest Road 50, a paved road in fairly good condition, reaches the upper division at Avery, 50 miles distant. To enter this road, drive north from St. Maries on State 3; just outside of town, cross the St. Joe River and turn east. From Avery, various Forest Roads lead into the Bitterroot and St. Joe Mountains and across the border into Montana. Before making the trip, inquire at the National Forest Service Center in St. Maries for information about the most rewarding places for birds and their accessibility by road or trail.

On the whole, the bird species in the upper division of St. Joe National Forest are identical to those in Heyburn State Park and nearby areas at low elevations, but the forests at high altitudes in the division attract additional species. Careful searching during June and early July in timbered places above 4,000 feet should yield the following breeding species: Northern Goshawk, Sharp-shinned and Red-tailed Hawks, Blue and Spruce Grouse, Vaux's Swift, Yellow-bellied Sapsucker (usually in aspen tracts), Black-backed Three-toed and Northern Three-toed Woodpeckers, Olive-sided Flycatcher, Gray Jay, Northern Raven, Clark's Nutcracker, Mountain Chickadee, Red-breasted Nuthatch, Brown Creeper, Winter Wren, Varied and Hermit Thrushes, Townsend's Solitaire, Golden-crowned Kinglet, Townsend's Warbler, Cassin's Finch, and Pine Grosbeak. In brushy places bordering lakes and streams above 4,000 feet, one should find the Wilson's Warbler and Lincoln's Sparrow. Other species are the North American Dipper along fast-flowing streams, the Rock Wren on rocky slopes near ridge tops, and the White-crowned Sparrow at the timber line.

SALMON

US 93 to Salmon | Salmon River Valley | Lemhi River Valley

Undoubtedly, **US 93 to Salmon,** a prosperous community just west of the Continental Divide in the southeastern Panhandle, is Idaho's longest and most scenic highway. From Hailey in the southern foothills of the Sawtooth Range, northward for some 200 miles, US 93 traverses a wilderness of lofty peaks, awesome canyons, mountain-walled lakes, and great coniferous forests; by countless switchbacks it climbs to heights of breath-taking grandeur and descends into deep valleys. Features that stand out in the memory of anyone who takes this route are the ascent of Galena Summit (elevation 8,701 feet), the highest point in Idaho accessible by car; the lovely drive through Challis National Forest with its fine stands of lodgepole pine, Douglas-fir, and Engelmann spruce; and the drive through Cronks Canyon and the Upper Salmon River Gorge. Practically every species of high-country bird in Idaho can be found in the early summer by pulling off the highway near suitable habitats and searching. A special bird to look for, especially near snowfields, is the Black Rosy Finch.

Two trips from Salmon are worth taking for a good variety of birds.

1. From Salmon, drive north on US 93 for 21 miles to North Fork; here turn left onto an unnumbered road which parallels the Salmon River downstream in a deep canyon of the **Salmon River Valley** for 50 miles. Only the first few miles are paved. A particularly good spot is Deadwater Slough, about 3 miles below North Fork. A widening of the River, the Slough is bordered by willows, cottonwoods, dogwoods, and other deciduous growth. Elsewhere on the more gradual canyon slopes are stands of conifers. Among the birds one is likely to see or hear on the trip are the White-throated Swift, Lewis' Woodpecker, Hammond's Flycatcher, Violet-green Swallow, Pygmy Nuthatch (in ponderosa pines), North American Dipper, Canyon and Rock Wrens, Hermit Thrush, Western Tanager, and Black-headed Grosbeak.

2. From Salmon, follow State 28 southeast through the **Lemhi River Valley** for 45 miles to Leadore. The Valley itself, wide and flat, has extensive stretches of riverside woodlands and thickets intermixed with irrigated fields and pastures. Some of the birds to expect in suitable situations are the Long-billed Curlew, Wilson's Phalarope, Short-eared Owl, Eastern and Western Kingbirds, Traill's and Western Flycatchers, Western Pewee, Gray Catbird, Veery, Mountain Bluebird, Warbling Vireo, Yellow-breasted Chat, Bobolink, Bullock's Oriole, Brewer's Blackbird, Lazuli Bunting, American Goldfinch, Green-tailed Towhee, and such sparrows as the Savannah, Vesper, Brewer's, and White-crowned.

SODA SPRINGS

Grays Lake | Grays Lake National Wildlife Refuge

This southeastern Idaho town, rich in early history and the site of many cold mineral springs, is also the take-off point for **Grays Lake,** which in the opinion of one Idaho ornithologist is 'possibly the outstanding area in the state for abundance and variety of birdlife.' Some 13,000 acres of the Grays Lake lakebed are currently encompassed by the **Grays Lake National Wildlife Refuge;**

eventually the Refuge will be enlarged to include the entire lakebed of 32,600 acres.

From Soda Springs, take State 34 north. This goes to the southwest side of the Lake, a distance of 20 miles. Here a road turns off left and goes north along the west side. State 34 swings eastward for 5 miles around the south end of the Lake to a junction with a road on the left that proceeds first northward on the east side, then around the north end of the Lake, in due course meeting the west-side road. Grays Lake is thus encircled by roads, providing many vantages for observation.

Grays Lake occupies a high plateau (elevation 6,400 feet), flanked on the east and north by the Caribou Mountains, with peaks rising as high as 10,000 feet, and on the west by the lower Little Valley Hills. Though a large body of water with a 41-mile shoreline, Grays Lake is nevertheless very shallow and to a large extent choked with bulrushes and other emergent plants. Numerous channels meander through the vegetation, and in spots there are open stretches of water.

Grays Lake is notable as a nesting site for about 250 pairs of Sandhill Cranes and as a staging area for migrating Sandhills, particularly in mid-September when the population peaks at 2,000 individuals. Among the other birds that are attracted to this great marshy area for nesting are seventeen species of ducks, including the Mallard, Gadwall, Common Pintail, Green-winged, Blue-winged, and Cinnamon Teal, Northern Shoveler, Redhead, and Ruddy Duck, and the following: Eared Grebe, Canada Goose, Northern Harrier, Virginia Rail, Sora, American Coot, Common Snipe, Willet, American Avocet, Wilson's Phalarope, Franklin's Gull, Forster's Tern, Short-eared Owl, Marsh Wren, and both Yellow-headed and Red-winged Blackbirds. In meadowlands adjacent to the Lake, the Long-billed Curlew, Bobolink, and Savannah Sparrow nest.

The mountain slopes bordering Grays Lake have a cover of lodgepole pine, limber pine, Douglas-fir, and aspen with intermittent grassy areas and patches of serviceberry, chokecherry, snowberry, wild rose, and other shrubby growth. Birds breeding in suitable habitats at one elevation or another are such kinds as the

Ruffed Grouse, Broad-tailed Hummingbird, Yellow-bellied Sapsucker, Traill's and Hammond's Flycatchers, Violet-green Swallow, Steller's Jay, Black-capped and Mountain Chickadees, Red-breasted Nuthatch, Rock Wren, Gray Catbird, Sage Thrasher, Swainson's Thrush, Mountain Bluebird, Ruby-crowned Kinglet, Warbling Vireo, Orange-crowned and Audubon's Warblers, Brewer's Blackbird, Cassin's Finch, Pine Grosbeak, Pine Siskin, American Goldfinch, Green-tailed Towhee, and several sparrows, the Vesper, Brewer's, White-crowned, and Fox among them.

Beginning in 1975, Grays Lake Refuge became the setting of a concerted effort by American and Canadian biologists to augment the small remaining population of wild Whooping Cranes by selecting breeding pairs of wild Sandhill Cranes on the Refuge and having them hatch and foster Whooping Crane chicks from eggs transplanted from a captive flock in Maryland and a wild flock in northwestern Canada. Bird finders contemplating a visit to the Refuge are advised to contact the Refuge Manager in advance (address: P.O. Box 837, 159 East Second Street, Soda Springs, ID 83276), as access to areas involved in the Sandhill Crane-Whooping Crane cross-fostering project is prohibited. The Manager may also suggest vantages from which, in late summer, one or more Whooping Cranes may be seen.

TWIN FALLS

Magic Reservoir | Macon Lake | Silver Creek Preserve

Centrally located in the Snake River Plain of southern Idaho, Twin Falls (elevation 3,746 feet) is an agricultural center surrounded by miles of gently sloping farmlands. For productive bird finding, there are three exceptionally good areas—the **Magic Reservoir, Macon Lake,** and **Silver Creek Preserve**—which lie north of the city.

From I 80N at the Twin Falls Interchange, exit north on US 93 and proceed to a point about 27 miles north of Shoshone. Here turn off northwest onto a gravel road and proceed for about 11 miles to a resort area midway up the west shoreline of the Magic Reservoir—an impoundment of the Wood River approximately 1.5

miles wide and 5 miles long. Drive east through this area to a shoreline road and follow it south about a mile to a public campground. On small islands and points of land southward between the campground and dam are readily observed nesting colonies of California and Ring-billed Gulls. Other birds nesting in their vicinity are Caspian and Black Terns and a variety of grebes and shorebirds.

Macon Lake, a unique pothole about 8 miles northwest of the Magic Reservoir campground, can be reached by backtracking to the main gravel road and continuing northwest. Having about 5 acres of surface with marshy borders, Macon Lake is in startling contrast to the surrounding sagebrush desert which features basaltic outcrops and lava outflows. When driving to the Lake, be alert for such resident or summer-resident birds as the Turkey Vulture, Ferruginous Hawk, Golden Eagle, Sage Grouse, Northern Raven, Sage Thrasher, and Brewer's Sparrow. The Lake itself attracts nesting birds that include the Northern Shoveler, Cinnamon Teal, Virginia Rail, and Black-necked Stilt.

The Silver Creek Preserve, property of The Nature Conservancy, may be reached by returning to US 93 and continuing north; turning east onto State 68 and going about 5 miles to the Gannett Intersection (marked by a sign); turning off south onto a gravel road and proceeding for over a mile; then, after crossing a small creek, driving easterly a half-mile on the only public road. This forms the southern boundary of the Preserve, a 500-acre tract of woods and marsh around the headwaters of spring-fed Silver Creek. A small nesting colony of Great Blue Herons occupies an aspen thicket in the center of the tract. Elsewhere in the tract are habitats for a diversity of nesting birds such as the American Bittern, Red-tailed Hawk, Northern Harrier, American Kestrel, Sandhill Crane, Sora, Long-billed Curlew, Eastern Kingbird, Marsh Wren, Audubon's Warbler, Yellow-breasted Chat, Bullock's Oriole, Western Tanager, American Goldfinch, and Oregon Junco.

After exploring the Silver Creek Preserve, the bird finder may wish to return to US 93 and continue north (*see under* **Salmon**) or drive east on State 68 to Carey, then north and east on US 20-26 to Craters of the Moon Natural Monument (*see under* **Carey-Arco**).

Iowa

SNOW GEESE, BLUE PHASE

Early March along the Missouri River in western Iowa marks the arrival of Snow Geese from their winter's stay in Louisiana. Snow still remains in patches on the bluffs, and ice has not yet disappeared from the streams and ponds when the first flocks appear overhead, pursuing an ancestral course that leads steadily northward—from the Missouri River at Sioux City to the Big Sioux River and thence, by way of the Red River of the North, into Canada. By mid-March the 'waveys' are moving by the thousands, their V-formations making zigzag patterns across the sky. Over certain areas, as if by prearrangement, they break ranks and descend rapidly on set wings, each bird wheeling dizzily, sometimes tumbling headlong or slipping sidewise, and finally alighting close to others to feed and rest. Such is the case at the Forney's Lake

Wildlife Area (*see under* **Sidney**) and the DeSoto National Wildlife Refuge (*see under* **Missouri Valley**) where, every spring, Snow Geese gather in such great numbers that they cover acres of ground. There are few spectacles like it in the United States.

From the Mississippi River on the east to the Missouri and Big Sioux Rivers on the west, Iowa is topographically much the same—a level to slightly rolling plain with only subtle differences here and there. Elevations vary from 477 feet at Keokuk in the southernmost point of the state to 1,670 feet near Sibley in the northwestern corner.

With a soil rich and arable, the Iowa scene is one of cornfields and other cultivated tracts, square or oblong with straight roads and highways running along section lines. Spaced almost evenly are sets of tree-sequestered farm buildings, each comprising a white house, red barn, and tall silo close to a pasture where dairy herds graze. Also spaced evenly, though farther apart, are the towns, each with its church spire and grain elevator. This monotony, albeit one of unqualified prosperity, is relieved only at irregular intervals by the wooded bluffs and river valleys. Some of the bluffs, such as those along the Mississippi in the northeast, are as high as 400 feet and are separated by richly forested ravines.

The greater part of Iowa was once prairie grassland, a vast biotic community which has all but disappeared save in a few state-owned and private preserves. Woodlands, made up chiefly of oak, hickory, maple, elm, ash, basswood, and cottonwood were confined to areas along rivers and streams, to the borders of lakes and sloughs, and—particularly in the eastern part of the state—to the plain as prairie groves. Where the grassland and woodlands came together, there were usually rank weeds, shrubby thickets, and low trees. Few of the original woodlands still exist. Like the prairie, the majority have been destroyed to make way for farming; others have long since been replaced by second- and third-growth timber.

Today, prairie birds reside in such places as fallow fields and grassy pastures, forest and forest-edge birds in timbered tracts, most of which are in the big river valleys. Throughout Iowa, the following birds breed regularly in the farmlands (fallow fields, pastures, wet meadows, shrubby areas, woodland borders, orchards, and dooryards) and woodlands including remnants of prairie groves:

FARMLANDS

Common Bobwhite
Mourning Dove
Red-headed Woodpecker
Eastern Kingbird
Eastern Phoebe
Horned Lark
Tree Swallow
Barn Swallow
House Wren
Gray Catbird
Brown Thrasher
Eastern Bluebird
Loggerhead Shrike
Yellow Warbler
Common Yellowthroat
Bobolink
Eastern Meadowlark
Western Meadowlark
Common Grackle
Northern Cardinal
Indigo Bunting
Dickcissel
American Goldfinch
Rufous-sided Towhee
Grasshopper Sparrow
Vesper Sparrow
Chipping Sparrow
Field Sparrow
Song Sparrow

WOODLANDS

Turkey Vulture
Red-shouldered Hawk
Broad-winged Hawk
Yellow-billed Cuckoo
Black-billed Cuckoo
Common Screech Owl
Barred Owl
Whip-poor-will
Yellow-shafted Flicker
Red-bellied Woodpecker
Hairy Woodpecker
Downy Woodpecker
Great Crested Flycatcher
Eastern Pewee
Blue Jay
Black-capped Chickadee
Tufted Titmouse
White-breasted Nuthatch
Wood Thrush
Blue-gray Gnatcatcher
Red-eyed Vireo
Warbling Vireo
Ovenbird
American Redstart
Baltimore Oriole
Scarlet Tanager
Rose-breasted Grosbeak

There are differences in breeding birdlife from border to border in Iowa but these are relatively slight, a situation expected in an area whose physiographic features show remarkable uniformity.

The Blue-winged Warbler breeds regularly only in eastern Iowa and the Prothonotary Warbler, Cerulean Warbler, Louisiana Waterthrush, and Kentucky Warbler reside as summer residents in eastern and southern Iowa; the Western Kingbird, Brewer's Blackbird, and Lark Sparrow nest mostly from the central part of the state westward; the Bell's Vireo breeds from border to border but is fairly common only in the western and southern sections; the Least Flycatcher and Savannah Sparrow are regular in summer in the extreme north, and the Acadian Flycatcher in the southern half of the state where the Carolina Wren and Northern Mockingbird are year-round residents. The Yellow-breasted Chat is fairly common as a nesting bird in the southern half of Iowa and northward in the valleys of the Mississippi and Missouri.

Extensive marshes with cattails, bulrushes, sedges, and other aquatic plants are mainly in the lake region of northwestern Iowa and on the bottomlands of the Mississippi and Missouri. In nearly all of Iowa's larger marshes the following birds breed regularly:

Pied-billed Grebe
American Bittern
Least Bittern
Virginia Rail
Sora
Common Gallinule
American Coot
Black Tern
Marsh Wren
Yellow-headed Blackbird
Red-winged Blackbird
Swamp Sparrow

The Mississippi and Missouri Valleys on the state's eastern and western boundaries, respectively, are the chief migration routes for birds breeding regularly in north-central North America. Although equally important in heaviness of traffic, the two routes nevertheless differ from each other with respect to the species using them. Such northern passerines as the Yellow-bellied Flycatcher, Winter Wren, Hermit Thrush, Veery, and various warblers—the Golden-winged, Northern Parula, Magnolia, Black-throated Green, Blackburnian, Chestnut-sided, and Canada—are much more abundant in the Mississippi Valley. The Connecticut Warbler follows this route, rarely the Missouri route. Whistling Swans more often show up along the Mississippi, but most other waterfowl appear more commonly along the Missouri. This is true not only of Snow Geese, whose spring movements provide the state's outstanding ornithological feature, but also of various ducks. In addition to waterfowl, other birds more frequent in the Missouri Valley than in the Mississippi include the Eared Grebe, American White Pelican, Northern Phalarope, Franklin's Gull, and Harris' Sparrow. In Iowa, the main migratory flights may be expected within the following dates:

Waterfowl: 10 March–20 April; 10 October–25 November
Shorebirds: 1 May–1 June; 1 August–1 October
Landbirds: 15 April–20 May; 25 August–20 October

Slate-colored Juncos and American Tree Sparrows are Iowa's most abundant winter visitants from northern climes, although Purple Finches and Pine Siskins are almost invariably numerous. Other northern visitants include the Rough-legged Hawk, Red-breasted Nuthatch, Brown Creeper, and, irregularly, the Evening Grosbeak, Common Redpoll, and Red and White-winged Crossbills. Where waterways remain ice-free with ample food supply, quite a few ducks—particularly Mallards, Common Goldeneyes, and Common Mergansers—pass the winter. Below the locks and dams on the Mississippi River, Bald Eagles catch fish in the open water and may stay in the immediate area all winter.

Authorities

Richard Antonnette, Charles C. Ayers, Jr., Margaret Brooke, Myrle M. Burk, Frieda and George Crossley, George E. Gage, Ruth C. Green, Darwin Koenig, Frederick Z. Lesher, Robert Moore, Robert L. Nickolson, Peter C. Petersen, Robert F. and Jean Vane.

Reference

Birding Areas of Iowa. Edited by Peter C. Petersen. Iowa Ornithologists' Union, 1979. Available from Librarian, 1560 Linmar Drive, Cedar Rapids, IA 52404.

ALGONA
Union Slough National Wildlife Refuge

Excellent for waterbirds, waterfowl, and open-country birds is **Union Slough National Wildlife Refuge** (2,155 acres) in north-central Iowa. Drive north from Algona on US 169 toward Bancroft for about 15 miles, turn east onto County A42 and drive 4 miles to headquarters in the Refuge. Here obtain leaflets about nature trails and an auto tour route.

Extending about 8 miles along Union Slough and Buffalo Creek, the Refuge embraces natural marsh habitat with open water—formerly drained—and about 600 acres of grassy upland with trees, mainly ash, cottonwood, and willow, bordering the streams.

In marsh habitat grow bulrushes, sedges, and other aquatic plants in rich profusion, thus providing ample nesting cover for such birds as Pied-billed Grebes, Least and American Bitterns, Virginia Rails, Soras, Common Gallinules, American Coots, Black Terns, Marsh and Sedge Wrens, Yellow-headed and Red-winged Blackbirds, and Swamp Sparrows. The more common breeding waterfowl include Mallards, Blue-winged Teal, and Wood Ducks. The Refuge is a haven in spring and fall for migrating Canada Geese, Greater White-fronted Geese, Snow Geese, and many ducks—mainly Mallards, Common Pintails, Green-winged and Blue-winged Teal, American Wigeons, Northern Shovelers, and Ruddy Ducks. Their peaks of abundance are in early April, October, and early November. Also, during the same season, Ring-billed and Franklin's Gulls appear in large numbers.

In addition to Ring-necked Pheasants, some of the open-country birds nesting on the Refuge are Horned Larks, Bobolinks, Western Meadowlarks, Dickcissels, and Grasshopper and Vesper Spar-

WESTERN MEADOWLARK

rows. Common Screech Owls and Great Horned Owls, both Yellow-billed and Black-billed Cuckoos, Red-headed Woodpeckers, Traill's Flycatchers, Warbling Vireos, and Baltimore Orioles are among the birds nesting in wooded areas.

BOONE
Ledges State Park

Ledges State Park in central Iowa, 6 miles south of Boone via State 164, is a pleasing contrast to the monotony of the surrounding farmland. Here the visitor drops down into a scenic wooded valley where Pease Creek descends rapidly through a ravine walled by huge sandstone ledges to join the Des Moines River within the Park's west boundaries. Giant maples, cottonwoods, and ashes shade the forest floor carpeted with wildflowers, and shrubby thickets abound in forest clearings. All parts of the Park's 900 acres are easily accessible by road, bridges, and foot trails.

Bird finding is at its best in late spring and early summer. One or more pairs of Red-tailed and Red-shouldered Hawks nest regularly. The chants of Whip-poor-wills can be heard at night. Eastern Phoebes find suitable nesting sites under bridges and overhanging ledges, Rough-winged Swallows in the tile drains emerging from bridge abutments. Along the tree-shaded edges of the Des Moines River, Wood Ducks, Green Herons, and Black-crowned Night Herons are frequent sights.

To see or hear many of the nesting passerines, park the car near the outlet of Pease Creek and walk upstream through the ravine. Be alert for Yellow-bellied Sapsuckers, Wood Thrushes, Veeries, Blue-gray Gnatcatchers, Bell's and Yellow-throated Vireos, Blue-winged and Cerulean Warblers, Louisiana Waterthrushes, Scarlet Tanagers, Rose-breasted Grosbeaks, and Indigo Buntings. A Yellow-throated Warbler is a possibility.

A flock of Turkey Vultures roosts in a stand of dead trees at the south end of the Park. Above the open grass-covered area just within the Park's east entrance, American Woodcock perform their crepuscular flight-songs during April and early May.

CEDAR FALLS–WATERLOO

George Wyth Memorial State Park | Island Park | Byrnes Park

These twin cities in the intensely cultivated region of east-central Iowa lie along the Cedar River, formerly bordered by oak, hickory, sugar maple, basswood, and butternut on the uplands, ash and silver maple on the lowlands. Where there are remnants of these woodlands in parks and preserves, bird finding is still productive, particularly during migration seasons and in winter.

The **George Wyth Memorial State Park** (400 acres), straddling both cities on the north side of the Cedar River, embraces typical woodland as well as much shrubbery and a small lake. Entrance is south from US 20 as it passes north of the Park in its east-west course. In late April and early May, again in September and early October, the trees often teem with transient kinglets, vireos, and warblers. The lake attracts various waterbirds and waterfowl in spring and fall.

Similarly rewarding for migrating small landbirds is **Island Park** (104 acres) in Cedar Falls, bounded on the south and west by the Cedar River and on the east by US 218 from which the Park is entered. Low-lying with an abundance of shrubby thickets in addition to woods, the Park provides good habitat for sparrows during migration as well as in winter.

Byrnes Park (170 acres) in south Waterloo, with main entrance from Fletcher Avenue, has extensive conifer plantations that induce finches, crossbills, redpolls, siskins, and other southbound finches to remain through winter.

CEDAR RAPIDS

Cedar Lake | Swan Lake | Palisades-Kepler State Park | Muskrat Slough

There are good opportunities for bird finding in and near Cedar Rapids, an industrial city in the rolling farmland of east-central Iowa.

Within the northeast city limits and close to the Cedar River is

Cedar Lake, covering about 100 acres. Numerous migrating waterfowl congregate here in spring and fall and many stay throughout winter. Because the Lake's waters are constantly used by a nearby power plant, the Lake does not freeze over. Even when air temperatures dip far below zero, flocks of Mallards, American Black Ducks, Lesser Scaups, and Common Goldeneyes, along with a few American Wigeons and Common Mergansers, usually stay all winter. Cedar Lake is approached from the center of the city by going east on First Avenue and turning north onto Thirteenth Street (Center Point Road) and continuing for 0.5 mile. The Lake will then appear on the left.

Sixteen miles south of the city, a marshy 44-acre slough called **Swan Lake** attracts many transient waterbirds, waterfowl, and shorebirds. The species at the Lake will depend on the particular phase the marsh cycle is in as well as on the water level. If vegetation is present across the Lake, Least Bitterns will be nesting in June and early July. Broods of Wood Ducks frequently appear. If vegetation has been destroyed by muskrats and the water level is low, Swan Lake becomes excellent for shorebirds in mid-May with Dunlins, Hudsonian Godwits, and Wilson's Phalaropes fairly dependable among the more common species. When there is higher open water, migrating waterfowl predominate.

Swan Lake may be reached from Cedar Rapids by driving south on I 380 for 13 miles (measured from US 30 Interchange), turning west onto County F28 and going west for 0.6 mile, north for 1.6 miles, then west again for 1.0 mile. Swan Lake is immediately adjacent to the Coralville Reservoir which affords excellent bird viewing in all seasons.

Palisades-Kepler State Park (688 acres), 12 miles southeast of Cedar Rapids, lies along the Cedar River where limestone cliffs rise from its banks some 30 to 75 feet. In these cliffs, especially on the west side of the River, Cliff Swallows have nesting colonies from mid-May through June. At the same time of year the oak-hickory shaded ravines that lead down to the River from the east side have breeding Red-bellied, Red-headed, Hairy, and Downy Woodpeckers. Pileated Woodpeckers nest along the River below the dam. Many passerines also breed in the Park including Wood Thrushes, Blue-gray Gnatcatchers, Cerulean Warblers, American

Redstarts, and Scarlet Tanagers. Blue-gray Gnatcatchers and Cerulean Warblers can usually be heard singing at the last turn before the River on the entry road. To reach the Park, drive south from the city on I 380, exit east on US 30 for 10 miles to the entrance.

Muskrat Slough, covering 366 acres about 27 miles east of Cedar Rapids, is a state-owned slough. On a much larger scale than Swan Lake, it attracts correspondingly greater numbers of transient waterbirds, waterfowl, and shorebirds. The surrounding shrubby habitat is fine for migrating small landbirds, especially sparrows. The Slough itself supports a colony of Yellow-headed Blackbirds. Both Marsh and Sedge Wrens are summer residents. In addition to all waterfowl species regularly passing through Iowa, Muskrat Slough attracts shorebirds galore, especially Lesser Golden Plovers. At the height of their migration, 5 May to 15 May, flocks of this species can be seen overhead flying north, feeding in plowed fields, or gathered around rain pools in numbers up to two hundred. The fields passed as one approaches Muskrat Slough in early summer are alive with Bobolinks and Dickcissels.

Muskrat Slough may be reached from Cedar Rapids by driving south on I 380, exiting east on US 30 through Mt. Vernon, Lisbon, and Mechanicsville. At a point 2.7 miles east of Mechanicsville, turn off north onto a road that goes 7.1 miles directly into the preserve.

CLINTON
Eagle Point Park

Commanding a magnificent view of the Mississippi River from the 200-foot bluffs on the north edge of Clinton is city-owned **Eagle Point Park** (121 acres), entered by turning east off US 67 (North 3rd Street). In late April and early May and in September, kinglets, vireos, warblers, and other transient passerines may be watched at treetop level from the brows of the bluffs. Although the Park is closed to cars in winter, a walk into the Park at that season to the stone tower may be well rewarded by the sight of Red-breasted Nuthatches, Purple Finches, and Pine Siskins sharing a nearby grove of pines with Barred Owls.

More rewarding for winter birds generally is a walk along the railroad and right-of-way below the Park between the east end of 32nd Avenue North (entered by turning east off US 67) and the east end of Deer Creek Road a mile farther north. The right-of-way, about 100 feet wide, is bordered on the west by the Park bluffs and on the east by the Mississippi. Two birds to be looked for especially are the Winter Wren among the exposed roots of the mature hardwoods at the foot of the bluffs, and the Carolina Wren, in the tumble of rocks below the bluffs or on the exposed ledges higher up. Red-bellied and Red-headed Woodpeckers are common. Occasionally a Gray Catbird, Brown Thrasher, or Rufous-sided Towhee may be spotted among the brush piles. In some winters, Evening Grosbeaks and Common Redpolls are attracted to the bluffs, where they feed on boxelder seeds and birch drupes.

COUNCIL BLUFFS
Lake Manawa State Park

Lake Manawa State Park (1,045 acres) encompasses Lake Manawa (660 acres), a meander cutoff of the Missouri River. Lying on flat, alluvial bottomland just south of Council Bluffs, the Lake may be reached by exiting south from I 80 on State 192 (South 4th Street), which leads to the Park and to a road encircling the Lake in the Park.

On any day from the first of March, when the Lake becomes free of ice, to June, and again from late August to December or January, when the Lake freezes over, one is certain to see a wide variety of waterbirds and waterfowl that stop to rest and feed during their migratory journeys. Among the species, many of them the same as in DeSoto National Wildlife Refuge (*see under* **Missouri Valley**), are the Common Loon, Horned, Eared, and Pied-billed Grebes and an occasional Western Grebe, Wood Duck, Redhead, Canvasback, Common Goldeneye, Ruddy Duck, Red-breasted Merganser, American Coot, numerous gulls—Herring, Ring-billed, Franklin's, and Bonaparte's—and the Forster's, Caspian, and Black Terns. The birds may be readily viewed from parking areas along the encircling road or by taking short walks to points of land that the road bypasses.

Well worth the effort is a walk out to Kaplan's Point, an extension into the Lake from the southeast side and marked by a sign. From the parking area, follow the path through mature woods—excellent for migrating small landbirds. Once at the Point, scan its marshy shores on either side for Yellow-headed Blackbirds in late April and early May, for Willets and other shorebirds in May, and for wading birds, including Great Egrets and Black-crowned Night Herons, in late summer and early fall.

DAVENPORT

Credit Island Park | Lock and Dam Number 14

In this urban area, consisting of Davenport and Bettendorf on the Iowa side of the Mississippi River and Rock Island and Moline on the Illinois side, the best year-round site for bird finding is **Credit Island Park** (420 acres; roughly 2 miles in length and 0.5 mile in width) in the Mississippi and readily accessible by a causeway.

Although Credit Island, which belongs to the city of Davenport, has been extensively developed for recreational purposes, the southern end has a fine, relatively undisturbed bottomland forest of silver maple, hackberry, swamp white oak, basswood, cottonwood, and willow. Here breeding birds include the Wood Duck, Barred Owl, Pileated Woodpecker, Red-bellied Woodpecker, Red-headed Woodpecker, Great Crested Flycatcher, Tufted Titmouse, White-breasted Nuthatch, Warbling Vireo, Prothonotary Warbler, and American Redstart. In late April and early May and in September as many as 23 species of warblers pass through the forest and in wooded tracts scattered among the recreational facilities.

During March and early April and occasionally in October and early November the shallow harbor off the north side of Credit Island has many surface-feeding ducks as well as Redheads, Ring-necked Ducks, Canvasbacks, Lesser Scaups, Ruddy Ducks, and Hooded and Red-breasted Mergansers, also Common Goldeneyes and Common Mergansers that may remain in winter as long as the water stays ice-free. Great Blue Herons and Great Egrets appear commonly along the shore. From Credit Island's south side, which faces the Mississippi channel, large numbers of Herring and Ring-billed Gulls and occasionally Caspian and Black Terns may be viewed in spring and fall.

Credit Island Park may be reached from I 280 west of Davenport by exiting east (marked north) on US 61 and proceeding 3.5 miles, then turning off right at a directional sign. In August and September, if the water level of the Mississippi is low, the exposed mud flats on the west side of the causeway to Credit Island attract numerous shorebirds such as the Semipalmated Plover, Greater and Lesser Yellowlegs, Solitary, Pectoral, Least, and Semipalmated Sandpipers, occasionally the White-rumped, Baird's, or Western Sandpiper. In winter, one or more Bald Eagles are often in sight on Pelican Island just east of the causeway.

More Bald Eagles can usually be seen in winter farther upriver below **Lock and Dam Number 14.** This may be reached from I 80 east of Bettendorf by exiting west (marked south) on US 67, proceeding 1.3 miles, and, after passing the first lock, turning off left at a directional sign to the access area. Look for the big birds below the Dam where they perch in trees on both the Iowa and Illinois sides of the River. Also look for Common Goldeneyes and Common Mergansers on the open water.

DES MOINES

Water Works Park | Greenwood and Ashworth Parks | Philip Jester County Park | Ankeny Ponds

Iowa's capital city lies in the central part of the state where the gently rolling agricultural land is broken up by the Des Moines River and its tributary, the Raccoon River. Three areas southwest of the business district offer the best opportunities for bird finding within the city limits.

One area is **Water Works Park** (2,000 acres), consisting of woods, a crabapple arboretum, conifer plantations, small ponds, and grassy lowlands bordering the Raccoon River. From I 235, take the 42nd or 31st Street Exit south, turn left onto Grand Avenue and, at the 17th Street traffic light, turn *sharply* right over the Raccoon River Bridge to Fleur Drive and follow it for 0.7 mile to the first traffic light. On the left is the entrance to Gray's Lake Park—worth a stop in spring and fall for ducks, gulls, and terns on the Lake. On the right is the entrance to Water Works Park. Turn in and proceed along some 8 miles of roads, stopping wherever there are promising sites for birds. In spring and fall, scan all ponds and settling basins for transient grebes, ducks, and shorebirds. In late spring and summer, investigate all woods for some of Iowa's more common passerines. In winter, visit the arboretum, whose fruits attract large flocks of American Robins, Cedar Waxwings, and Purple Finches, and inspect the conifers, which are frequented in some years by both Red and White-winged Crossbills.

The other two areas in the city are contiguous **Greenwood and Ashworth Parks** (70 and 65 acres), which are, essentially, a wooded continuation north from Water Works Park. From I 235, again take the 42nd or 31st Street Exit south, but turn right onto Grand Avenue, left onto 45th Street, and then immediately right up the hill to the Art Center parking lot. Leave the car and first check the wooded hill south of the Center. This is excellent for resident woodpeckers, transient warblers, and numerous summer residents such as the Rose-breasted Grosbeak. Then look over the lagoon for transient warterfowl, walk past the swimming pool, and follow the path that leads south into steep, wooded ravines, which are rewarding for birds in any season. Broad-winged Hawks have nested here. The path comes out on a railroad right-of-way. Along its shrub-bordered edge, look for Yellow Warblers, Common Yellowthroats, Yellow-breasted Chats, and Indigo Buntings.

There are two highly recommended sites for bird finding outside the city. One is **Philip Jester County Park** (846 acres), 19 miles northwest of the business district, on the west side of Saylorville Lake, an impoundment of the Des Moines River. From I 235 west of Des Moines, exit north on I 35 for 4 miles, exit right on State 141 north for 7 miles, then turn off right and proceed 3 miles to the Park entrance. Drive through the Park, bearing left at intersections. In open areas during summer, listen for Vesper, Grasshopper, and Field Sparrows. Where the road bends left past the golf course, leave the car, walk by a picnic stove, and take a path leading down to the Lake through woods in which summer-resident birds include Great Crested Flycatchers, Wood Thrushes, Yellow-throated Vireos, Cerulean Warblers, Ovenbirds, and Scarlet Tanagers. After completing the drives in the Park, return to just inside the entrance gate and turn left to the Lake, about 0.5 mile down the hill. Preferably, leave the car and walk, looking and listening for birds in the woods along the way and then scanning the Lake and shore for wading birds, gulls, and terns.

The other site is **Ankeny Ponds** northeast of the city. At the junction of I 235 and I 35 northeast of the business district, go north on I 35 for 5.2 miles to the second Ankeny Exit. Here turn right onto N. E. 94th Avenue and proceed 4 miles to N. E. 56th Street. The grassy fields along the way are summer habitats for Bobolinks, Dickcissels, and Grasshopper Sparrows. At 56th Street, turn left (north) and, in about a mile, start watching on the left side of the road for marshy areas. These attract in May, if the season is wet, many transient shorebirds which often include Lesser Golden Plovers, Greater and Lesser Yellowlegs, Pectoral Sandpipers, and sometimes Hudsonian Godwits. All farm ponds should be viewed for ducks, American Coots, and Common Snipes. The deeper ponds with marshy edges may have nesting Yellow-headed Blackbirds.

DUBUQUE

Eagle Point Park | Linwood Cemetery | Mt. Calvary Cemetery | John Deere Levee | Mud Lake Park | White Pine Hollow Forest Preserve

In the environs of this city are several places for good bird finding northeast of the business district which lies along the Mississippi River at the foot of high bluffs.

Eagle Point Park (133 acres), reached from US 52 by turning off east onto 20th Street to Rhomberg Avenue, then (before crossing High Bridge) left on Shiras Road, is an oak-wooded area on bluffs overlooking the Mississippi and Lock and Dam Number 11.

Among the wide variety of birds breeding in the Park are Red-bellied and Red-headed Woodpeckers, Tufted Titmice, Wood Thrushes, Blue-gray Gnatcatchers, and Scarlet Tanagers. From the tops of the bluffs during late April and early May, and again in September, when small landbirds are migrating, one may readily view many kinglets, vireos, and warblers below at treetop level. In winter the water below the Dam, which remains ice-free, should be scanned, either from the Park, or from the river levee, for Common Goldeneyes, Common Mergansers, and Red-breasted Mergansers. Look for a few Bald Eagles in riverbank trees on the opposite shore where they pass the time between flights for fishing.

Linwood Cemetery (140 acres), reached from US 52 by exiting east on 22nd Street and going north on Windsor Avenue, has dense stands of both conifers and hardwoods with shrubby edges that yield woodland birds in any season. **Mt. Calvary Cemetery** (64 acres), bordering Linwood on the north, has denser stands of conifers well worth exploring in the colder months for daytime-roosting owls.

John Deere Levee, reached from US 52 north by exiting east on 32nd Street and its continuation, Peru Road, for 2 miles, and turning right across railroad tracks to a power-plant intake structure. Park here and walk out on the Levee which extends into the Mississippi about 0.5 mile. In early spring and fall, as well as during winter when there is open water, both surface-feeding and diving ducks are offshore, and invariably there are gulls in view. Except during winter and midsummer, there may be Bonaparte's Gulls and Forster's, Caspian, and Black Terns. If the water level is low in May and in late August and September, the exposed mud flats attract many shorebirds—Semipalmated Plovers, Ruddy Turnstones, Whimbrels, Willets, Greater and Lesser Yellowlegs, Pectoral and Least Sandpipers, Dunlins, Short-billed and Long-billed Dowitchers, Semipalmated Sandpipers, and others.

Mud Lake County Park (20 acres), reached by driving north on US 52 for 5 miles, then turning right at a directional sign and following more such signs for 3.5 miles, encompasses a small bay off the Mississippi, a stretch of open river, and a cattail marsh. Least Bitterns, Virginia Rails, Soras, and Marsh Wrens are among the species nesting in the marsh. Great Egrets and other wading birds commonly frequent the area in late summer and early fall.

White Pine Hollow Forest Preserve (800 acres), far to the northwest of Dubuque, is rewarding for a rich variety of woodland birds. Drive north and west on US 52 for 26 miles to Luxemburg; continue west on State 3 for 0.75 mile, then turn off right and proceed 2 miles to the entrance sign where a foot trail begins. Embracing rough terrain with deep ravines walled by cliffs, the Preserve features deep woods with the largest stands of virgin white pine remaining in Iowa. Among the many birds nesting regularly are the Ruffed Grouse, Pileated Woodpecker, Acadian Flycatcher, Blue-winged and Cerulean Warblers, Louisiana Waterthrush, and Kentucky Warbler.

LANSING
Upper Mississippi River Wildlife and Fish Refuge

The bottomlands of the Mississippi River from Lansing north to the Minnesota line are included in the great **Upper Mississippi River Wildlife and Fish Refuge** (headquarters address: Box 226, Winona, MN 55987), a maze of channels, small islands, ponds, sloughs, marshes, and wooded swamps.

In spring and fall, transient Canada Geese and several species of ducks—Mallards, Common Pintails, Blue-winged Teal, Ring-necked Ducks, and Lesser Scaups—are common to abundant. Whistling Swans in small flocks linger on isolated stretches of water for a few days in early spring. Breeding birds include the following: in marshes, the Northern Harrier, Sora, American Coot, Black Tern, Marsh Wren, Common Yellowthroat, and Swamp Sparrow; in bottomland woods and wooded swamps, the Double-crested Cormorant, Great Blue and Green Herons, Great Egret, Black-crowned Night Heron, Wood Duck, Red-tailed Hawk, Great Horned and Barred Owls, Prothonotary Warbler, and American Redstart.

State 26 going north from Lansing to New Albin, a distance of 11 miles, skirts the Mississippi bottomlands—the Refuge is on one side and high rocky bluffs on the other. Yellow-breasted Chats nest in thickets along the highway and sometimes show up on utility wires. During migration in May, Connecticut Warblers may be found on the wooded and brushy banks of the Upper Iowa River where State 26 crosses the River just south of New Albin.

MARQUETTE
Yellow River State Forest

Unexcelled in northeast Iowa for birds in timbered uplands back from the Mississippi bottomlands is **Yellow River State Forest,** encompassing some 6,000 acres of hilly country in separate units. One, highly recommended for bird finding, is the 4,000-acre Paint Creek Unit, which may be reached as follows: From State 76, about 10.5 miles north and west of Marquette and 4.5 miles from its junction with State 364, turn off north (right) on to a gravel road and drive 2 miles to the entrance. Continue on the road to Forest headquarters for a map and information about hiking trails.

The Paint Creek Unit has willows, cottonwoods, and boxelders standing along creeks, other trees such as maples, oaks, and hickories covering slopes as well as the tops of ridges where there are also aspens. Red cedars grow on bluff outcroppings. Here and there are plantations of conifers including red and white pines. In a few areas the timbered cover is interrupted by meadows and old fields.

Among the many bird species breeding in the Unit are the Turkey Vulture, Cooper's, Red-shouldered, and Broad-winged Hawks, Ruffed Grouse, Wild Turkey, American Woodcock, Whip-poor-will, Pileated Woodpecker, Yellow-bellied Sapsucker, Acadian and Least Flycatchers, Blue-gray Gnatcatcher, Blue-winged and Cerulean Warblers, Louisiana Waterthrush, Kentucky Warbler, Yellow-breasted Chat, Scarlet Tanager, and the Savannah and Henslow's Sparrows.

MISSOURI VALLEY
DeSoto National Wildlife Refuge

North of Council Bluffs and Omaha, Nebraska, **DeSoto National Wildlife Refuge** embraces 7,800 acres of the Missouri River bottomlands including 750-acre DeSoto Lake, a meander cutoff of the Missouri. Headquarters at the north end of the Refuge may be reached by exiting west from I 29 at Missouri Valley on US 30 for 6 miles.

DeSoto Lake, shaped like a horseshoe, has long sand bars extending into its shallower water and steep banks overlooking its deeper water. Back from the Lake are drainage ditches, several small marshes, and about 2,800 acres of forest—primarily of cottonwood with some willow and boxelder—interspersed with fields and croplands.

The stellar ornithological feature of the Refuge are the hordes of Snow Geese and some fewer Canada Geese that stop off in their spring and fall migrations, resting on DeSoto Lake and feeding on the Refuge's 3,000 acres of croplands specially maintained for them. Vast numbers of ducks also congregate in spring and fall. Although Mallards predominate, Common Pintails, Blue-winged Teal, and American Wigeons are impressively abundant and American Black Ducks, Gadwalls, Green-winged Teal, Northern Shovelers, Ring-necked Ducks, Lesser Scaups, Buffleheads, and Common Mergansers are notably common. In spring the numbers of geese peak in mid-March, ducks in late March and April. In fall, both geese and ducks appear in more spectacular concentrations, the geese peaking in mid-November, the ducks later in the month.

Along with the transient waterfowl in spring and fall, small flocks of migrating American White Pelicans show up in April and early May and again in October and early November. Many Bald Eagles are present in winter and may be observed in trees near the north end of the Lake where the small numbers of wintering Mallards and other waterfowl gather when water elsewhere on the Lake freezes over.

In suitably wooded and brushy habitats on the Refuge, breeding birds include the Red-tailed Hawk, Yellow-billed and Black-billed Cuckoos, Great Horned and Barred Owls, Red-headed Woodpecker, Eastern Kingbird, Gray Catbird, Brown Thrasher, Wood Thrush, Bell's and Warbling Vireos, American Redstart, Orchard and Baltimore Orioles, Rose-breasted Grosbeak, Indigo Bunting, Rufous-sided Towhee, and Field Sparrow. In fields, the Dickcissel is a common summer resident.

OTTUMWA
Greater Ottumwa Park | YMCA Camp Arrowhead | Lake Wapello State Park

This prosperous meat-packing city lies along the Des Moines River in southeastern Iowa, about 75 miles from the Mississippi River and 35 miles from the Missouri line.

Within the city is **Greater Ottumwa Park** (360 acres), south of the Des Moines River. At the point where US 63 (Wapello Street) coming south joins US 34 going east-west, continue south on the Wapello Street Extension (also called Ferry Street Extension), which soon crosses the Park, passing small bodies of water and long lagoons on either side. From the extension, take any of the asphalt roads leading into the Park. These give ready access to sites for wading birds, geese and ducks, shorebirds, gulls, and terns in spring, late summer, and fall. From mid-April to mid-May, transient warblers and other small passerines pass through the Park woods, sometimes in impressive numbers.

About 4 miles east of Ottumwa, a sign on the south side of US 34 indicates a road to the **YMCA Camp Arrowhead,** 3 miles distant. In the camp's vicinity are wooded ravines, mature stands of timber, open brushy and grassy areas, and a small lake—a wide variety of habitats for a wide variety of birds. Among the species breeding regularly are the Red-tailed Hawk, Barred Owl, Chuck-will's-widow, Red-bellied Woodpecker, Eastern Pewee, Wood Thrush, White-eyed and Yellow-throated Vireos, warblers including the Worm-eating, Blue-winged, Kentucky, and Hooded, and both Scarlet and Summer Tanagers. Bird finders are welcome to the Camp, provided they first make known the purpose of their visit.

Lake Wapello State Park (1,143 acres), southwest of Ottumwa, is surrounded by rolling, wooded hills and embraces Lake Wapello, an artificial body of water covering 287 acres. Many geese and ducks stop on the Lake during their migrations in early spring and fall. American White Pelicans occasionally appear in spring, usually in early May. The southeastern end of the Lake has good marshy habitat for nesting bitterns, rails, Marsh Wrens, and Red-winged Blackbirds. Among the passerines nesting in the area are the Northern Mockingbird, Orchard and Baltimore Orioles, and Scarlet and Summer Tanagers. Conifer plantations in the Park provide winter shelter for Long-eared and Saw-whet Owls. The Park may be reached from Ottumwa by driving south for 16 miles on US 63, then west on State 273 for 10 miles.

SIDNEY
Waubonsie State Park | Forney's Lake Wildlife Area

In extreme southwestern Iowa, high on loess bluffs overlooking the Missouri River bottomlands from the east, is scenic **Waubonsie State Park** (1,208 acres). To reach the Park from Sidney, drive south on US 275 for 4 miles, west on State 2 for 2 miles, then south for 0.5 mile on State 239, which dead-ends at the Park. To reach the Park from I 29 which runs north-south west of the Park, exit on State 2 for 5 miles, then go south on State 239.

From the Park's main parking area close to the ranger station, walk to the top of the nearest bluff for a remarkable view—of Nebraska, across the Missouri River, 9 miles to the west; of Missouri, 4 miles to the south; and, if one has a good eyesight (or imagination), of Kansas, 40 miles to the southwest.

Seven miles of foot trails lead over the crests of the neighboring bluffs, mostly grass-covered, and down 300 feet through wooded ravines. Various shrubs, such as dogwood, coralberry, and wild plum, grow commonly on the slopes of bluffs; redbuds and pawpaws are abundant in ravines—conspicuously so in May when they come into bloom.

Birds of special interest in the Park from the middle of May through July are both the Scarlet and Summer Tanagers, as well as the Wood Thrush, Blue-gray Gnatcatcher, Bell's Vireo, Kentucky Warbler, Yellow-breasted Chat, Orchard Oriole, Indigo Bunting, and Lark and Field Sparrows. Look or listen for the tanagers and Kentucky Warbler in the ravines. The Bell's Vireo prefers the plum thickets and a brushy area just outside the entrance gate. The Lark and Field Sparrows have territories on the grassy, brush-scattered ridges.

To see one of Iowa's ornithological spectacles—the spring concentrations of migrating Snow Geese—visit **Forney's Lake Wildlife Area** comprising 1,800 acres of which 400 make up a permanent refuge. This may be reached by driving north on I 29 from the State 2 Interchange for 15 miles to Bartlett, then exiting east to the northwest side of the Area and turning left to the Lake within the Area.

An old oxbow of the Missouri River, Forney's Lake is actually more marsh than lake. Here the Snow Geese begin appearing in early March with the retreat of winter and reach peak abundance by the middle of the month. Throughout their stay the handsome birds are restless and noisy. Early in the morning, large numbers fly off to feed on waste grain in nearby croplands; in the late afternoon, especially at sunset, they return to pass the night. All day, small groups depart and return—the shuttling back and forth between foraging and resting is almost constant. The great aggregation of Snow Geese tends to eclipse other waterfowl present at the time: small flocks of Canada Geese, and scattered numbers of Greater White-fronted Geese and ducks, which include hordes of Mallards along with impressive numbers of Gadwalls, Common Pintails, Green-winged and Blue-winged Teal, American Wigeons, Northern Shovelers, and a few diving ducks—Redheads, Ring-necked Ducks, Lesser Scaups, Buffleheads, and Common Mergansers.

Forney's Lake is rewarding for other birds later in the season. Small flocks of American White Pelicans often stop off in their migration. Among birds that remain to nest in the Lake's marshy periphery are Least and American Bitterns and Yellow-headed Blackbirds. Great Egrets and Black-crowned Night Herons frequent the Lake during the late summer.

SIOUX CITY

Logan Park Cemetery | Stone State Park | Brown's Lake | Bigelow Park

This large city covers some 45 square miles of bottomlands and bluffs along Iowa's northwestern boundary where the Big Sioux and Missouri Rivers meet and where, across the boundary, South

Dakota meets Nebraska. In the vicinity are wooded tracts, oxbow lakes (meander cutoffs), and great stretches of farmlands that were once wetlands and prairie—a wide variety of physiographic features that are in turn attractive to a wide variety of birds.

Logan Park Cemetery and **Stone State Park,** both in the northwestern part of the city, are convenient for good bird finding the year round. The Cemetery, occupying 100 acres of rolling terrain, may be reached from I 29 by exiting north on Nebraska Street and continuing north to its dead-end, here turning left onto Stone Park Boulevard and watching for the entrance on the left in the 4000 block. From the entrance, winding drives lead to all parts of the Cemetery. Although it is extensively landscaped, the introduction of many trees and shrubs, native as well as exotic, has provided food and cover for many woodland and forest-edge birds. The prevalence of pines, cedars, spruces, and other conifers has induced the Pine Siskin to become a permanent resident and the Red Crossbill a winter visitant. Numerous passerine species, including Swainson's and Gray-cheeked Thrushes and the Lincoln's Sparrow, pass through in migration (late April and early May, September and early October); other species from the north country, such as the Slate-colored Junco, American Tree Sparrow, and Harris' Sparrow, remain through the winter.

Stone State Park (1,200 acres) embraces loess bluffs and valleys along the east bank of the Big Sioux and may be reached from I 29 by exiting north on Riverside Boulevard (State 12) for 3 miles. The valleys are heavily wooded with ash, hawthorn, basswood, boxelder, cottonwood, maple, and oak and are carpeted with wildflowers in spring; on the more exposed slopes, wild plum, sumac, and other shrubs form dense thickets. All productive bird habitats are readily accessible by foot trails and bridle paths. Three species of owls—Common Screech, Great Horned, and Barred—and the Red-bellied Woodpecker are permanent residents. Among the many summer residents are the Turkey Vulture, Red-tailed Hawk, Broad-winged Hawk (occasionally), Great Crested Flycatcher, Wood Thrush, Bell's Vireo, Orchard Oriole, Scarlet Tanager, Blue Grosbeak, Indigo Bunting, Rufous-sided Towhee, and Lark Sparrow.

Southeast of Sioux City on the flat Missouri bottomlands are vast grainfields, patches of willows and shrubs, groves of cottonwoods, hackberries, maples, basswoods, and ashes, and oxbow lakes. A good site for bird finding is state-owned **Brown's Lake,** an oxbow, with **Bigelow Park** (24 acres) on the Lake's east side. From Sioux City, drive south on I 29 for 15 miles, then exit west at Salix to the Park. Here take a gravel road which runs around the southern edge of the Lake to the west side, where there are cottonwoods, wild plums, and introduced white pines—habitat in early May for many migrating warblers and other northbound small landbirds.

Brown's Lake hosts a good representation of migrating waterfowl and waterbirds in spring and fall. Mallards, Gadwalls, Common Pintails, and Blue-winged Teal are among the commoner ducks. Eared Grebes and sometimes small flocks of American White Pelicans stop off as do many gulls—Herring, Ring-billed, and Franklin's. The Wood Duck is a summer resident. In some of the marshy borders, Yellow-headed Blackbirds have small nesting colonies in early summer.

Kansas

BY ROBERT M. MENGEL, CHARLES A. ELY, MAX C. THOMPSON, AND MARVIN SCHWILLING

UPLAND SANDPIPER

Nearly rectangular except for its short northeastern border along the Missouri River, Kansas comprises 82,276 square miles, extending some 410 miles from Missouri to Colorado and just over 200 miles north to south. Elevation rises steadily westward at about eight feet to the mile, from 743 feet in the southeast to 4,135 feet at a point near the Colorado line. The geographical center of the contiguous 48 states is about 50 miles northwest of Salina. The entire state lies within the Great Plains which slope gently from the Rocky Mountains to the Mississippi River. The rivers flow eastward, with historically evocative names like Saline, Smoky Hill, Republican, Arkansas, and Cimarron. (On the Smoky Hill, in 1874, O. C. Marsh, harassed by hostile Indians, made the first discoveries of the toothed birds *Hesperornis* and *Ichthyornis* of the Cretaceous Niobrara Chalk whose outcrops are conspicuous in Gove County.) Across these shallow, meandering streams came the great cattle drives of the 1880s and 1890s to railheads like Dodge City and Abilene, and the same streams have served as avenues of exploration eastward and westward for both men and birds.

Although a great tableland in the center has given Kansas a reputation for unrelieved flatness—interminable distances with only an occasional grain elevator to break the horizon—in fact most of the state is gently rolling.

Originally, except for the extreme east, Kansas was almost entirely prairie, tall grass (bluestems, switchgrass) in the east, through mixed prairie, to short grass (buffalo grass, grama) in the west. This prairie has largely given way to agriculture and extensive tree-planting.

Extending from eastern farmland and forest to western wheat land and range, Kansas predictably supports a large avifauna consisting of a comparatively few very common species and many uncommon, local, or rare ones—because of the great extent of a few simple habitats and the patchy nature of others. More than 410 species have been reliably recorded and more than 175 known to breed. This avifauna is essentially eastern in eastern Kansas, grading to dilute western in the west—especially the southwest. Eastern and western representatives of various barely, or not quite, differentiated species overlap—perhaps as a result of recent disturbance—and hybridize to varying degrees. Scattered outliers from both east and west breed in pockets beyond their continuous ranges.

Most of the several recognized physiographic and vegetational subdivisions of the state are not now conspicuously bounded and the latter have been obscured by agricultural disturbance. The changes in birdlife are therefore for the most part subtle and more quantitative than qualitative.

Eastern Kansas was originally a mosaic of forest and prairie, the extent of forest being debated by ecologists. Today the mosaic is of forest and farmland, with some prairie remnants. Forest is best represented along river valleys in those bottomlands spared from the plow and on the rougher north-facing slopes above the floodplains. Upland forests consist of oaks, hickories, elm, hackberry, ash, red cedar, buckeye, basswood, black cherry, redbud, and others. Cottonwood, sycamore, and willows predominate in lowland, riparian forests. Upland forest is nowhere so tall or floristically complex as in the Appalachian and Ozarkian areas to the east.

Diversified farms of a few hundred acres or less are devoted to woodlots, pasture, and corn, wheat, alfalfa, soybeans, and milo.

The eastern flavor of the avifauna derives from various typical—or exclusively—deciduous-forest species, and from a few of more open habitats. Listed below are representative and regularly breeding species. Those marked with an asterisk continue somewhat farther west than the others, which become rare or drop out in central Kansas.

Red-shouldered Hawk (*local, mainly southeastern*)
Broad-winged Hawk (*local*)
* Greater Prairie Chicken (*local*)
* Upland Sandpiper
Barred Owl
* Chuck-will's-widow
Whip-poor-will
Pileated Woodpecker (*rare, increasing southward*)
Acadian Flycatcher (*southeast*)
* Eastern Pewee
* Carolina Chickadee (*southern Kansas*)
Tufted Titmouse
Carolina Wren
Wood Thrush
White-eyed Vireo
Yellow-throated Vireo
Northern Parula Warbler
Cerulean Warbler (*local*)
Ovenbird (*local, northeast*)
Louisiana Waterthrush
Kentucky Warbler
* Eastern Meadowlark
* Baltimore Oriole (*hybridizes with Bullock's in west*)
Scarlet Tanager (*local, mainly northeast*)
Rose-breasted Grosbeak
* Indigo Bunting
Rufous-sided Towhee (black-backed form)

West of the first two tiers of eastern counties (50 to 60 miles), forest becomes restricted to watercourses, sheltered and favorable sites in ravines, and north-facing slopes.

This is particularly true of the Flint Hills, a rolling upland 50 to 75 miles wide extending nearly from the northern to the southern border. A flinty chert poor for crop culture is responsible for the preservation of a large area of rich prairie. Very green in spring, purplish red in autumn, these hills afford wide vistas and are among the richest grazing lands in the world. They harbor populations of Greater Prairie Chickens (uncommon, local, or absent elsewhere) and Upland Sandpipers, which are probably now as numerous as ever in historic times. Rough-legged Hawks are numerous in winter, greatly outnumbering the Red-tails of more wooded areas, and Golden Eagles are regular at this season. Sprague's Pipits, Smith's Longspurs, and many Chestnut-collared Longspurs are transients.

The Flint Hills are traversed east-west by I 70 west of Topeka, by the Kansas Turnpike from Topeka to Eldorado, and by I 35 from Ottawa to Emporia. Another fine drive is provided by State 177 from Manhattan south to Eldorado. In summer, along any of the lesser Flint Hills roads, Upland Sandpipers may often be seen standing on fence posts and utility poles. Listen also for their penetrating 'wolf whistle' song, usually delivered aloft, April through June.

Prairie chickens are not easily seen. One must approach ranchers in the better areas (e.g., near Manhattan, Council Grove, Emporia, Florence) and ask for information and permission. Look for areas with hilly prominences, wide expanses of bluestem, and scattered plots of food crops. The birds often fly overland early and late in the day and are then most readily seen.

West of the Flint Hills the terrain is variously rolling or comparatively flat and the country gradually assumes a western aspect. Increasingly numerous are large wheat farms and cattle ranches. Woody vegetation is scattered, although more abundant than originally. Plantings are found about towns (elm, evergreens, cottonwood, exotics), ranch houses (juniper, Norway spruce, elms, Russian olive), and in many shelterbelts (largely Osage orange) planted in the 'dust bowl' of the 1930s. Many areas have sweeping views across broad valleys and eroded upland where sage is prominent. On these dry, windy plains the nights are cool even in the hottest summer.

A few points distinguish southern Kansas. Here the Greater Roadrunner is found in low density statewide, and the Scissor-tailed Flycatcher, local northward, is a conspicuous roadside species. The Painted Bunting, local northeast to Lawrence, is common in brushy situations and the Blue Grosbeak more numerous than elsewhere. Most notably, from near Wellington, Sumner County, west, the Mississippi Kite is numerous and readily seen in summer, with outposts north to Barton and Ellis Counties. These kites nest semicolonially in shelterbelts, city parks, and riparian forest, feeding daily over the prairies, and are particularly numerous along the Medicine Lodge River, Barber County, and about Meade County State Park.

From Medicine Lodge almost to Coldwater, Comanche County, and south to the Oklahoma line, the Cedar Hills Prairie occupies an area of rough 'breaks' in bright red shales, eroded into minor canyons and buttes with many cacti, dotted with juniper and sage, and decidedly of southwestern aspect. Here, in 1885, N. S. Goss found Black-capped Vireos breeding, and ornithologists have sought them in vain ever since. These areas are crossed by US 160.

The High Plains of western Kansas, north or south, from roughly Hays, Ness City, Dodge City, and Meade west, are typically western plains. Straggling Prairie Falcons are present at all seasons. Golden Eagles nest at a few remote points. In the southwest, White-necked Ravens are locally common, building their large nests of barbed wire and other exotic materials. They should be watched for on utility lines near the Colorado border. The following western species breed (those marked with an asterisk also nest farther east, locally in central Kansas, some of them rarely eastward):

Ferruginous Hawk (*local*)
Lesser Prairie Chicken (*southwest*)
Scaled Quail (*southwest*)
Mountain Plover (*rare, local*)
Long-billed Curlew (*rare*)
* Burrowing Owl
* Poor-will
Ladder-backed Woodpecker (*southwest*)
* Say's Phoebe
* Black-billed Magpie
White-necked Raven
* Rock Wren

Sage Thrasher (*rare, local*)
Bullock's Oriole (*hybridizes with Baltimore*)
Black-headed Grosbeak
Lazuli Bunting
* Lark Bunting
Cassin's Sparrow (*southwest*)

The following species are among those which breed throughout Kansas:

OPEN AREAS

Swainson's Hawk (*local in east*)
Northern Harrier (*local in east*)
Mourning Dove
Eastern Kingbird
Western Kingbird (*local in east*)
Scissor-tailed Flycatcher (*rare in north*)
Horned Lark
Barn Swallow
Northern Mockingbird
Loggerhead Shrike
Bell's Vireo (*local in west*)
Common Yellowthroat
Western Meadowlark (*local in east*)
Blue Grosbeak (*local in north*)
Dickcissel (*local in west*)
Grasshopper Sparrow

WOODED AREAS

Common Bobwhite
Yellow-billed Cuckoo
Yellow-shafted Flicker
Red-bellied Woodpecker
Red-headed Woodpecker
Hairy Woodpecker
Downy Woodpecker
Great Crested Flycatcher (*rare in west*)
Eastern Phoebe (*rare in west*)
Blue Jay
White-breasted Nuthatch (*rare in west*)
House Wren
Bewick's Wren (*rare in north*)
Gray Catbird
Brown Thrasher
Eastern Bluebird
Blue-gray Gnatcatcher (*rare in west*)
Red-eyed Vireo
Warbling Vireo
Common Grackle
Orchard Oriole (*rare in west*)
Northern Cardinal (*rare in west*)
American Goldfinch (*rare in west*)
Field Sparrow (*local in west*)

Kansas lies athwart the diffuse central flyway. Heavy flights of shorebirds stop at a few favored areas, with good flights of waterfowl, especially geese. In 1950 a canoe was a novelty; now one may buy a yacht in several cities, attesting the proliferation of reservoirs, 20 or more large ones (*see under* **Lawrence, Manhattan**). With many smaller ones, these have greatly augmented the resting areas for transient waterbirds (*see also under* **Great Bend, Pleasanton**). Loose flocks of Franklin's Gulls wheeling overhead or feeding in fields are common in April and early May and in September and October. The migration of passerines is conspicuous in spring, diffuse in autumn. Parulids are scarce by eastern standards but the Tennessee, Orange-crowned, and Nashville Warblers are very numerous. In central and western Kansas where wooded habitats are scattered, concentrations of kinglets, thrushes, vireos, parulids, and fringillids can be amazing. The Clay-colored Sparrow is a common transient, and the mysterious Baird's Sparrow appears occasionally among the casualties at western television towers.

Main migration flights may be expected within the following dates:

Waterfowl: 1 March–10 April; 15 October–10 December
Shorebirds: 25 April–25 May; 5 August–10 October
Landbirds: 10 April–10 May; 5 September–1 November

Winter bird populations are fairly rich, especially in mild winters; Christmas counts commonly range from 50 to 70 species at eastern localities, 35 to 50 westward. Lapland Longspurs appear in great flocks on open lands, along with other longspurs and Horned Larks. Common winter birds include the Brown Creeper, Golden-crowned Kinglet, Slate-colored (including Oregon) Junco, American Tree Sparrow, White-crowned Sparrow, and Song Sparrow. Much of the winter range of the Harris' Sparrow, one of the more numerous winter birds, falls within Kansas. In irruption years many Red Crossbills frequent the conifers in Kansas towns. Other northern and western birds irregular in winter are: Bohemian Waxwing, Townsend's Solitaire (rare), occasional Scrub and Pinyon Jays; and Clark's Nutcrackers. The Mountain Bluebird is regular in the west, ranging eastward rarely as far as Lawrence. The same is true of the Red-shafted Flicker and hybrids.

Authorities

Ivan L. Boyd, Jean W. Graber, Richard R. Graber, Eugene R. Lewis, Edmund F. Martinez, Sebastian T. Patti, Harrison B Tordoff, John L. Zimmerman.

Reference

A Directory to the Birds of Kansas. By Richard F. Johnston. University of Kansas, Museum of Natural History (Lawrence, KS 66045), Miscellaneous Publication No. 41, pp. 1–67, map. 1965.

ARKANSAS CITY
Chaplin Nature Center

Arkansas City is about 50 miles from Wichita and can be reached by US 77. The **Chaplin Nature Center,** run by the Wichita Audubon Society, is on the Arkansas River and preserves rich stands of mature cottonwoods and some oaks. The Center is reached by going west from Arkansas City on US 166 and then 2 miles north. It is developing extensive nature trails through woodlands, prairie, and streamside. More than 200 species of birds have been recorded. Pileated Woodpeckers nest on the reservation.

ELKHART
Cimarron National Grassland | Point-of-Rocks | Cimarron River

Elkhart, Morton County, in extreme southwestern Kansas is of interest because of its large, nearby tracts of sand-sage grassland or short-grass prairie. The **Cimarron National Grassland,** administered by the U.S. Forest Service, encompasses all of Morton County. The Grassland has been restored since the 1930s, when much of the prairie was lost to the wind through mismanagement. Much of this important community has also been lost to agricultural uses. Access to the present Grassland is unrestricted. Forest Service personnel are helpful and knowledgeable about local birdlife. Excellent maps are available at the Service headquarters in Elkhart.

Of primary interest is the Lesser Prairie Chicken, and Morton County is one of the better places anywhere to see this local and diminished species. The chickens are south of the Cimarron River, which bisects the County. The Forest Service sets up blinds on booming grounds from March until May and personnel are helpful in assisting visitors to find these birds. Other species typical of the sand-sage grassland include: Rough-legged and Ferruginous Hawks, Prairie Falcon, McCown's and Chestnut-collared Longspurs (winter); Long-billed Curlew, Burrowing Owl, Lark Bunting, and Cassin's Sparrow (summer); White-necked Raven (mainly summer); and Scaled Quail (especially about abandoned dwellings).

The rocky slopes of the Cimarron River known as the **Point-of-Rocks,** 7 miles north and 2 miles west of Elkhart by State 27 and 51, provide one of the few areas in Kansas occupied by the Brown Towhee and Rufous-crowned Sparrow. Rock Wrens are present in summer.

The cottonwood-willow-tamarisk floodplain of the **Cimarron River** provides habitat for such common breeding species as the Common Bobwhite, Wild Turkey, Mississippi Kite, Poor-will, Black-billed Magpie, Warbling Vireo, Yellow Warbler, and Bullock's (probably with some Baltimore admixture) and Orchard Orioles. Ladder-backed Woodpeckers and Ash-throated Flycatchers are also present. The floodplain is bordered by intermittent bluffs, in places forming shallow canyons. Drive about 8 miles north of Elkhart on State 27, or north for a comparable distance from Rolla on State 51 or from Hugoton (Stevens County) on US 270. One may walk from any of these roads along the riverbed, April through June providing the best bird finding.

The man-made impoundments 8 miles north and 5 miles east of Elkhart on State 27 and 51 should be checked for migrating waterfowl and marsh birds in spring and fall. Providing the only water for miles, these ponds have an oasis effect.

The Soil Service headquarters, 2.5 miles north of Elkhart on State 27, has planted conifers and provides good shelter for birds in winter.

Unexpected species often appear in Morton County because of its geographic location, and the only Kansas records of several western species are from this locality, where numerous others were first recorded. The area has the most complete blend in Kansas of eastern and western avifaunas.

EMPORIA
Flint Hills National Wildlife Refuge

For good all-round bird finding, especially in spring and fall, go to the **Flint Hills National Wildlife Refuge** (18,500 acres, established 1966) in southeastern Kansas. The Refuge is convenient to I 35 and the Kansas Turnpike, which meet in Emporia. The Refuge lies within the flood-pool of the John Redmond Reservoir in the Neosho River Valley. In migration and winter the Refuge has large populations of migratory aquatic birds including many ducks and geese. Passerine populations are also varied and extensive. Information and permission should be obtained at Refuge headquarters at the west edge of Hartford.

Pileated Woodpeckers reside in the timber near the west end of the Reservoir and Greater Prairie Chickens frequent the prairie on the north side of the Neosho River between I 35 and Hartford.

Drive 10 miles east from Emporia on I 35 and turn south by county road to Neosho Rapids (about 3 miles), continuing southeast and south on county roads about 5 miles to Hartford.

GREAT BEND
Cheyenne Bottoms State Wildlife Management Area | Quivira National Wildlife Refuge

The finest site in Kansas for aquatic birds and, at the peak of migration, probably the best over-all, is **Cheyenne Bottoms State Wildlife Management Area,** approximately 7 miles northeast of Great Bend in central Kansas. This is one of the premier bird-finding spots in the midwest, routinely visited by bird finders from many states. In late April or early May it is not difficult to record 125 or more species daily, more than 50 of them waterbirds, from an Area list exceeding 320.

Originally Cheyenne Bottoms was a great marsh and wet

meadow occupying a natural depression of some 40,000 acres. The Area was developed in the early 1950s (dedication 1958) as a resting and feeding area for migrant waterfowl with controlled public hunting. Water was diverted from the great bend of the Arkansas River to flood five immense, shallow pools (less than 2 feet deep) separated by dikes which support good sand roads. The usual water area is about 12,500 acres. A large, central pool of 3,300 acres is a permanent refuge, surrounded by four additional large pools, three of which are dotted with grids of 167 permanent concrete duck blinds each. Out of hunting season these provide resting places for cormorants, pelicans, eagles, and other birds. Here and there are areas of marsh growth—spartina grass, cattails, and bulrushes—and peripherally there are great expanses of wet meadow, sandy plain, and, to the northwest, swampy woods. Flooded ditches favored by American Bitterns parallel the roads leading into the Bottoms, and shallow ponds are scattered in nearby pastures in and out of state ownership.

Breeding waterfowl include the Canada Goose, Mallard, Gadwall, Common Pintail, Blue-winged Teal, Northern Shoveler, Wood Duck, Redhead, Canvasback, and Ruddy Duck. Other nesting birds are the Pied-billed Grebe, Double-crested Cormorant, Little Blue Heron, Cattle Egret (first nested in 1973), Snowy Egret, Least and American Bitterns, King and Virginia Rails, Sora, Common Gallinule, American Coot, Snowy Plover, American Avocet, Wilson's Phalarope, and Forster's, Little, and Black Terns. Since 1965, steadily increasing and breeding regularly have been White-faced Ibises, commonly seen in flocks of five to twenty in wet, extensive grassy areas well back from the pools.

Yellow-headed Blackbirds are conspicuous near the dikes. Increasing since the late 1960s, and now conspicuous, are many of the Great-tailed Grackles that have been breeding in south-central Kansas in recent years.

During migration, many additional waterfowl appear. Greater White-fronted and Snow Geese are regular. American White Pelicans number up to 15,000 in migration, with non-breeders summering. Sandhill and occasionally Whooping Cranes pass through in March and October, with large concentrations of Canada Geese in late fall and winter, when many Ferruginous Hawks and Prairie and Peregrine Falcons are seen.

COMMON PINTAIL

The edges of the large pools provide extensive mud flats for all kinds of shorebirds, supplemented by wet meadows, and the Bottoms list is virtually that of central North America, including rare vagrants. Fall migration is even more notable than that of spring. The Hudsonian Godwit is common, the Marbled much less numerous, although regular. Staples are the thousands of Long-billed Dowitchers, Wilson's Phalaropes, American Avocets, Semipalmated Sandpipers (with a few Westerns), and Baird's Sandpipers (mainly spring). Black-necked Stilts are rare but regular.

In migration periods the woods, particularly a roadside park 1.0 mile west of headquarters, can provide fine concentrations of landbirds.

No permission is required to search for birds on the refuge. Advice should be sought, however, as to areas closed during hunting season. Refuge headquarters is open normally Monday through Friday, but may occasionally be closed due to work requirements. Maps and an authoritative list of species recorded and their status are available upon request. Camping is permitted free at the nearby roadside park (picnic tables and rest buildings only; no water).

The refuge headquarters may be reached in several ways. (1) From Great Bend, at the intersection of US 56 and 281, drive north 5 miles on US 281, thence 2 miles east by marked gravel road; or drive east from Great Bend for 3 miles on US 56, thence northeast about 6 miles on US 156 to a state roadside park. Just short of the park turn left and continue west by gravel road for 8 miles through the Bottoms. (2) Eastbound on I 70, turn south on US 281 at Russell and drive about 30 miles to Hoisington (which, May to August, has a few Mississippi Kites; others are in the Great Bend parks). From Hoisington, continue south 4 miles and turn east 2 miles (as above, from Great Bend). (3) Westbound on I 70, past the outskirts of Ellsworth, turn southwest onto US 156 and proceed about 35 miles, past the Claflin turnoff, to the same state roadside park mentioned above, thence west. Watch for the large colony of Great Blue Herons nesting in the tall cottonwoods along a creek just south of the road for the first few miles after leaving I 70. This is the land of the 'postrock,' where the observant visitor will note many remaining limestone fence posts cut by early settlers as a substitute for scarce wood.

Alternatively, the visitor coming from I 70 may proceed east from Hoisington or west from Claflin on State 4, in each case about 6 miles to the store at Redwing. A magnificent view of the Bottoms is afforded from a prominence a few miles west of Claflin on this route. From Redwing, turn south on a gravel road into the Bottoms and take two successive right turns to refuge headquarters, about 8 miles in all. Each of these accesses provides abundant bird finding and all should be traversed on a thorough visit.

The major gravel roads into and about the central pool are good all-weather roads. Back roads may be difficult or impassable in wet weather, even for vehicles with four-wheel drive.

The massasauga (pygmy) rattlesnake is common. These little snakes frequent the roadsides and dry, thick, grassy areas just back from the water; reasonable caution is prudent.

Burrowing Owls are regular near Great Bend and usually a colony or two is somewhere nearby in association with a prairie dog 'town.' Inquire at refuge headquarters; local dog towns are ephemeral, owing to intermittent poisoning by land owners.

During invasion winters, Snowy Owls find the barren white wastes of Cheyenne Bottoms congenial, with 7 to 10 present on occasion. Bald Eagles are regular in winter as long as water is open.

In nearby Stafford County is 8,760-acre **Quivira National Wildlife Refuge** (established 1955). More diversified than Cheyenne Bottoms, this migratory waterfowl Refuge contains rangeland, farmland, sandhills, and two salt marshes near its northern and southern boundaries. Snowy Plovers are very numerous and the area has the largest breeding colony of Little Terns in Kansas. Black-necked Stilts nested in 1979. The over-all representation of waterfowl and shorebirds is much the same as that of Cheyenne Bottoms (*which see*) but numbers average lower. Much of the Refuge is inaccessible and often off limits to disturbance of any kind. Permission to enter must be obtained.

From Great Bend, drive east on US 56 to Ellinwood, thence south on blacktop county road approximately 20 miles, passing the Hudson turnoff. Turn east about 6 miles north of Stafford and drive about 5 miles to Refuge headquarters.

HAYS

The region around Hays in west-central Kansas, about a 75-minute drive from Cheyenne Bottoms and Great Bend, is representative of the mixed-grass prairie of central Kansas. Most of the area is rangeland or cultivated (food cereals) but plantings, ponds, and three streams provide enough local habitats to allow a list of more than 270 species within 15 miles of Hays. Bird finding is usually unremarkable, often hampered by high winds, and generally best in the early morning.

The three best spots close by are at the southwest quarter of Hays. From I 70, take Exit 157 and proceed south on Business I 70 (Alternate US 183) following conspicuous highway signs.

To visit Fort Hays State University, drive 2.3 miles to the FHSU exit and proceed about 0.8 mile to the northwest edge of the campus. At the Biology office in Albertson Hall (southwest corner of campus, observatory on roof) one can usually get information on local bird finding. The campus and adjacent areas of town are especially good for birds during migration.

To visit the Hays parks, follow Alternate US 183 from I 70 for 3.5 miles to the stop light at State 274. Turn left, park, and walk Big Creek both directions from Main Street.

To visit the Fort Hays Experiment Station, turn right onto State 274 and proceed less than 0.1 mile, then turn left onto the Station grounds, which are very well landscaped and especially good for wintering birds and transients. Some of the breeding species in these areas are the Yellow-shafted Flicker, Eastern and Western Kingbirds, Blue Jay, Northern Mockingbird, Gray Catbird, Brown Thrasher, Warbling Vireo, Orchard and Baltimore Orioles, and Black-headed Grosbeak.

Prairie birds, such as Lark Buntings and Lark Sparrows, can be observed along any county roads in the area. Poor-wills reside about rocky slopes. Among local specialties are the Mississippi Kite (one pair nests regularly in the Lincoln Draw area of North Hays) and Rock Wren (outcrops near Saline River). To reach the latter locality, follow US 183 north from Hays. About 10.5 miles north of the I 70 overpass, turn left on to a chalk road which roughly parallels US 183. Drive almost 2 miles and stop just before the road drops into the Saline River Valley. Park on the left near some oil tanks and investigate the nearby ravines.

Numerous small ponds provide excellent bird finding during migration. Most of these are near the Smoky Hill River and are on private land, so one should obtain permission before bird finding. One of the best is on the Harold Kraus farm, 2 miles west and 3 miles south of Antonino. (Antonino is about 5 miles south and 3 miles west of Hays.)

LAWRENCE

Perry Reservoir | Clinton Reservoir | Lakeview | Kaw Valley Fish Farms | University of Kansas Natural History Reservation | Baldwin Woods | Martin Park

This university town is about 35 miles west of Kansas City, Kansas, on the Kansas Turnpike (toll section of I 70) and US 24 and 40. Several good bird finding areas are nearby.

Perry Reservoir, approximately 12,000 acres, is reached (1) by traveling about 14 miles west on US 24 from the East Lawrence Exchange of the Kansas Turnpike to Perry and north 3 miles by county road to the dam (road marked) and Corps of Engineers headquarters, or (2) by driving west from Lawrence on US 40 (2 miles south of the West Lawrence Exchange of the Kansas Turnpike) to the Lecompton-Perry Road (County 1029), thence north to Perry and proceed as before. Good roads around the Reservoir lead to picnic and campgrounds, boat landings, etc., and afford excellent views. Woodland and edge birds are numerous. Information on trails is available at headquarters. The lake sometimes attracts many waterfowl in migration, especially when cold weather suppresses boating. Often seen are Canada, Greater White-fronted, and Snow Geese, with some American White Pelicans, a few Double-crested Cormorants, and occasional Western Grebes. Herring and Ring-billed Gulls are regular; Franklin's and Bonaparte's less frequent. Common ducks include Mallards, Lesser Scaups, Common Goldeneyes, Buffleheads, Common Mergansers, and sometimes others. When waterfowl are numerous, a few Bald Eagles are usually present.

Clinton Reservoir, 7,000 acres, scheduled for completion of facilities about 1981 (filling 1978–80), will be similar and promises to be even better for waterbirds. Travel west on Sixth Street (US 40) in Lawrence to the overpass of the West Lawrence Turnpike Interchange (Iowa Street) or to this point from that exchange, thence west for 4 miles on US 40 and south for 2 miles to Clinton Dam on a marked road (or west from the intersection of 23rd and Iowa Streets, Lawrence, on the Clinton Parkway). There will be roads all around the lake when completed.

Lakeview, an old oxbow of the Kansas River (locally the Kaw), is good for shorebirds and ducks in April and May when the water level is right. From Lawrence, drive west on US 40 (Sixth Street) to Kasold Drive and north on Kasold, which becomes County 1041. Follow this about 4 miles to Lakeview. Becoming good gravel, this road continues northwest, unmarked, on the south side of the River to Lecompton, where it joins County 1029 (*see above*) as the second road south of the Lecompton Bridge. It traverses forest where Chuck-will's-widows and Whip-poor-wills are common in summer. In late winter a few Bald Eagles are usually perched in large cottonwoods along the River.

The **Kaw Valley Fish Farms,** a series of large shallow ponds, often attract many ducks and shorebirds, including occasional rarities. April (ducks) and May and September (shorebirds) are best. Drive 0.5 mile east on US 24 and 40 from their intersection just north of the East Lawrence Exit of the Kansas Turnpike, turn left at the Lawrence Airport, and continue north for 1.0 mile to the ponds.

Just to the east of the fish ponds is the **University of Kansas Natural History Reservation,** approximately 600 acres of forest undisturbed since 1947, and the adjacent Robinson and Nelson experimental ecology tracts. These have a variety of breeding birds including the Yellow-billed Cuckoo, Great Crested Flycatcher, Eastern Pewee, Black-capped Chickadee, Tufted Titmouse, White-breasted Nuthatch, Gray Catbird, Brown Thrasher, Blue-gray Gnatcatcher, Wood Thrush, White-eyed, Bell's, Warbling, and Red-eyed Vireos, Kentucky Warblers, Scarlet and Summer Tanagers, Northern Cardinals, and Indigo Buntings. Outside the Reservation are grasslands and hedges frequented by Eastern and Western Kingbirds, Eastern Meadowlarks, and Grasshopper and Field Sparrows. Drive 1.5 miles east of the intersection of US 24 and 40 (*see just above*), then nearly 2 miles north on a gravel road to a sign that marks the entrance. Permission should be obtained from the Division of Biological Sciences in Snow Hall at the University of Kansas.

Similar permission should be obtained to visit the University's Breidenthal, Rice, and Wall research tracts in **Baldwin Woods,** in a forested valley 14 miles south-southeast of Lawrence. In Lawrence, go east on 23rd Street to Haskell Avenue, turn south, and follow County 1055 all the way. Here the finest relatively undisturbed forests in the area support summer-resident Whip-poor-wills, Chuck-will's-widows, Wood Thrushes, Yellow-throated Vireos, Northern Parula Warblers, Louisiana Waterthrushes, and Kentucky Warblers. Cerulean Warblers, Ovenbirds, Scarlet and Summer Tanagers, and Rose-breasted Grosbeaks are sometimes present. Nearby shrubby areas and fields may have Bewick's Wrens, Bell's Vireos, Dickcissels, Blue Grosbeaks (rare), and Painted Buntings (regular about many of the newer Lawrence suburbs west of Iowa Street and south of Sixth Street).

Just south of Baldwin Woods a well-marked road turns east 1.0 mile to Douglas County State Lake, which is good for waterfowl in season. Upland Sandpipers are sometimes seen in the open areas along this road.

Martin Park is 2 miles east of Lawrence, with an exceptionally fine stand of oak-hickory forest. Drive north on Iowa Street to Peterson Road, west for 1.7 miles, and north on a marked gravel road. The small Park is on the left a few hundred feet from this point and is excellent for Wood Thrushes, Louisiana Waterthrushes (along the creek just to the west), and Kentucky Warblers. Chuck-will's-widows, Whip-poor-wills, and Rose-breasted Grosbeaks are common as are various parulids and other passerines in migration.

MANHATTAN

Tuttle Creek Reservoir | Konza Prairie Research Natural Area | Agronomy Research Farm | Pottawatomie State Lake No. 2

Manhattan is in the northern Flint Hills, 8 miles north of I 70 on State 177, also on US 24. A variety of good bird-finding areas is nearby.

Tuttle Creek Reservoir averages 1.0 mile wide and stretches north more than 30 miles toward the Nebraska border. There are large concentrations of waterfowl on this body of water during migration and in winter, but the best year-round bird finding is in the River Pond State Park below Tuttle Creek Dam, 5 miles north of Manhattan on US 24 and State 177. Besides the typical forest-edge birds of eastern Kansas, Western Kingbirds, Scissor-tailed Flycatchers, and both meadowlarks breed within the park. Bank Swallows nest in the sandy bluffs across from the swimming beach, and in the small cattail marsh, behind several beaver dams along the western edges of the campground, Least Bitterns have nested. From November to early April, Bald Eagles frequent the entire Reservoir; several are usually in the large cottonwoods along the old river channel south of the state park.

Seven miles southwest of Manhattan is the headquarters of the **Konza Prairie Research Natural Area,** a 13-square-mile tract of largely virgin tall-grass prairie administered by the Division of Biology, Kansas State University. Prior arrangements to enter Konza Prairie Area must be made with the Division of Biology office in Ackert Hall on the K-State campus, but it is well worth the effort since not only the birds but also the vegetation exemplify those of the true prairie. Leave Manhattan southward on State 177 and immediately after crossing the Kansas River turn right onto McDowell Creek Road and continue for about 6 miles. Just after passing a side road on the right going to the Ashland community is the gate across the road into Konza Prairie. Present during the breeding season are Greater Prairie Chickens, Upland Sandpipers, Poorwills, and a few Henslow's Sparrows, with more abundant Eastern Meadowlarks, Dickcissels, and Grasshopper Sparrows. In the riparian woods along the main creek flowing down from the upland prairie are species typical of the deciduous forest coves of the Flint Hills: Yellow-billed Cuckoo, all resident woodpeckers, Great Crested Flycatcher, Eastern Pewee, Black-capped Chickadee, Tufted Titmouse, White-breasted Nuthatch, Carolina Wren, Blue-gray Gnatcatcher, Red-eyed and Warbling Vireos, Northern Parula Warbler, Louisiana Waterthrush, and Summer Tanager. In the brushy mixture of dogwoods, buckbrush, and juniper below the steeper outcrops of limestone, the Bewick's Wren, Bell's Vireo, Blue Grosbeak, and Lark Sparrow nest. While passerine numbers and diversity are very low in the winter on the prairie, raptors are numerous, including an occasional Prairie Falcon or Peregrine as well as small flocks of Short-eared Owls. Also in winter large concentrations of Greater Prairie Chickens feed in some of the cultivated plots near the entrance road and in fields along McDowell Creek Road.

Upon leaving Konza Prairie, it is often worthwhile to take the Ashland Road passed before, but continue on following the signs to the USDA Plant Materials Laboratory and **Agronomy Research Farm** close by the Kansas River. Birds characteristic of riparian forest reside in the riverbottom woods, and the orchards and mature windbreaks of the Research Farm offer good bird finding at all seasons except late summer.

During waterfowl migration, **Pottawatomie State Lake No. 2,** east of Manhattan, is excellent for ducks, particularly the divers. Although powerboats are restricted on the Lake, it is still necessary to arrive before the fishermen for the greatest variety. Even in the afternoon, however, there are still small groups of Redheads, Ring-necked Ducks, Canvasbacks, scaups, and Buffleheads in the middle of the Lake and several species of surface-feeding ducks along less accessible Lake margins. The brushy edges of the two streams feeding into the Lake from the north are particularly good for transient thrushes and sparrows in the spring and a variety of wintering species. Drive east from Manhattan on US 24. About 1.5 miles after crossing the Blue River, turn north onto a blacktop road that soon becomes gravel and continue for another 1.5 miles. At the top of the hill turn left and go 1.25 miles and turn right rather than dropping back down into the valley of the Blue River. The entrance to the park is about 1.0 mile from this junction.

OXFORD
Slate Creek Salt Marsh

Oxford lies on the west bank of the Arkansas River in southern Kansas, about 30 miles from Wichita, and can be reached by US 160. The Oxford Mill area is a major bird-finding site. Heading east, turn left on the east side of town just before crossing the bridge. Drive through the park and on for 1.0 mile. The Mill is situated on a millrace which attracts many passerines and some waterfowl. Another mile north a dam across the Arkansas River diverts water to the millrace. Bald Eagles winter here. Breeding birds include the Wild Turkey, Pileated Woodpecker, Summer Tanager, and Indigo and Painted Buntings.

Slate Creek Salt Marsh (1,000 acres) lies 8 miles south of Oxford on the county road and 0.5 mile west. The side road is unimproved and should not be traveled in bad weather. The road is passable through the Marsh only by four-wheel-drive vehicles. One may park near the highway and walk to the Marsh. This is privately owned but bird finding from the road into it is public and can be quite good. A gun club owns the large lake there and bird finders should not visit it during duck season. Birds in migration include many ducks and most of the shorebird species that appear in Kansas. King Rails breed in the Marsh.

PHILLIPSBURG
Kirwin National Wildlife Refuge

Visitors to Phillipsburg and east-west travelers on US 36 or US 24 in northwestern Kansas, or north-south transients on US 183, may enjoy good bird finding for waterfowl, shorebirds, and other species at 10,755-acre **Kirwin National Wildlife Refuge.** Embracing 5,000-acre Kirwin Reservoir and diverse farmlands and ranchlands, the Refuge was established in 1954 to provide resting and feeding areas for migrating American White Pelicans, Greater White-fronted Geese, Sandhill Cranes, and many ducks and shorebirds. March-May and September-November are best, but some waterbirds and waterfowl are present at all times. Up to 75,000 Mallards and 3,000 Canada Geese winter on the Refuge, where more than 200 species of birds have been recorded. The area is popular with fishermen and there are boating and camping facilities. Information and permission to look for birds should be sought at Refuge headquarters near Kirwin. Drive south 5 miles from Phillipsburg to Glade on US 183 and east about 10 miles to Kirwin on State 9, following signs to Refuge headquarters.

PLEASANTON
Marais des Cygnes State Wildlife Management Area

In extreme east-central Kansas, about 28 miles north of Fort Scott, or 5 miles north of Pleasanton, is the **Marais des Cygnes State Wildlife Management Area.** US 69 passes through the 6,646-acre Management Area. It is on the floodplain of the Marais des Cygnes River about 5 miles west of the Missouri line and roughly midway between the Nebraska and Oklahoma lines. Here the River opens into a broad floodplain between Ozark-type hills. The River is deep and narrow and its banks have built up from centuries of flooding as much as six feet above the level of the plain. The vegetation is largely oak-hickory. That of the bottomland is dense, with the greatest number of tree species found anywhere in Kansas. The hills bordering the River are forested with sugar maple and at least nine species of oaks, which color beautifully in early October.

With the oaks and other mast trees—pecan, shagbark, shellbark, and bitternut hickories, and black walnut—there is a rich population of gray and fox squirrels. Whitetail deer are also numerous. The valley was originally rich in fur-bearers and waterbirds, and the early French trappers called it 'marais des cynges,' marsh of the swans.

The Area is owned and managed by the Kansas Fish and Game Commission which completed acquisition in 1953 and began development in 1955. Numerous units from a few to 600 acres now provide 1,800 acres of marsh managed to provide rest and food for migratory waterfowl and regulated public hunting.

Spring and fall concentrations of waterfowl are impressive, sometimes reaching 150,000. Over 70,000 ducks normally winter. Thousands of Snow Geese, Common Pintails, and Blue-winged Teal stop during migration. Mallards, American Wigeons, Northern Shovelers, and Lesser Scaups are also common, as are herons, egrets, and numerous species of shorebirds. Wood Ducks nest in natural cavities and nest boxes. More than two hundred species of birds have been recorded.

Where US 69 crosses the southern boundary, turn east and south onto an unimproved trail and proceed for almost 0.5 mile, then turn east for another 0.5 mile to an old abandoned farmstead. Park here and walk northeast into an orchard of crabapple and haw trees to an excellent parulid site. (A 9 May visit revealed twenty-one species including the Black-and-white, Prothonotary, Golden-winged, Northern Parula, Yellow, Magnolia, Black-throated Green, Chestnut-sided, Bay-breasted, Blackpoll, Ovenbird, Northern and Louisiana Waterthrushes, Kentucky, Yellow-breasted Chat, Wilson's, and Canada.) Continue on north-northeast to the heavy, mature woods along Muddy Creek (Hagie Woods), where Red-shouldered Hawks and Pileated Woodpeckers reside as well as Eastern Pewees, chickadees, titmice, and nut-hatches.

Continuing north on US 69, cross the Marais des Cygnes River at the small town of Trading Post and turn west onto a gravel road and go past the manager's residence. When the road forks, at 0.25 mile, keep right and continue north along the east edge of Wood Duck Lake (Unit B), stopping often to look and listen. Prothonotary Warblers nest abundantly in the dead snags in the marsh. Kentucky Warblers nest along the riverbank. Wood Thrushes can be heard. Eastern Pewees and Summer Tanagers are abundant, Ruby-throated Hummingbirds are common, and Scarlet Tanagers often nest about 100 yards south of the open field west of the road. Pass the Area headquarters, where maps and current information may be obtained, before returning to the highway. Working days are Monday through Friday, but headquarters may be closed when personnel are afield.

Great Blue Herons nest in a colony almost at the north edges of the Management Area where Middle Creek flows across the eastern border. As it flows west farther south, Middle Creek re-enters the Area. This is a remote spot difficult to reach even on foot. Yellow-crowned Night Herons are regular, scattered nesters along the many creeks.

Gravel all-weather roads almost completely border Burr Oak Lake (Unit A), Flathead Lake (Unit G), the west edge of Wood Duck Lake (Unit B), and the south edge of Unit E, providing easy access to much excellent bird-finding habitat. Cormorants, herons, Bald Eagles, and Ospreys are often seen on the many dead snags in the larger marsh pools.

TOPEKA

Lake Shawnee | Gage Lake | Silver Lake | Shawnee State Fishing Lake

This capital city, just east of the Flint Hills in eastern Kansas, has several good bird-finding areas convenient to downtown. One of the better year-round areas is the 600-acre woodland park that includes 400-acre **Lake Shawnee.** Southeast of the city, the park is reached by driving south on Kansas Avenue and turning left onto 29th Street, which leads to the entrance. Circling the Lake is a drive with many pulloffs for observations.

The Lake is bordered by a few patches of cattails and, on higher ground, by fields, groves of oaks, and evergreens. Bird finding is best in April and early May and from early September through November. In migration, Common Loons (rarely), Eared Grebes,

Pied-billed Grebes, American White Pelicans, and Double-crested Cormorants appear at times along with other birds often including the Mallard, Common Pintail, Blue-winged Teal, Northern Shoveler, Redhead, Canvasback, Lesser Scaup, Common Merganser, Killdeer, Lesser Yellowlegs, Least Sandpiper, and Ring-billed and Franklin's Gulls. Regular are such landbirds as the Red-winged and Brewer's Blackbirds, American Tree Sparrow (in winter), Field Sparrow, and Harris' Sparrow (September to May).

Mallards usually winter, sometimes with Common Mergansers and American Coots. Great Blue, Green, and Little Blue Herons and Great Egrets are present at times in summer and early fall.

Gage Park (146 acres), west of the city on 10th Avenue, is a recreational area with open, grassy spots, many oak trees, and clusters of ornamental shrubs. Parulids and other small passerines are numerous in early May. The area is good in winter for Yellow-shafted (and occasional Red-shafted) Flickers, Red-bellied Woodpeckers, Black-capped Chickadees, Tufted Titmice, Red-breasted Nuthatches, Golden-crowned Kinglets, Cedar Waxwings, Northern Cardinals, Purple Finches, and Pine Siskins.

Silver Lake, 12 miles west of the city on US 24, is an old oxbow of the Kansas River. Its 180 acres are shallow and marked with many dead willow stumps and emergent aquatic plants which attract many breeding Red-winged Blackbirds. Prothonotary Warblers nest in the large willows, cottonwoods, and elms along the shore when suitable cavities are available. From late April through May, and from late July through September, Silver Lake is one of the most productive spots in the Topeka area for shorebirds, provided mud flats are exposed.

Shawnee State Fishing Lake is approximately 6 miles north and 4 miles west of the city limits. Drive north on US 75 and turn left onto a marked county road. This more than 100-acre body of water is good in migration for waterfowl. Poor-wills, Horned Larks, and, irregularly, Henslow's Sparrows nest on the adjacent prairie tracts where Greater Prairie Chickens are often seen.

For Perry Lake to the northeast, *see under* **Lawrence.**

WICHITA

Oak Park | Sim Park | Wichita-Valley Center Floodway

The award-winning city parks of Wichita are a haven for woodland birds as the city is surrounded by open farmland. The Arkansas and Little Arkansas Rivers join within the city limits, and several nearby bodies of water attract many waterbirds and waterfowl. Cheney Reservoir, approximately 10 miles west on US 54 (thence north to the Reservoir) is attractive to wintering waterfowl. Many sand pits on the western edge of the city attract waterfowl and many nesting passerine birds. Great-tailed Grackles nest in various parts of the city; the best area is Midcontinent Airport just off US 54 on the western edge of Wichita.

Oak Park (about 40 acres) occupies a point of land formed by a bend in the Little Arkansas River and may be reached by going north on Broadway and turning left onto 11th Street. Arriving at the Little Arkansas River, turn right just before crossing the bridge. Park in a designated area and walk. The eastern half of the Park has a rather dense woods of oak, black walnut, and cottonwood. The western half is more open. Bird finding is best in April and May when many parulids are in migration. The following species are numerous or regular. *Year round:* Common Screech Owl, Red-bellied Woodpecker, White-breasted Nuthatch, Northern Mockingbird, Brown Thrasher (rare in winter), Cedar Waxwing, Red-winged Blackbird, and Rufous-sided Towhee (spotted form in winter). *Migration and winter:* Mallard, Slate-colored (including Oregon) Junco, Harris', White-crowned, White-throated, Lincoln's, and Song Sparrows. *Migration and summer:* Green and Black-crowned Night Herons, Mississippi Kite, Eastern and Western Kingbirds, Blue-gray Gnatcatcher, Orchard and Baltimore Orioles, Rose-breasted Grosbeak. *On migration:* Chipping and Clay-colored Sparrows.

Sim Park (183 acres), on the east bank of the Arkansas River, is reached by continuing on 11th Street 1.0 mile west of Oak Park. This is open prairie with some thickets on sand dunes by the River. Conifers attract northern finches in winter. Several birds here are not found in Oak Park. Throughout the year there are Common Bobwhites and Eastern and Western Meadowlarks.

Transients and breeding species include the Great Blue Heron, Mississippi Kite, Yellow-billed Cuckoo, Yellow-shafted Flicker, Scissor-tailed Flycatcher, Brown Thrasher, Bell's Vireo, and Solitary Vireo (transient only).

The **Wichita-Valley Center Floodway** on the west side of Wichita is a diversion ditch for the Arkansas River in flood. It generally has water in it and is a good place to find waterbirds. Shorebirds appear in migration and a few may winter.

For other bird-finding sites within an hour's travel from Wichita, *see under* **Arkansas City** *and* **Oxford.**

Louisiana

FULVOUS WHISTLING DUCKS

The great John James Audubon once declared Louisiana his 'favorite portion of the Union.' In its vast coastal marshes and deep woods, he found a richness of birdlife that inspired him to draw birds in greater number and excellence. Today when the bird finder explores the coastal marshes, areas that ornithologists recognize as having some of the largest bird populations in the United States, and works along the forest byways of West Feliciana Parish, he will fully understand why Louisiana meant so much to Audubon.

Louisiana lies entirely on the Coastal Plain. Lower Louisiana is a strip of lush coastal marshes, 10 to 40 miles wide, lying along the Gulf of Mexico. Upper Louisiana is an upland area of low rolling hills and bottomlands. The highest point is Driskill Mountain at

535 feet, near Liberty Hill in northwestern Louisiana. Bordering the Mississippi River, which cuts through the state, is a broad bottomland, averaging 50 miles in width, which roughly divides the upland area into the Uplands of the Florida Parishes, north of Lake Pontchartrain and east of the Mississippi, and the West Louisiana Uplands, west of the Mississippi and north of the coastal marshes. Apart from the Mississippi, the most important watercourses are the Red River, flowing into the Mississippi from the northwest, the Atchafalaya River running south from the juncture of the Mississippi and Red Rivers, and the Sabine River, forming much of the western boundary and emptying into the Gulf.

Both the Uplands of the Florida Parishes and the West Louisiana Uplands have, along their southern extremities, extensive flats forested with longleaf and slash pines, often mixed with stands of hardwoods, principally black and post oaks and mockernut hickory. Characteristic nesting birds are Red-cockaded Woodpeckers, Brown-headed Nuthatches, Pine Warblers, Prairie Warblers, and Bachman's Sparrows. The moist alluvial bottomlands of the Mississippi and of the other large rivers have remnants of the once dense forests. In lower or backwater areas where the soil is poorly drained, overcup oak, water locust, cypress, and tupelo grow commonly, but on higher, well-drained areas the common trees are sweetgum, green ash, swamp red oak, willow oak, water oak, elm, and pecan. Some of the birds breeding regularly in forests throughout the state are listed below.

Mississippi Kite
Red-shouldered Hawk
Broad-winged Hawk
Yellow-billed Cuckoo
Common Screech Owl
Barred Owl
Chuck-will's-widow (*mainly upland forests*)
Yellow-shafted Flicker
Pileated Woodpecker
Red-bellied Woodpecker
Hairy Woodpecker
Downy Woodpecker
Great Crested Flycatcher
Acadian Flycatcher
Eastern Pewee
Blue Jay
Fish Crow
Carolina Chickadee
Tufted Titmouse
White-breasted Nuthatch (*mainly upland forests*)
Wood Thrush
Blue-gray Gnatcatcher

Yellow-throated Vireo
Red-eyed Vireo
Warbling Vireo
Prothonotary Warbler
Swainson's Warbler
Northern Parula Warbler
Yellow-throated Warbler
Louisiana Waterthrush
Kentucky Warbler
Hooded Warbler
American Redstart
Baltimore Oriole
Summer Tanager

Upper Louisiana was originally forested in its entirety, except near the western boundary where the surface supported prairie vegetation, but today more than half of upper Louisiana is occupied by settlements and farmlands (fields, brushy areas, thickets, orchards, and dooryards), where the following birds breed regularly:

Common Bobwhite
Mourning Dove
Red-headed Woodpecker
Eastern Kingbird
Barn Swallow
Carolina Wren
Northern Mockingbird
Gray Catbird (*mainly northern Louisiana*)
Brown Thrasher
Eastern Bluebird
Loggerhead Shrike
White-eyed Vireo
Common Yellowthroat
Yellow-breasted Chat
Eastern Meadowlark
Orchard Oriole
Common Grackle
Northern Cardinal
Blue Grosbeak
Indigo Bunting
Painted Bunting
Dickcissel
Rufous-sided Towhee
Chipping Sparrow
Field Sparrow

For the bird finder the outstanding attractions of upper Louisiana are the beautiful Feliciana Parishes north of Baton Rouge. A typical spot, once a favorite of Audubon, is St. Francisville, on the Mississippi. Near this quiet community there are unfrequented roads, small fields, and thicketed ravines, together with several ante-bellum plantations noted for their deep woods and dignified houses shaded by moss-hung trees and surrounded by lawns with flowering shrubbery and lovely gardens. In such a setting, where habitats are many and varied, birdlife is unusually rich.

The coastal marshes of lower Louisiana are a flat plain of rushes, cord grasses, bulrushes, and quillreeds interrupted by a maze of shallow lakes, ponds, bayous, bays, and man-made canals. Scrubby trees, such as willows and hackberries, find foothold only where there are small ridges of sand and shells. Near the coast the water is generally salt or brackish; farther inland it gradually becomes fresh. Except where the Mississippi Delta spreads out into the Gulf, the seaward side of the marshes is protected by a barrier of sand beaches that rise to crests called cheniers (locally pronounced *shin-years*), so named because some support groves of live oaks. A few 'land islands' a mile or more in diameter rise out of the marshes between the cheniers and the interior uplands. These are usually bordered by cypress and tupelo and have dry-land hardwoods growing in their interiors. Among the regularly nesting birds of the coastal marshes, including the beaches, tree-covered ridges, cheniers, and land islands, are the following:

Olivaceous Cormorant
American Anhinga
Great Blue Heron
Green Heron
Little Blue Heron
Cattle Egret
Great Egret
Snowy Egret
Louisiana Heron
Black-crowned Night Heron
Yellow-crowned Night Heron
Least Bittern
American Bittern
White-faced Ibis
White Ibis
Roseate Spoonbill
Fulvous Whistling Duck
Mottled Duck
King Rail
Clapper Rail
Purple Gallinule
Common Gallinule
Willet
Black-necked Stilt
Forster's Tern
Marsh Wren
Red-winged Blackbird
Great-tailed Grackle
Boat-tailed Grackle
Seaside Sparrow

The coastal marshes undoubtedly constitute Louisiana's major ornithological attraction. For fine examples, *see under* **Jennings, Lake Charles,** *and* **Thibodaux.**

Offshore in the Gulf are sand and shell islands, covered with grasses and fringed with mangrove thickets. Notable are the large islands of the Timbalier and Dernieres groups westward from the Delta and of the Chandeleur group northeastward. Birds breeding regularly on these islands—and occasionally on undisturbed beaches on the mainland—are the following:

American Oystercatcher (*Chandeleurs only*)
Wilson's Plover
Laughing Gull (*mainly Chandeleurs*)
Little Tern
Royal Tern
Sandwich Tern
Caspian Tern
Black Skimmer

Almost none of the islands is readily accessible. Although the Timbalier and Dernieres groups have fine nesting colonies of terns and skimmers, their distance from ports, coupled with the scarcity of craft for hire at reasonable prices, makes them impossible to reach by ordinary means. The islands of the Chandeleur group, some of which have all the bird species mentioned above, are more accessible—if the bird finder does not object to the great expense of chartering a powerboat at Gulfport, Mississippi, the nearest port.

The bottomlands of the Mississippi comprise Louisiana's chief migratory route for waterbirds, waterfowl, shorebirds, and landbirds in their journeys to and from northern latitudes. In the fall it becomes a veritable bottleneck, where birds from both northeastern and north-central North America converge in their southward flight. During these journeys, impressive numbers of grebes, wading birds, ducks, gulls, and terns linger on bayous and meander cutoffs, while tremendous waves of small landbirds move through the hardwoods.

By far the most remarkable aspect of migration in Louisiana is the spring return of landbirds, primarily passerines, across the Gulf. If at the time of migration the weather in lower Louisiana is warm, or mild, the birds with northern destinations proceed inland some distance before coming to land, or else settle down in the deeply wooded bottomlands in such a widely dispersed fashion that they are not readily detected. The coastal area thus appears to be an ornithological 'hiatus.' But if, as a result of inclement weather, a cold front or 'norther' sets in, the birds are forced to 'fall out' and to accept the first land reached, usually the islands and cheniers. The coastal area is then flooded with birds. As soon as the weather clears or mild temperatures resume, the birds disappear. These spring fall-outs are frequently observed at various localities, notably at Grand Isle (*see under* **Thibodaux**) and in Johnsons Bayou Woods (*see under* **Lake Charles**). Heavy fall-outs usually take place between 1 April and 10 May.

The following migration timetable applies chiefly to birds passing through Louisiana:

Shorebirds: 15 March–15 May; 15 August–20 October
Landbirds: 15 March–5 May; 15 September–10 November

During winter the coastal marshes of lower Louisiana teem with waterfowl. West of the Delta, thousands upon thousands of Snow Geese, probably the bulk of the North American population, pass the winter. In addition, vast numbers of Greater White-fronted Geese remain through the winter even though many may continue into southern Texas and Mexico. Grebes, cormorants, wading birds, ducks, gulls, and terns abound along the entire coast. The wintering populations of shorebirds are always impressive, but the species present vary greatly, making it impossible to say with certainty which species one may see in appreciable numbers. Almost invariably the coastal area hosts representatives of a few species from western United States, providing an unexpected element for the bird finder. Upper Louisiana is the wintering ground of many passerine birds, including the Hermit Thrush, Golden-crowned and Ruby-crowned Kinglets, Solitary Vireo, several warblers, and numerous fringillids. Where there are small lakes, reservoirs, bayous, and meander cutoffs, waterbirds and ducks frequently stay from fall to spring.

Authorities

Bruce Crider, Kathleen S. Harrington, Horace H. Jeter, Buford 'Mac' Myers, Robert J. Newman, Robert D. Purrington.

Reference

Louisiana Birds. By George H. Lowery, Jr. 3rd ed. Baton Rouge: Louisiana State University Press. 1974.

ABITA SPRINGS
Area for Red-cockaded Woodpeckers

The Florida Parishes north of Lake Pontchartrain have pine flats, which extend for 30 to 50 miles inland from the Lake, and pine uplands which stretch beyond the flats east and north to the Mississippi line and northwestward to the Feliciana Parishes. Throughout these pine woodlands, Brown-headed Nuthatches and Bachman's Sparrows are year-round residents as are Red-cockaded Woodpeckers, although uncommon and local. Where the pine woods are interrupted by fields of broom sedge and other uncut grasses, Grasshopper, Henslow's, and Le Conte's Sparrows often reside in winter. To find these birds and others, the following trip is suggested.

From I 12, take Exit 65 north on State 59 for 5 miles to Abita Springs; turn off east (right) onto State 36 and drive about 4 miles to the Mid-South Turf Farm, marked by a sign; here turn south (right) onto a paved road, which, after a mile or so, becomes dirt and sometimes impassable in wet weather. At this point, the road reaches a grove of mature slash pines—habitat for Red-cockaded Woodpeckers. (*See under* **Fluker-Greensburg** for methods of finding them.) Bachman's Sparrows are also present and may be easily located in the breeding season by their distinctive, ringing song. Other birds in this area all year are Common Bobwhites, Red-headed Woodpeckers, Blue Jays, Brown-headed Nuthatches, Carolina Wrens, and Pine Warblers. In summer there are Summer Tanagers, Orchard Orioles, and Blue Grosbeaks, possibly Prairie Warblers.

Backtrack to State 36 and continue east, passing other stands of mature pines which may also have Red-cockaded Woodpeckers. About 2.5 miles from the Turf Farm, turn off south (right) onto State 1088. In the next 3 miles, especially in the first mile, the highway passes open pine woods with scattered brush—prime habitat for Bachman's Sparrows. In late spring and summer, when they are singing, stop and listen for them.

During the trip in winter, all fields passed that contain broom sedge—or better still, broom sedge mixed with other tall, fine grasses—may hold that elusive trio, the Grasshopper, Henslow's, and Le Conte's Sparrows. They are difficult to flush and identify as they flutter up at one's feet, then immediately drop back out of sight and run ahead under cover. By flushing them repeatedly, sometimes they will pause on a perch long enough for identification. Some of the birds flushed may prove to be Sedge Wrens or Savannah Sparrows as both species are common winter residents in the same habitat.

BATON ROUGE

University and City Park Lakes | Highland Road Park | Bayou Manchac

Both a capital city and a university city, Baton Rouge on the east side of the Mississippi River offers several good areas for birds.

University and City Park Lakes and two adjacent smaller lakes, all on the southeast edge of the Louisiana State University Campus, have over 8 miles of shoreline worth investigating in any season. From I 10 in Baton Rouge, take Exit 156B south on Dalrymple Drive which leads to the Campus.

In winter, when bird finding is best, expect to see Pied-billed Grebes, American Coots, and large numbers of ducks, chiefly Ring-necked Ducks, Lesser Scaups, Canvasbacks, and Ruddy Ducks, and smaller numbers of others. A few Wood Ducks are present the year round, as are Great and Snowy Egrets and Louisiana and Black-crowned Night Herons, although they are more abundant in spring and summer. In the tree-shaded residential districts encircling the Lakes, regular winter visitants include, besides a few Solitary Vireos, such birds as Purple Finches, American Goldfinches, and White-throated Sparrows. American Crows, Fish Crows, Carolina Chickadees, Carolina Wrens, Northern Mockingbirds, Northern Cardinals, and Rufous-sided Towhees are permanent residents.

For woodland and field birds, visit **Highland Road Park** southeast of the city. Go south on I 10; turn off right (west) at Exit 166 and follow Highland Road for 2.5 miles to the Park entrance on the left. Leave the car in the parking lot and explore the heavily wooded area behind it. (Caution: The area is designated as an archery range.) Among the breeding birds are Red-shouldered Hawks, Barred Owls, Red-bellied and Downy Woodpeckers, Acadian Flycatchers, Wood Thrushes, Blue-gray Gnatcatchers, White-eyed Vireos, and Swainson's, Kentucky, and Hooded Warblers. In winter, Sedge Wrens frequent the wet, grassy stretches near Highland Road and a wide variety of sparrows, the nearby fields and bordering brushy thickets.

For birds in bayou country, visit **Bayou Manchac** southeast of the city. This is an old distributary of the Mississippi, now cut off by levees, that was part of the route followed by William Bartram when he traveled through the South in the 1770s. Bayou Manchac Road, a shell road, lined with moss-draped cypresses and live oaks entwined with lush vines, parallels the old waterway and occasionally passes fields and weedy fencerows on the side opposite the waterway. Along the Road in spring and summer some of the birds to watch or listen for are the Prothonotary, Northern Parula, Yellow-throated, and Kentucky Warblers, Orchard Oriole, Summer Tanager, and Indigo and Painted Buntings. Likely to be seen at any time are the Cattle Egret, White Ibis, Black Vulture, Mississippi Kite, Broad-winged Hawk, and Pileated Woodpecker.

To reach Bayou Manchac Road, go south from the city on I 10; turn off left at Highland-Perkins Exit (Exit 166) and start east on Highland Road; take the first right, which is State 427. Follow it south until, at 3.1 miles, it swings sharply east; here turn off acutely right onto State 928 and follow it back over I 10. Bayou Manchac Road begins immediately to the right at the bottom of the overpass.

FLUKER-GREENSBURG
Area for Red-cockaded Woodpeckers

A reliable Louisiana area for Red-cockaded Woodpeckers is in the Uplands of the Florida Parishes, reached from I 55 by taking the Fluker-Greensburg Exit (Exit 53) west on State 10 toward Greensburg to a triangular median space just beyond the pass of State 10 over I 55. Within this space, search the pines for the woodpeckers' nesting holes, always high up in *live* pines and easily recognized in the breeding season by a whitish resin or pitch surrounding the openings and streaking the trunks below. On locating such a hole, withdraw a short distance and watch for the occupant adults as they arrive and depart. In winter, the woodpeckers use the holes for roosting at night; hence they can sometimes be seen entering the holes in the late afternoon and leaving in the early morning.

Farther west on State 10 are pine woods on the right, accessible by several narrow dirt roads but unsafe for conventional cars. The woods and open brushy areas are worth investigating on foot in any season. Year-round residents include—besides Red-cockaded Woodpeckers—Common Screech Owls, Red-headed Woodpeckers, Brown-headed Nuthatches, Pine Warblers, and Bachman's Sparrows. Among the summer residents are Prairie Warblers (common), Eastern Bluebirds, Gray Catbirds, and Orchard Orioles.

For another area for Red-cockaded Woodpeckers, *see under* **Abita Springs.**

HENDERSON
Upper Atchafalaya Basin

As the principal distributary of the Mississippi River's flood waters in Louisiana, the Atchafalaya River begins near Simmesport southwest of the confluence of the Mississippi and Red Rivers and flows 140 miles southward to the Gulf of Mexico by way of Morgan City. The vast basin through which the Atchafalaya courses—well over 17 miles wide at its narrowest point—encompasses 13,400,000 acres of wetlands, bottomlands, cypress swamps, lakes, and bayous—all in all an area holding enormous wildlife resources.

Interstate 10 from Henderson east to Ramah crosses a 17-mile segment of the **Upper Atchafalaya Basin.** Even from his moving car, one obtains a sampling of its birdlife: in spring and summer months, American Anhingas circling overhead; wading birds—Little Blue Herons, Cattle, Great, and Snowy Egrets, and Louisiana Herons—flying along side or across the highway; and Mississippi Kites and perhaps a Swallow-tailed Kite sailing above treetop level.

Anyone tantalized by the sampling of birds from his car (stopping is inadvisable), and wanting to see more of the Basin and its birds, may take the Henderson-Cecilia Exit (Exit 115) south, turn right onto State 347, then turn immediately left onto State 352 and follow it through Henderson to Bayou Peyronnet (2.8 miles from Exit 115). At this point, drive up the road ascending the levee and follow the shell road on the levee north for 17 miles through the Basin to US 190.

In addition to the birds already mentioned, some of the other birds to look for: *In any season:* Red-shouldered Hawks and woodpeckers—Pileated, Red-bellied, Hairy, and Downy. *In summer:* Wood Storks occasionally and Broad-winged Hawks, Wood Thrushes, White-eyed Vireos, Prothonotary and Kentucky Warblers, Summer Tanagers, and Indigo and Painted Buntings. *In winter:* a few accipiters, Red-tailed Hawks, Yellow-bellied Sapsuckers, Tree Swallows flying low over water, and, in suitably grassy or brushy situations, such sparrows as the Savannah, White-throated, Lincoln's, Swamp, and Song.

JENNINGS
Lacassine National Wildlife Refuge

In southwestern Louisiana the **Lacassine National Wildlife Refuge** embraces 31,125 acres of marshland. Although once consisting almost entirely of salt and brackish marshes with rushes and grasses constituting the principal vegetation, the Refuge now has a 16,000-acre pool of fresh water and a maze of man-made canals. The Refuge is open to the general public from 15 March through 15 October.

All parts of the Refuge must be explored by boat; only headquarters, on the banks of the Mermentau River, is accessible by car. To reach headquarters from I 10, take the Jennings Exit (Exit 64) south on State 26 for 11 miles to the town of Lake Arthur; turn right (west) onto State 14 and drive 7 miles to State 3056 (Lowry Road); turn left onto it and proceed 5 miles south to headquarters. Here obtain a checklist of birds and inquire about the possibility of viewing birds from a boat.

The Refuge hosts huge concentrations of wintering waterfowl including one of the largest wintering populations of Greater White-fronted Geese anywhere. Other wintering waterfowl in impressive numbers are Snow Geese, Mallards, Gadwalls, Common Pintails, Green-winged and Blue-winged Teal, American Wigeons, Northern Shovelers, and Ring-necked Ducks. Common among the nesting waterfowl are Fulvous Whistling Ducks and Mottled Ducks. Besides them, among the nesting birds are Olivaceous Cormorants, American Anhingas, Cattle Egrets and many other egrets, herons, White-faced and White Ibises, King Rails, Purple and Common Gallinules, and Forster's Terns.

LAKE CHARLES

Gum Cove Area | Sabine National Wildlife Refuge | Mud Lake | Holly Beach | Johnsons Bayou Woods | Cameron Channel Area

Cameron Parish in extreme southwestern Louisiana, the state's primary area for bird finding, is one of the very few places in Louisiana where the Gulf of Mexico may be reached by road. Here are miles of beaches and mud flats, marshes, many shallow lakes, and the famous cheniers, site of the 'fall-outs' of returning trans-Gulf passerines in spring.

From I 10 west of St. Charles, take Exit 20 (Sulphur Exit) and proceed south on State 27 toward the Gulf, 35 miles distant. For a suggested side trip to the **Gum Cove Area,** turn off west, after 5 miles, onto State 108 and proceed for about 9 miles, then turn off south onto the Gum Cove Road. Along the way, watch fence posts and mounds of dirt for Black Francolins, which were introduced in the early 1960s and have become well established. Watch the sky also for soaring Turkey and Black Vultures and the possiblity of a Crested Caracara among them. In winter, large flocks of Greater White-fronted Geese, with a few Canada Geese interspersed, are often in view not far from the roadside. In spring, scan the adjacent ricefields for a few transient Hudsonian Godwits. After about 5.5 miles from State 108, Gum Cove Road comes to a small ferry across the Intracoastal Canal. If francolins and caracaras have been missed so far, one is more likely to see them by crossing the Canal and continuing south for several miles until the pavement ends.

Backtrack to State 27 and continue south on this highway which passes through Hackberry and soon begins traversing the **Sabine National Wildlife Refuge.** Stop at headquarters on the left, 28 miles from I 10, for desired information about the Refuge and a checklist of Refuge birds. In winter, the live oaks and shrubby growth around headquarters usually provide habitat for one or more Merlins and/or Vermilion Flycatchers.

The Sabine Refuge comprises 142,846 acres of coastal marshland cut up by bayous, man-made canals, alligator ponds, and potholes. Calcasieu Lake, 14 by 18 miles, bisects the eastern section of the Refuge, and Sabine Lake, 7 by 14 miles, forms the western boundary. Both Lakes are brackish and surrounded by great stretches of three-square bulrushes and cord grasses. In the interior of the Refuge, away from the brackish water of the Lakes, is marsh containing mostly fresh water, with vegetation such as twig rushes, freshwater bulrushes, and quillreeds. Here and there are a few 'islands' covered with scrub willows, chinaberries, and hackberries.

Hordes of waterfowl winter on the Refuge from mid-October until mid-March, reaching peak numbers in December and January. Chief among the hordes are geese—Snow and Greater White-fronted. Others in abundance include Mallards, Gadwalls, Common Pintails, Green-winged and Blue-winged Teal, American Wigeons, and Northern Shovelers. Mottled Ducks and some Blue-winged Teal reside on the Refuge the year round. In the interior of the Refuge are nesting colonies of waterbirds—Olivaceous Cormorants, Great Blue Herons, Cattle, Great, and Snowy Egrets, Louisiana Herons, Yellow-crowned and Black-crowned Night Herons, White-faced and White Ibises, and Roseate Spoonbills.

Although most of the Refuge is accessible only by boat, practically all the bird species regularly breeding on, or otherwise frequenting, the Refuge may be viewed feeding in the marshes along State 27 as it passes through the Refuge. Huge flocks of wintering Snow Geese are always in evidence. From the right (west) side of State 27, 3.7 miles south of headquarters, the Refuge's nature trail leads past marshes and ponds, terminating after a mile at a tower for a fine view of the area. During the warmer months, a walk along the trail will very likely yield Least Bitterns and Purple Gallinules. There will be many alligators.

South from the Refuge, State 27 soon passes close to **Mud Lake** on the left (east). Stop at the Mud Lake Overlook, a widened area for parking, and scan the Lake for waterbirds and wading birds in any season, grebes, American White Pelicans, and waterfowl in winter.

Continue south on State 27 from Mud Lake to the town of Holly Beach; turn off right (west) on State 82 for a mile; stop either on a wide shoulder of the highway's south side, or in the paved parking area close to **Holly Beach** on the Gulf. During a walk along the Beach, some of the birds to expect on the Beach itself, the pilings, and offshore waters: *In any season:* Herring, Ring-billed, and Laughing Gulls, Gull-billed, Forster's, Common, Royal, and Caspian Terns, and Black Skimmers. *In summer:* Wilson's Plovers, Little, Sandwich, and Black Terns. *In spring:* many transient shorebirds including Semipalmated, Piping, and Black-bellied Plovers, Ruddy Turnstones, Red Knots, Least and Semipalmated Sandpipers, and Sanderlings. *In fall, occasionally in spring:* Franklin's Gulls. *In winter:* Horned and Eared Grebes, and many ducks, among them Greater and Lesser Scaups, Ruddy Ducks, Red-breasted Mergansers, and scoters.

Continue west on State 82. When, beyond the point where it diverges inland from Holly Beach, some good cheniers begin on the left. Any roads leading into them are worth investigating from late March through mid-May. **Johnsons Bayou Woods,** south of the 14-mile marker in the diffuse town of Johnsons Bayou, is an especially productive chenier. After a heavy fall-out, most of the parulid species in eastern United States are represented along with other passerines such as Great Crested Flycatchers, *Empidonax* flycatchers, White-eyed, Yellow-throated, Red-eyed, and Warbling Vireos, Scarlet and Summer Tanagers, Rose-breasted and Blue Grosbeaks, and Painted and Indigo Buntings. Western Kingbirds, Olive-sided Flycatchers, and Philadelphia Vireos are some of the passerines that are apt to be more common in fall. During winter there are often vagrant species from as far west as the Pacific states.

Backtrack on State 82 to the town of Holly Beach and continue east on State 82, after State 27 joins it, passing marshes and paralleling a roadside canal. Along the way, shell roads, built by oil companies, lead off into the marshes. If the gates are open and not posted, the roads are worth driving for views of ibises, spoonbills, Willets, Clapper Rails, and numerous other marsh-preferring birds.

About 5 miles from Holly Beach, follow the highway as it swings left and take the ferry across the Calcasieu River Ship Canal to the **Cameron Channel Area.** Go into the town of Cameron. Be alert for both Great-tailed and Boat-tailed Grackles in the tree-shaded residential districts. From the main road through town, watch on the right for the Cameron Parish Road 3143 and take it 3 miles south to the mouth of the Calcasieu River. Locally known as the East Jetty Road, this is a rewarding drive through marshy areas. If Clapper Rails have been missed, look for them here. Seaside Sparrows are usually abundant and often perch in full view on fences. Sedge Wrens and Sharp-tailed Sparrows are frequent winter visitants in the salt grasses.

East Jetty Road ends at the Gulf shoreline with East Jetty adjacent on the right. In the vicinity at low tide, Reddish Egrets and Black Skimmers are often present as are hundreds of gulls and terns and most of the other birds that show up in season at Holly Beach. Probably the shore in the vicinity of East Jetty is the most reliable spot on the Louisiana coast in spring and fall for seeing three of the larger shorebirds—the Long-billed Curlew, Whimbrel, and Marbled Godwit.

NEW IBERIA
Avery Island

From the New Iberia-Delcambre Exit on US 90, about 20 miles south of Lafayette in south-central Louisiana, continue south on US 90 for 1.9 miles and take the Avery Island Exit right on a service road which doubles back to State 329; turn left onto State 329 and proceed 5 miles to a toll gate marking the entrance to privately owned **Avery Island.**

Circular in shape, with a diameter of approximately 3 miles, Avery Island is in reality a salt dome rising from the coastal marshes and cypress swamps. On the 'Island' are several woodland areas and a development comprising homes, schools, shops, a pepper sauce factory (offers tours), a salt mine, and an oil field. Of special interest to the bird finder are the Jungle Gardens and Bird City.

Jungle Gardens consists of more than 250 acres surrounding the home of the late Edward Avery McIlhenny, ornithologist. In addition to native flora, the area contains an extraordinary array of exotic plants, ranging from immense palms and giant bamboos to huge beds of camellias, azaleas, and chrysanthemums. Birds such as Tufted Titmice, Carolina Wrens, Northern Mockingbirds, Yellow-breasted Chats, Orchard Orioles, Boat-tailed Grackles, and Northern Cardinals are common.

Bird City, in the southern section of Jungle Gardens, is a 35-acre pond bordered by buttonwood trees and willows where 30,000 or more egrets—Cattle, Great, and Snowy—and Louisiana Herons nest from late March to September. Other residents of Bird City in smaller numbers are American Anhingas, Great Blue, Green, and Little Blue Herons, and Purple and Common Gallinules.

The pond and its waterbird colony were developed by E. A. McIlhenny, neither having existed prior to 1893. As the colony increased in size, the pond was enlarged. When the breeding population became so enormous as to cause a shortage of nesting sites and nesting materials, elevated platforms of bamboo were set up in the pond as supports for nests, and carloads of twigs were gathered from distant points and heaped on specially constructed benches, from which they could be gathered by birds at will. The colony soon doubled in size. From a watchtower erected in the pond, one may view the colony: birds nesting in the bordering trees and shrubs, the egrets standing out as conspicuously as patches of snow; birds nesting on the platforms, a setting so unnatural as to seem ludicrous; birds resting; birds fishing in the shallows; birds skirmishing; birds displaying; birds incessantly arriving and departing. Such is Bird City!

NEW ORLEANS
City Park | Lakefront | Pearl River Wildlife Management Area

Hemmed in by marshland, New Orleans lies mostly between the south shore of shallow Lake Pontchartrain and the Mississippi River. In adjacent wetlands are numerous bayous and lakes where waterbirds are abundant the year round and waterfowl are common in fall, winter, and spring.

City Park (1,500 acres) in north-central New Orleans is a convenient area for a variety of birds in season. To reach it, take the Canal Street Exit south from I 610 and turn left onto City Park Avenue and proceed to the entrance. The several lagoons in the Park attract moderate numbers of wintering waterbirds and waterfowl including Pied-billed Grebes, Ring-necked Ducks, Lesser Scaups, and American Coots, occasionally Common Goldeneyes and Buffleheads. Bayou St. John, paralleling Wisner Boulevard along the eastern edge of the Park, attracts the same wintering species and, in addition, Common Loons. Two spots are worth investigating for migrating passerines, especially on days after the passage of a cold front. One is along Stadium Drive in the Park between Roosevelt Mall and Marconi Drive; the other is around Popp Memorial Fountain just northeast of the Marconi Drive underpass of I 610.

North of City Park is the **Lakefront**—the name given to the Lake Pontchartrain shore from Municipal Yacht Harbor on the west to Seabrook Bridge over the Industrial Canal on the east. Wisner Drive north from City Park terminates on Lakeshore Drive, which, as its name correctly indicates, parallels the Lakefront. During winter, from vantages almost anywhere along the Drive, one may view good numbers of ducks, gulls, and terns. Lesser Scaups, Red-breasted Mergansers, Herring, Ring-billed, Laughing, and Bonaparte's Gulls, and Forster's, Royal, and Caspian Terns may be expected. Often there are Common Loons and Horned Grebes; and there is always the possibility of uncommon or rare ducks and gulls.

The **Pearl River Wildlife Management Area** (27,000 acres) on the Pearl River bottoms along the Louisiana-Mississippi border northeast of New Orleans and east of Lake Pontchartrain, offers many nesting birds typical of southeastern Louisiana's extensive tupelo-cypress swamps. Drive northeast from New Orleans on I 10. When it turns east about 16 miles from the city's limits, continue northeast on I 59 for 6 miles, cross the West Pearl River, and then take the Honey Island Swamp Exit (Exit 5B), the old roadbed of US 11. This proceeds east into the Area and ends at East Pearl River which forms the state border. The roadbed—paved, narrow, and tree-lined—traverses densely wooded swamp. Because water levels in the swamp are always high, one cannot leave the roadway, but he may nevertheless park indiscriminately for observations.

Among the more common nesting birds to look or listen for along the road are the Red-shouldered Hawk, Yellow-billed Cuckoo, Pileated, Red-bellied, Hairy, and Downy Woodpeckers, Eastern Kingbird, Great Crested Flycatcher, Acadian Flycatcher (very common), American Crow, Carolina Wren, Wood Thrush, White-eyed and Red-eyed Vireos, Prothonotary, Swainson's, Northern Parula, Kentucky, and Hooded Warblers, American Redstart, and Summer Tanager.

Two gravel roads lead south off the pavement. The first is the better for bird finding. On its left, just after it begins, is a marked nature trail. Farther along the road, which is about 3 miles in length, be alert for all the bird species mentioned above and watch the sky for such birds in flight as the American Anhinga and both the Swallow-tailed and Mississippi Kites. Listen for Barred Owls. Just before the road ends, a trail leads off to the right for about 0.5 mile to a pond where Green Herons, Yellow-crowned Night Herons, and other wading birds reside.

An especially good spot for Swallow-tailed Kites is at the end of the paved road on the East Pearl River and over the field to the north. Wild Turkeys sometimes show up in early morning along the edges of the field. Both Indigo and Painted Buntings are common in brushy places. With luck, one may see a Pileated Woodpecker flying across the River.

ST. FRANCISVILLE

Audubon State Commemorative Area

It was here in West Feliciana Parish on the banks of the Mississippi that John James Audubon found his 'happy land.' For five months, beginning 18 June 1821, he lived at Oakley, the James Pirrie plantation, and studied birds nearby in 'the most beautiful of all Louisiana's beautiful woods.' Birdlife was rich in variety and abundance, inspiring him to produce some of his finest paintings. Among the birds he depicted, the Carolina Paroquets and Ivory-billed Woodpeckers have long since vanished and many other species have suffered sharp reductions in their populations; nevertheless the St. Francisville area is still one of the finest in Louisiana for landbirds, both breeding and transient.

The Oakley plantation house in **Audubon State Commemorative Area** still stands today. The plantation—the house and a hundred encompassing acres—was purchased by the state and restored to resemble as closely as possible the Oakley that Audubon knew in 1821. Three-storied and with quaint shuttered porches, the house is surrounded by moss-hung live oaks, yellow poplars, and southern magnolias. Tours of the house are available.

Across the road from the house are picnicking facilities in a park-like area with tall trees and dense shrubbery—ideal for birds the year round. Breeding birds include the Red-bellied Woodpecker, Acadian Flycatcher, Eastern Pewee, Carolina Chickadee, Tufted Titmouse, Carolina Wren, Wood Thrush, Blue-gray Gnatcatcher, the Northern Parula, Yellow-throated, Kentucky, and Hooded Warblers, Summer Tanager, and Northern Cardinal. In winter, Yellow-bellied Sapsuckers join the resident woodpeckers along with an occasional Red-breasted Nuthatch and Brown Creeper. Hermit Thrushes, Solitary Vireos, Orange-crowned and Myrtle Warblers, American Goldfinches, and White-throated Sparrows may be expected.

The Commemorative Area may be reached from St. Francisville by driving 1.0 mile south on US 61, then left (east) onto State 965 and going 3 miles to the entrance on the right.

SHREVEPORT
Cross Lake

West of this northwestern Louisiana city, in agricultural country, is **Cross Lake,** a large impoundment about 10 miles in length, with an irregular shoreline. The northern side of the Lake is very productive ornithologically. Here, bordering the Lake at different locations, are cypresses, willows, mud flats, grassy marshes, and wooded swamps. Back from the Lake, on higher ground, are abandoned fields, open shrubby areas with thick tangles, and woods composed of oaks, and pines, both shortleaf and loblolly. The entire area is readily accessible as follows:

From I 20 in Shreveport, exit north on US 171 (Hearne Avenue), then turn off west onto State 173 (Blanchard Road); continue 2 miles and turn off left onto North Lakeshore Drive and follow it along the Lake. After about 10 miles, watch for West Lakeshore Drive, a side road on the left, and follow it down to its end. Backtrack to North Lakeshore Drive and continue about 2 miles to its end at the Blanchard-Furrh Road. Here turn left (west) and go about 1.35 miles, entering on the right the Walter B. Jacobs Memorial Nature Park, 160-acre preserve with nature trails leading from a parking area.

From the long list of birds one is likely to see or hear during late spring and summer in suitable locations reached on the above trip, the following are selected to show the great diversity of species: American Anhinga, Little Blue Heron, Wood Duck, Black Vulture, Red-shouldered Hawk, Yellow-billed Cuckoo, Barred Owl, Chuck-will's-widow, Pileated and Red-bellied Woodpeckers, Acadian Flycatcher, Eastern Pewee, Fish Crow, White-breasted and Brown-headed Nuthatches, Carolina Wren, Brown Thrasher, Wood Thrush, Eastern Bluebird, Loggerhead Shrike, White-eyed and Yellow-throated Vireos, nine warblers (the Black-and-white, Prothonotary, Swainson's, Northern Parula, Yellow-throated, Pine, Louisiana Waterthrush, Kentucky, and Hooded), Orchard Oriole, Summer Tanager, Blue Grosbeak, and Indigo and Painted Buntings.

Among the winter birds are many White-throated Sparrows and hordes of Red-winged Blackbirds, Common Grackles, and Brown-headed Cowbirds. Other birds to be found regularly in winter are Yellow-bellied Sapsuckers, Eastern Phoebes, Brown Creepers, Winter Wrens, Hermit Thrushes, Golden-crowned and Ruby-crowned Kinglets, Solitary Vireos, Orange-crowned and Myrtle Warblers, Rufous-sided Towhees, Slate-colored Juncos, and Field,

Fox, and Song Sparrows. Common Loons, Eared Grebes, American Coots, Canvasbacks, Ruddy Ducks, and Ring-billed Gulls commonly pass the winter on the Lake.

THIBODAUX
Grand Isle

A narrow strip of land 7 miles long and lying 40 miles west of the Mississippi Delta, **Grand Isle** is the only island on the Gulf Coast of Louisiana accessible by car. After a spring 'norther' it is an excellent area to observe fall-outs of passerines returning from across the Gulf. Countless numbers of flycatchers, vireos, parulids, and fringillids collect in the live oaks and oleanders and swarm about the many dwellings, even entering through open windows. One has to see the spectacle to appreciate it fully.

Grand Isle is reached from Thibodaux by driving south on State 1 to its end on the island. For most of the distance of 78 miles, the highway traverses marshes and, for the first 40 miles or so, also parallels Bayou Lafourche, providing views of herons, egrets, gulls, terns, and other waterbirds. Past Golden Meadow, the highway begins passing more ornithologically productive marshes. In ditches beside the highway, one may see numerous rails and gallinules. During winter an inspection of grassy areas should yield Sedge Wrens and Sharp-tailed Sparrows.

About 5 miles south of Leeville, where State 1 crosses Bayou Lafourche and bears sharply left, turn off right onto State 3090 (Fourchon Road) which leads directly south to the Gulf. As the Road passes extensive mud flats, park the car and either walk out onto them, or scan them, for nesting Black-necked Stilts and Little Terns. Where the Road passes large expanses of water, again park the car and scan them in any season for American White Pelicans and in winter for many ducks. In spring and fall, look over their shallow borders for Dunlins, Willets, Marbled Godwits, and other transient shorebirds such as dowitchers and phalaropes. At the end of the Road, one may drive west on the beach all the way to the jetties at the mouth of Bayou Lafourche, with the expectation of

MAGNIFICENT FRIGATEBIRD

seeing the same bird species as on Holly Beach (*see under* **Lake Charles**).

Backtrack to State 1 and continue east, crossing the bridge to Grand Isle. The beach in the state park at the far end of the island is good for the smaller shorebirds in season as well as for gulls and terns anytime of the year. Its stellar bird attraction, however, is probably the presence of Magnificent Frigatebirds from late June through September.

In the bays to the north and northwest of Grand Isle are many small mangrove islets, some of which are nesting sites for Green and Little Blue Herons, Reddish Egrets, Louisiana and Black-crowned Night Herons, and both White-faced and a few Glossy Ibises. (This is the only known area in the state where both of these dark ibises are known to nest side by side.) To view the aggregations, one should find a fisherman willing to take him by boat closely around some of the sites.

The spring fall-outs of passerines, similar to those in Johnsons Bayou Woods (*see under* **Lake Charles**), are best observed in the town of Grand Isle by wandering down the tree-lined side streets to the left of the main highway as it passes eastward down the island. One must stay in the streets as there are few places not in somebody's dooryard. Cemetery Lane, about midway down the island, is a famous vantage for seeing heavy fall-outs.

Minnesota

COMMON LOON

Scattered throughout Minnesota, though more concentrated in the northern part, are 15,291 lakes. These bodies of water, together with three river systems and their permanent expansions, cover 5,637 square miles, providing more water surface than in any other state. Little wonder that Minnesota's outstanding ornithological attraction are hosts of waterbirds and waterfowl from Common Loons and American White Pelicans to Red-breasted Mergansers and Franklin's Gulls.

In general, Minnesota has a gently rolling surface with an average elevation above sea level of only 1,200 feet, but its four corners show considerable variation: in the northeast, rugged hills that are practically mountains; in the northwest, flat, monotonous prairie; in the southeast, high hills and deeply cut ravines; in the southwest, an undulating, sometimes flat prairie, interrupted by a hilly upland called the Coteau des Prairies and deep valleys. The highest elevation in the state, 2,301 feet, is reached at the summit of Eagle Mountain in the Misquah Hills of the extreme northeastern part; the lowest elevation, 602 feet, is the 200-mile strip of land bordering Lake Superior.

Despite the fact that its elevation is relatively low, Minnesota gives rise to three great drainage systems of North America. In Itasca State Park the Mississippi begins its southward course, soon joined by the Minnesota River from the west and the St. Croix from the north; in the northeast, the St. Louis River and numerous streams cascade eastward into Lake Superior of the St. Lawrence River system; along the western boundary, the Red River of the North rises to flow northward into the Hudson Bay outlet.

Throughout Minnesota, except in the southeastern section, are large marshes with aquatic plants—i.e., emergent bulrushes, cattails, quillreeds, and sedges; submerged pondweeds, bladderworts, and duckweeds—which provide ideal conditions for breeding waterfowl and for many varieties of marsh-loving birds. King Rails and Common Gallinules nest in large marshes in the southern part of the state. In wet grassy lowlands near the marshes and lakes of central and western Minnesota, a few Wilson's Phalaropes nest. Birds in the larger marshes of western Minnesota may include the Red-necked and Horned Grebes (mainly in northwestern Minnesota) and the Eared Grebe, Western Grebe, Franklin's Gull, and Forster's Tern. The following birds breed regularly in nearly all large marshes throughout the state:

Pied-billed Grebe
American Bittern
Least Bittern (*except northeastern Minnesota*)
Virginia Rail
Sora
American Coot
Black Tern
Marsh Wren
Yellow-headed Blackbird
Red-winged Blackbird
Swamp Sparrow

In Minnesota's most scenic country—the Lake Superior shore (commonly called the North Shore), the gorge of the St. Croix River, and the bluffs overlooking the Mississippi south of Red

Wing—are bold escarpments that were formerly sites of Peregrine Falcon aeries. Perhaps the most picturesque site of all is Gwinn's Bluff, which towers 550 feet above the Mississippi, a short distance southeast of Winona.

Minnesota has parts of three great biological realms: the northern coniferous forest in the northeast and north-central sections of the state; the open prairie along the western boundary and in the southwestern section; the deciduous forest, a belt interposed between the coniferous forest and the prairie grassland, extending diagonally from the southeastern corner of the state to within 100 miles of the northern boundary. Transition areas where these realms come together provide some of the finest places for bird finding.

The coniferous forest has fared better in Minnesota than in states farther east. Since much of the land on which it exists is unsuited for agriculture, only fires and lumbering have brought devastation, and this has been offset by extensive reforestation and the early reclaiming of lands by government agencies. Even today there are a few virgin stands of pine. The principal trees of the upland coniferous forest are white pine, red pine, jack pine, white spruce, and balsam-fir; of the lowlands and bogs, black spruce, tamarack, and white cedar. Where the forest was burned or cut, birches, aspens, red maples, and other deciduous trees have succeeded the conifers. Breeding birds characteristic of the coniferous forest are the following:

Northern Goshawk
Spruce Grouse
Yellow-bellied Sapsucker
Black-backed Three-toed Woodpecker
Yellow-bellied Flycatcher
Olive-sided Flycatcher
Gray Jay
Boreal Chickadee
Red-breasted Nuthatch
Brown Creeper
Winter Wren
Hermit Thrush
Swainson's Thrush
Ruby-crowned Kinglet
Golden-crowned Kinglet
Solitary Vireo
Nashville Warbler
Northern Parula Warbler
Magnolia Warbler
Black-throated Blue Warbler
Myrtle Warbler
Black-throated Green Warbler
Blackburnian Warbler

- Northern Waterthrush
- Mourning Warbler
- Canada Warbler
- Purple Finch
- Pine Siskin
- Red Crossbill
- White-winged Crossbill
- Slate-colored Junco
- White-throated Sparrow

The deciduous forest is continuous northward from the southeastern corner of the state, except where it is divided by an irregular peninsula of prairie extending from the southern boundary near Albert Lea and Austin north to the Mississippi Valley between Red Wing and St. Paul. Expectedly, the deciduous belt shows changes in composition along its northward course. In the southeastern corner, river birch, swamp white oak, white oak, black sugar maple, and Kentucky coffeetree grow on the Mississippi Valley bottomlands, black oak, honey locust, shellbark hickory, and black walnut predominate on the adjacent bluffs and uplands. Birds that regularly nest here and not much farther north in the state are the following:

- Red-shouldered Hawk
- Red-bellied Woodpecker
- Tufted Titmouse
- Blue-gray Gnatcatcher
- Prothonotary Warbler
- Blue-winged Warbler
- Cerulean Warbler
- Louisiana Waterthrush

Farther north, across the prairie peninsula, from the vicinity of Faribault north to the vicinity of St. Cloud, is the 'Big Woods' section of the deciduous belt, once a dense forest of hardwoods—sugar maple, basswood, red oak, and American and slippery elms—but now reduced to patches here and there. From St. Cloud north, the deciduous belt is less imposing, with elm, boxelder, bur oak, aspen, and black cherry prominent in upland areas that have not been cleared for farming, and willow and cottonwood in the lowland areas such as bottomlands along streams. On the prairie side of the deciduous belt are prairie groves—i.e., islands of deciduous trees separated from the main belt by prairie—and bottomland forests—peninsulas of deciduous trees extending into the prairie along rivers and streams. Usually the tree composition of these groves and forests is similar to that of the adjacent parts of the deciduous belt.

The original prairie region of Minnesota was open grassland with only scattered shrubs, but nearly all of it has been turned into farmland and is no longer recognizable. A few Swainson's Hawks and Burrowing Owls occasionally reside during the summer in the extreme western part of the state. In the northwestern part, in the Red River Valley, there are suitable, undisturbed areas where the Marbled Godwit, Sprague's Pipit, Baird's Sparrow, Le Conte's Sparrow, Sharp-tailed Sparrow, and Chestnut-collared Longspur may reside in the summer, though not so abundantly as farther west in North Dakota.

Among the birds that nest regularly in the deciduous forests throughout Minnesota (including prairie groves and bottomland woods) and on Minnesota farmlands (fields, wet meadows, brushy lands, woodland borders, orchards, and dooryards) are the following:

DECIDUOUS FORESTS

Broad-winged Hawk (*mainly northern Minnesota*)
Ruffed Grouse
Yellow-billed Cuckoo (*mainly southern Minnesota*)
Black-billed Cuckoo
Common Screech Owl
Barred Owl
Whip-poor-will
Yellow-shafted Flicker
Hairy Woodpecker
Downy Woodpecker
Great Crested Flycatcher
Least Flycatcher
Eastern Pewee
Blue Jay
Black-capped Chickadee
White-breasted Nuthatch
Wood Thrush
Yellow-throated Vireo
Red-eyed Vireo
Warbling Vireo
Ovenbird
American Redstart
Baltimore Oriole
Scarlet Tanager
Rose-breasted Grosbeak

FARMLANDS

Common Bobwhite (*southeastern Minnesota*)
Mourning Dove
Red-headed Woodpecker
Eastern Kingbird
Western Kingbird (*central and western Minnesota*)
Eastern Phoebe
(Prairie) Horned Lark
Tree Swallow

FARMLANDS (*Cont.*)

Barn Swallow
House Wren
Gray Catbird
Brown Thrasher
Eastern Bluebird
Loggerhead Shrike
Yellow Warbler
Common Yellowthroat
Bobolink
Eastern Meadowlark (*mainly east-central Minnesota*)
Western Meadowlark
Orchard Oriole (*southern Minnesota*)
Common Grackle
Northern Cardinal (*central and southern Minnesota*)
Indigo Bunting
Dickcissel (*southern Minnesota*)
American Goldfinch
Rufous-sided Towhee
Savannah Sparrow
Grasshopper Sparrow
Henslow's Sparrow (*southern Minnesota*)
Vesper Sparrow
Chipping Sparrow
Clay-colored Sparrow
Field Sparrow (*southern Minnesota*)
Song Sparrow

Minnesota has two heavily used migration routes: the Mississippi Valley and the Red River Valley. In spring, waterbirds, waterfowl, shorebirds, and passerine birds move up the Mississippi to the vicinity of Minneapolis and St. Paul, where the traffic divides: one part of it going north along the St. Croix River Valley and thence to the vicinity of Duluth and along the north shore of Lake Superior; the other part continuing to follow the Mississippi but eventually veering off westward. Waterfowl use the Red River route more commonly than the Mississippi route. This is particularly true of Whistling Swans and geese. American White Pelicans usually follow the Red River route, rarely the Mississippi route. In their northward flight, the majority of birds taking the Red River route first fly up the Missouri River Valley and the Big Sioux River Valley, thence to the Red River and directly northward. Apparently fall migration takes place along the same routes, although Whistling Swans tend to go southward in a dispersed manner, rarely concentrating anywhere. A spectacular feature of migration that bears no relation to these routes is the passage of diurnal predators around the western extremity of Lake Superior during the late summer and fall. On a given day from the bluffs overlooking the city of Duluth one may view vast numbers heading southward.

The following timetable indicates the periods of heaviest migratory movements in Minnesota:

NORTHERN MINNESOTA:
Waterfowl: 1 April–1 May; 1 October–15 November
Shorebirds: 1 May–1 June; 1 August–15 September
Landbirds: 15 April–1 June; 15 August–15 October

SOUTHERN MINNESOTA (including Minneapolis and St. Paul):
Waterfowl: 20 March–1 May; 1 October–1 December
Shorebirds: 20 April–20 May; 10 August–20 September
Landbirds: 15 April–15 May; 1 September–1 November

Minnesota winters are generally severe with several periods of sub-zero temperatures, but they do not have the extreme cold and dryness of the western plains states at the same latitude. Regular winter visitants include the Rough-legged Hawk, Horned Lark (northern races), Northern Shrike, Common Redpoll, Lapland Longspur, and Snow Bunting. Winter visitants appearing irregularly are the Bohemian Waxwing, Evening and Pine Grosbeaks, and Red and White-winged Crossbills. In southern Minnesota the Brown Creeper, Pine Siskin, Slate-colored Junco, and American Tree Sparrow pass the winter, sometimes commonly, in sheltered areas, but the majority of individuals continue southward. Occasionally visiting during winter the woods, bogs, or open country in the extreme northern part of the state are the Snowy Owl, Hawk Owl, and Great Gray Owl.

Authorities

Ron E. Adams, W. J. Breckenridge, K. R. Eckert, Bradley D. Ehlers, Janet C. Green, Pershing B. Hofslund, Roger Holmes, Robert B. Janssen, Federick Z. Lesher, Ron E. Miles, David F. Parmelee, Orwin A. Rustad, Gary H. Swanson, Sarah S. Vasse, Richard and Gloria Wachtler, David K. Weaver.

References

Minnesota Birds: Where, When, and How Many. By Janet C. Green and Robert B. Janssen. Minneapolis: University of Minnesota Press, 1975.

A Birder's Guide to Minnesota. By K. R. Eckert. Distributed by the James Ford Bell Museum of Natural History, Minneapolis, MN 55455.

BAUDETTE
'Big Bog Country' | Red Lake Wildlife Management Area

About 365 miles north of Minneapolis and St. Paul, between Baudette on the Canadian border and the Red Lakes to the south, is the **'Big Bog Country,'** a level wilderness of seemingly endless proportions. Much of the wilderness is typical 'quaking' bog with a sphagnum mat on which there are 'islands': some of densely growing black spruce and tamarack with some red cedar; others of aspens, red pines, and jack pines. Here and there are wide stretches of sedges and deciduous shrubs.

State 72, reached from State 11, 2 miles east of Baudette, passes south through many miles of Big Bog Country. In June and early July, pay special attention to the islands on both sides of State 72, as soon as it begins, for Connecticut and Palm Warblers. Listen and look for Connecticut Warblers on all the islands, Palm Warblers particularly on the aspen-pine islands.

The **Red Lake Wildlife Management Area,** embracing some 427,570 acres of Big Bog Country, offers nearly all the birds characteristic of the northern coniferous forest. From Baudette, drive west on State 11 to Roosevelt, a distance of 20 miles; turn south onto State Forest Service Road and continue for about 15 miles to Norris Camp in the Area. This is an information center, office of the supervisor, and main entrance to 225 miles of service roads through the Area. Although permission to enter the Area and use the roads is not required, the bird finder is nonetheless advised to notify the supervisor of his intentions and determine which roads are the most productive ornithologically. There are no public accommodations, but camping is permitted in specified places. Never venture from any of the roads without a compass.

Among the breeding birds that should be looked for especially are Gray Jays near frequently used campsites; Red Crossbills in red pines; Spruce Grouse, Yellow-bellied Flycatchers, Boreal Chickadees, Ruby-crowned Kinglets, and Northern Waterthrushes primarily in spruce-tamarack islands; Black-backed Three-toed Woodpeckers where there are dead trees and stumps. The Red-tailed and Broad-winged Hawks and Great Horned Owl are the most common raptors, the Northern Goshawk, Northern Harrier, and American Kestrel less so. The Merlin and Hawk Owl are always possibilities.

In winter the Hawk Owl is more likely, along with other visitants such as the Great Gray Owl, Boreal Owl, and Northern Raven.

CALEDONIA
Beaver Creek Valley State Park

An enchanting break in the farmlands of extreme southeastern Minnesota is **Beaver Creek Valley State Park** (615 acres). Here a stream wends its way through a deep ravine whose steep 300-foot slopes support a mature forest of maple, basswood, and walnut interrupted by impressive outcroppings of sandstone and limestone. The forest is of special interest ornithologically as the chief nesting habitat of the Acadian Flycatcher in Minnesota.

To reach the Park from Caledonia, drive west 0.5 mile on State 76; where it turns north, continue west 4 miles on County 1 to the entrance. At headquarters, obtain a map showing the trails which meander up the ravine slopes. Recommended for the Acadian Flycatcher is the Switch Back Trail. Among other birds to note along the way are the Blue-gray Gnatcatcher, Blue-winged Warbler, and Cerulean Warbler calling or singing high in the trees.

CORRELL
Pelican Island

Many American White Pelicans and few numbers of Double-crested Cormorants nest on **Pelican Island** in Marsh Lake, an impoundment of the Minnesota River within the 28,000-acre Lac Qui Parle Wildlife Management Area in southwestern Minnesota. Less than an acre in size, the Island may be readily seen from the north shore of Marsh Lake, reached from Correll on State 7 by a public access road leading 1.25 miles south. The birds are best viewed in early summer, at which time there are large numbers of young and considerable movement of the attending adults to and from the Island.

DETROIT LAKES
Tamarac National Wildlife Refuge

From this community in west-central Minnesota, drive 9 miles east on State 34 to intersecting County 29; turn left and drive north for another 9 miles to the headquarters of **Tamarac National Wildlife Refuge.** Inquire here about nature trails and roads leading to the best opportunities for viewing birds.

This preserve of 42,484 acres has vast, level uplands, 24 lakes, many marshes, and wet lowlands. Since it lies in a region of transition between the western prairies, the northeastern coniferous forest, and the southern hardwoods, it attracts birds characteristic of all three natural realms.

On trips along Refuge roads in June and July, one can be almost certain to see broods of such waterfowl as the Canada Goose, Mallard, Blue-winged Teal, Wood Duck, Ring-necked Duck, and Common Goldeneye. From early October to freeze-up, many geese and thousands of ducks stop during their southward flight. Lesser Scaups are remarkably abundant from mid-October to mid-November.

DULUTH
Hawk Ridge Nature Reserve | Minnesota Point | Sax-Zim Bog | Gooseberry Falls State Park | North Shore of Lake Superior | Grand Marais Harbor

This port city, at the westernmost extremity of Lake Superior, offers an unexcelled opportunity to watch southbound migrating hawks in impressive numbers as they veer westward around Lake Superior rather than flying directly south across its watery expanse.

The southbound raptors may be viewed from the bluffs that overlook Duluth from the west. The birds tend to fly low—sometimes they can be looked down upon—and frequently close enough for ready identification. Peak flights of the more common species may be expected as follows: American Krestrels, the last week of August through September; Sharp-shinned Hawks,

Northern Harriers, and Ospreys, September through mid-October; Broad-winged Hawks in September; Northern Goshawks, Rough-legged Hawks, and Golden and Bald Eagles, mid-October through late November. The ideal time is mid-September when, on a clear day with strong westerly winds, Broad-winged Hawks pass in spectacular numbers along with a good many Sharp-shinned Hawks and Northern Harriers.

Hawk Ridge Nature Reserve (300 acres), high on the bluffs, has excellent vantages for watching the flights. Drive north through Duluth on US 61 (Superior Street, then London Road), left on 45th Avenue East for 1.0 mile, left again on Glenwood Street for 0.8 mile, then sharp right on Skyline Parkway for 1.0 mile to the Reserve.

For bird finding at Duluth in winter as well as in spring and fall, **Minnesota Point** is highly productive. A 6-mile sand spit extending into Lake Superior and separating it from Superior Bay, the Point is severed from the mainland by the Duluth Shipping Canal over which the Aerial Lift Bridge passes. Although the Point is heavily developed with homes and industries for over half its length, it has sandy beaches, a 100-acre landscaped park at the end of the road, and 1.0 mile of naturally wooded dunes beyond the road. Superior Bay remains partly ice-free through most of the winter owing to winter shipping of taconite and coal.

To reach Minnesota Point, turn east off US 61 (Superior Street) in downtown Duluth on Lake Avenue and proceed on it across the Aerial Bridge; then take Minnesota Avenue which passes down the Point as far as 43rd Street, where the Recreation Center begins. Along the way, turn off on any of the streets, to the Lake on the left, the Bay on the right, for scanning open water; the area around Hearding Island on the Bay side at 19th to 23rd Streets is the most productive. From 43rd Street, one may take the Bayside Boulevard to Sky Harbor Airport and from there walk to the tip of the Point.

The beaches and open grassy areas near the Recreation Center and Sky Harbor Airport are especially good for shorebirds. Through May, and from early August to mid-October, look not only for the expected varieties of shorebirds—Semipalmated Plovers, Lesser Golden Plovers, Black-bellied Plovers, Ruddy Turnstones, Greater and Lesser Yellowlegs, Pectoral Sandpipers, Least Sandpipers, Dunlins, Semipalmated Sandpipers, and Sanderlings—but also for Whimbrels, Red Knots, White-rumped Sandpipers (spring), Baird's Sandpipers, Short-billed and Long-billed Dowitchers, Stilt Sandpipers, Buff-breasted Sandpipers (fall), and an occasional Western Sandpiper.

Offshore birdlife warrants attention as there is always a chance in fall, winter, and spring of a rare loon, diving duck, jaeger, or gull. Wintering gull concentrations usually contain a few Glaucous Gulls. In May there are large numbers of transient Bonaparte's Gulls. Snowy Owls are a strong possibility on the Lake or Bay ice in midwinter. In both spring and fall, hawks, particularly Sharp-shinned, migrate along the dunes. The Recreation Center is invariably rewarding for the warbler migration in May and for northern tundra birds—Lapland Longspurs and Snow Buntings—in fall.

Accessible within 50 miles of Duluth are many extensive spruce-tamarack bogs. Recommended is the **Sax-Zim Bog,** reached by driving north from Duluth on US 53, west on County 133 which passes at once along the south side of the Bog for 6 miles, then north on County 7 through the Bog for 16.5 miles to County 27. Go west on County 27 for 10 miles along the north edge of the Bog, then south on County 5 to County 133 again. County 7, County 5, and the roads connecting them traverse fine habitats for boreal-bog birds: Gray Jays and Boreal Chickadees the year round; Yellow-bellied Flycatchers, Palm Warblers, and Lincoln's Sparrows in late spring and summer; many hawks in fall and early winter and possibly Hawk Owls, Great Gray Owls, and Northern Shrikes in winter.

The scenic **North Shore of Lake Superior,** northeast from Duluth for 150 miles to Canada, is ornithologically worthwhile in any season. From Duluth, take the North Shore Highway (State 61) along the Lake for 21 miles to Two Harbors. Along the way in winter, be alert for a Boreal Owl, Northern Ravens, Bohemian Waxwings, Pine Grosbeaks, and Common Redpolls. Stoney Point, 11 miles along the North Shore Highway from Lester River in Duluth, is a good place for rare vagrants and to watch migration in the fall.

Gooseberry Falls State Park (638 acres), 15 miles northeast of

Two Harbors on US 61, features two waterfalls of the turbulent Gooseberry River and the bold rocky shore of Lake Superior. Footpaths lead through woods containing birch, aspen, and alder and some stands of balsam-fir and white cedar, sometimes mixed with spruce and pine. Among the birds regularly breeding are the Red-breasted Nuthatch, Swainson's Thrush, Veery, Golden-crowned Kinglet, at least a dozen warblers—Black-and-white, Nashville, Magnolia, Myrtle, Black-throated Green, Blackburnian, Chestnut-sided, Ovenbird, Mourning, Common Yellowthroat, Canada, American Redstart—Purple Finch, and White-throated Sparrow.

Grand Marais Harbor at Grand Marais, on US 61, 105 miles from Duluth, invariably has a good variety of wintering ducks that includes Mallards, American Black Ducks, Common Goldeneyes, Oldsquaws, and Common Mergansers, as well as rarities from northern climes. The birds may be viewed easily from either the city park and power plant on the west side of the Harbor, or from the Coast Guard Station on the east side.

ELY
Superior National Forest

In the northeastern corner of Minnesota, called the Arrowhead Country because of its shape, are nearly 4 million acres comprising **Superior National Forest.** Within its boundaries are more than 1,200 lakes and tremendous stretches of splendid forests. For 108 miles along its northern border, which is also the International boundary, a million acres of wilderness are reserved as the Boundary Waters Canoe Area. Across the border Canada has set aside a great roadless area named Quetico Provincial Park. Travel by canoe is the method of transportation in both the American and Canadian areas, thus keeping them in primitive condition.

Much of the terrain in Superior National Forest is rugged, in some places so hilly as to be virtually mountainous. The lakes, crystal clear and cool, have shorelines varying from ledges and loose rock to sandy beaches, and are linked by rivers and streams that greatly facilitate travel by canoe. The forests are primarily coniferous. In bogs they consist of black spruce and tamarack; in upland areas they comprise white, red, and jack pines, white spruce, and balsam-fir mixed with white birch, aspen, sugar maple, yellow birch, and other deciduous stands, especially on ridges near Lake Superior. Where there is muskeg country—and there is much of it—deciduous shrubs such as Labrador tea and leather-leaf grow abundantly. Birdlife of Superior National Forest is strongly typical of northern coniferous forests.

A wide choice of travel awaits the bird finder. He may go on foot or by car, canoe, snowmobile, or skis, depending on his inclinations, financial resources, physical stamina, and available time. Prior to his trip he should write the Forest Service Office, Superior National Forest (address: Box 338, Duluth, MN 55801) for a recreation map, visitor regulations, and tourist information.

Ely, an iron-mining and summer-resort town 120 miles northeast of Duluth, is the most practicable place from which to begin ornithological explorations. Leading from Ely into the National Forest are well-marked routes that usually have picnic and camping facilities. With a recreation map at hand for reference, take the following trip by car for several bird specialties.

Drive southeast from Ely on State 1 to the South Kawishiwi River Recreation Site. During the nesting season the woods in the area are worth investigating for Black-backed Three-toed Woodpeckers as well as for such north-country birds as Yellow-bellied Flycatchers, Olive-sided Flycatchers, Red-breasted Nuthatches, Winter Wrens, and numerous warblers. As many as twenty-four species of warblers breed in Superior National Forest including the Tennessee, Cape May, Bay-breasted, and Connecticut.

Continue southeast and then east on State 1, passing the junction with County 2 on the right, to Isabella; here turn left and continue for 1.0 mile on National Forest Road 172, left again onto NFR 369 which passes north toward Sawbill Landing through coniferous bogs, where Spruce Grouse are a good possibility. The black spruce-jack pine stand at the junction of 172 and 369 is a good spot to listen for the Cape May and Bay-breasted Warblers during the breeding season.

Return west to State 1 from Sawbill Landing via NFR 173, 373, and then 173 again. The bog at the junction of NFR 173 and State

1 is a typical nesting habitat for Yellow-bellied Flycatchers, Boreal Chickadees, and Connecticut Warblers. Both the Black-backed and Northern Three-toed Woodpeckers have been found here in winter.

Reference

Birds of the Superior National Forest. Compiled by Karl P. Siderits. Eastern Region, Forest Service, U. S. Department of Agriculture. Available from U. S. Government Printing Office. 1978.

FARIBAULT

Nerstrand Woods State Park | Heron Island

A fine remnant of the 'Big Woods' and excellent for woodland birds is **Nerstrand Woods State Park** (587 acres), 13 miles northeast of Faribault and 80 miles south of Minneapolis. To reach the Park from Faribault, exit east from I 35 on State 60 through the city. After crossing the bridge over a river, turn left onto County 20, then right onto County 27, and finally left onto County 40 to the Park entrance with headquarters and parking lot.

The forest, much of it in virgin condition, is entirely deciduous,

PILEATED WOODPECKER

the predominant trees being sugar maple, basswood, American and slippery elms, and red oak. Besides the majestic stands of timber, the Park's scenic attractions are a rolling terrain and Prairie Creek with its two waterfalls. Trails from the parking lot, leading down to and along Prairie Creek, offer the best bird-viewing opportunities. During the second week of May at the peak of the warbler migration it is not unusual to see as many as seventy-five species. Breeding birds include the Red-tailed Hawk, Pileated and Red-bellied Woodpeckers, Blue-gray Gnatcatcher, Veery, Cerulean Warbler, and Scarlet Tanager.

Ten miles northwest of Faribault is a heron colony on **Heron Island,** a county-owned bird sanctuary in General Shields Lake. Exit west from I 35 on State 21 and continue through Shieldsville to the Lake on the left just beyond the town. Farther along, on the northwest shore, is county-owned McCullough Park where boats may be rented.

Heron Island, the only island in the Lake, looms high near the far shore, its surface of 6.4 acres covered entirely by trees, mainly elms, maples, and boxelders. The colony, consisting mostly of Great Blue Herons with small numbers of Great Egrets and Black-crowned Night Herons, occupies the tops of the tallest trees. Any time from mid-May to mid-July is best for viewing the colony because there are young in the nests during this period, consequently more activity. Although the Island is posted against landing, the colony can be closely observed from a boat.

FRONTENAC STATION

Frontenac Area | Frontenac State Park

About 60 miles southeast of Minneapolis-St. Paul, on a widening of the Mississippi called Lake Pepin, lies the **Frontenac Area,** famed as a site for the spring warbler migration. Between 15 May and 20 May when the migration usually peaks, it is possible to record as many as thirty species of parulids, together with a wide representation of other birds. Take the following trip.

From Frontenac Station, proceed southeast on US 61 to the sign 'Villa Maria' at the intersection of County 2 going north. Stop on the left to explore the wayside rest area for warblers and Blue-gray Gnatcatchers. Go north on County 2, stopping frequently. After about a mile, where County 2 turns sharply left, take the gravel road that angles down to the right. Follow it through the Methodist Campus and park. North of the buildings between the road and Lake Pepin is a small wooded depression—the 'mudhole'—which is one of the best spots for warblers, particularly waterthrushes, as well as for thrushes, vireos, and Baltimore Orioles. Continue through the woods to the Lake for a possible Pileated Woodpecker. Return to the gravel road and walk north, checking the ravine on the left for more warblers and Rose-breasted Grosbeaks.

Come back to County 2 and drive west on it, watching for Indigo Buntings. Continue on County 2 through the town of Frontenac and then turn off north on the entrance road to **Frontenac State Park** (562 acres). This passes through open country for such birds as Red-tailed Hawks, Eastern Kingbirds, Eastern Bluebirds, Bobolinks, both Eastern and Western Meadowlarks, and Vesper, Clay-colored, and Field Sparrows.

Come back again to County 2 and follow it southwest. Just before reaching Frontenac Station on US 61, County 2 traverses a marshy area and open water, attractive to herons, ducks, Virginia Rails, Soras, American Coots, and transient shorebirds.

LUVERNE
Blue Mounds State Park

A veritable oasis for birds in the plowed and grazed prairie of extreme southwestern Minnesota is **Blue Mounds State Park** (1,300 acres). Here are two small lakes connected by a little stream thickly bordered by woods, a large tract of prairie with quartzite outcroppings, and, along the eastern border of the Park, quartzite cliffs 40 to 100 feet high and heavily wooded at their base.

From I 90, exit north on US 75 for 0.1 mile to Luverne and continue north on US 75 for 4.5 miles to the Park entrance on the right. At headquarters, obtain a map and drive eastward on a half-mile road that gives access to the wooded stream on the left and the prairie (where there is a small herd of bison) on the right, and that dead-ends near where the cliffs begin to rise. In suitable habitat during the late spring and summer, look for such birds as the Green Heron, Swainson's Hawk, Gray Partridge, Upland Sandpiper, Western Kingbird, Traill's Flycatcher, Orchard Oriole, Blue Grosbeak, and Grasshopper Sparrow.

The Park offers the best bird finding from 1 April to 15 June and from 15 August to 1 November when many other species are passing through in migration. At these times of year a walk along the cliffs can be especially rewarding for vireos, warblers, and other passerines at the wooded base, for hawks on top.

McGREGOR
Rice Lake National Wildlife Refuge

In east-central Minnesota the **Rice Lake National Wildlife Refuge** is an inviting wilderness of 20,296 acres, 4,500 of which are water surface. Surrounding Rice Lake are extensive wet lowlands with islands of tamarack; in the Lake itself are large beds of wild rice and luxuriant growths of pondweeds, bulrushes, and cattails.

Common Loons reside regularly in summer as do Pied-billed Grebes, American Bitterns (probably Least Bitterns and Virginia Rails), Soras, Marsh Wrens, and Sedge Wrens. Canada Geese, Mallards, American Black Ducks, Common Pintails, Blue-winged Teal, American Wigeons, Wood Ducks, and Hooded Mergansers breed, but waterfowl are in greater abundance during migration, particularly in October, when the population is swelled by southbound Snow Geese, Ring-necked Ducks, and Lesser Scaups among other species.

The upland portions of the Refuge support stands of mixed hardwoods, chiefly maple and ash, and some stands of spruce and fir. Here the Ruffed Grouse is common. Where there are open brushy areas the Sharp-tailed Grouse resides in small numbers.

To reach the Refuge, drive south from McGregor on State 65 for 5 miles to the very small town of East Lake; then turn west onto an unnumbered dirt road which goes directly to headquarters on the north side. Several trails leading away from headquarters provide excellent opportunities for watching the warbler migration in May.

MINNEAPOLIS—ST. PAUL

Theodore Wirth Park | Eloise Butler Wildflower and Bird Sanctuary | Thomas Sadler Roberts Bird Sanctuary | Lake Harriet | Carver Park Reserve | Lowry Nature Center | Sunny Lake Refuge | 'Shorebird Management Area' | Fred E. King Waterfowl Sanctuary | Black Dog Lake | Minnesota River Bottom

The Twin Cities have many parks, lakes, and tree-shaded byways for good bird finding. Probably **Theodore Wirth Park,** a 681-acre tract 0.5 mile wide and 2.30 miles long about 3 miles west of downtown Minneapolis, offers the greatest variety of landbirds. This is due largely to varied habitats since the Park's rolling terrain includes woods, open spaces, glens, springs, open water (Theodore Wirth Lake, Birch Pond, and Brownie Lake), rivulets, and a meandering stream. Oaks thrive on the uplands, boxelders, ashes, and willows on the lowlands. Numerous introduced conifers such as jack pine and Douglas-fir, together with many deciduous shrubs, cover much of the area. In winter the Park is attractive to birds, owing to certain springs that remain open in freezing temperatures, evergreens and shrubs which hold their fruit in cold weather, and feeding stations that supplement the natural food supply.

Part of the Minneapolis Park System, Theodore Wirth Park is conveniently reached from the west side of the city by exiting east from I 494 on State 55 (Olson Memorial Highway) for 7.3 miles, then turning south onto Theodore Wirth Parkway through the Park. Although almost any section of the Park is productive for bird finding, a recommended starting point is near Theodore Wirth Lake where the Parkway is crossed by Glenwood Avenue. Footpaths from parking facilities here lead to all the better sites for birds, especially the **Eloise Butler Wildflower and Bird Sanctuary,** a 60-acre parcel within the Park south of Glenwood Avenue. The Sanctuary has parking facilities for immediate access.

The best time for bird finding is early May when the Park's birdlife is vastly augmented by northbound transients. Among them are twenty or more warblers, the Tennessee, Orange-crowned, Myrtle, and Chestnut-sided being the most common.

Hermit Thrushes and White-throated and Fox Sparrows abound; Swainson's and Gray-cheeked Thrushes as well as Veeries are well represented.

Summer-resident birds include the Green Heron, Wood Duck, Great Crested Flycatcher, Eastern Pewee, Yellow-throated, Red-eyed, and Warbling Vireos, Rose-breasted Grosbeak, and Indigo Bunting. Northern Cardinals are year-round residents as are Pileated, Hairy, and Downy Woodpeckers, Black-capped Chickadees, and White-breasted Nuthatches. Species showing up in the winter, some more regularly than others, are the Red-breasted Nuthatch, Bohemian and Cedar Waxwings, Evening Grosbeak, Purple Finch, Pine Grosbeak, Common Redpoll, Pine Siskin, Red and White-winged Crossbills (particularly attracted to seeds in the Douglas-fir cones), and Slate-colored Junco.

Especially good for thrushes, warblers, and other passerines during spring and fall migrations is the **Thomas Sadler Roberts Bird Sanctuary** in Lyndale Park, part of the Minneapolis Park System, on the north side of Lake Harriet about 5 miles southwest of downtown Minneapolis. From I 494 south of the city, exit north on I 35W for 1.6 miles to the State 190 Exit from the left lane; go northward 2.5 miles on State 190 (becomes Lyndale Avenue South), turn west (left) onto West 46th Street and proceed 0.5 mile, then turn north (right) on East Lake Harriet Boulevard and go 0.4 mile to the stop sign. Here jog left to remain on the Boulevard and reach the formal Lake Harriet Rose Garden just 0.1 mile distant. Immediately past the Garden, turn right on Roseway Road and leave the car in the parking lot across the Road from the Garden. The Roberts Sanctuary, entered off the north end of the parking lot, is a fenced area of 31 acres with a tree- and shrub-bordered path and a marsh through which a stream passes.

Before or after visiting the Roberts Sanctuary, a walk or drive around **Lake Harriet** can be profitable for viewing transient diving ducks such as Common and Red-breasted Mergansers. Black-crowned Night Herons are frequent around the Lake.

Rewarding for birds in any season is Hennepin County Park Reserve District's **Carver Park Reserve,** a 3,700-acre open space park about 25 miles southwest of Minneapolis. The gently rolling land is dotted with lakes, marshes, swamps, old fields, and mature forests. Areas of outstanding interest are **Lowry Nature Center, Sunny Lake Refuge,** a **'Shorebird Management Area,'** and the **Fred E. King Waterfowl Sanctuary.**

Lowry Nature Center maintains a feeding station the year round. Often in view from the building are Pileated and Red-bellied Woodpeckers, Harris' and White-throated Sparrows in spring and fall, and occasionally Northern Shrikes and various owls in winter. A system of trails leads from the Center. *Maple Trail* goes through a mature forest, where Scarlet Tanagers nest, to Sunny Lake Refuge (partly ice-free in winter) where there are resident Canada Geese (the giant subspecies) and captive Trumpeter Swans. Least Bitterns are summer residents in the bordering marsh. *Tamarack Trail* passes meadows with summer-resident Bobolinks to a tamarack swamp where, from a boardwalk, one may observe American Woodcock flight-singing at dusk in early spring and later hear or see Traill's Flycatchers and Veeries.

An unofficial 'Shorebird Management Area,' adjacent to the Lowry Nature Center entrance, features ponds with manually controlled water levels. During late April, May, August, and September, the mud flats attract transient Common Snipes, Solitary Sandpipers, Greater and Lesser Yellowlegs, Pectoral, Baird's, and Least Sandpipers, Short-billed and Long-billed Dowitchers, Stilt, Semipalmated, and Western Sandpipers, and Wilson's Phalaropes. In March and April, middle pond supports a good population of both surface-feeding and diving ducks. In open country adjacent to the ponds are Tree Swallows, Eastern Bluebirds, Bobolinks, Eastern and Western Meadowlarks, Savannah Sparrows, and, in winter and early spring, Short-eared Owls.

The Fred E. King Waterfowl Sanctuary is accessible by a hiking trail from the Park Reserve's Parley Lake Tent Campground. The Sanctuary's main lake has many shallow bays, marshy areas, and mud flats. Besides Common Loons, which have nested, are Canada Geese, surface-feeding ducks, and numerous summer visitants, or transients in spring or fall, such as Great Egrets, Herring, Ring-billed, Franklin's, and Bonaparte's Gulls, and Forster's and Common Terns. As in the Shorebird Management Area, the mud flats attract many shorebirds.

The bird finder visiting Carver Park Reserve should go first to the Lowry Nature Center. To reach it from I 494, exit west on State 5 for 12.3 miles through the village of Victoria, then turn north (right) onto County 11 and proceed 1.6 miles to the Nature Center entrance gate on the east side of the road. At the Center, obtain directions to the aforementioned areas and inquire about any birds of special interest currently available for viewing.

Excellent for over-wintering waterfowl and an occasional gull from northern climes is **Black Dog Lake** south of the Twin Cities along the Minnesota River. Because of the warm water effluent from the adjacent Black Dog Power Plant, about 250 acres of water stay open during the winter. Great numbers of Mallards and a sizable flock of Common Goldeneyes consequently remain even through the coldest months. A good representation of other waterfowl stop off temporarily in spring migration. Great Blue Herons and Great Egrets frequent the marshy edge of the Lake in the late spring and summer.

To reach Black Dog Lake from I 494, exit south on I 35W for 4.4 miles to Exit 4B (113th Street); go east under I 35W on Black Dog Road which runs between Minnesota River on the left and Black Dog Lake on the right. The eastern end of the Lake is accessible for good viewing by driving through the Power Plant site.

Bird finding is highly productive during the migration periods along the **Minnesota River Bottom** west of Black Dog Lake. Here there is a fine variety of habitats—shallow lakes, tributary creek bottoms, springs, swamps, marshes, and wet meadows. On the north side of the River there are heavily wooded hills and grassy bluffs called 'goat prairies.' Take the following loop tour, stopping at areas that seem promising.

From I 494, exit south on I 35W for 5.9 miles to Exit 3B (State 13 South) and proceed west on State 13 to State 101, then continue west on State 101 to Shakopee (about 12.3 miles from I 35W). Here go north on US 169 across the Minnesota River and east 4.4 miles; bear off right on River View Road to the intersection with County 18 which heads north to I 494, 4.7 miles distant.

River View Road has several overlooks for scanning the River Bottom and its lakes including Grass Lake where Great Blue Herons and Great Egrets come from their nesting colony at nearby Blue Lake to feed and rest. From State 101 south of the River there is a public access to Rice Lake whose mud flats attract transient shorebirds.

ORTONVILLE
Big Stone National Wildlife Refuge

Big Stone National Wildlife Refuge, lying close to the South Dakota border, attracts a fine variety of breeding waterbirds and waterfowl largely because some 4,000 of its 10,800 acres consist of open water and marsh resulting from impoundment of the Minnesota River. Both Western and Pied-billed Grebes nest as do Least and American Bitterns, Northern Harriers, Virginia Rails, Soras, American Coots, Franklin's Gulls, Forster's and Black Terns, and the following ducks: Mallard, Gadwall, Common Pintail, Green-winged and Blue-winged Teal, American Wigeon, Northern Shoveler, Wood Duck, Redhead, Canvasback, Ruddy Duck, and Hooded Merganser. Double-crested Cormorants and a few Great Blue Herons and Great Egrets colonize a small island in one of the pools.

To reach the Refuge from Ortonville, just inside the Minnesota border, drive southeast on US 75 for 2 miles, then turn off right to the Refuge and take the 5.4-mile auto tour route which offers excellent views of the marsh and its birds. To see the cormorant-heron-egret colony, return to US 75 and continue east 4 miles; turn off right and proceed 2 miles to the Refuge gate, enter, and drive to the shore of the pool where the island will come into view to the southwest.

PARK RAPIDS
Itasca State Park | Upper Rice Lake

Itasca State Park, midway between Park Rapids and Bemidji and 225 miles north of Minneapolis and St. Paul, enfolds 32,820 acres of natural beauty and charm. Within its boundaries are more than one hundred lakes, numerous bogs, magnificent virgin stands of red and white pines, and great stretches of young forests. To reach the Park from Park Rapids, drive north on US 71 for 17 miles to the South Entrance at left. Obtain a map here showing the auto drives and foot trails, then proceed north on Park Drive for 3 miles to the Contact Station (headquarters) near Douglas Lodge on the eastern arm of Lake Itasca.

The largest of the lakes in the Park and the source of the Mississippi River, Lake Itasca attracts many birds. Common Loons, Great Blue Herons, and Ospreys fish in its waters; a pair of Bald Eagles rears its young each year in an aerie near the western arm across from Chambers Creek; ducks, among them Wood Ducks and Hooded Mergansers, frequent Floating Bog Bay and the entire western shore.

Itasca Park yields a remarkable mixture of landbirds because it lies at a juncture of the northern coniferous forest, southeastern deciduous forest, and western prairie grassland. In addition to the primeval stands of pines are extensive stands of black spruce and tamarack in bogs, jack pine, balsam-fir, white spruce, willow, aspen, white birch, basswood, and sugar maple on the uplands. Thus it is possible on any given day in early summer to encounter both Gray Jays and Black-billed Magpies, or Yellow-throated and Solitary Vireos, or even Golden-winged and Connecticut Warblers.

Among the several trips for birds that one may take in the Park, the following three are recommended for the widest variety between mid-May and mid-July.

1. *Wilderness Drive* which goes west from near the Contact Station and eventually swings north and finally east, terminating after 15 miles at the Headwaters Interpretive Center near the north boundary. Stop now and then to look for Scarlet Tanagers, Rose-breasted Grosbeaks, Indigo Buntings, and Evening Grosbeaks. At the Wilderness Sanctuary parking area, leave the car and carefully search the general area. Both Pine and Black-throated Green Warblers are common in the towering pines. Be alert for Black-backed Three-toed Woodpeckers, Gray Jays, and Brown Creepers. Always possibilities here and elsewhere along the Drive are the Spruce Grouse, Saw-whet Owl, both Red and White-winged Crossbills, and all three accipiters—Northern Goshawk, Sharp-shinned Hawk, and Cooper's Hawk.

2. *Schoolcraft Trail* which goes south from Headwaters Interpretive Center on the west side of Lake Itasca for slightly over a

mile. In addition to the Broad-winged Hawk, Pileated Woodpecker, Olive-sided Flycatcher, and Red-breasted Nuthatch, as many as fifteen species of warblers may be heard or seen. French Creek Bog, just south of Hill Point at the terminus of the Trail, is especially productive.

3. *La Salle Trail* which goes east from the University of Minnesota Forestry and Biological Station for 1.5 miles to the La Salle Bog in the northeast corner of the Park. In the Bog, listen and watch for the Yellow-bellied Flycatcher, Boreal Chickadee, Winter Wren, Golden-crowned Kinglet, Tennessee, Nashville, and Blackburnian Warblers, and, in the spruces or tamaracks festooned with *Usnea* lichen, the Northern Parula Warbler. On the edges of the Bog and the nearby slopes where there is open secondary woods with shrubby spots are such birds as the Ruffed Grouse, American Woodcock, Whip-poor-will, Hermit Thrush, Golden-winged, Chestnut-sided, and Mourning Warblers, Ovenbird, Rufous-sided Towhee, and White-throated Sparrow.

The Biological Station, within the Park on the east side of Lake Itasca, is reached from Park Drive either by proceeding north from the Contact Station for 3 miles, or south from the Headwaters Interpretive Center for 2.5 miles. Ornithologists at the Station are helpful in pinpointing the most likely sites for particular species.

Upper Rice Lake, north of Itasca State Park, is notably attractive to breeding waterbirds and waterfowl. To reach it, leave the Park via the North Entrance from Park Drive to the community of Lake Itasca; continue north 9.5 miles on County 2; then turn west on County 36 to the Lake 3 miles distant.

Numerous pairs of Red-necked Grebes may be seen nesting offshore. Other birds to look for are Common Loons, Ring-necked Ducks, and Common Goldeneyes on the open water, American and Least Bitterns and Marsh and Sedge Wrens in the marshy borders.

ROTHSAY
Rothsay Wildlife Management Area

For a fine remnant of virgin prairie with its associated birdlife, visit the **Rothsay Wildlife Management Area** (2,947 acres) in northwestern Minnesota during spring and early summer. From I 94 between Fergus Falls and Moorhead, exit west to Rothsay; just south of town, take County 26 west for 4 miles, then turn north into the Area for over a mile.

The Area is very flat with some open water and minor depressions supporting emergent aquatic plants and sedges. Most of the Area, however, contains short-grass prairie with springs and ground seeps that keep it wet and long ago saved it from cultivation. No 'sea of grass' as prairies are sometimes described, it has a rich flora of some one hundred or more herbaceous species that typified the prairie of western Minnesota before settlement. Seeing this unspoiled prairie when many of the plants are producing showy flowers, one can readily understand what the first settlers meant by 'the blooming prairie.'

In April, Sandhill Cranes stop in the Area during their northbound migration and Greater Prairie Chickens perform their booming displays at sunrise. Both species are shy but may sometimes be observed closely from the entrance road, provided one stays in his car. Breeding birds to be looked for later in spring and early summer include the Upland Sandpiper, Marbled Godwit, Short-eared Owl, Sedge Wren, Bobolink, Brewer's Blackbird, Henslow's Sparrow, and possibly the Le Conte's Sparrow.

ST. CHARLES

Whitewater State Park | Whitewater Wildlife Area

Typical of the upper Mississippi bottomland country are **Whitewater State Park** and **Whitewater Wildlife Area,** midway between Winona and Rochester in southeastern Minnesota.

From I 90, exit north on State 74 for 0.1 mile to St. Charles. After crossing US 14 in St. Charles, continue north on State 74 for 6 miles to the Park. Its 1,600 acres include a deep, broad valley, through which State 74 descends, with precipitous walls ascending 300 to 400 feet above the valley floor. Whitewater River, in most places a shallow stream, passes through the valley. Although the surrounding country is undulating farmland, the Park is densely forested with mature, mixed hardwoods—mostly oak—and some white pine and red cedar. Certain parts of the Park have been developed for recreational purposes—golfing, picnicking, camping, and swimming—but this does not interfere with good bird finding.

From alder thickets in the early spring American Woodcock emerge for their crepuscular flight-songs. Later in the spring and in the early summer, Whip-poor-wills utter their nighttime chants from the woods; Turkey Vultures ride on warm air currents rising from the valley floor; Cliff Swallows sometimes nest in a colony under a highway bridge; Blue-winged Warblers call or sing frequently from brushy growth on slopes below the bluffs; and, rarely, a Louisiana Waterthrush may appear along the edge of a secluded stream. Footpaths lead into the deep woods where the Ruffed Grouse and Pileated Woodpecker reside the year round and the Wood Thrush, Cerulean Warbler, and Scarlet Tanager are common summer residents.

Continue north from the Park on State 74. Soon after passing through the small village of Elba, State 74 enters the 31,000-acre Whitewater Wildlife Area. Travel this road slowly; stop and look frequently. In winter, Golden Eagles soar above the bluff tops on the east side of Whitewater River; Bald Eagles perch near the ponds west of the road, presumably with an eye on the hundreds of Mallards that winter on the ponds. Red-shouldered Hawks hunt along the river bottom in winter and spring.

At the large, open intersection of State 74 and County 30 in the village of Beaver, turn east up a long valley to the boundary of the Wildlife Area, then return. Watch and listen carefully along this road for introduced Wild Turkeys. Near the bottom of the bluffs search during the late spring and summer for the Yellow-breasted Chat in brush bordering the fields.

Continue north on State 74 to US 61 along the Mississippi River. Between Beaver and US 61, look for wintering eagles. The Dorer Pools and their immediate surroundings near the north end of the Wildlife Area have breeding habitats for ducks, rails, and passerines including the Tufted Titmouse, Blue-gray Gnatcatcher, Bell's Vireo, Cerulean Warbler, and Yellow-breasted Chat.

THIEF RIVER FALLS
Agassiz National Wildlife Refuge

A must on the itinerary of any bird finder visiting northwestern Minnesota is **Agassiz National Wildlife Refuge,** embracing 61,487 acres of uplands, marsh, and open water, 24 miles northeast of Thief River Falls. To reach the Refuge, drive north from this community on State 32; at 0.5 mile north of the village of Holt, turn east onto County 7 and proceed on an improved road for 11 miles to Refuge headquarters on the left. Stop here for information on a self-guided, 4-mile auto drive over roads through the Refuge.

The upland terrain of the Refuge is flat with varied wildlife habitats. These include many abandoned farmlands reverting to tree growth; scattered groves of aspen, ash, and elm, with shrubby undergrowth; large areas of willows; and a 4,000-acre block of spruce-tamarack bog which is part of the National Wilderness Preservation System. Ruffed Grouse and Sharp-tailed Grouse are well represented, the latter frequenting the boundaries near farmlands during their 'dancing' season in April. Bobolinks reside in uncultivated fields beginning in May. Cliff Swallows establish small colonies under bridges and eaves of buildings, and many Tree and Barn Swallows find suitable nest sites.

The entire marsh, occupying more than 36,000 acres, is exceedingly rich in aquatic plants—cattails, quillreeds, and bulrushes in large contiguous areas; submerged pondweeds, bladderworts, and waterweeds—as well as in fishes and fresh-water invertebrates. Waterbirds and waterfowl are consequently in abundance.

The breeding season in June and July is an ideal time for visiting the marsh. Chief among the attractions is the presence of five species of grebes—Red-necked, Horned (uncommon), Eared (uncommon), Western, and Pied-billed. Canada Geese, Mallards, Gadwalls, Blue-winged Teal, American Wigeons, Northern Shovelers, Redheads, Ring-necked Ducks, Canvasbacks, and Ruddy Ducks nest commonly, American Coots in vast numbers. Double-crested Cormorants and Black-crowned Night Herons colonize on an island in Agassiz Pool, the widest extent of open water; and Great Blue Herons and possibly a few Great Egrets colonize in a remote spruce-tamarack bog within the Wilderness Area. Conspicuous are Franklin's Gulls passing to and from their colonies in the marsh and Forster's and Black Terns in flight over the marsh in which they nest or over open water in which they search for food.

With the exception of Yellow-headed and Red-winged Blackbirds which are conspicuous by their sheer numbers, most of the other birds may be observed only by careful watching or listening: American Bitterns, Virginia Rails, Soras, and Marsh Wrens in the cattails; Sedge Wrens, Savannah and Swamp Sparrows, and, less commonly, Le Conte's and Sharp-tailed Sparrows in wet grassy places fringing the marsh.

In late summer, when the water level is lower, transient shorebirds in great number and variety appear on exposed mud flats and surrounding shallows. The Semipalmated and Black-bellied Plovers, Solitary Sandpiper, Greater and Lesser Yellowlegs, Pectoral, Baird's, and Least Sandpipers, Short-billed and Long-billed Dowitchers, Stilt, Semipalmated, and Western Sandpipers, Marbled Godwit, and Wilson's and Northern Phalaropes are the most regular among the thirty species listed for the Refuge.

In early spring when the ice breaks up and in fall until freeze-up, the Refuge is alive with transient waterfowl: Whistling Swans, Snow Geese among the hordes of Canada Geese, and impressive numbers of Canvasbacks, Lesser Scaups, and Buffleheads.

WAUBUN
Waubun Marsh

Of all the prairie marshes in Minnesota, **Waubun Marsh** in the northwest is the most dependable for Yellow Rails in the breeding season. From the community of Waubun, 25 miles north of Detroit Lakes and 9 miles south of Mahnomen, drive south on US 59 for 2 miles; just before passing the well-marked Mahnomen-Becker County Line, turn off west onto an unimproved and unnumbered road. This almost immediately crosses a railroad grade, bends southward, and then goes west, reaching within 1.5 miles a vast grassy wetland—Waubun Marsh.

Since Yellow Rails, always a challenge to see and nearly always impossible to flush, are ordinarily located by their *ticking* calls, the ideal time to visit Waubun Marsh is on a warm windless evening in June or early July when they are at their noisy best. But there are other birds worth observing here too. Therefore, arrive at the Marsh well before sundown with ample time to (1) find singing Le Conte's and Sharp-tailed Sparrows as well as Sedge Wrens and (2) explore the drier prairie south of the Marsh, reached by backtracking more than a mile on the entrance road and taking the first road south, for Greater Prairie Chickens, Upland Sandpipers, Marbled Godwits, and Grasshopper Sparrows. While listening for Yellow Rails in the evening, provided it is early in the breeding season and calm, one may hear a few prairie chickens booming in the distance.

WHEATON
Mud Lake | Lake Traverse

A great ornithological spectacle is the swan and goose migration in spring (late March to early May) at **Mud Lake** and **Lake Traverse,** long, narrow reservoirs forming Minnesota's western boundary. On a day in mid-April, at the peak of migration, one may see thousands of Whistling Swans, Canada Geese, Snow Geese, and scattered numbers of Greater White-fronted Geese. Several hundred American White Pelicans, in addition to thousands of ducks, swell the throng.

Both Lakes, separated by a dike, have marshy shores rising to cultivated land on the Minnesota side and high, grassy bluffs on the South Dakota side. To reach points closely overlooking their open water and bordering marshes: (1) Drive north from Wheaton on County 9 and turn west on State 236 which soon provides good views of Mud Lake as it crosses a dike to South Dakota. (2) Drive south from Wheaton on State 27 and turn west onto State 117 for good views of Mud Lake on the north and Lake Traverse on the south as it proceeds to South Dakota on a dike between the Lakes. (3) Backtrack and continue south on State 27 which parallels Lake Traverse and overlooks it from certain spots on its way to Browns Valley.

If possible, plan a trip to the Lakes so as to be present near evening when flocks of geese start coming in from grainfields, where they have been feeding during the day, to settle on the water for the night.

ZIMMERMAN
Sherburne National Wildlife Refuge

Forty miles northwest of Minneapolis and St. Paul, the **Sherburne National Wildlife Refuge** embraces 30,552 acres of woodland, shrubby swamps, wet meadows, fields, and open water, including Rice Lake. From US 169, 4 miles north of Zimmerman, turn west onto County 9 and proceed 5.5 miles to headquarters in the Refuge.

The variety of habitats in the Refuge attracts a corresponding variety of birds such as the Common Loon, Wood Duck, Red-tailed Hawk, Ruffed Grouse, Sora, Black Tern, Black-billed Cuckoo, Common Screech Owl, Sedge Wren, Veery, Warbling Vireo, Bobolink, Rose-breasted Grosbeak, and Indigo Bunting. Most of the habitats are accessible from the *Mahnomen Wildlife Trail* which leads south from headquarters and goes over a marsh by boardwalk to an observation tower overlooking Rice Lake, and from the *Blue Hill Hiking Trail* which starts west of headquarters.

Two ornithological features of the Refuge are a few breeding Sandhill Cranes (two nests in 1976) and American Woodcock (two broods in 1976). Woodcock arrive in late March and flight-sing in subsequent evenings through early spring. Inquire at headquarters about sites from which their performance may be observed.

Missouri

GREATER PRAIRIE CHICKEN

In spring days long ago the Missouri prairie resounded at sunup to the low, vibrant 'booms' of the Greater Prairie Chicken. The setting for these sounds was a grassy knoll or meadow. Here were the assembled males, each vigorously maintaining his territory by a succession of antics: quick flights by each bird straight upward, followed, after a drop to the ground, by cackles and baleful squawks; charges, as one bird rushed after another; direct encounters, as two birds attacked each other by slapping their wings or by striking with their bills and feet. The principal feature, however, was the booming display, which was climaxed by rapidly stamping the feet, bringing the pinnated neck feathers to a forward position, inflating the bright orange air sacs, and producing the three-syllabled doleful sound, *old-mul-doon*. The booming of this fine bird may still be heard in Missouri, but in few areas only. Both species and habitat have been greatly depleted by man.

The prairie is one of the three distinctive physiographic regions in Missouri, occupying most of the state north of the Missouri River and a part southwest of it—roughly a section west of a line drawn from Boonville on the Missouri southward through Versailles, Clinton, Appleton City, El Dorado Springs, Greenfield, and Joplin to the western boundary. Throughout, the Missouri prairie is level to rolling country, dissected by numerous rivers and their tributaries. Originally grassland, with deciduous trees along watercourses, or in isolated groves, it is now the principal farming district. Most of the remaining Greater Prairie Chickens reside where there are extensive acreages of native prairie or grassland undisturbed from year to year (*see under* **Clinton**). Other birds sharing the environment with Greater Prairie Chickens include the Upland Sandpiper, Horned Lark, Eastern Meadowlark, and Grasshopper Sparrow.

The Ozarks, the second region, occupy the rest of the state south of the prairie except the southeastern lowlands, which is the third region, extending as far north as Cape Girardeau on the Mississippi and west to points near Poplar Bluff and Naylor. In general, the Ozarks are a broad upland, or plateau, interrupted by innumerable wide valleys, though in several places there are many knobs and peaks with intervening narrow, deep valleys. Few localities have elevations exceeding 1,400 feet. The St. Francis Mountains, in the eastern Ozarks, have a landscape that is rugged, even mountainous. Here Taum Sauk Mountain reaches 1,772 feet, the highest point in the state.

Another scenic locality, made famous through the novel *The Shepherd of the Hills* by Harold Bell Wright, is north, east, and west of Lake Taneycomo in the southwestern Ozarks (*see under* **Branson**). The Ozarks are for the most part tree-clad, though summits of certain knobs and ridges are naturally treeless, or 'bald,' while other areas have been cleared for settlement and agricultural pursuits. The forests vary in character. Most are quite open and some are actually interrupted by prairie-like glades, often with stands of red cedar. The timber, predominantly oaks—red, black, post, and blackjack—and hickory, is medium-sized at lower elevations, but becomes stunted as higher elevations are approached. In some parts of the Ozarks, on the lower slopes and valley floors, there are forests of shortleaf pine mixed with bur oak, black oak, walnut, sycamore, black gum, hackberry, boxelder, black cherry, and other deciduous trees. Some of the birdlife characteristic of Ozark forests may be observed along the Karkaghane Scenic Drive in Mark Twain National Forest (*see under* **Salem**).

The southeastern lowlands, which includes the Bootheel that projects into Arkansas, represent a widening of the Mississippi bottomland south of Cape Girardeau. With an elevation less than 400 feet above sea level and only 10 to 20 feet above the Mississippi, its numerous shallow ponds, meander cutoffs, and bayous increase many times their size by flooding early in the year, yet are almost always dry in late summer. Dense bottomland forests and swamps containing trees of immense size once stood here and still do in certain places as in Big Oak Tree State Park (*see under* **Charleston**). Sycamore, cottonwood, bald cypress, tupelo, sweetgum, magnolia, tulip tree, pecan, persimmon, catalpa, sassafras, various oaks, and other trees characteristic of southern lowlands are common.

Expectedly this region attracts birds of southern affinities. Among breeding species are the Little Blue Heron, Cattle Egret, Great Egret, Snowy Egret, Yellow-crowned Night Heron, Black Vulture, Mississippi Kite, and Swainson's Warbler.

Throughout Missouri the following birds breed regularly in the forests (including bottomland woods, the Ozark forests, and remnants of prairie groves) and farmlands (fallow fields, pastures, wet meadows, fencerows, brushy draws, woodland borders, orchards, and dooryards):

FORESTS

Broad-winged Hawk
Yellow-billed Cuckoo
Common Screech Owl
Barred Owl
Chuck-will's-widow (*southern Missouri*)
Whip-poor-will
Yellow-shafted Flicker
Pileated Woodpecker
Red-bellied Woodpecker
Hairy Woodpecker
Downy Woodpecker
Great Crested Flycatcher
Acadian Flycatcher
Eastern Pewee
Blue Jay

Black-capped Chickadee (*northern and central Missouri*)
Carolina Chickadee (*southern Missouri*)
Tufted Titmouse
White-breasted Nuthatch
Wood Thrush
Blue-gray Gnatcatcher
Yellow-throated Vireo
Red-eyed Vireo
Warbling Vireo
Prothonotary Warbler
Worm-eating Warbler (*southern Missouri*)
Ovenbird
Louisiana Waterthrush
Kentucky Warbler
American Redstart
Baltimore Oriole
Scarlet Tanager (*northern Missouri*)
Summer Tanager
Rose-breasted Grosbeak (*except extreme southern Missouri*)

FARMLANDS

Common Bobwhite
Upland Sandpiper (*prairie region*)
Mourning Dove
Red-headed Woodpecker
Eastern Kingbird
Eastern Phoebe
Horned Lark
Barn Swallow
House Wren (*except southern Missouri*)
Bewick's Wren (*chiefly Ozarks*)
Carolina Wren
Northern Mockingbird
Gray Catbird
Brown Thrasher
Eastern Bluebird
Loggerhead Shrike
White-eyed Vireo
Bell's Vireo
Yellow Warbler
Prairie Warbler (*southern Missouri*)
Common Yellowthroat
Yellow-breasted Chat
Eastern Meadowlark
Western Meadowlark (*northwestern Missouri*)
Orchard Oriole
Common Grackle
Northern Cardinal
Blue Grosbeak (*western and southern Missouri*)
Indigo Bunting
Dickcissel
American Goldfinch
Rufous-sided Towhee
Grasshopper Sparrow
Vesper Sparrow (*northern Missouri*)
Chipping Sparrow
Field Sparrow
Song Sparrow (*northern Missouri*)

The Mississippi River on the eastern boundary and the Missouri River, which flows across the state from the western boundary, bring through the state birds that habitually follow these great river valleys during migration. On their bottomlands and on adjacent areas, large numbers of birds stop off: waterbirds, waterfowl, and shorebirds on the countless sand and mud bars, lakes, ponds, sloughs, meander cutoffs, and marshes; landbirds in weed patches, brushy thickets, and woods. Various geese, ducks, gulls, and terns—also American White Pelicans in northwestern Missouri—are the most conspicuous transients, but careful inspection of thickets and woods is almost certain to reveal in the proper season impressive waves of small landbirds. The bottomlands of the Missouri north of Kansas City and of the Mississippi are, in spring and fall, among the best places for bird finding in the state. As in Iowa, a number of northern species of passerines—for example, the Yellow-bellied Flycatcher, Winter Wren, Gray-cheeked Thrush, Philadelphia Vireo, Bay-breasted Warbler, and Canada Warbler—move more commonly along the Mississippi in eastern Missouri than they do farther west. Conversely, a few species such as the Orange-crowned Warbler and Harris' Sparrow are more common in migration in the western part of the state. The following timetable gives the dates within which one may expect the main spring and fall migratory flights:

Waterfowl: 1 March–10 April; 15 October–10 December
Shorebirds: 25 April–25 May; 5 August–10 October
Landbirds: 10 April–10 May; 5 September–1 November

Missouri is a common wintering ground for Slate-colored Juncos and American Tree Sparrows and is frequented in winter by Evening Grosbeaks, Purple Finches, Common Redpolls, Pine Siskins, and Red Crossbills. In southern Missouri, especially in the southeastern lowlands, the following birds winter commonly: Red-breasted Nuthatch, Brown Creeper, Golden-crowned Kinglet, Myrtle Warbler, White-crowned and White-throated Sparrows. Large aggregations of waterfowl, mainly Canada Geese and Mallards, with fewer numbers of other species, pass the winter on bodies of water where the food supply is sufficient. Bald Eagles are all-winter visitants along rivers and reservoirs that contain fish, their favorite food.

Authorities

Richard A. Anderson, Paul E. Bauer, Donald M. Christisen, D. Wayne Davis, William H. Elder, Nathan Fay, James Haw, Paul L. Heye, James C. Irvine, Larry T. Keck, Floyd R. Lawhon, Edward J. and Evelyn McCrae, Earl S. McHugh, David Plank, James Rathert, Simon Rositzky, Katherine A. White.

BRANSON
Shepherd of the Hills Country

In the Ozarks of southwestern Missouri, Power Site Dam impounds the waters of the White River to form Lake Taneycomo, a narrow, S-shaped reservoir about 15 miles long, in the heart of **Shepherd of the Hills Country.** The nearby bluffs and rocky ravines are rather openly forested with oak and hickory. In the rugged hill country beyond, the oak-hickory forests are often interrupted by grass-covered glades, sometimes supporting brush and scattered stands of red cedar.

Among the regularly breeding birds in forests and shrubby areas are the Wild Turkey, Yellow-billed Cuckoo, Great Crested Flycatcher, Carolina Chickadee, Bewick's Wren, Wood Thrush, White-eyed Vireo, Black-and-white Warbler, Ovenbird, Yellow-breasted Chat, Orchard Oriole, Summer Tanager, Blue Grosbeak, Indigo Bunting, and Rufous-sided Towhee. In lowland areas along watercourses are such summer residents as the Prothonotary Warbler, Northern Parula Warbler, and Louisiana Waterthrush. Prairie Warblers, Field Sparrows, and a few Bachman's Sparrows reside in the glades during summer; American Robins and flocks of Cedar Waxwings frequent the cedar glades during winter.

Branson, a resort center on Lake Taneycomo, is a good point from which to begin bird finding since highways lead from here along the Lake and into the hill country. State 76 northwest from Branson goes through typical hill country where one may turn off and look for birds. Inspiration Point on State 76, 7 miles west from Branson, is worth a stop for a fine view of the Ozarks from an elevation of 1,341 feet. Visit the following sites for birds south and southwest of Branson:

School of the Ozarks Campus on Lake Taneycomo, reached by driving 4.5 miles south from Branson on US 65. This is rewarding for viewing the warbler migration in late April and early May and offers good chances of seeing Wood Ducks and Greater Roadrunners in any season.

Table Rock Dam, which holds back the waters of the White River to form Table Rock Lake; reached by continuing south on US 65 for about 2 miles, then turning off west onto State 165 which, after about 7 miles, crosses the Dam. This is a good vantage for observing birds in winter: on the lakeside, Horned Grebes, Bald Eagles, and both Herring and Ring-billed Gulls; below the Dam, Buffleheads, Hooded Mergansers, and other waterfowl. Canada Geese reside in the general area the year round. *Table Rock State Park,* traversed by State 165 before it reaches the Dam, has nature trails worth walking for woodland birds in any season.

Whenever traveling along the above highways, be alert for occasional Scissor-tailed Flycatchers on fences and utility wires.

BROOKFIELD
Swan Lake National Wildlife Refuge

In north-central Missouri the **Swan Lake National Wildlife Refuge** embraces 10,670 acres lying on the bottomlands of the Grand River. About 5,000 acres consist of open water and marshes impounded by levees, the remainder of croplands and heavy timber, principally oak, hickory, pecan, walnut, maple, and sycamore.

Nearly 200,000 Canada Geese pass the winter on the Refuge, arriving steadily after mid-September, reaching maximum numbers by late December, and leaving by the first of April. During October and November, some 20,000 Snow Geese and a few hundred Greater White-fronted Geese stop off in their southbound migration, as do about 250,000 ducks, chiefly surface-feeders—Mallards, Gadwalls, Common Pintails, Green-winged and Blue-winged Teal, and Northern Shovelers. In spring, fewer Snow and Greater White-fronted Geese and ducks stop off although there are usually more diving ducks represented than in fall.

Small flocks of American White Pelicans show up from late August to mid-October and again during April and early May.

Many Bald Eagles are invariably present through the winter along with Red-tailed and Rough-legged Hawks, Northern Harriers, an occasional Short-eared Owl, large mixed flocks of 'blackbirds,' and an abundance of Slate-colored Juncos, American Tree and White-crowned Sparrows, and other fringillids. Harris' Sparrows often stay in and along the rows of multiflora rose.

Nesting in suitable habitats on the Refuge are Green Herons, Mallards, Blue-winged Teal, a few Wood Ducks, American Coots, Whip-poor-wills, Great Crested Flycatchers, Sedge Wrens, Bell's Vireos (in thickets along levee roads), Northern Parula Warblers, Indigo Buntings, Dickcissels, and Field Sparrows.

To reach the Refuge from Brookfield, proceed west on US 36 for 5 miles, exit south (left) on State 139 to Sumner, a distance of 12 miles, then turn off south onto Recreational Access Road and proceed 1.0 mile to headquarters on the west side. The Refuge and its auto road are closed to the public from 1 October to 10 March, but many of the birds may be readily seen during that period from the perimeter of the Refuge or from the observation tower at headquarters.

CAPE GIRARDEAU

I. R. Kelso Wildlife Sanctuary | Trail of Tears State Park

On the north edge of the city, which overlooks the Mississippi at a point about 150 miles downstream from St. Louis, is the **I. R. Kelso Wildlife Sanctuary** (57 acres), operated by Southeast Missouri State University. It may be reached by driving north from Cape Girardeau on State 177 to the city limits.

Running through the Sanctuary is Juden Creek, bordered by patches of weeds and briars, willow thickets, and stands of cottonwood and sycamore. From this stream a wooded slope rises to a high ridge where there are dense woods. The Sanctuary is excellent for the more common birds characteristic of the southeast Missouri countryside. Some of the species definitely breeding are the Wood Duck, Pileated Woodpecker, Eastern Phoebe, Acadian Flycatcher, Blue Jay, Carolina Chickadee, Tufted Titmouse, Carolina Wren, Wood Thrush, Eastern Bluebird, Blue-gray Gnatcatcher, Prothonotary, Northern Parula, and Yellow-throated Warblers, and Louisiana Waterthrush. Visitors are welcome to use the trails leading to a variety of productive bird habitats.

Overlooking the Mississippi, about 12 miles north of Cape Girardeau on State 177, is **Trail of Tears State Park** (3,000 acres). Its limestone bluffs provide good vantages for viewing migrating waterfowl during the spring and fall and for Canada Geese, Bald Eagles, and an occasional Golden Eagle during the winter months.

CHARLESTON
Big Oak Tree State Park

An excellent example of floodplain forest in the bottomlands of the Mississippi in southeast Missouri just northeast of the Bootheel is embraced by **Big Oak Tree State Park** (1,004 acres). Much of the forest contains virgin timber, the rest well-developed second growth. Some of the commonest trees are oak, cottonwood, cypress, ash, shagbark hickory, sweetgum, tupelo, and persimmon. Many attain giant size—for example, the famous 'Hunter's Oak' (a bur oak) for which the Park was named stood 143 feet tall and had a circumference of 23 feet 5 inches. The epiphytic fern, *Polypodium*, grows on the trunks and branches of some of the larger trees. Twenty-two acres of the Park comprise Big Oak Lake, an impoundment for fishing.

As many as 65 species of birds may be observed in the Park during the breeding season, including the Little Blue Heron, Mississippi Kite, Fish Crow, and Swainson's Warbler, as well as the following: Wood Duck, Yellow-billed Cuckoo, Pileated, Red-bellied, and Red-headed Woodpeckers, Great Crested and Acadian Flycatchers, Carolina Chickadee, White-breasted Nuthatch, Carolina Wren, Wood Thrush, Blue-gray Gnatcatcher, Yellow-throated and Red-eyed Vireos, Prothonotary, Worm-eating, Northern Parula, Cerulean, Yellow-throated, and Kentucky Warblers, Orchard Oriole, and Summer Tanager.

The passage of passerine birds through the area in April, early May, and September is heavy—which is to be expected in view of the Park's location with respect to the Mississippi flyway. The abundance of Northern Waterthrushes is frequently impressive. Regular among the wintering species are the Yellow-bellied Sapsucker, Brown Creeper, Hermit Thrush, Golden-crowned and Ruby-crowned Kinglets, Myrtle Warbler, and White-crowned, White-throated, and Fox Sparrows.

The Park, about 21 miles south of Charleston, may be reached by exiting from I 57 at Charleston on State 105 south to the northeast edge of East Prairie, then bearing left on State 102 which continues south to the entrance.

CLINTON
Taberville Prairie Refuge | Schell-Osage Wildlife Area

Here in west-central Missouri are the remaining Greater Prairie Chickens in the state. For their booming in spring the birds prefer native grassland, unmowed and ungrazed, although they will accept permanent pastures, meadows, and other areas where grasses are short. Year after year they use the same booming grounds and begin returning to them in January, reach peak activity by the first two weeks of April, and cease appearing by the end of June. Best dates for observation are between the second week of March and the third week of April when they boom regularly each day: from half an hour before sunrise until two or three hours later, and from an hour before sundown until dusk.

The **Taberville Prairie Refuge,** owned by the state and administered by the Missouri Conservation Commission, and adjacent private lands offer opportunities for watching booming activities. From Clinton, drive south on State 13 for 9 miles and west on State 52 for 8 miles to Montrose; stay on State 52 south for 5 miles until it turns west, then continue south on County A for 2 miles and County H for 7.5 miles to the Refuge entrance and parking lot on the left. Here inquire at the information booth about booming grounds on which activities may be watched from the car.

There are booming grounds on privately owned land outside

that Refuge that may be viewed from intersecting county roads which run north-south and east-west. To see the activities the bird finder must arrive at the viewing site *before* the first light of day, stay in the car, and keep quiet. At the slightest disturbance the birds will disperse and not reappear that day.

The Taberville Refuge embraces about 1,200 acres of rolling grassland, rocky outcrops, oak groves, and streams with deep pools. A self-guided nature trail, winding 2.25 miles through the Refuge, introduces the principal features of native Missouri prairie. Besides permanent-resident Greater Prairie Chickens, some of the birds one is likely to hear or see include: Northern Harriers, Horned Larks, and Eastern Meadowlarks all year; Upland Sandpipers, Scissor-tailed Flycatchers, Grasshopper Sparrows, and Henslow's Sparrows during nesting season; Short-eared Owls and flocks of Smith's Longspurs during winter.

From the Taberville Refuge one may continue south on County H through Taberville for 5 miles, turn left onto County Y and proceed 2.5 miles, and then follow directional signs to headquarters of the **Schell-Osage Wildlife Area** (8,633 acres). Within its boundaries are flooded bottomland timber along the Osage River, marshes, open-water impoundments, and prairies—a rich assortment of habitats for waterbirds, waterfowl, and landbirds that include wintering Bald Eagles. Little Terns are often visitants in the late summer.

COLUMBIA

Ashland Wildlife Research Area | Little Dixie Lake Wildlife Area

A favorite locale for Columbia bird watchers is the **Ashland Wildlife Research Area,** a 2,250-tract of wild land which the University of Missouri administers as an experimental wildlife area. Habitats include grassland, woodland edge, oak-hickory woods, and pine plantations. Virgin sugar maples stand along Brushy Creek at the south end of the Area. This variety of habitats attracts a consequent variety of birds. Some of the species breeding in the Area, beginning in May, are the Broad-winged Hawk, Ruffed Grouse,

Wild Turkey, Prothonotary, Worm-eating, and Blue-winged Warblers, Prairie Warbler (in the younger pine plantations), Kentucky Warbler, Summer Tanager, Blue Grosbeak, and Lark Sparrow.

The Ashland Area may be reached from Columbia by driving 16 miles south on US 63 to the Ashland-Guthrie Exit, then east 3.4 miles on a road (at first blacktop then gravel) going south through the Area.

The best place for waterbirds and waterfowl within a 50-mile radius of Columbia is the **Little Dixie Lake Wildlife Area** (673 acres including Dixie Lake of 205 acres). Drive 9 miles east from Columbia on I 70; at the Millersburg Exit, turn off south onto County J and proceed 4 miles, then go east for 0.25 mile to the entrance on County RA. There are parking areas on three sides of the Dixie Lake.

At the north end of the Lake are several beaver dams and moist, partly open areas where American Woodcock perform their evening flight-songs in March and April.

KANSAS CITY

Swope Park | Horseshoe Lake | Sugar Lake | Lewis and Clark State Park | La Benite County Park | Lake Jacomo County Park | James A. Reed Memorial Wildlife Area

Within the southeastern limits of this big city lies **Swope Park** (1,346 acres), a recreational center of exceptional beauty. Rocky hills drop steeply into deep ravines and the valley of the Blue River; oaks and hickories cover the slopes, and stately sycamores stand in the lowland areas. Landscaped gardens, two golf courses, and two lakes—the Lagoon for boating and the Lake of the Woods for fishing—add variety to the setting.

Despite the Park's use for many recreational purposes, bird finding is fairly good in any season. Turkey Vultures are common in summer and roost among remote limestone outcroppings. Wood Ducks and Northern Parula Warblers nest in woods along the River, Yellow-billed Cuckoos, Acadian Flycatchers, Wood Thrushes, Yellow-throated Vireos, and Scarlet and Summer Tanagers on forested hillsides. Barred Owls and Whip-poor-wills are vociferous in evenings during the early summer. Many small landbirds pass through the Park in spring and fall, Harris' Sparrows in April and October, numerous warblers, such as the Tennessee, Orange-crowned, Nashville, Myrtle, Blackpoll, and Wilson's, in early May and September.

Swope Park is readily accessible from I 435 east of the Park by exiting west on Gregory Boulevard which soon crosses its southern part. Here stop at the Nature Center on the right for information on the best sites for birds in season.

Because of its situation in the Missouri Valley—on the Missouri River where it turns eastward from the state's western boundary—Kansas City is in the traditional path of hosts of waterbirds, waterfowl, and shorebirds during their spring and fall migrations. Numerous sites where these birds stop off to rest and feed are northwest of the city and may be reached by leaving the city northward on I 29 and then exiting west on State 45. This highway follows up the Missouri River over flat marshy bottomlands intersected by streams, backwaters, meander cutoffs, and lakes whose water levels fluctuate seasonally, being especially high in March and April. The following three sites, passed by State 45, are particularly recommended:

Horseshoe Lake, about 20 miles northwest of Kansas City, is a meander cutoff where ducks and shorebirds gather in notable numbers, the former in March, October, and November, the latter in late April, May, late August, and September. Among the more common shorebirds are the Semipalmated Plover, Solitary Sandpiper, Greater and Lesser Yellowlegs, Pectoral Sandpiper, Baird's Sandpiper, Least and Semipalmated Sandpipers, and Hudsonian Godwit. For the best vantage, turn off State 45, 1.0 mile south of Farley, and cross the railroad trucks.

Sugar Lake, southwest of the junction of State 45 and US 59 and about 44 miles from Kansas City, is an irregularly shaped, shallow body of water connected to the Missouri River at its westernmost end. From late February to early April, and again in late October and November, the 400-acre surface of Sugar Lake is alive with ducks: Mallards, Gadwalls, Common Pintails, Green-winged and Blue-winged Teal, American Wigeons, Northern Shovelers, Ring-necked Ducks, Lesser Scaups, and Ruddy Ducks. In early March great numbers of Canada Geese, Snow Geese, and fewer numbers of Greater White-fronted Geese may be seen on the Lake, or flying to and from feeding grounds elsewhere. In April and September, flocks of American White Pelicans stop off during their migrations, as do Forster's and Caspian Terns.

About a mile south of its junction with US 59, State 45 passes the easternmost arm of Sugar Lake and the entrance road to **Lewis and Clark State Park,** which occupies 60 acres back from the Lake's shore. Both this arm and the Park have good vantages for observation. Other vantages are on the northern shore, parts of which are near the left side of US 59 going west from its junction with State 45.

On the eastern outskirts of Kansas City are the following three areas that yield a great variety of birds:

La Benite County Park (500 acres) northeast of the city on the south side of the Missouri River. From I 435, exit east on US 24 for 6.3 miles, then exit north on State 291 until it crosses the Missouri over Liberty Bend Bridge. Leave the car here and walk east through the Park on trails that parallel the River.

La Benite Park is rewarding in early May when the cottonwoods and willows along the River abound with migrating warblers. Here common summer residents include Yellow-billed Cuckoos, Warbling Vireos, Yellow Warblers, American Redstarts, Orchard and Baltimore Orioles, and Rose-breasted Grosbeaks. In adjacent open areas a few Dickcissels and Lark Sparrows nest.

Lake Jacomo County Park (4,394 acres) east of the city. From I 435, exit east on I 70 for 7.2 miles, exit south for 4 miles on State 291 to Woods Chapel Road for 2 miles to the entrance.

Lake Jacomo (940 acres of surface water) is excellent in October and November for transient Common Loons and Horned Grebes, as well as for the same variety of ducks appearing concurrently on Sugar Lake (*see above*). In winter the several stands of conifers ringing the Lake are invariably worth investigating for Great Horned and Long-eared Owls, Red-breasted Nuthatches, and Red Crossbills. In the same season all brushy habitats should be explored for northern fringillids such as Slate-colored Juncos and American Tree, Harris', White-crowned, and White-throated Sparrows.

The oak-hickory woods on uplands around the Lake produce many woodpeckers in any season, migrating kinglets, vireos, and warblers in late April, early May, and September. Both Red-eyed and Warbling Vireos, Kentucky Warblers, and American Redstarts are among the many summer residents.

The **James A. Reed Memorial Wildlife Area** (1,500 acres) southeast of the city; administered by the Missouri Conservation Commission. From I 435, exit east on US 50, passing Lee's Summit, for 12.9 miles, then turn south onto Ranson Road and go 1.0 mile to the entrance and headquarters on the left. Inquire here about roads and trails in the Area.

Although a popular recreation center, especially in the warmer months, the Reed Wildlife Area is nonetheless a 'true island of wildlife' in a suburban setting. Several small lakes bring numerous waterfowl and wading birds in spring and fall. Canada Geese nest in the Area and may be viewed in any season.

Small woodlots and fields together with their brushy edges provide desirable habitats for many different birds. One or two Sharp-shinned and Cooper's Hawks are almost a certainty in the introduced stands of conifers near headquarters and in the southeastern section of the Area. Northern sparrows, including the Savannah, Le Conte's, and Vesper, are a specialty in October, November, March, and April; other fringillids stay through the winter, attracted to feeding stations maintained by the Area's personnel. Among summer residents are the Eastern Pewee, Eastern Bluebird, Blue-gray Gnatcatcher, Indigo Bunting, Dickcissel, and Field Sparrow.

PUXICO
Mingo National Wildlife Refuge

For productive bird finding anytime of year in southeast Missouri, go to **Mingo National Wildlife Refuge** just west off State 51, 1.5 miles north of Puxico. Here some 16,000 of the Refuge's 21,646 acres are swamp, formed in an ancient channel of the Mississippi River and flanked by low hills and bluffs. Much of the Refuge is heavily timbered with stands of cypress and sweetgum in the swamp and various oaks on higher ground. The visitor center at the Refuge has interpretive services and directions to the Boardwalk Trail through the swamp and Bluff Trail leading along the uplands.

Spring is delightful in Mingo. Wildflowers begin blooming in March at the base of the bluffs and in April are blooming all over. By late April the first of the northbound warblers appear and from then on warblers increase until, in early May, waves of them may be noted from the Boardwalk and Bluff Trails. Among the warblers arriving to nest are the Prothonotrary, Northern Parula, Cerulean, Yellow-throated, and Kentucky as well as a few Louisiana Waterthrushes.

Besides nesting warblers, the Refuge holds a wide variety of other breeding birds including Green and Little Blue Herons, Wood Ducks, Hooded Mergansers, Red-tailed and Red-shouldered Hawks, Yellow-billed Cuckoos, Chuck-will's-widows, Barred Owls, Pileated and Red-headed Woodpeckers, Acadian Flycatchers, Carolina and Sedge Wrens, White-eyed and Warbling Vireos, Orchard Orioles, and Indigo Buntings.

Waterfowl migration from northern climes begins with Green-winged and Blue-winged Teal in late September. Mallards become the most common ducks as fall advances but Gadwalls, Common Pintails, American Wigeons, Northern Shovelers, and Ring-necked Ducks attain great abundance by the end of October when the duck population reaches its maximum of about 100,000. As the duck population dwindles in November, several thousand Canada Geese and fewer Snow and Greater White-fronted Geese arrive. By mid-December the numbers of waterfowl decline to a wintering population consisting mainly of a few thousand Mallards and Canada Geese.

An impressive number of Bald Eagles arrives in November and stays through the winter, feeding on fish or incapacitated waterfowl and perching in the tall cypresses. Wild Turkeys, although present in the area the year round, gather during the winter in large flocks which one may readily watch as they forage in the open croplands maintained by the Refuge.

ST. JOSEPH
Squaw Creek National Wildlife Refuge | Honey Creek Wildlife Area

Farther northwest of this northwestern Missouri city that overlooks the Missouri Valley from the east is **Squaw Creek National Wildlife Refuge.** Except for a small area on hilly bluffs, most of this 6,886-acre Refuge is on the bottcmlands of the Missouri River Valley. About half is open water and marsh; the remainder consists of woodlands and croplands.

The Refuge's greatest ornithological attraction begins in early March when tens of thousands of Snow Geese stop off on their way north as do smaller numbers of Canada Geese and a few Greater White-fronted Geese. No sooner has the goose migration passed its peak after mid-March when, almost as an anticlimax, come hordes of northbound ducks—Mallards, Gadwalls, Common Pintails, Green-winged and Blue-winged Teal, American Wigeons, Northern Shovelers, Lesser Scaups, and fewer numbers of American Black Ducks, Redheads, Ring-necked Ducks, Canvasbacks, Buffleheads, Ruddy Ducks, and mergansers.

During April and early May other northbound birds stop off: hundreds of American White Pelicans and Franklin's Gulls; Pied-billed Grebes and American Coots; huge flocks of Yellow-headed Blackbirds; and numerous shorebirds, particularly Semipalmated and Lesser Golden Plovers, Common Snipes, Solitary Sandpipers, Greater and Lesser Yellowlegs, Pectoral, Least, and Semipalmated Sandpipers as well as others such as Black-bellied Plovers, Willets, White-rumped and Baird's Sandpipers, Hudsonian Godwits, and Wilson's Phalaropes.

The marshes with their rich growth of cattails and bulrushes are breeding habitat for American Bitterns, King Rails, Soras, Black Terns, Marsh Wrens, and Yellow-headed and Red-winged Blackbirds, and feeding areas for summer-resident Great Blue and Green Herons, Great Egrets, Yellow-crowned and Black-crowned Night Herons, and Wood Ducks.

Raptors regularly breeding in the Refuge are the Red-tailed Hawk, Northern Harrier, American Kestrel, and Common Screech, Great Horned, and Barred Owls.

The congregations of transient waterfowl and pelicans in the fall, October through mid-December, are usually as impressive as in the spring. A great many Canada Geese along with vast numbers of Mallards stay through the winter. The most notable among the winter visitants are Bald Eagles, often two hundred or more, and a few Golden Eagles. Other winter visitants include Rough-legged Hawks and Short-eared Owls.

To reach the Refuge from St. Joseph, drive north on I 29 for 30 miles and exit west on US 159 for 2 miles to headquarters.

Almost midway between St. Joseph and the Squaw Creek Refuge is **Honey Creek Wildlife Area** comprising 1,448 acres of timbered hilly country that is ideal for woodland passerines in spring migration. From I 29, 17 miles north of St. Joseph, exit east at Fillmore and immediately turn west over I 29; once on the other side, turn left onto a graveled road (Honey Creek Area directional sign here) which enters the Area after 1.4 miles and for the next 1.5 miles goes along the top of a ridge, providing excellent vantages for watching birds, sometimes at treetop level. As many as twenty-five species of warblers can be identified at the height of their migration in early May.

ST. LOUIS

Forest Park | Creve Coeur Lake and Memorial Park | St. Charles Airport | Dardenne Marshes | Rockwoods Reservation | August A. Busch Wildlife Area | Missouri Botanical Garden Arboretum

From the Mississippi River just south of the points where the Illinois River enters from the east and the Missouri River from the west, this enormous city and its numerous suburbs spread out westward over many miles of flat to gently undulating terrain. Despite the dense settlement and the vigorous industrial activity, there exists in this region, even within the limits of the city itself, an array of good bird-finding areas—lakes, ponds, lagoons, and marshes for migrating waterbirds, waterfowl, and shorebirds; forested tracts for many different kinds of landbirds.

The best year-round area in the city is **Forest Park** (1,293 acres), one of the largest natural parks in any United States city. About 2 miles west of downtown St. Louis, it extends from Kingshighway Boulevard on the east to Skinker Boulevard on the west and from Lindell Boulevard on the north to the Daniel Boone Expressway on the south. A group of small lakes and connecting waterways, open fields, and hilly woodland offer a variety of habitats. The spot of special interest is a sanctuary consisting of an undisturbed, unlandscaped tract of hardwood forest back of the City Art Museum. From I 270, take the Boone Expressway (US 40) east as far as Forest Park Exit (Hampton Avenue); here turn north and proceed 0.2 mile, passing the zoo; turn left from Concourse Drive onto Washington Drive and go 0.3 mile; turn left onto Government Drive; then right onto Fine Arts Drive to the parking area behind the Art Museum. Proceed on foot south and west on trails through the sanctuary and along open water.

On a single morning in late April and early May it is possible to record as many as seventy-five resident and transient species including the Pied-billed Grebe, Blue-winged Teal, Wood Duck, American Coot, Solitary Sandpiper, Lesser Yellowlegs, Red-bellied Woodpecker, Yellow-bellied Sapsucker, Great Crested Flycatcher, Rough-winged Swallow, Carolina Chickadee, Tufted Titmouse, Brown Creeper, Northern Mockingbird, Wood and Swainson's Thrushes, Blue-gray Gnatcatcher, Philadelphia Vireo, Worm-eating and Magnolia Warblers, Louisiana Waterthrush, Baltimore Oriole, Rufous-sided Towhee, and White-crowned and Lincoln's Sparrows. Thirty species of warblers may be identified at the peak of their migration. On the last two Sundays in April and the first two in May, the St. Louis Audubon Society conducts bird walks, starting back of the Art Museum at 8:00 a.m.

Perhaps the best year-round area for bird finding just outside the city is **Creve Coeur Lake and Memorial Park** (1,260 acres), reached by exiting west from I 270 onto Dorsett Road for 1.5 miles and turning right onto Marine Avenue, which winds down for 0.4 mile to the parking area near the boat ramps on the northeast side of the Lake.

Situated on the bottomlands of the Missouri River, shallow Creve Coeur Lake, 2 miles long, has muddy shores bordered by willow thickets, shrubs, and tall rushes, overlooked from the eastern side by a high bluff well covered by woods. March, October, and November are the best months for transient ducks; April, early May, late August, and September for transient waterbirds and shorebirds. Some of the species that may be expected from time to time are the Common Loon, Horned Grebe, Yellow-crowned Night Heron, Pectoral, Least, and Semipalmated Sandpipers, Franklin's and Bonaparte's Gulls, and Forster's and Common Terns.

From the above-mentioned parking area, walk south along the base of the steep bluff past a waterfall to deep woods where the Prothonotary Warbler nests. The Lake at this point is very shallow with mud flats attractive to waterbirds, ducks, and shorebirds.

Return to the car and backtrack about 0.3 mile up Marine Avenue and turn right to the Greensfelder Memorial on top of the bluff. Park here and walk north from the Memorial area along the crest of the bluff through woods that are excellent for observing migrating kinglets, vireos, and warblers, often at treetop level. From the Memorial area one may drive across Marine Avenue and follow the Park road as it winds over the top of the bluff through mature woods where there are Pileated Woodpeckers and numerous other birds in any season.

North of St. Louis the narrow strip of lowland between the Mississippi and Missouri Rivers at their confluence is flat and often marshy with farmlands, grassy meadows, and shallow lakes. The extent to which it is covered by water depends on rainfall. The entire area is excellent in fall, winter, and spring for a multitude of birds from waterbirds and waterfowl to passerines. Although much of the area is owned by hunting clubs and closed to public access, a great many birds may be seen from State 94, which traverses the area between St. Charles and West Alton, or, better still, from side roads leading off State 94. Two areas are particularly good for birds:

St. Charles Airport, reached from State 94 north of Boschertown by turning northwest onto State Secondary B and going 2 miles. From the road beside the Airport or by walking along the edge of the grassy runways, look for, *in winter*, Horned Larks (northern races), Le Conte's Sparrows, and Lapland Longspurs; *in late winter and early spring*, Smith's Longspurs; *in early spring*,

Lesser Golden Plovers; *in both spring and fall,* Sprague's Pipits and numerous fringillids. Upland Sandpipers arrive in April; some may remain to nest along with (Prairie) Horned Larks and Grasshopper Sparrows.

Dardenne Marshes, reached by proceeding west from St. Charles Airport on SS B for about 7 miles, then north (right) on SS C. Owned by a duck club and fenced, the Marshes are closed to public access, but permission to enter may be obtained, except during the hunting season, at the clubhouse on the Mississippi River near the terminus of SS C. Many unimproved roads—passable only in dry weather—run through the Marshes, crossing small streams and skirting muskrat ponds, partly submerged woodlands, and open grassland. In winter the area is usually highly rewarding for birds of prey—Red-tailed and Rough-legged Hawks, Bald Eagles, Northern Harriers, American Kestrels, and Short-eared Owls—and other birds such as Loggerhead Shrikes, Rusty Blackbirds, and Swamp Sparrows.

While in this general area north of St. Louis, one should always watch for Eurasian Tree Sparrows. In fall and winter they are usually in flocks, often in association with House Sparrows. Look for them in or near hedgerows and brushy thickets, or in farmyards.

Twenty miles west of St. Louis the Missouri Conservation Commission maintains the **Rockwoods Reservation** (3,200 acres), a section of very rough country with steep hills and narrow valleys, for the most part forested with oak and hickory. Excellent for warblers during the height of their migration in early May, it is equally good for birds of prey: *nesting* Turkey Vultures, Cooper's, Red-tailed, and Broad-winged Hawks, American Kestrels, Great Horned Owls, and occasionally Long-eared Owls where there are pine groves; *migrating* Sharp-shinned Hawks.

To reach the Rockwoods Reservation, exit west from I 270 on I 40 for 11.7 miles; exit north on State 109 for 4.5 miles; then turn west (left) onto Glencoe Road and proceed 1.5 miles to the natural history museum in the Reservation. Inquire here about nature trails that will lead to the best vantages for birds.

Twenty-five miles west of the city the Missouri Conservation Commission also supervises the **August A. Busch Wildlife Area.**

This may be reached by exiting west from I 270 on US 40 for 18.9 miles; exiting left on State 94 for 1.1 miles, turning right onto State Secondary D and proceeding 1.75 miles to the headquarters area, and then right again for a short distance to headquarters and Shop Lake.

The Busch Wildlife Area's 7,500 acres include bottomland woods along Dardenne Creek, many man-made lakes of varying size, brushy fields, and some croplands. The lakes attract numerous ducks in migration; Shop Lake, which rarely freezes over, holds a few ducks and American Coots all winter. Wood Ducks nest commonly as do Canada Geese. Northern Parula and Cerulean Warblers, and occasionally Yellow-throated Warblers, nest in the woods along Dardenne Creek; Yellow-breasted Chats, Orchard Orioles, and a few Blue Grosbeaks in the more open shrubby areas.

For small landbirds the year round, one of the outstanding places in the St. Louis region is the **Missouri Botanical Garden Arboretum** (1,600 acres), 30 miles southwest of the city. Exit west from I 270 on I 44 for 23 miles; at Gray Summit, exit south on State 100 for a short distance and drive into the Arboretum through its stone gate and park. All parts of the area are accessible by footpaths.

Several hundred acres of the Arboretum are Meramec River bottomland with a small lake, open fields, and woodlands in which many wildflowers, shrubs, and trees—including large stands of conifers—have been introduced. During evenings in early spring, American Woodcock may be heard flight-singing above fields. Pileated Woodpeckers are permanent residents in the deeper woods; Blue-winged and Prairie Warblers, Blue Grosbeaks, and Lark Sparrows nest in brushy places. In winter the conifers attract for food or shelter a variety of birds such as Red-breasted Nuthatches, Cedar Waxwings, Myrtle Warblers, Pine Siskins, and sometimes Red Crossbills.

Across the Mississippi in Illinois are several excellent areas for birds. For information, *see under* **East St. Louis** *and* **Grafton** in the Illinois chapter of *A Guide to Bird Finding East of the Mississippi*, second edition, the eastern counterpart of this book.

No account of bird-finding opportunities in the St. Louis region

EURASIAN TREE SPARROW

would be complete without calling further attention to the Eurasian Tree Sparrow mentioned in a foregoing paragraph. This species, native in the British Isles, Europe, and northern Siberia, was introduced in the St. Louis region in 1870. At that time about 20 individuals were released in a residential section on the south side of the city. By 1877 a small colony had become permanently established in the same neighborhood. Despite the subsequent intrusion of the House Sparrow and its tendency to bully its Old World colleague, the species prospered and soon appeared around the city and across the Mississippi in Illinois. Today the species resides in many places within a radius of 100 miles from St. Louis.

Reference

A Guide to Finding Birds in the St. Louis Area. By Richard A. Anderson and Paul E. Bauer. St. Louis: Webster Groves Nature Study Society, 1968.

SALEM
Karkaghne Scenic Drive

For forest birds in the Ozarks of south-central Missouri, take the **Karkaghne Scenic Drive** in Mark Twain (formerly Clark) National Forest east of Salem. There are a few overlooks with wide views of the Ozarks but most of the drive is enclosed by dense forest, primarily hardwoods with some pine.

Before taking the trip, obtain a map of the National Forest from the office of the United States Forest Service on the south edge of Salem, one block south of the intersection of State 19, 32, and 72. Then from this intersection, proceed east on State 32-72 for 12 miles; continue east on State 32, after State 72 turns off right, for 14 miles; soon after passing through the village of Boss, turn off south (right) onto State KK and proceed 9 miles to an intersection with graveled Forest Service Road 2233. Turn east (left) onto FS 2233, the start of Karkaghne Scenic Drive, and begin looking for birds in the woods along the road and in adjacent steep hollows for the next 10 to 12 miles. Two right side roads, FS 2235 to Cooks Spring and FS 2236 to Sutton Bluff, offer similar opportunities.

Red-shouldered and Broad-winged Hawks may be expected. Both the Chuck-will's-widow and Whip-poor-will are summer residents. Among the other birds breeding in suitable forest cover are the Pileated and Red-bellied Woodpeckers, Great Crested Flycatcher, Eastern Pewee, Acadian Flycatcher, Carolina Chickadee, Wood Thrush, Yellow-throated and Red-eyed Vireos, Black-and-white, Worm-eating, Yellow-throated, and Pine Warblers, Ovenbird, Kentucky Warbler, and Scarlet and Summer Tanagers.

SULLIVAN
Meramec State Park

A fine spot in which to find small landbirds in east-central Missouri is **Meramec State Park** (7,153 acres), one of the major recreational areas in the state, just east of Sullivan and reached from I 44 by exiting south on State 185 to the entrance. A public playground as well as a forest preserve and game refuge, it consists largely of an unbroken tract of rolling deciduous woodland with several miles of nature trails. Among the natural features are many springs, more than a dozen caves—including Fisher's Cave which is open to the public during the summer months—and the clear waters of the Meramec River tumbling swifly through a narrow valley.

Both the Pileated and Red-headed Woodpeckers nest in the Park. Wild Turkeys breed in the valleys, as do Northern Parula and Cerulean Warblers and Yellow-breasted Chats. During the first half of May many warblers—e.g., Blue-winged, Tennessee, Nashville, Myrtle, Blackpoll, and Northern Waterthrush—pass through in migration. In winter the brushy and grassy bottomlands are attractive to great numbers of fringillids such as Slate-colored Juncos, and American Tree, White-throated, Fox, Swamp and Song Sparrows.

Montana

WESTERN GREBES

In May the Western Grebes return to their favorite marshes in the prairie lakes of Montana and at once begin their courtship displays. From then until July they perform, usually in twos, their varied and fantastic antics—'water skimming,' 'weed tricks,' 'head-waggling bouts,' 'penguin struts,' 'ghost dives,' 'swan gliding,' and 'habit preening.' A description would require paragraphs, at best inadequate. To be fully appreciated, these antics must be seen—and they may be readily observed by any bird finder who will take the time and make the effort to visit such a spot as Medicine Lake in the Medicine Lake National Wildlife Refuge (*see under* **Culbertson**) in late May or June, select a vantage point overlooking the marsh, and simply watch.

Montana has two major natural divisions: the Prairie Region, roughly the eastern two-thirds of the state, and the Mountain Region, the remaining western third. The meeting of the two divisions is along a line running diagonally across the state in a northwest-southeast direction.

The Prairie Region lies on the Missouri Plateau, a part of the Great Plains, with elevations ranging from 1,900 feet in the northeast to 4,500 feet at the edge of the Mountain Region. Eastward across the Prairie Region flow Montana's principal eastern rivers, the Missouri and the Yellowstone, to their confluence just across the North Dakota line. The valleys of these great waterways and their main tributaries are deep and broadly terraced, in places severely dissected by erosion, forming typical badlands, or flanked by high ridges, sometimes showing bold cliffs and striking formations such as pinnacles and 'tables.' Where the terrain has a rough character of this sort, birds such as Say's Phoebes, Bank and Cliff Swallows, Rock Wrens, and perhaps a Prairie Falcon or Golden Eagle may be expected. Beyond the valleys are the smooth rolling plains that dominate the topography of the Prairie Region; yet the plains do not extend for many miles without being interrupted by flat-topped buttes and gravelly mesas of varying heights. In the west, interruptions are even more pronounced, for here small, isolated mountains groups—for example, the Little Rocky, Big Snowy, Bearpaw, Highwood, and Crazy Mountains—rise 3,000 to 4,000 feet above the plains in widely separated localities.

The plains of the Prairie Region, though modified in varying degrees by cultivation or grazing, are mainly grassland, but the grass associations differ according to conditions of soil or moisture.

The higher plains, including most hills and mesas, are relatively dry and thus support a short-grass association. Since they are generally unsuited to crop production, they are used to a great extent for grazing cattle and show little change. Favorite habitats for both Horned Larks and McCown's Longspurs, the higher plains are also, if not too greatly disturbed, the most attractive areas in the state for breeding Long-billed Curlews.

The lower plains, having greater moisture, were originally long-grass prairie, but are now largely agricultural in character. Nevertheless there are meadows, pastures, and haylands that have grasses, either native or introduced, which provide cover for species of birds—the Bobolink, Savannah Sparrow, Chestnut-collared Longspur, and others—that may have typified the birdlife of the long-grass prairie over the state.

In scattered localities of the Prairie Region are stretches of sagebrush where Sage Grouse, Clay-colored Sparrows, and Brewer's Sparrows regularly reside. Groves of cottonwood, aspen, boxelder, and other deciduous trees together with thickets of willow and various shrubs—for example, silverberry, buffaloberry, snowberry, and wild rose—grow along the watercourses and in sheltered coulees (trench-like ravines); shrubby thickets also grow in draws (shallow ravines) and occasionally spread out to some extent on the adjacent prairie. These wooded environments, especially when isolated by broad expanses of prairie, often contain surprisingly large numbers of birds and varieties of species. Some of the higher hills, mesas, buttes, and ridges have a cover of juniper which, in the southern part of the Prairie Region, holds small nesting populations of Pinyon Jays. On the prairie-surrounded mountains in the western Prairie Region, the vegetation and associated birds are identical with those of the Mountain Region.

No discussion of the Prairie Region and its birds would be adequate without mentioning the thousands of small man-made farm ponds and their attractiveness to nesting ducks. Although often quite small, the ponds average about two broods of ducks per acre of water surface with the more mature ponds being the most productive. Mallards, Common Pintails, Blue-winged Teal, and American Wigeons are the species most commonly represented, Gadwalls and Northern Shovelers less commonly, and Green-winged Teal and diving ducks occasionally.

Because of the great variation in the physiography of the Prairie Region, there is no bird species that resides everywhere; yet a great many species range widely. Listed below are species that breed more or less regularly throughout the Prairie Region in open country (prairie grasslands, haylands, pastures, fallow fields, meadow-like lowlands, brushy places, woodland borders, orchards, and dooryards) and in wooded areas (along watercourses and in coulees). Species marked with an asterisk breed also at higher altitudes in the Mountain Region.

OPEN COUNTRY

Swainson's Hawk
Ferruginous Hawk
Northern Harrier
Sharp-tailed Grouse
Sage Grouse
Mountain Plover
Long-billed Curlew
Upland Sandpiper
Mourning Dove
Burrowing Owl
Red-headed Woodpecker
Eastern Kingbird
Western Kingbird
Say's Phoebe
Horned Lark
* Tree Swallow
Barn Swallow
Black-billed Magpie
House Wren
Gray Catbird
Brown Thrasher
* Mountain Bluebird
Loggerhead Shrike
Yellow Warbler
Common Yellowthroat
Yellow-breasted Chat
Bobolink
Western Meadowlark
Common Grackle
Lazuli Bunting
* American Goldfinch
Rufous-sided Towhee
Lark Bunting
Savannah Sparrow
Grasshopper Sparrow
Baird's Sparrow (*eastern Prairie Region*)
Vesper Sparrow
* Chipping Sparrow
Clay-colored Sparrow
Brewer's Sparrow
Song Sparrow
McCown's Longspur
Chestnut-collared Longspur

WOODED AREAS

* Cooper's Hawk
Black-billed Cuckoo
Common Screech Owl
* Red-shafted Flicker
* Hairy Woodpecker
* Downy Woodpecker
* Dusky Flycatcher
* Western Pewee
* Black-capped Chickadee
* White-breasted Nuthatch
Red-eyed Vireo
Warbling Vireo
American Redstart
Bullock's Oriole
Black-headed Grosbeak

The Mountain Region embraces various ranges in the Rocky Mountains, nearly all of which trend in a northwest-southeast direction. The Prairie Region meets the easternmost ranges—the

Lewis, Big Belt, Absaroka, Beartooth, and the northern tip of the Big Horn—along their eastern foothills. A few of the ranges of the Mountain Region are separated by broad valleys, but most of them are close together, separated, sometimes indistinctly, only by narrow valleys or even by canyons. The broad valleys are generally flat to rolling and are typically treeless except near streams. It is thus not surprising to find here many bird species characteristic of the open country in the Prairie Region.

Elevations in the Mountain Region begin at 1,800 feet in the Kootenai River Valley, the lowest point in the state in the extreme northwest, and reach 12,799 feet on the summit of Granite Peak, the highest point near Yellowstone National Park in the southeast. The elevations of the valley floors usually range between 3,000 and 5,000 feet. Only a few mountains exceed 11,000 feet, but there are many that reach above 10,000. The Continental Divide zigzags northward over several of the ranges, passing west of many of the highest points including Granite Peak.

The higher ranges in the Mountain Region are notably rugged, with craggy ridges and peaks; those in Glacier National Park (*see under* **East Glacier Park**) are the most extreme in this respect and thus the most scenic. Coniferous trees characterize the forests, although cottonwoods, aspens, alders, willows, and other deciduous growth stand in the lower valleys and ravines, usually near streams, or, in the case of aspens, in groves on the lower slopes.

East of the Continental Divide, in the foothills and mountains up to about 6,000 feet in the south and 5,000 feet in the north, are forests of ponderosa pine and Douglas-fir and, not infrequently, belts of juniper. Between 6,000 and 8,500 feet in the south and between 5,000 and 7,500 feet in the north, lodgepole pine becomes the predominant forest growth, but there are also stands of Douglas-fir, Engelmann spruce, and limber pine. As the elevation increases, subalpine fir appears.

West of the Continental Divide, where there is greater humidity, the forests show a somewhat different character, especially in the northwest. Here western white pine, western larch, western red cedar, western hemlock, and grand fir are common, sometimes at elevations much lower than 5,000 feet. Above 8,500 feet in the southern Mountain Region and above 7,500 feet in the northern, begins the stunted or subalpine forest consisting mainly of subalpine fir. This continues to the timber line at 9,000 feet in the south and 8,000 feet in the north. Where the mountains reach far above the timber line an alpine region exists. In parts of Glacier National Park the various forest belts described above are, as a rule, correspondingly lower, with the result that the alpine region sometimes begins at 6,500 feet.

Since most of the main mountain ranges in both the Mountain and Prairie Regions are so close together as to be practically continuous, the species of birds inhabiting them are much the same throughout; but there are nonetheless a few differences between the ranges owing to latitude, elevation, humidity, vegetation, and other factors, and this accounts for the presence or absence of certain species. For example, the Varied Thrush evidently resides only in the northern mountains west of the Divide, the Black Rosy Finch and Green-tailed Towhee in the southern mountains, and the Pinyon Jay in the juniper belt east of the Divide. Listed below are birds that breed regularly in the forests of the major mountain ranges.

Northern Goshawk
Sharp-shinned Hawk
Blue Grouse
Spruce Grouse (*western ranges*)
Ruffed Grouse
White-tailed Ptarmigan (*above timber line in northern ranges*)
Northern Pygmy Owl
Pileated Woodpecker (*northern ranges*)
Yellow-bellied Sapsucker (*generally below 7,500 feet*)
Williamson's Sapsucker
Black-backed Three-toed Woodpecker
Northern Three-toed Woodpecker
Hammond's Flycatcher (*western ranges at higher elevations*)
Western Flycatcher (*western ranges at lower elevations*)
Olive-sided Flycatcher
Violet-green Swallow
Gray Jay
Steller's Jay
Northern Raven
Clark's Nutcracker
Mountain Chickadee
Red-breasted Nuthatch
Brown Creeper
Winter Wren (*northwestern ranges*)
Hermit Thrush
Swainson's Thrush

Townsend's Solitaire
Golden-crowned Kinglet
Ruby-crowned Kinglet
Water Pipit (*above timber line*)
Solitary Vireo
Orange-crowned Warbler (*in deciduous thickets*)
Audubon's Warbler
Townsend's Warbler (*northwestern ranges*)
Northern Waterthrush
MacGillivray's Warbler (*in deciduous thickets below 7,500 feet*)
Wilson's Warbler (*in deciduous thickets above 7,500 feet*)
Western Tanager
Cassin's Finch
Pine Grosbeak (*subalpine forest*)
Pine Siskin
Red Crossbill
Oregon Junco
White-crowned Sparrow
Fox Sparrow
Lincoln's Sparrow (*in deciduous thickets*)

Contributing immeasurably to Montana's attractiveness for bird finding are the prairie lakes, most of which are in the northern Prairie Region. Usually shallow and quite marshy, with an abundance of cattails and quillreeds, they frequently hold an impressive breeding population of waterbirds, waterfowl, Yellow-headed Blackbirds, and Red-winged Blackbirds. In their immediate vicinities, Willets, American Avocets, and Wilson's Phalaropes often nest. The larger the lakes, the more open water they have and the more likely they are as breeding areas for Western Grebes, California and Ring-billed Gulls, Common Terns, and, sometimes American White Pelicans and Double-crested Cormorants. Among the larger prairie lakes notable for variety of birds are Medicine Lake and Lake Bowdoin (*see under* **Culbertson** *and* **Malta,** *respectively*) and Freezeout Lake, 35 miles west of Great Falls along US 89.

In the Mountain Region lie many bodies of water that hold a good array of birds, though they do not have the immense populations of the prairie lakes. The beautiful Red Rock Lakes in the extreme southwest (*see under* **Monida**) have nesting Trumpeter Swans, and the reservoirs in the Ninepipe and Pablo National Wildlife Refuges (*see under* **Ronan**) have nesting all five species of grebes and numerous ducks. In the northwest are glacial lakes of varying size; most are deep and bordered by willow thickets and coniferous forests. Flathead Lake (*see under* **Polson**) is the largest, but not so productive ornithologically as the smaller, more remote lakes farther north. Several in Glacier National Park, away from disturbances by man, offer the best chances of seeing broods of Common and Barrow's Goldeneyes and Common Mergansers.

In the spring and fall, great numbers of transient Canada Geese, ducks, and shorebirds pass through the Prairie Region, stopping on the larger lakes for feeding and resting. Medicine Lake and Lake Bowdoin are two of the best places for seeing large aggregations. Thousands of Whistling Swans and Snow Geese usually visit Freezeout Lake in April and again in November. Fairly impressive numbers of waterfowl and shorebirds also visit suitable bodies of water, such as the reservoirs in the Ninepipe and Pablo National Wildlife Refuges, of the Mountain Region. The wooded areas in all the river valleys throughout the state show spring and fall movements of small landbirds, though nowhere is the number of birds particularly great. Some of the species appearing in the valleys of the eastern Prairie Region are the Tennessee Warbler, Orange-crowned Warbler, Myrtle Warbler, Blackpoll Warbler, Slate-colored Junco, American Tree Sparrow, Harris' Sparrow, and White-throated Sparrow. In the Mountain Region, most of the landbirds moving through in migration are northern representatives of the same species that breed in the Region itself; hence there are few different species to be observed in the spring and fall. The main migration flights in Montana take place within the following dates:

Waterfowl: 20 March–25 April; 15 September–15 November
Shorebirds: 1 May–1 June; 1 August–15 September
Landbirds: 15 April–1 June; 15 August–1 October

As in North Dakota, the Prairie Region offers generally unproductive bird finding in the winter. The low temperatures, the persistent snow, and the lashing of bitter winds make conditions unfavorable for all but the hardiest of bird species—and the hardiest of bird finders. For a day's search over a wide stretch of plains and in a well-wooded river valley or coulee, several Rough-legged Hawks, a few flocks of Common Redpolls, Lapland Longspurs, and

Snow Buntings, perhaps a small flock of American Tree Sparrows, one or two Northern Shrikes, and a half-dozen or so permanent-resident species are considered good results. The Mountain Region, particularly that part west of the Continental Divide, which enjoys the milder climate of the Pacific slope, has additional birds in winter. These include Mallards, Common Pintails, Common and Barrow's Goldeneyes, and a few other ducks on spring-fed bodies of water, North American Dippers near the shallow rapids of open streams, and Townsend's Solitaires, Bohemian Waxwings (often large numbers), Pine Grosbeaks, and Evening Grosbeaks in forested valleys.

Authorities

Winston E. Banko, Mrs. Robin Boyd, Helen Carlson, Clifford V. Davis, Beatrice D. FitzGerald, C. J. Henry

BILLINGS
Rimrocks | Coulson Park

This large city, lying in the valley of the Yellowstone River at 3,200 feet elevation, is on the western edge of the prairie and near an eastern spur of the Rockies, the Beartooth Mountains. The two best areas for bird finding are the **Rimrocks** and **Coulson Park** north of the city.

Starting at the airport, drive east on State Secondary 318 for 1.0 mile, then turn off right onto Black Otter Trail which winds along the top of the Rimrocks. In late spring and summer, watch for White-throated Swifts and Barn and Cliff Swallows in flight along the face of the Rimrocks and for Rock Wrens perched on the roadside barriers. Other birds to look or listen for are Canyon Wrens, Western Meadowlarks, Rufous-sided Towhees, and Lark and Vesper Sparrows. In winter, Townsend's Solitaires may be found.

Continue east on Black Otter Trail; turn right onto State Secondary 318, left at the traffic light onto US 87, right at the next traffic light and go to the end of the street, left onto Billings Bench Boulevard for 0.5 mile, and then right onto a gravel road entering Coulson Park. Leave the car in a parking lot at the bottom of the hill.

Coulson Park borders the Yellowstone River. Even though within metropolitan Billings, the Park's 120 acres remain purposely 'unimproved' and accessible only on foot.

Nesting birds include: *in the woods near the parking lot*, the Downy Woodpecker, Western Pewee, Black-capped Chickadee, and Bullock's Oriole; *in or about the marsh near the entrance*, the Sora, Common Yellowthroat, Red-winged Blackbird, and Song Sparrow; *in the interior*, the American Kestrel, Common Screech Owl, Red-headed Woodpecker, House Wren, Gray Catbird, Brown Thrasher, Cedar Waxwing, Red-eyed and Warbling Vireos, Yellow Warbler, Yellow-breasted Chat, and Black-headed Grosbeak. Winter birds usually include the Rough-legged Hawk, Red-breasted Nuthatch, and American Tree Sparrow, occasionally the Bohemian Waxwing and Northern Shrike.

A walk along the banks of the Yellowstone should yield: *in summer*, the Double-crested Cormorant, Great Blue Heron, Killdeer, Spotted Sandpiper, California, Ring-billed, and Franklin's Gulls, and four swallows, the Violet-green, Tree, Bank, and Rough-winged; *in winter*, Canada Geese, Mallards, Common and Barrow's Goldeneyes, and Common Mergansers on the open water.

An ideal time for bird finding in Coulson Park is from mid-April to mid-May and in September when there are many migrating small landbirds such as Swainson's Thrushes, warblers—Orange-crowned, Audubon's, Blackpoll, Northern Waterthrush, Wilson's, and American Redstart—and sparrows—Clay-colored, Brewer's, Harris', White-crowned, and White-throated.

BOZEMAN
Fish Cultural Development Center

There are probably few places in Montana holding a wider variety of landbirds in so small an area as the vicinity of the **Fish Cultural Development Center** northeast of Bozeman on Bridger Drive. Its richness of birdlife is not surprising because of the varied environment. Here, meeting the farmlands of the Gallatin Valley, are mountain forests of spruce, Douglas-fir, and pine, ravines with tangles of wild rose, snowberry, and dogwood, and a swift stream that is bordered by chokeberry, serviceberry, willow, and hawthorn and overhung by rocky cliffs. An early morning trip to this spot in May, June, or July should yield forty or more species.

The Center is easy for newcomers to find owing to the proximity of Montana State University's giant block-letter M, a conspicuous landmark on the slope visible northeast of the city. To reach the area, exit north from I 90 on US 10 north, then immediately turn right onto Griffin Drive and soon bear left onto Bridger Drive (State Secondary 293), proceeding 5 miles to a sign marking the Center on the right. Drive in, park the car, and walk south through the grounds and out the stone exit gate.

From late May to late June one can expect the Calliope Hummingbird on the terminal twigs of shrubs in the area. Turn left and follow the road along Bridger Creek which is flanked by trees and shrubs where such birds breed as the Hammond's Flycatcher, Western Pewee, Black-capped Chickadee, Veery, Warbling Vireo, Audubon's and MacGillivray's Warblers, American Redstart, Black-headed Grosbeak, and Pine Siskin.

To the right, rising abruptly from the road, is a knoll covered sparingly by coniferous trees and junipers. Here look or listen for the Clark's Nutcracker, Swainson's Thrush, Townsend's Solitaire, Ruby-crowned Kinglet, Western Tanager, Lazuli Bunting, Evening Grosbeak, Cassin's Finch, both Green-tailed and Rufous-sided Towhees, White-crowned Sparrow, and other nesting species. The jutting face of rock to the north of the knoll is occupied by Cliff Swallows and, occasionally, Violet-green and Rough-winged Swallows. At the point where the road crosses Bridger Creek the Spotted Sandpiper and North American Dipper are often in view along the Creek.

CULBERTSON
Medicine Lake National Wildlife Refuge

In northeastern Montana, **Medicine Lake National Wildlife Refuge** is excellent for waterbirds, waterfowl, and birds frequenting upland prairie. Of the Refuge's 31,457 acres, about 40 per cent consists of 8,700-acre Medicine Lake, the main body of water, and smaller lakes and ponds—all bulrush-bordered impoundments of what was once entirely marsh. The rest of the Refuge is largely grassy upland with stretches of such shrubby growth as snowberry, buffaloberry, and silverberry. Cottonwoods, ashes, and willows along streams and on islands in Medicine Lake constitute the only notable tree growth.

Waterbirds and waterfowl abound during the nesting season from April to mid-July. Outstanding are the large numbers of Western Grebes and their unforgettable courtship displays early in the season as they establish colonies among the bulrushes bordering Medicine Lake. Also in large numbers are American White Pelicans, Double-crested Cormorants, Great Blue and Black-crowned Night Herons, California and Ring-billed Gulls, and Common Terns which have colonies on islands in Medicine Lake. Waterfowl nesting in greatest numbers are Canada Geese, Mallards, Gadwalls, Common Pintails, Blue-winged Teal, American Wigeons, Northern Shovelers, Redheads, Canvasbacks, and Ruddy Ducks. Other common breeding birds associated with the water and marshy areas include the Eared Grebe, American Bittern, Northern Harrier, Sora, American Coot, Spotted Sandpiper, Willet, American Avocet, Wilson's Phalarope, Marsh Wren, Yellow-headed Blackbird, and Savannah Sparrow.

In late August and September the bird populations of the Refuge are vastly augmented with the arrival of Franklin's Gulls from miles around and transient shorebirds, the more common being Greater and Lesser Yellowlegs, Least Sandpipers, Long-billed Dowitchers, and Semipalmated Sandpipers. From late September to mid-November the populations are augmented further by thousands of southbound waterfowl and frequently by hundreds of Sandhill Cranes, always with the chance of one or more Whooping Cranes.

Residing in the grasslands are Sharp-tailed Grouse. For watching their 'dances' in late April and May, ask at headquarters about the best vantages. Among other breeding grassland birds are the Upland Sandpiper, Marbled Godwit, Burrowing Owl, Horned Lark, Sprague's Pipit, Bobolink, Lark Bunting, Grasshopper and Baird's Sparrows, and Chestnut-collared Longspur. A few McCown's Longspurs are a possibility in or near the Refuge.

The Medicine Lake Refuge is crossed by State 16 about 24 miles north of Culbertson and 23 miles south of Plentywood. Headquarters lies on the Refuge, 2 miles east of State 16 from which it is reached. Most parts of the Refuge are accessible by car except in wet weather. At headquarters, inquire about taking the 18-mile, self-guided auto tour.

EAST GLACIER PARK
Glacier National Park

The 1,583 square miles of **Glacier National Park** in northwestern Montana have some of the most exceptional scenery in the Rocky Mountains. Here, through prolonged effects of glaciation, are immense peaks with steeply cut walls, lofty cirques holding the remnants of great glaciers, broad U-shaped valleys separated by high sharp-edged ridges, and 200 or more blue-green lakes, some in deeply curved basins and others held by morainic dams. Many streams, arising partly from melting glaciers and snowfields, speed down to the valleys, frequently plunging during their course in foaming cascades.

The mountains in Glacier Park run in a northwest-southeast direction for 45 miles. The easternmost front on the Great Plains, the westernmost slope down to the Flathead Valley. Southward over the mountains meanders the Continental Divide, thus making a natural separation of the Park into east and west sides. Elevations in the Park range from 10,488 feet at the summit of Mt. Cleveland, the highest point, down to 4,834 at St. Mary Lake on the east side and 3,153 feet at Lake McDonald on the west side.

Coniferous trees characterize the forests. Those on the east side are mainly lodgepole pine and Douglas-fir, giving way higher up to Engelmann spruce, subalpine fir, and whitebark pine; those on the west side, where there is ample moisture, are predominantly western white pine, western hemlock, western larch, grand fir, and western red cedar. Only in a few places are there deciduous trees—for example, near lakes and streams are cottonwood and willow thickets and on some of the slopes on the east side are groves of aspen.

Because of the generally high terrain, all forests are restricted to the lower valleys and slopes. Much of the Park, including the higher valleys, is therefore above timber line and is subalpine to alpine in character. Where there is soil, mosses and hardy wildflowers grow in profusion. In early July, following the slow retreat of the last snowbanks, wildflowers are at their best, the blooms of the bear grass, shooting star, glacier lily, larkspur, gentian, carpet pink, Indian paintbrush, and others giving lavish color to benches, gentle slopes, and valley meadows.

Complementing the aforementioned attractions are the mammals. So much of the Park is open country that a visitor cannot be within its boundaries very long before viewing some of the bigger species: mountains goats and bighorn sheep moving sure-footedly along the ledges of cliffs at dizzy heights; moose feeding in boggy creek bottoms; mule and whitetail deer grazing in the forest openings; and perhaps a coyote skulking over a meadow, or a black bear patrolling a stream. The smaller mammals are remarkably numerous, particularly the marmots and pikas on the high talus slopes and the squirrels and chipmunks at the timber line and below. Always a special delight to the visitor are the handsome golden-mantled squirrels, many of which are very tame and confidently solicit food.

Amid the superlative scenery, the richness of the flora, and the abundance of mammals, birdlife seems eclipsed. The truth is, owing to the high elevation coupled with the wide extent of treeless terrain, birds are limited in numbers and variety. A few species such as Red-tailed Hawks, Golden Eagles, Ospreys, Rufous and Calliope Hummingbirds, and Mountain Bluebirds range widely and a few others such as Water Pipits and MacGillivray's and Townsend's Warblers confine themselves to specific habitats, but in no case is any species remarkably abundant save perhaps Pine Siskins. Moreover, no bird family is represented by any notable number of species save the grouse family, an exception that provides an outstanding ornithological feature. In all, there are five species: the Sharp-tailed Grouse (a few) in burned-over areas and brushy ravines on the east side; the Ruffed Grouse in the aspen groves on the east side; the Spruce Grouse in the pine-fir forests on the west side; the Blue Grouse in all coniferous forests

SPRUCE GROUSE

from about 5,000 feet elevation to the timber line; and the White-tailed Ptarmigan above the timber line.

Glacier Park, unlike most other National Parks, has limited access to cars. It is crossed only by the spectacular Going-to-the-Sun Road, from the St. Mary entrance on the east side across the Continental Divide to the West Glacier entrance on the west, a distance of 50 miles. There are two other roads, both on the east side, entering separately and running short distances to Many Glacier and Two Medicine, respectively. Glacier Park is primarily a 'trail park,' most of its primeval wilderness being accessible only by trails. For the visitor who prefers traveling on foot or riding horseback, there are over 1,000 miles of trails into the 'back country.'

Embracing, as it does, so much high country, the Park has a relatively brief season each year—15 June to 15 September—at which time its concessions—hotels and cabins—are open. For several weeks before and after the regular season there are accommodations on privately owned land in the Park and outside. A map of the Park showing the principal roads and trails, as well as campgrounds and other facilities, may be obtained at any one of the entrance checking stations. Naturalists at the ranger stations in Many Glacier, Two Medicine, St. Mary, Avalanche Creek, and Apgar are available for inquiry about natural-history features and interpretive services.

Despite the general scarcity of birds in Glacier Park, there are nonetheless localities as rewarding as in any area of the northern Rocky Mountains. Below are directions to several of the more promising spots and some of the birds to be expected in the early summer.

From the town of East Glacier Park, just east of the easternmost extremity of the Park itself, drive northward on US 89 (Blackfeet Highway), passing the turnoffs to Two Medicine and St. Mary entrances to the town of Babb, a distance of 38 miles. Leave the numbered highway here and proceed 8.5 miles west, on the Park road, to the Many Glacier entrance; obtain a map of the Park at the checking station and continue 4 miles to Many Glacier, at the foot of Swiftcurrent Lake (elevation 4,878 feet), which has a hotel, cabins, and campground.

A few Gray Jays frequent the campground for handouts. Chipping Sparrows and White-crowned Sparrows are common in the vicinity. Well worth following is a trail going west from the campground along the shore of Swiftcurrent Lake, across Swiftcurrent Creek, and then south and around Lake Josephine. Keep an eye on the open water for Barrow's Goldeneyes and Common Mergansers. In the bordering willow thickets, be alert for Veeries, Northern Waterthrushes, and MacGillivray's and Wilson's Warblers. And expect to see Steller's Jays and Oregon Juncos at any time along the way.

Back from the shore of Swiftcurrent Lake are patches of dense conifers and intervening open spaces scattered with aspen that should be investigated for Ruffed Grouse, Hammond's Flycatchers, Red-breasted Nuthatches, Swainson's Thrushes, Ruby-crowned Kinglets, and Audubon's and Townsend's Warblers.

From the southern end of Lake Josephine a trail leads up to Grinnell Lake (about 3 miles distant) to Grinnell Glacier (6 miles distant) over which Golden Eagles sometimes soar. For the scenery alone, this trip is rewarding.

Having explored the Many Glacier area, return to the Blackfeet Highway outside the Park and go south 8 miles to the St. Mary entrance, then begin a trip westward across the Park on the Going-to-the-Sun Road.

West from the St. Mary entrance the Road soon passes close to the north shore of St. Mary Lake, where both Common and Barrow's Goldeneyes, Common Mergansers, and California Gulls may be readily viewed from the car. Other birds that may be noted are American Kestrels and at least four swallows—Violet-green, Tree, Barn, and Cliff—flying low over the water or adjacent fields.

At Logan Pass (elevation 6,664 feet), 18 miles from the St. Mary entrance, the Road crosses the Continental Divide. Except for a few stunted conifers, the environment at this point is typically subalpine with wildflowers luxuriantly carpeting terraced benches. Special birds here are the White-tailed Ptarmigan, Clark's Nutcracker, Water Pipit, and Gray-crowned Rosy Finch. Failing to see these species here, try the Hidden Lake Trail. This leads south from the Pass across similar environment. After 2 miles, however, it goes through a small pass and comes to the edge of a cliff overlooking Hidden Lake, 500 feet below at 6,375 elevation. Listen carefully for songs of the Hermit Thrush rising from among the subalpine firs near the Lake.

From Logan Pass the Road descends by hairpin turns into McDonald Valley and follows down McDonald Creek. Stop where there is rapid water to look for Harlequin Ducks and North American Dippers, and watch the sky for swifts that can be either Black or Vaux's.

From Avalanche Creek Campground, 15.5 miles from Logan Pass on the left side of Going-to-the-Sun Road, take the 3.5-mile trail up to Avalanche Lake. The ascent is gradual through a magnificent stand of red cedar and hemlock with some lodgepole pine, Douglas-fir, and Engelmann spruce. There is no better place in Glacier Park for Spruce Grouse than this particular forest. Seldom is the trail very far from the deep, cool gorge through which Avalanche Creek rushes noisily—and North American Dippers are a near certainty.

Avalanche Lake (3,905 feet elevation), a small body of milky colored water, is bordered by a dense growth of deciduous shrubs and, farther back, by mixed conifers and deciduous trees. A high cirque, with cascades spilling over its face, looms in the distance. Swifts commonly forage for insects in the air above Avalanche Lake.

Among the birds that one can hear or see on the entire trip from Avalanche Campground to the Lake are the Rufous and Calliope Hummingbirds, Yellow-bellied Sapsucker, Black-backed Three-toed Woodpecker, Hammond's and Olive-sided Flycatchers, Black-capped and Chestnut-backed Chickadees, Brown Creeper, Winter Wren, Varied and Swainson's Thrushes, Solitary, Red-eyed, and Warbling Vireos, Audubon's and Townsend's Warblers, Northern Waterthrush, American Redstart, Western Tanager, Cassin's Finch, Pine Grosbeak, Pine Siskin, Oregon Junco, and Fox Sparrow.

Soon after Avalanche Creek Campground the Going-to-the-Sun Road leaves McDonald Creek and parallels for about 10 miles the east shore of Lake McDonald, the Park's largest lake, where a few Eared Grebes reside as well as goldeneyes, Common Mergansers, and California Gulls. Occasionally a Bald Eagle makes an appearance. At the south end of Lake McDonald the Road passes through Apgar to West Glacier, site of Park headquarters and the west entrance to the Park.

Although Bald Eagles are always a possibility during the summer, the time to see them is in the fall when large numbers gather along McDonald Creek from near its entry into Lake McDonald and for about 5 miles below the Lake. The big birds—as many as 377 counted on a single day in 1975—are here to feed on salmon, dead or dying after spawning. Golden Eagles join them, and later in the season and early winter Snowy Owls arrive for a share of the harvest.

Reference

Birds of Glacier National Park. By Lloyd P. Parratt. West Glacier: Glacier Natural History Association. 1964

GREAT FALLS

Benton Lake National Wildlife Refuge

Drive north from Great Falls in west-central Montana on US 87 for 2 miles, turn off left onto State Secondary 225 (Bootlegger Trail) and continue north for 12 miles, then turn onto the road at left which leads to headquarters of the **Benton Lake National Wildlife Refuge.**

The Refuge (12,383 acres) comprises marshy Benton Lake and adjacent rolling grasslands.

In spring and fall Benton Lake is host to hordes of migrating waterfowl. The largest numbers appear in the fall, peaking from September to mid-November. Many waterfowl also nest; thus one may see commonly in June and early July numerous broods of such species as the Canada Goose, Mallard, Gadwall, Common Pintail, Green-winged and Blue-winged Teal, American Wigeon, Northern Shoveler, Redhead, Canvasback, Lesser Scaup, and Ruddy Duck.

In diked units of Benton Lake where bulrushes, spike rushes, and cattails thrive are nesting colonies of Eared Grebes and Franklin's Gulls. Other birds nesting in the same habitat are the Northern Harrier, Sora, American Coot, Black Tern, and Yellow-headed and Red-winged Blackbirds. Willets, American Avocets, Wilson's Phalaropes, and Savannah Sparrows find suitable nest sites on the Lake's periphery.

The Refuge offers a 14-mile, self-guided auto tour that passes over dikes for viewing all of the aforementioned species. Inquire at headquarters about a road to the grassy uplands which are worth exploring for a variety of breeding birds that will include the Long-billed Curlew, Upland Sandpiper, Marbled Godwit, Burrowing Owl, Horned Lark, Lark Bunting, Grasshopper, Baird's and Vesper Sparrows, and McCown's and Chestnut-collared Longspurs.

HARLOWTON
Long-billed Curlew and Mountain Plover Country

Harlowton in central Montana lies in short-grass prairie country attractive as nesting habitat to Long-billed Curlews and a few Mountain Plovers.

Long-billed Curlews may be readily observed along US 191 from about 20 miles north to 25 miles south of Harlowton. They arrive in April, nest in late May and June, and congregate in large flocks before leaving in early September. Wherever they nest they are invariably vociferous when disturbed. If the bird finder as much as walks over the general nesting area, the birds fly toward him, then circle and alight on a nearby knoll, meanwhile pouring out an endless series of shrill whistles and angry rattles.

To observe Mountain Plovers, drive west on US 12 for 12 miles; opposite State Secondary 296 that goes south to Twodot, turn north onto a gravel road, locally called Olaf Road. For the next 7.5 miles, look for Mountain Plovers. They arrive soon after mid-April, have broods by mid-June, and leave by the first of August. Very often the birds appear on the sides of the road or even on the road ahead of the car. Sometimes they are remarkably tame, allowing close observation.

The smaller prairie birds may be seen by the thousands in spring and early summer, particularly along US 191 north from Harlowton to Judith Gap. Horned Larks, Lark Buntings, and McCown's Longspurs fly up from the roadsides ahead of the car. North from Judith Gap to Moore, where the highway crosses wheat country, diurnal predators appear in wider variety than in most other parts of the state. Such species as the Red-tailed and Ferruginous Hawks, Golden Eagle, Northern Harrier, Prairie Falcon, and American Kestrel may be expected. Say's Phoebes and Mountain Bluebirds are frequent around unoccupied buildings; the former are also frequent near bridges, as are Barn and Cliff Swallows. Spotted Sandpipers and Wilson's Phalaropes usually reside around the borders of the numerous farm ponds, which are nesting habitats as well for a few ducks and, occasionally, a pair of Eared Grebes. Savannah and Vesper Sparrows are often common in grassy areas adjoining the ponds. Where the highway crosses streams that are brush-bordered, Black-billed Magpies, American Goldfinches, and Common Yellowthroats are in evidence.

HELENA

Lake Helena | MacDonald Pass

This capital city (elevation 4,157 feet), built around famous Last Chance Gulch of the Gold Rush days, is well up in the foothills of Mt. Helena. To the northeast, broad Prickly Pear Valley (Helena Valley) descends gradually to the Missouri River some 10 miles away; beyond the Missouri rise the Big Belt Mountains. Several lakes lie along the Missouri within 25 miles of the city. About 15 miles west of Helena looms the Continental Divide.

Of special interest because of its nesting Canada Geese, as well as its many loons, grebes, ducks, and transient shorebirds, is **Lake Helena,** formed by the backing up of the Missouri into the mouth of Prickly Pear Creek. Approximately 3 miles long and 1.5 miles wide, this body of water is very shallow and produces an abundance of cattails and bulrushes. Although the geese may be seen or heard from almost any spot along the shore, their preferred breeding habitat is in the southern half of the area. One may reach the northern side of Lake Helena by driving north from Helena on I 15 for 8 miles, then exiting east 4 miles on State Secondary 453, which soon bears south and crosses the Lake near its eastern end.

US 12 west from I 15 at Helena goes over **MacDonald Pass** (elevation 6,325 feet) on the Continental Divide. As the highway ascends to the Pass it leads up through a fine coniferous-forest belt to conditions typically subalpine. From parking places along the way one may walk into adjacent stands of lodgepole pine, Engelmann spruce, and Douglas-fir and search for such birds as Spruce Grouse, Black-backed Three-toed and Northern Three-toed Woodpeckers, Gray Jays, Mountain Chickadees, Red-breasted Nuthatches, Brown Creepers, Audubon's and Wilson's Warblers, and Cassin's Finches. Near the Pass, Clark's Nutcrackers, Oregon Juncos, and, occasionally, Pine Grosbeaks are observed.

MALTA

Bowdoin National Wildlife Refuge | Little Rocky Mountains | Charles M. Russell National Wildlife Range

No bird finder passing through northeast-central Montana in spring, summer, or fall should miss visiting **Bowdoin National Wildlife Refuge** (15,437 acres) for its remarkable variety and large numbers of breeding and transient birds.

The Refuge is centered on Lake Bowdoin. Fed by surface drainage, it has strongly alkaline water, the level of which, until stabilized by damming, fluctuated seasonally. Now open water and marsh regularly cover 4,300 acres. Since the water averages between 4 and 5 feet in depth and most of the shore rises abruptly, there is relatively little marsh vegetation; but in two arms of the Lake—the southeastern and southwestern—the water is sufficiently shallow and the incline of the shore is gradual enough to afford an extensive growth of bulrushes, cattails, and sedges. In the main part of the Lake where the two arms meet are several small islands, the largest—Woody Island—with a surface of about 5 acres. Though bordered by a few marshy patches, their interiors are generally barren save for scattered clumps of shrubs—e.g., greasewood—and grasses.

Birds regularly breeding from April to mid-July in the Lake Bowdoin marshes include the Eared Grebe, Black-crowned Night Heron (in colonies), American Bittern, Northern Harrier, Sora, American Coot, Franklin's Gull (in colonies), Black Tern, Marsh Wren, Yellow-headed and Red-winged Blackbirds, and the following ducks: Mallard, Gadwall, Common Pintail, Green-winged and Blue-winged Teal, American Wigeon, Northern Shoveler, Redhead, Canvasback, Lesser Scaup, and Ruddy Duck. Many Canada Geese nest, sometimes on muskrat houses and on shoulders of the Refuge roads. Killdeers, Willets, American Avocets, Wilson's Phalaropes, and Savannah Sparrows reside on the edges of marshes or very near them.

Contributing in large measure to Lake Bowdoin's ornithological distinction are the impressive colonies of American White Pelicans, Double-crested Cormorants, Great Blue Herons, California and Ring-billed Gulls, and Common Terns that settle on the islands, particularly Woody and Pelican Islands. Unless the season is late, the hatching period begins the first week of June and is over by the last week.

The prairie uplands in the western part of the Refuge attract nesting Sharp-tailed and Sage Grouse, Long-billed Curlews, Marbled Godwits, Sprague's Pipits, Baird's, Clay-colored, and a few Brewer's Sparrows, and large numbers of Horned Larks, Lark Buntings, and Chestnut-collared Longspurs. Bank, Rough-winged, and Barn Swallows nest locally as do both Eastern and Western Kingbirds, Brown Thrashers, and Bullock's Orioles.

From September to mid-November southbound geese and

WILSON'S PHALAROPE

ducks congregate by the thousands on the Refuge, and Sandhill Cranes, sometimes as many as 10,000, stop to feed and rest.

Headquarters of the Refuge, overlooking the southeastern arm of Bowdoin Lake, may be reached from Malta by driving 7 miles east on Old US 2, then turning off right onto the entrance road where the buildings are already in view ahead. Most of the Refuge is accessible by car except during wet weather. At headquarters, inquire about taking the 15-mile, self-guided auto tour around Lake Bowdoin and about the best vantages for seeing different birds and the nesting colonies.

For bird finding 'off the beaten parth,' go to the **Little Rocky Mountains,** 35 miles or more southwest of Malta via US 191.

The Little Rockies are the easternmost mountain range in the northern half of Montana, separated from the major Rocky Mountains to the west by nearly 200 miles and surrounded in all directions by dry plains. Although a small range—about 15 miles long (north–south) and 10 miles wide—it nonetheless rises sharply from the plains, attaining altitudes of 3,500 to 5,500 feet.

The principal forest cover consists of ponderosa pines at higher elevations and lodgepole pines with aspens and other deciduous growth on the lower slopes and in the lower canyons along the few creeks, which are often dry during the summer months.

Toward the close of the last century the Little Rockies were the scene of productive goldmining, but the boom days were over by 1900. Such place names on the map—Zortman, Landusky, Hays, and Lodgepole—are today practically ghost towns.

Birdlife is essentially characteristic of the Rocky Mountains. What makes the Little Rockies exciting ornithologically is the surprising number of species that breed in this 'island' despite the gap of prairie between it and the main range.

The following trip to the Little Rockies is suggested. From Malta, take US 191 southwest for 55 miles, then turn north onto State Secondary 376. After about 6 miles turn off right at the directional sign to Landusky and Hays. The road, like all roads in the Little Rockies, is dirt-surfaced and unnumbered. Between Landusky and Hays the road goes over the highest pass accessible by car. At Hays, a paved road returns to State Secondary 376. Begin bird finding anywhere after leaving the highway by parking the car and hiking over old mine roads which wind through the area, up mountain slopes and down into canyons. Herewith is an indication of the great variety of breeding birds.

In the higher pine-clad reaches of the Little Rockies are the Clark's Nutcracker, both Black-capped and Mountain Chickadees, Red-breasted Nuthatch, both Hermit and Swainson's Thrushes, and the White-crowned Sparrow. In the lower canyons with suitable cover are both Dusky and Western Flycatchers together with such species as the Western Pewee, Violet-green Swallow, Veery, Orange-crowned Warbler, Ovenbird, and MacGillivray's Warbler. Common wherever there are stands of pine are the Audubon's Warbler, Western Tanager, Pine Siskin, and Oregon Junco.

The Townsend's Solitaire frequents rocky, brush-covered slopes as does the Rock Wren, although the latter is not restricted to woody growth.

In the foothills along tree- and brush-lined stream beds is a wide assortment of birds including the Eastern Kingbird, Black-billed Magpie, Gray Catbird, Yellow-breasted Chat, Lazuli Bunting, and Rufous-sided Towhee.

American Kestrels are common but most raptors are relatively few. However, one may see Golden Eagles on the western side of the range, particularly from the road between Landusky and Hays. Peregrine Falcons have nested near the mouth of Peoples Creek south of Hays.

After exploring the Little Rocky Mountains, return to US 191 and continue south. About 34 miles from Hays and just before reaching the Robinson Bridge across the Missouri River, turn east off US 191 onto a dirt road and go 6 miles to the Slippery Ann Wildlife Station, subheadquarters of the **Charles M. Russell National Wildlife Range.** In its vicinity are Wild Turkeys, Mountain Plovers, Burrowing Owls, and Poor-wills.

MILES CITY

Tongue River | Pumpkin Creek | Miles City National Fish Hatchery

Miles City in southeastern Montana lies in the heart of the 'boots and saddle country,' characterized by rolling grasslands and sagebrush flats, buttes, canyons, and coulees, river bottoms, willow thickets, and occasional sloughs.

For a sample of the country, exit from I 94 on US 312 south from Miles City along the **Tongue River** and then **Pumpkin Creek** over terrain featured by red buttes and canyons. For the first 15 miles the highway traverses an area particularly good for breeding Sage Grouse, Burrowing Owls, Horned Larks, Sage Thrashers, Lark Buntings, and Lark Sparrows. Pinyon Jays may be expected on the juniper-clad ridges to the east of the highway and Prairie Falcons in flight almost anywhere.

The vicinity of the **Miles City National Fish Hatchery** is well worth a visit for waterbirds, waterfowl, and shorebirds in season. Exit from I 94 to Miles City and take US 10 west and south; about 4 miles out of town, turn right onto a road that passes under an arch marked 'U.S. Range Livestock Experiment Station' and continue for 1.3 miles north and west; then turn right and follow the only road 0.7 mile to a railroad track but do not cross it; instead, turn left (west) and go to a group of white buildings belonging to the Hatchery.

Leave the car at the Hatchery and proceed on foot between large rearing ponds to grassy sloughs and meadows, where there will be a variety of such summer-resident birds as grebes, American Bitterns, Soras, American Coots, Common Snipes, and Yellow-headed Blackbirds.

About a mile farther west are the sand and gravel bars of the Yellowstone River, which, during May and September, attract numerous shorebirds, including Solitary Sandpipers, Greater and Lesser Yellowlegs, Pectoral and Baird's Sandpipers, and Northern Phalaropes. The cottonwood groves and brushy thickets that flank the River are likely nesting habitat for Mourning Doves, Red-headed Woodpeckers, Western Pewees, Black-capped Chickadees, Mountain Bluebirds, Warbling Vireos, Yellow Warblers, Yellow-breasted Chats, and Lazuli Buntings.

MISSOULA
Pattee Canyon

Due partly to its relatively low altitude (3,223 feet) and partly to its location on the Pacific slope of the Continental Divide, this city enjoys early springs, late falls, and mild winters. Consequently in undisturbed outskirts some bird species reside that may be rare or even absent east of the Divide.

The best area for birds near Missoula is **Pattee Canyon** southeast of the city. Two or three early-morning hours of searching here in April, May, or June should yield a good list of landbirds. From I 90, exit south on US 12; turn off left onto Higgins Avenue and proceed to the outskirts of the city, and where Higgins Avenue becomes South Higgins Avenue and goes right, turn left onto Pattee Canyon Drive. After passing through a small residential area and skirting Mt. Sentinel which bears on its north slope the University of Montana's big letter M, the Drive leads into the Canyon.

Once in the mouth of the Canyon, where there are deciduous trees and thick clumps of shrubs, search for such birds as Lewis' Woodpeckers, Varied Thrushes, Veeries, MacGillivray's Warblers, and Bullock's Orioles. Then continue up the Canyon.

Stop and explore the Lolo National Forest Picnic Area where

Steller's Jay may be expected. Listen for Swainson's Thrushes.

Above the Picnic Area, Pattee Canyon Road winds up through coniferous woods to the saddle in the vicinity of which are many high-country birds including Clark's Nutcrackers, Winter Wrens, Evening Grosbeaks, Pine Siskins, and Oregon Juncos. Calliope Hummingbirds are not uncommon here; a Ruffed Grouse may sometimes drum; with luck, a Northern Pygmy Owl may be discovered among the lower branches of evergreens.

MONIDA
Red Rock Lakes National Wildlife Refuge

Situated in the east end of the isolated, 6,600-foot Centennial Valley of southwestern Montana, the **Red Rock Lakes National Wildlife Refuge** (40,223 acres) enjoys a mountain-rimmed setting of great charm. Immediately to the east and south rises the Centennial Range along whose crests at 10,000 feet runs the Continental Divide; to the north begin the rolling foothills of the Gravelly Range.

The Refuge features 14,000 acres of open water—primarily three large lakes, Upper Red Rock, Lower Red Rock, and Swan, with bulrush islands—and extensive sedge meadows. The rest of the Refuge consists mainly of dry, undulating terrain, on which sagebrush thrives, and the lower slopes of the Centennial Mountains, which support open stands of pine, fir, and aspen.

The United States Fish and Wildlife Service established the Refuge in 1935 to save the Trumpeter Swan from extirpation in northwestern United States. The setting of the Refuge was ideal for the purpose. Not only did it embrace ample nesting habitat and food sources remote from human disturbances; it also, thanks to warm springs flowing into the Lakes, maintained open water through the winter—an important factor since the Trumpeter Swan in the contiguous northwestern states rarely migrates far from its breeding grounds. By 1972 as many as 250 Trumpeters—one-third of the population outside Alaska—were nesting in the Refuge, and the population has remained about the same ever since.

TRUMPETER SWAN

Until these splendid birds have their cygnets, by the end of June, viewing them is allowed only from knolls overlooking the Lakes. Thereafter through August many Trumpeters with broods may be observed closely at the Shambow Display Pool, an inlet to Upper Red Rock Lake.

Bear in mind that the Refuge has much to offer besides the Trumpeters. Other waterfowl breeding regularly include Canada Geese and at least fifteen species of ducks. American White Pelicans are present from spring to fall but do not nest in the Refuge. Eared Grebes, Sandhill Cranes, Willets, American Avocets, Wilson's Phalaropes, and Forster's Terns find suitable nesting habitat in the wetlands, Long-billed Curlews on the adjacent uplands. In the sagelands, Sage Grouse reside the year round, Sage Thrashers and Brewer's Sparrows in the warmer months. Mountain Bluebirds are numerous, their turquoise adding to the riotous hues of blooming wildflowers and the red and yellows of the mountains and foothills that rim the Valley.

Allow time while in the Refuge to climb the forested slopes on the south side of the Refuge, where there are both Blue and Ruffed Grouse and vantages for scanning the Centennial Valley and watching for soaring buteos—the Red-tailed, Swainson's, and Ferruginous Hawks—sometimes at eye level.

Thousands of geese and ducks stop on the Lakes temporarily during the southward passage in September and early October as do hundreds of Whistling Swans later in October.

In addition to birds, one is almost certain to see moose foraging around the Lakes and along creeks—a favorite spot is Tom Creek—and bands of pronghorns grazing in the sagelands.

Refuge headquarters at Lakeview on the south side of the Refuge may be reached by exiting from I 15 at Monida and driving east on a dirt road through the Centennial Valley for 28 miles, or by turning off US 191 to Henrys Lake (10 miles west of West Yellowstone, Montana) and proceeding west on a dirt road for 30 miles through Red Rock Pass. Both roads are generally impassable for ordinary cars until mid-May, and any time of the year following heavy rains, even in summer. Inquire about current road conditions at Monida or Henrys Lake. There are two campgrounds in the Refuge but no accommodations.

POLSON
University of Montana Biological Station

At Yellow Bay on the east shore of Flathead Lake (elevation 3,000 feet) is the **University of Montana Biological Station,** reached from Polson by driving east and north on State 35.

The Station grounds proper comprise 70 acres, forested principally with ponderosa pine, Douglas-fir, grand fir, and western larch. Where the timber has been removed there are deciduous shrubs and a few cherry orchards. Directly east of the Station are the Mission and Swan Ranges, which rise to altitudes of about 10,000 feet.

Some of the breeding birds residing regularly in the vicinity of the Station are the Cooper's Hawk, Osprey, Ruffed Grouse, Spotted Sandpiper, Northern Pygmy Owl, Pileated Woodpecker,

Yellow-bellied Sapsucker, Black-backed Three-toed Woodpecker, Hammond's Flycatcher, Steller's Jay, Northern Raven, Mountain Chickadee, Red-breasted Nuthatch, Winter Wren, Varied Thrush, Swainson's Thrush, Ruby-crowned Kinglet, Solitary Vireo, Audubon's, Townsend's, and MacGillivray's Warblers, American Redstart, Western Tanager, Black-headed and Evening Grosbeaks, Lazuli Bunting, Pine Siskin, American Goldfinch, Red Crossbill, and Oregon Junco.

The Biological Station is in operation from mid-June to mid-August. Stop in and inquire about the latest information on bird finding in the area.

RONAN

Ninepipe and **Pablo National Wildlife Refuges** | **National Bison Range**

In northwestern Montana the **Ninepipe** and **Pablo National Wildlife Refuges** lie at 3,000 feet elevation on the gently rolling terrain of the Flathead Valley between the Mission and Cabinet Ranges. Both Refuges are essentially irrigation reservoirs, Ninepipe with 2,000 acres of water and Pablo with 2,500. Ninepipe is the better of the two for birds because of its more marshy character.

Waterfowl, sometimes in numbers as high as 200,000, frequent the Refuges in migration from late September to mid-November. Of the waterfowl that nest in the Refuges from April to mid-July, the most common are Canada Geese, Mallards, Gadwalls, Common Pintails, Green-winged and Blue-winged Teal, American Wigeons, Northern Shovelers, Redheads, Ruddy Ducks, and Common Mergansers.

All five species of grebes nest, the Horned and Western more commonly. Other breeding birds include the American Bittern, Northern Harrier, Sora, American Coot, Common Snipe, American Avocet, Wilson's Phalarope, California and Ring-billed Gulls, Forster's and Black Terns, Marsh Wrens, and three blackbirds—Yellow-headed, Red-winged, and Brewer's.

The Ninepipe Refuge is reached from Ronan by driving 5 miles south on US 93, then turning right (west) onto State Secondary 212 and proceeding 1.8 miles. At the Refuge directional sign, turn south onto the road that leads along the top of a high dike overlooking the reservoir. To reach the Pablo Refuge from Ronan, drive north for 8.3 miles on US 93 and then, turning off left (west), on an unnumbered road for 1.5 miles. At the Refuge directional sign, drive up and along the dike for good views of the reservoir.

While in this area the bird finder may enjoy touring the **National Bison Range.** From the Ninepipe Refuge, return to State Secondary 212 and follow it west and then south for 12 miles to the Range entrance and headquarters at Moiese.

The Bison Range covers 18,541 acres, largely grassland, over which several hundred bison or buffalo roam free as do elk, whitetail deer, mule deer, pronghorns, and bighorn sheep. A 19-mile self-guided auto tour of the Range begins and ends at headquarters. Since bison are apt to be belligerent, one must stay in or near his car. Nevertheless there is ample opportunity along the way to see a variety of open-country birds and always a very good chance of sighting one or more Golden Eagles over the higher parts of the Range.

SIDNEY
Fox Lake

A good spot for breeding ducks in prairie surroundings is **Fox Lake,** half an hour's drive from Sidney, a trading center near the North Dakota line. About a mile south of Sidney on State 16, turn west onto State 200 and continue straight west about 26 miles, passing the town of Lambert. Fox Lake, a widening of Fox Creek, will appear on the left just west of town.

Approximately 1.5 miles long and 0.25 mile wide, Fox Lake is shallow—not more than 3 feet deep anywhere—and in the summer is a green stretch of waving bulrushes, with cattails in abundance along its treeless margin. Mallards, Gadwalls, Common Pintails, Green-winged and Blue-winged Teal, American Wigeons, and Northern Shovelers are the ducks most likely in considerable numbers. Rails, blackbirds, and other marsh-loving birds are numerous.

In suitable upland situations to the north and south of Fox Lake, where the prairie is rolling, are such summer-resident birds as Burrowing Owls, Horned Larks, Sprague's Pipits, Baird's Sparrows, and McCown's and Chestnut-collared Longspurs.

STEVENSVILLE
Lee Metcalf National Wildlife Refuge

Along the west bank of the Bitterroot River in western Montana, beginning about 2 miles north of Stevensville and extending north for approximately 5 miles, is the 2,800-acre **Lee Metcalf National Wildlife Refuge.** To reach it from US 93, turn off east across the Bitterroot River to Stevensville and proceed through the north end of town; then turn left onto the first public farm road and go north about 1.5 miles to the Refuge's south boundary. From here a road leads through the Refuge, passing several ponds, marshes, and wet meadows and giving access to foot trails through timbered areas including bottomland woods.

The Refuge is situated in the sandy Bitterroot Valley between the Sapphire Mountains on the west and the Bitterroot Range on the west and south. Although several peaks within 20 miles of the Refuge reach 8,000 feet or more, the elevation of the Refuge is only 3,300 feet. Winters are consequently milder than in most parts of the state, with the result that such birds as Great Blue Herons, Canada Geese, many ducks, American Coots, Mourning Doves, American Robins, and Red-winged Blackbirds stay through the winter months.

The tall-timbered areas, consisting largely of ponderosa pine and cottonwood, hold a variety of breeding woodpeckers—the Red-shafted Flicker, Pileated, Lewis', Yellow-bellied Sapsucker, Hairy, and Downy. Their nesting cavities are often used in later seasons by Tree Swallows, both White-breasted and Pygmy Nuthatches, and Mountain Bluebirds. Elsewhere in the Refuge in suitable habitat the Wood Duck, Hooded Merganser, and Virginia Rail nest as does the Rufous Hummingbird.

WHITEHALL
Lewis and Clark Cavern State Park

In the Tobacco Root Range in southwestern Montana, about halfway between Bozeman and Butte, lies **Lewis and Clark Cavern State Park** (2,770 acres), containing one of the biggest limestone caves in the United States. The Park may be reached by exiting from I 90 at Whitehall and proceeding east on US 10 for 13.5 miles and turning north to the entrance road.

In this picturesque area are opportunities for both sight seeing and bird finding in the summer months. Persons wishing to visit Lewis and Clark Cavern should continue on the entrance road for 4 miles to the parking area near the Cavern entrance at the base of a high cliff. Bird finders should stop at the picnic area and campground off the entrance road midway between the highway and Cavern entrance. This gives access to several bird habitats: *to the south,* the Jefferson River with adjacent slopes sparsely covered with cactus, yucca, mountain juniper, and sagebrush; *to the west,* deeply eroded gullies and rocky outcrops; *to the north,* within a stone's throw, a spruce-fir forest at the mouth of a small canyon that winds invitingly back into the limestone hills of the Tobacco Root Range.

During May, June, and July, an exploration of the above habitats should yield a great many bird species including the Turkey Vulture, Northern Goshawk, Ferruginous Hawk, Golden Eagle, Rufous and Calliope Hummingbirds, Say's Phoebe, Western Pewee, Violet-green Swallow, Gray and Steller's Jays, Northern Raven, Pinyon Jay, Clark's Nutcracker, Rock Wren, Sage Thrasher, Mountain Bluebird, Townsend's Solitaire, Ruby-crowned Kinglet, Lazuli Bunting, Cassin's Finch, Pine Siskin, Green-tailed and Rufous-sided Towhees, Oregon Junco, and White-crowned Sparrow.

Nebraska

SANDHILL CRANES

Spring in the Platte River Valley means to bird finders one of the country's greatest ornithological shows—an enormous gathering of Sandhill Cranes. At no other point in their long migration through the interior of the continent do these spectacular birds stop in such numbers over such an extended period of time; and at no other point can they be watched more conveniently. Thousands upon thousands roost at night on the sand bars of the broad, shallow Platte River and feed during the day on the level agricultural lands nearby (*see under* **Kearney**). If disturbed during their daytime foraging, they frequently circle skyward, soaring on set wings like Turkey Vultures, and sometimes disappear from view. Occasionally some flocks boldly appear near farmyards, where their lanky forms and long, jerky strides are in comical contrast to the squat shapes and brisk steps of domestic fowl.

Nebraska is a roughly rectangular part of the interior plains, sloping gradually upward from an elevation of 825 feet in the Missouri Valley to 5,300 feet in the Wildcat Hills near the Wyoming line. The North Platte River, entering from Wyoming, joins the South Platte River from Colorado to become the Platte River which loops eastward across the state to enter the Missouri just south of Omaha. Various other rivers, such as the Niobrara, Elkhorn, Loup, and Blue, drain from west to east, eventually joining either the Platte or the Missouri.

Eastern Nebraska—approximately the eastern third of the state from the Missouri River westward for 125 miles—comprises bottomlands and bluffs along the Missouri and low, rolling hills to the west. On the bottomlands, besides cultivated areas, are cottonwood stands, willow thickets, and here and there small, water-filled meander cutoffs and marshes. The neighboring bluffs and their intervening hollows support woodlands, usually open and sometimes extensive, in which shagbark hickory, bur oak, red oak, walnut, green ash, and basswood are among the principal trees. Bordering the woods are low thickets of wild plum, chokecherry, and sumac. West of the bluffs are rolling hills and wide valleys through which streams wend their way in broad, sand-bedded channels. This undulating country was formerly tall-grass prairie, treeless except for cottonwoods, willows, boxelders, silver maples, sycamores, and other deciduous growth lining the watercourses and reaching back into sheltered draws; but it has since become converted into agricultural lands and the once limited tree growth has been augmented by orchards, groves, and other plantations. Birdlife of eastern Nebraska is eastern in its composition, containing a number of species at the western limits of their ranges. Listed below are birds that breed more or less regularly in eastern Nebraska and, in some cases (marked by an asterisk), nest in restricted areas farther west.

Red-shouldered Hawk
Barred Owl
Whip-poor-will
* Yellow-shafted Flicker
Red-bellied Woodpecker
Great Crested Flycatcher
* Eastern Phoebe
Eastern Pewee
Tufted Titmouse
* Wood Thrush

Blue-gray Gnatcatcher
White-eyed Vireo (*southeast only*)
* Bell's Vireo
Yellow-throated Vireo
Cerulean Warbler
Louisiana Waterthrush (*southeast only*)
Kentucky Warbler (*southeast only*)
Scarlet Tanager
Summer Tanager (*southeast only*)
* Rose-breasted Grosbeak
* Indigo Bunting

Central and western Nebraska, the remaining two-thirds of the state, lies on the Great Plains. Generally higher, more arid, and more sharply dissected than the rolling hills of eastern Nebraska with which they merge imperceptibly, the Great Plains show considerable variation with respect to physiography. Two regions, in fact, are so distinctive as to warrant separate descriptions.

North of the Platte and North Platte Rivers is a 24,000-square-mile region of the Great Plains known as the Nebraska Sandhills. Bounded roughly on the north by the Niobrara River and stretching westward from eastern Nebraska to within 50 miles of the Wyoming line, the Sandhills are countless numbers of dunes 200 or more feet in height with inclines varying from gradual to steep. For the most part they are fixed by a sod of tall and short grasses and a sprinkling of cacti and yuccas; thus they are used extensively as grazing lands for cattle. But in places overgrazing and wind action have broken the sod; the wind has then blown out the sand, leaving depressions called blowouts. From these the sand has drifted over wide areas. Between some of the hills are pockets with small trees such as hackberries, and thickets of wild plum; between others are broad meadows in which rushes and sedges thrive. Here and there, usually between the hills, are groups of ranch buildings around which a few trees have been planted for shelter. Scattered throughout the Sandhills are numerous lakes, the majority clustered in the north-central and western sections. Relatively shallow, frequently supporting cattails, bulrushes, and other marsh vegetation, and occasionally bordered by willows and shrubs, these bodies of water are highly important ornithologically, for they attract a wide variety of breeding and transient waterbirds, waterfowl, and shorebirds. The bird finder interested in exploring the Sandhills should take State 2 between Grand Island and Alliance; this runs through the heart of the region. If he wishes to investigate certain of the more productive lakes and at the same time see typical sections of the Sandhills, he should take one or more of the trips described elsewhere (*see under* **Ogallala, Oshkosh,** *and* **Valentine**).

In the extreme northwest corner of Nebraska the Great Plains drop down to the Missouri Plateau—a northern, topographically different section of the Great Plains—in a north-facing escarpment, the Pine Ridge, which is cut by canyons and eroded near the base to form a strip of badlands. Entering Nebraska from the Wyoming line some 15 miles south of South Dakota, the Pine Ridge curves southward through points south of Crawford and Chadron, then swings northeastward to enter South Dakota near White Clay. South of Crawford and Chadron, it rises almost 1,000 feet in height and has generally sharp slopes that include vertical walls; to the west and east it gradually becomes lower and less abrupt. Over the Pine Ridge, except on the steepest slopes, are stands of ponderosa pine, and near the rim is a sparse growth of juniper. In the deeper canyons are cottonwood, aspen, boxelder, and other deciduous trees, together with various shrubs such as chokecherry, skunk-brush, and buckbrush. This rough country and its fairly extensive cover of conifers brings into the state many birds of western affinities. Thus bird finding here (*see under* **Chadron** *and* **Crawford**) will yield results not to be matched anywhere else in the state.

The rest of Nebraska's Great Plains—those north of the Niobrara River, west of the Sandhills, and south of the Platte and North Platte Rivers—consist of slightly uneven tablelands and valley lowlands. On higher ground, wheat, corn, and other grains are cultivated, unless conditions are too arid, in which case the natural cover of short grasses—chiefly grama and buffalo—is retained and used for grazing. Crops such as sugar beets, potatoes, and alfalfa are produced in the valleys, particularly where irrigation has been undertaken. Westward the terrain is increasingly broken by ridges, buttes, and canyons. The Wildcat Hills in extreme western Nebraska (*see under* **Scottsbluff**) typify the rougher areas. Throughout this remaining part of the Great Plains, deciduous trees and shrubs grow mainly in valleys, sheltered draws, and canyons, and to some extent about farmyards, ranch houses, and settlements; in the western part, ponderosa pine and juniper are scattered over the slopes of ridges, buttes, and canyons.

On the Great Plains of Nebraska, including the Sandhills and Pine Ridge, the birds listed below breed regularly. Rarely, if at all, do they nest in eastern Nebraska. Birds marked with an asterisk reside mainly in the extreme western part of the state.

Ferruginous Hawk
Sharp-tailed Grouse
Long-billed Curlew (*mainly in the Sandhills*)
Poor-will
Red-shafted Flicker
Say's Phoebe (*uncommon eastward*)
Western Pewee
* Violet-green Swallow
Black-billed Magpie
* Pinyon Jay
* Rock Wren
Mountain Bluebird
Audubon's Warbler
Bullock's Oriole
* Western Tanager
Black-headed Grosbeak
Lazuli Bunting
Lark Bunting
* White-winged Junco
* Brewer's Sparrow
* McCown's Longspur
* Chestnut-collared Longspur

Throughout Nebraska the following birds breed more or less regularly in wooded areas, or in open country (natural grasslands, agricultural lands, meadow-like lowlands, brushy places, woodland borders, orchards, and dooryards):

WOODED AREAS

Cooper's Hawk
Yellow-billed Cuckoo
Black-billed Cuckoo
Common Screech Owl
Hairy Woodpecker
Downy Woodpecker
Blue Jay
Black-capped Chickadee
White-breasted Nuthatch
Red-eyed Vireo
Warbling Vireo
Ovenbird (*less common westward*)
American Redstart
Baltimore Oriole (*less common westward*)

OPEN COUNTRY

Swainson's Hawk
Northern Harrier
Common Bobwhite (*uncommon northwestward*)
Upland Sandpiper
Mourning Dove
Burrowing Owl
Red-headed Woodpecker
Eastern Kingbird
Western Kingbird
Horned Lark
Barn Swallow
House Wren
Northern Mockingbird (*mainly southern Nebraska*)
Gray Catbird
Brown Thrasher
Eastern Bluebird
Loggerhead Shrike
Yellow Warbler
Common Yellowthroat
Yellow-breasted Chat
Bobolink
Eastern Meadowlark
Western Meadowlark
Common Grackle
Orchard Oriole
Northern Cardinal
Blue Grosbeak
Dickcissel
American Goldfinch
Rufous-sided Towhee
Grasshopper Sparrow
Vesper Sparrow
Chipping Sparrow
Field Sparrow

Spring arrives in this prairie state with a burst. Rivers rise, flooding the bottomlands; temporary ponds and streams appear everywhere. Hordes of waterbirds and waterfowl come as soon as the ice goes out, and shorebirds show up later to take full advantage of muddy places left by the receding waters. Flycatchers, kinglets, vireos, and warblers throng the wooded borders of streams, lakes, and sloughs, or in canyons and city parks. Since these environments are often widely separated by vast stretches of open country, especially in the western part of the state, the birds tend to 'bunch up,' frequently in impressive numbers. Fall comes to Nebraska in a more leisurely manner, and the bird migration is seldom so spectacular, most birds passing through the state in a more dispersed and less hurried fashion. The principal migration flights in Nebraska take place within the following dates:

Waterfowl: 10 March–20 April; 10 October–25 November
Shorebirds: 1 May–1 June; 1 August–1 October
Landbirds: 15 April–20 May; 25 August–20 October

During winter, the more common visitants in sheltered areas are Slate-colored Juncos, Oregon Juncos (better represented in western Nebraska), American Tree Sparrows, and Harris' Sparrows. In open country, one may expect to see one or more Rough-legged Hawks and perhaps a flock or two of Lapland Longspurs. Wherever there is open water on the larger lakes and rivers, one may usually count on a few ducks, particularly Common Goldeneyes and Common Mergansers, and sometimes several Belted Kingfishers.

Authorities

John C. W. Bliese, R. G. Cortelyou, Doris Gates, Ruth C. Green, Ralph Harrington, Margaret Morton, Kenneth Robinson, Roger S. Sharpe, Mary M. Tremaine, Roy J. and Maud Witschy, C. Fred Zeillemaker.

Reference

A Preliminary List of the Birds of Nebraska and Adjacent Plains States. By Paul A. Johnsgard. Published by the author, School of Life Sciences, University of Nebraska, Lincoln, NB 68588. 1980.

BURWELL
Prairie Chicken Booming Grounds

To see Greater Prairie Chickens on their booming grounds, take State 11 north from Burwell through eastern sandhill country for 18.5 miles, then turn left onto the county road and go 0.5 mile. The grounds are in view on the right. Or, continue on State 11 for 2 more miles to a point just south of the Garfield-Holt County line. The grounds are in view on the left. (For other grounds in Nebraska, *see under* **Valentine.**)

Display performances begin in early March, reach peak intensity in the second and third weeks of April, begin declining soon thereafter, and are over by early June. To observe activities, which get under way at daybreak and last about two hours, the bird finder should arrive before daybreak and should remain in his car because the birds will disperse at the slightest disturbance and not return for the rest of the day.

CHADRON
Chadron State Park

The bird finder approaching northwestern Nebraska from the east on heavily traveled US 20 will notice that the undulating grasslands become increasingly interrupted by buttes, rough gullies, and steep-walled canyons, and that ponderosa pine begins to dot various slopes, especially those on the Pine Ridge to the south. Arriving in the northwestern corner, which is much like Wyoming to the west and South Dakota to the north, he is likely to hear in the nesting season his first Poor-will and catch his first glimpses of the Violet-green Swallow, Pinyon Jay, Lazuli Bunting, and other western species; at the same time he may see the last of such eastern species as the Orchard Oriole and Indigo Bunting. If the bird finder wishes to investigate this mixture of eastern and western birds, as well as see other birds regularly breeding in Nebraska's northwest corner, a good opportunity awaits him in **Chadron State Park** (850 acres), 9 miles south of Chadron on US 385.

Almost in the center of a long, narrow Pine Ridge, the Park embraces rough country with a conglomeration of buttes, canyons, and bluffs. Ponderosa pine grows in the higher areas. Along Chadron Creek, which traverses the east side of the Park and around two small ponds, are deciduous trees—cottonwoods, ashes, boxelders, and willows—and low shrubs. Birds nesting in the Park include the species mentioned above and the following: Red-tailed Hawk, Prairie Falcon, Red-headed Woodpecker, Say's Phoebe, Western Pewee, Red-breasted Nuthatch, Rock Wren, Brown Thrasher, Mountain Bluebird, Red-eyed Vireo, Black-and-white Warbler, Ovenbird, Yellow-breasted Chat, American Redstart, Black-headed Grosbeak, Pine Siskin, Rufous-sided Towhee, and Lark Sparrow. Other birds present in summer and probably nesting are the Pygmy Nuthatch and Red Crossbill. Sighting a Golden Eagle is always a possibility.

The Park is open the year round. An auto road winds through the Park from the highway, and several trails lead through canyons and up the highest bluffs.

CRAWFORD
Fort Robinson State Park

Fort Robinson State Park (20,000 acres), crossed by US 20, 2 miles west of Crawford in the Pine Ridge of northwestern Nebraska, is much larger with greater multiple use than Chadron State Park (*see under* **Chadron**). Like Chadron Park, it has buttes and canyons studded with ponderosa pine and deciduous trees lining waterways—habitats attracting practically the same species of birds. Additionally, the Park has a colony of White-throated Swifts about 6 miles west of headquarters (inquire here for directions to the site) and extensive grassland for breeding birds such as Sharp-tailed Grouse, Bobolinks, both Eastern and Western Meadowlarks, Lark Buntings, Grasshopper and Vesper Sparrows, and both McCown's and Chestnut-collared Longspurs.

WHITE-THROATED SWIFTS

HALSEY
Nebraska National Forest, Bessey Division

A unique project, the creating of a forest, was begun in 1902 in the **Nebraska National Forest, Bessey Division** (90,000 acres) which lies 3 miles west of Halsey in central Nebraska. The area selected was in typical grass-covered, sandhill terrain, devoid of trees and shrubs except in pockets between hills and on the borders of the Loup and Dismal Rivers and other streams. Planting was started with jack pine and ponderosa pine. Since 1902 the original trees have matured and additional varieties of both deciduous and coniferous trees have been planted. The forested area, now 30,000 acres, is reached from Halsey on a well-marked road.

Although for miles over the treeless sandhill terrain that surrounds the Forest such birds as the Greater Prairie Chicken, Sharp-tailed Grouse, Upland Sandpiper, Burrowing Owl, Horned Lark, Eastern and Western Meadowlarks, and Grasshopper and Lark Sparrows still breed regularly, the Forest has attracted great numbers of woodland and forest-edge birds—some occupy the river valleys in the east and others are common in the wooded canyons in the west.

Birds that breed in the created forest and its edges include the American Kestrel, Great Horned Owl, Blue Jay, Black-capped Chickadee, House Wren, Gray Catbird, Brown Thrasher, Wood Thrush, Eastern Bluebird, Loggerhead Shrike, Bell's Vireo, Black-and-white Warbler, Yellow-breasted Chat, American Redstart, Orchard Oriole, Scarlet Tanager, Northern Cardinal, Black-headed Grosbeak, Blue Grosbeak, and Rufous-sided Towhee.

KEARNEY
Platte River Valley | Lillian Annette Rowe Bird Sanctuary

From 1 March to 10 April, Sandhill Cranes may be observed almost anywhere along the Platte and North Platte Rivers from Grand Island to Lewellen. Fairly large numbers concentrate in the **Platte River Valley** between the small towns of Odessa and Elm

Creek, and also east of Kearney for about 7 miles. Look for the birds, often in big flocks in the previous year's cornfields, in alfalfa fields, around haystacks, and on pasture lands. Another good site is a recently plowed field. Now and then the bird finder may catch sight of a few individuals, or perhaps an entire flock, performing their peculiarly stiff-legged bounding 'dances.' In that case, slow down and stop the car, stay in it and watch. If the bird finder is unusually lucky, he may spot one or more of the few remaining Whooping Cranes (*see under* **Port Lavaca, Texas**) which sometimes linger in the Platte River Valley during their journey north to Canada.

To find flocks of Sandhill Cranes west of Kearney, drive on US 30 (avoid I 80), which parallels the Platte River on the north and goes through Odessa and Elm Creek. West of Odessa, start searching from dirt roads between the highway and the River. Some of the byways run north and south; others run east and west connecting them. All pass farmlands that are likely sites for cranes. (For a part of the North Platte River Valley where Sandhill Cranes may be searched for in a like manner, *see under* **North Platte.**)

To see equally good concentrations of cranes east of Kearney, drive south on State 44, cross the Platte River, and continue south about a mile. Then turn east onto State Link 50A and work the roads that cross this highway at mile intervals, both northward and southward about a mile each way. Although many cranes can be seen from Link 50A, the side roads are better for viewing because they are almost devoid of traffic.

Where Link 50A dead-ends in 7 miles on State 10, continue straight ahead on a gravel road for 2 miles, then turn north and go about 4 miles to the **Lillian Annette Rowe Bird Sanctuary** (1,821 acres). Established by the National Audubon Society primarily for Sandhill Cranes, the Sanctuary extends along the Platte River, including much of the riverbed, for 2.5 miles. Permission to enter must be obtained well in advance of a visit from the Sanctuary Department, National Audubon Society, 950 Third Avenue, NY 10022.

LINCOLN
Wilderness Park | Pioneers Park | Chet Ager Nature Center | Branched Oak Lake | Pawnee Lake | Wagon Train Lake | Stagecoach Lake

In this capital city and its outskirts are several parks that offer good bird finding in any season. Two, both on the south side of the city, are considered the best.

One is **Wilderness Park** (1,455 acres), a mixed deciduous woodland bordering Salt Creek for 7 miles, beginning in the city and extending south of its limits. Through the Park from north to south run a bridle path and a hiking trail, along which opportunities for observing birds are ideal. Waves of warblers, including the more common Tennessee, Orange-crowned, and Myrtle, pass through the Park in late April and early May and in September. Among the breeding birds, besides permanent-resident Common Screech, Great Horned, and Barred Owls, Red-bellied, Hairy, and Downy Woodpeckers, Tufted Titmice, White-breasted Nuthatches, and Northern Cardinals, are such species as the Yellow-billed and Black-billed Cuckoos, Gray Catbird, Brown Thrasher, Wood Thrush, Eastern Bluebird, Orchard and Baltimore Orioles, Rose-breasted Grosbeak, Indigo Bunting, and Rufous-sided Towhee. In winter the Park is a haven for Slate-colored Juncos and sparrows—American Tree, Harris', White-crowned, Song, and others.

The Park is accessible from at least four main entrances where there are parking areas and signs marking paths and trails. Herewith are directions to the northernmost and southernmost entrances: (1) From US 77-State 2, exit west on South Street; turn left onto Park Boulevard at Gooch Mill (in view on the right); turn right onto West Van Dorn Street across Salt Creek; then turn immediately south to the entrance on the right. (2) From State 2, 1.0 mile after US 77 separates right, turn off right onto 27th Street for 5 miles, then turn right onto West Saltillo Road where the entrance is immediately on the right.

The other is **Pioneers Park** (600 acres), west of Wilderness Park and reached by following West Van Dorn Street west from the turnoff to Wilderness Park (*see above*) for 1.5 miles, then turning south onto Coddington Avenue and proceeding 0.2 mile to the entrance on the right. The Park's extensive plantings of pine and other conifers, now mature, are choice sites for Long-eared Owls in winter and frequently attract, either in winter or during migration, Red-breasted Nuthatches, Brown Creepers, Golden-crowned Kinglets, Purple Finches, and Pine Siskins.

Incorporated in Pioneers Park is the 55-acre **Chet Ager Nature Center**, embracing a variety of bird habitats—a small grove of pines ('Pine Knoll'), ponds, old fields, low prairie, and deciduous woods. At the Center's lodge and activities buildings, open daily all year, obtain a map of trails and a checklist of birds. Especially good for warblers in migration is the deciduous woods bordering Haines Branch, a creek along the south side of the Center. Haines Creek remains ice-free in winter; one may see Mallards and Belted Kingfishers here even in the severest of weather.

A boon to bird finding in the Lincoln area is a series of eight major lakes west and south of the city, created by federal and state agencies for flood control and recreational purposes. Known as the Salt-Wahoo Creeks Watershed, these bodies of water and their peripheral wetlands vastly increased both transient and breeding bird populations. All the lakes have public access and excellent vantages for observation. The following four lakes are particularly recommended.

Branched Oak Lake (1,800 acres of surface). Drive north from Lincoln on I 180; continue from its terminus north and west on US 34 for 4 miles; turn north onto State 79 (Valparaiso Road) and drive 5 miles; then turn west onto Raymond Road and proceed 3 miles to the road circling the Lake. Being the largest of the lakes, it accommodates the largest concentrations of waterfowl. Among the vanguards in March, as soon as the water is ice-free, are Canada and Snow Geese, a few Greater White-fronted Geese, sometimes a pair or two of Whistling Swans, and many Mallards and Common Mergansers. Very soon the numbers of waterfowl greatly increase with Gadwalls, Common Pintails, Green-winged and Blue-winged Teal, American Wigeons, Northern Shovelers, Redheads, Ring-necked Ducks, Canvasbacks, Lesser Scaups, Buffleheads, and Ruddy Ducks. Some of the Canada Geese, Mallards, Blue-winged Teal, and a few late-arriving Wood Ducks remain to nest. Northbound American White Pelicans appear in April and early May.

The Lake has extensive mud flats where, in May, late August, and September, migrating shorebirds congregate. Among the species are the Semipalmated and Lesser Golden Plovers, Ruddy Turnstone, Common Snipe, Willet, Greater and Lesser Yellowlegs, Pectoral, White-rumped, and Baird's Sandpipers, Dunlin, Short-billed and Long-billed Dowitchers, Stilt Sandpiper, Marbled and Hudsonian Godwits, Sanderling, American Avocet, and Wilson's and Northern Phalaropes. The Lake also has beds of cattails in which Marsh Wrens and Yellow-headed Blackbirds nest and weedy or grassy borders which are nesting habitat for Sedge Wrens and Savannah Sparrows as well as good cover in late summer and early fall for such sparrows as the Henslow's, Le Conte's, and Sharp-tailed.

Pawnee Lake (740 acres). Drive west on US 6 (O Street) to Emerald which is 6 miles west from 1st Street; here turn off north, passing over I 80, and go 2 miles, then west 1.5 miles to the road giving access to the Lake's east side. Though less than half the size of Branched Oak, Pawnee is nonetheless equally good for its variety of waterfowl and shorebirds. Unlike Branched Oak, it is surrounded primarily by prairie grassland with Eastern and Western Meadowlarks, Dickcissels, and Grasshopper Sparrows among the summer residents.

Wagon Train Lake (315 acres). Drive south and east on State 2 to South 56th Street; turn onto it and go south for 4 miles; then turn east onto Saltillo Road and continue 0.75 mile to Hickman Road (South 68th Street) and follow it south for 5 miles to the north edge of the community of Hickman; from here proceed 2 miles east to the Lake. Besides attracting impressive concentrations of transient waterfowl and shorebirds, Wagon Train is notable for Double-crested Cormorants that perch on dead trees at the north end and for an occasional Common Loon in migration. Much of the Lake is bordered by a rich growth of cattails and other aquatic plants where Great Blue and Green Herons stalk their prey and Virginia Rails and Soras nest.

Stagecoach Lake (120 acres). Drive 1.5 miles south of Hickman (*see above*) and turn west onto Panama Road from which in the next mile there are entrances to the Lake north of the road. Despite its small size, Stagecoach is remarkably good for transient diving ducks including Red-breasted Mergansers. Willows, cottonwoods, and other woody growth around the Lake are favored by Wood Ducks for nesting and often abound with migrating kinglets, vireos, and warblers in spring and fall.

NORTH PLATTE

North Platte River Valley | Whitehorse Creek | Twin Lakes | Whitehorse Marsh | Jackson Lake | Ambler Lake | Lake Maloney | Interstate Lake

This city, inside the angle formed by the confluence of the North and South Platte Rivers, has in its general vicinity some of the state's most productive areas for bird finding in spring. From early March to the last of May the Rivers and their tributary streams, the valley lowlands and adjacent sandhills, and the many small ponds, marshes, and wet meadows attract hordes of migrating waterbirds, waterfowl, shorebirds, and landbirds. Wooded tracts, limited for the most part to the borders of watercourses, are oases for masses of tree-loving birds.

One of the best areas for flocks of migrating Sandhill Cranes is the **North Platte River Valley** between North Platte and Hershey. This may be traversed by driving west from North Platte on US 30 for about 3 miles from the fairgrounds at the west edge of the city, north on a gravel road for 1.0 mile, then west on a road which eventually meets another road going south to US 30 at Hershey. Along the way, cranes may be viewed by the thousands from 1 March to 10 April on farmlands in situations similar to those in the Platte River Valley (*see under* **Kearney**).

For spring bird finding in the environs of North Platte, the following two loop trips are suggested, both starting from North Platte.

1. Take US 30 east 1.0 mile from the city limits, crossing the North Platte River Bridge. Cliff Swallows nest under it. (In winter, Bald Eagles perch on dead trees to the left of the Bridge.) Continue east 1.0 mile from the Bridge, then turn off US 30 over railroad tracks onto a northbound road and proceed 2 miles. Where it crosses **Whitehorse Creek,** look for Upland Sandpipers

and Bobolinks in the adjacent meadows and grasslands on higher ground beyond. Great Blue Herons have a nesting colony on the ranch to the right of the road.

Turn left (west) and go 1.0 mile to the small **Twin Lakes** on the right where Wood Ducks may be sighted. Continue straight west for another 2 miles, watching for Long-billed Curlews which reside in this area.

Turn right (north) and go about 2 miles on US 83 which traverses wet meadows; turn left onto State 97 and drive 3 miles to **Whitehorse Marsh**—a widening of Whitehorse Creek. Stop here for a variety of ducks and numerous marsh dwellers including American Bitterns and Northern Harriers. Another mile beyond, ask permission at the farm on the right to look for birds from a small dam that forms **Jackson Lake,** a farm pond on the right of the road. Besides ducks, there are such birds in the nearby marsh and vicinity as Marsh Wrens, Common Yellowthroats, Yellow-headed Blackbirds, and Savannah and Swamp Sparrows.

On leaving the farm entrance, turn right and drive 0.5 mile, park the car, and walk a short distance north past a little cemetery. On the right, about 0.25 mile away in a pasture, is a 3-acre woodlot, mostly boxelders, where Great Horned Owls, Yellow-shafted Flickers, Red-headed Woodpeckers, Black-capped Chickadees, Eastern Bluebirds, and other common birds reside.

Backtrack on State 97 to a point just south of Whitehorse Marsh and here turn off right onto a side road and proceed straight south, past numerous farm ponds, marshy spots, and wet meadows for viewing birds that may have been missed so far on the trip; then work east and south on roads to US 83 which enters North Platte via the North Platte River Bridge.

2. Follow US 83 north for 28 miles, then turn left onto State 92 and go through the small town of Stapleton. While crossing the treeless, rolling sandhills country in the last 20 miles before turning onto State 92, watch for Burrowing Owls, Horned Larks, Loggerhead Shrikes, Western Meadowlarks, Lark Buntings, Vesper and Lark Sparrows, and Chestnut-collared Longspurs. Three miles west of Stapleton, State 92 passes along the north side of **Ambler Lake.** Mainly a large marsh containing cattails, sedges, bulrushes, fringed by cottonwoods and willows, it attracts many

migrating waterfowl and shorebirds and holds an impressive population of nesting birds that includes Pied-billed Grebes, Mallards, Blue-winged Teal, Virginia Rails, Soras, American Coots, Marsh Wrens, and Yellow-headed Blackbirds.

Continue from Ambler Lake on State 92 for 18 miles through more sandhill country, then turn south to North Platte on State 97 to US 83, passing on the way some of the sites pointed out in Loop Trip 1.

Lake Maloney, one in a chain of reservoirs south of North Platte, is excellent for viewing spring transient waterbirds and shorebirds. Drive south on US 83 to the University of Nebraska Station, turn west and go 1.0 mile, then follow a well-traveled road south, west, and south again. Every spring American White Pelicans, Double-crested Cormorants, and herons appear here. When the water is low, the exposed mud flats abound with transient shorebirds, including Lesser Yellowlegs, Pectoral, Baird's, and Least Sandpipers, Short-billed and Long-billed Dowitchers, Semipalmated Sandpipers, Marbled Godwits, and American Avocets.

A good site for spring transient ducks, as well as waterbirds, is reached from North Platte by driving west on US 30 for 15 miles to Hershey, turning off south over I 80, taking the first road to the left, and turning in at the **Interstate Lake** of 15 acres on the left. This is a favorite stop-over for grebes—Horned, Eared, and Western—and diving ducks—Redhead, Canvasback, Lesser Scaup, Bufflehead, and others.

OGALLALA

Lake McConaughty | Lake Ogallala | Eagles Canyon

Lake McConaughty, a huge reservoir formed by the Kingsley Dam across the North Platte River, is 6 miles north of Ogallala in open, rolling country of corn and wheat fields and rangelands. In the immediate surroundings of the reservoir the vegetation consists largely of grasses and forbs with occasional stands of cottonwoods and willows. Below the Kingsley Dam is a cattail marsh and a small lake, **Lake Ogallala,** with marshy edges and a border of willows, false indigos, and a few cottonwoods. South of both bodies of water is a number of 'breaks' or canyons in which there is a rich growth of juniper, chokecherry, buckbrush, and wild rose. North of the Lakes the land gradually rises to sandhill country.

Although the peak periods for birds, in numbers of species and individuals, are during migration from March through May and from August to the last of October, this is exciting country for bird finding in any season. Since there are a number of country roads leading to the lakes which, because of the open setting, are easy to find, the bird finder may have a very successful day by just following one road after another. A trip touching only the high spots is outlined below.

Drive north from Ogallala on State 61 to the Kingsley Dam, an excellent vantage for viewing transient Horned, Eared, and Western Grebes, American White Pelicans, Canada and Snow Geese, and ducks—Mallards, Gadwalls, Common Pintails, Green-winged and Blue-winged Teal, American Wigeons, Northern Shovelers, Redheads, Ring-necked Ducks, Canvasbacks, Lesser Scaups, and others. Below the overhanging concrete spillway of the Dam a great many Cliff Swallows begin nesting in June.

At the north end of the Dam, take the first turn to the right, which leads down to Lake Ogallala and the cattail marsh. Here nesting birds include American Bitterns, Marsh Wrens, and Red-winged Blackbirds. A small, marshy island in the Lake supports a colony of Black-crowned Night Herons. On the Lake in winter, especially if Lake McConaughty freezes over, are vast numbers of ducks, Mallards, Common Goldeneyes, and Common Mergansers predominating. The breaks south of Lakes Ogallala and McConaughty give shelter in winter to Rough-legged Hawks, Bald Eagles, American Robins, American Goldfinches, Pine Siskins, and American Tree Sparrows. In spring, the canyons are alive with migrating sparrows together with American Kestrels and warblers—Yellow, Myrtle, Audubon's, and Common Yellowthroats. Among nesting birds are the Prairie Falcon, Say's Phoebe, Bank and Cliff Swallows, Black-billed Magpie, Blue Grosbeak, and probably the Golden Eagle and Lazuli Bunting.

Return to State 61 and drive north through a typical stretch of the Nebraska Sandhills. During wet years there are many temporary ponds along this highway which attract transient geese, ducks, and shorebirds for resting and feeding. Beginning in late May, look for nesting Long-billed Curlews near the ponds. Other breeding shorebirds around the ponds are Common Snipes, Willets, Wilson's Phalaropes, and a few American Avocets. Summer-resident landbirds to observe along the highway are the Burrowing Owl, Eastern and Western Kingbirds, Western Meadowlark, Brewer's Blackbird, Dickcissel, Lark Bunting (very common), and Grasshopper and Vesper Sparrows.

After investigating this sample of sandhill country, return south on State 61 and, just before crossing the railroad tracks on the north side of the Kingsley Dam, turn right to Lewellen on State 92 which parallels the tracks and the north shore of Lake McConaughty. Small roads and trails lead south from the highway to points on the Lake that are good for observing transient waterfowl. Other birds to watch for are Great Blue Herons and Great Egrets.

Along the North Platte River between the west end of the Lake and Lewellen are numerous sites where many hundreds of Sandhill Cranes stop off during their migrations in spring and fall. They are best seen from mid-March to early April and in September and October, either on the sand bars in the River or feeding in the adjoining corn and wheat fields.

At Lewellen, turn left, cross the River, and return to Ogallala on US 26, or, after crossing the River, take the first left turn and follow the River back to the west end. From here proceed along the shore until the road turns right and returns to State 26 through **Eagles Canyon,** attractive to many birds, probably including a nesting pair or two of Prairie Falcons.

OMAHA
Fontenelle Forest Nature Center

The outstanding site for bird finding adjacent to Nebraska's largest city is the **Fontenelle Forest Nature Center** (1,300 acres), overlooking the Missouri River about 2 miles south of the city limits in the suburb of Bellevue. From I 80, exit south on US 73-75 for 3.5 miles, crossing the city line, to Southroads; here at the traffic signal, turn east (left) onto Camp Brewster Road and go 2 blocks to its dead-end; then turn right (south) onto Bellevue Boulevard and go one block to the sign indicating the entrance to the Center headquarters.

Fontenelle Forest consists largely of bur oak and shagbark hickory covering the tops of steep bluffs, and basswood, red oak, green ash, and black walnut standing in the ravines that reach down to the Missouri bottomlands where, outside the Center, are extensive marshes and streams lined with cottonwood, mulberry, boxelder, and willow. All these natural features, inside and outside the Center, are accessible by self-guided nature trails, described in a booklet available at headquarters.

Among the birds nesting in Fontenelle Forest and its environs are the following: Green Heron, Wood Duck, Red-tailed Hawk, American Kestrel, Common Screech and Barred Owls, Common Nighthawk, Red-bellied Woodpecker, Red-headed Woodpecker, Great Crested Flycatcher, Eastern Pewee, Blue Jay, White-breasted Nuthatch, Gray Catbird, Brown Thrasher, Wood Thrush, Bell's, Yellow-throated, and Warbling Vireos, Prothonotary Warbler, Ovenbird, Kentucky Warbler, American Redstart, Baltimore and Orchard Orioles, Scarlet Tanager, Rose-breasted Grosbeak, Indigo Bunting, American Goldfinch, Rufous-sided Towhee, Field Sparrow. Bird finding is especially productive during migration from mid-April to mid-May and from mid-September to mid-October when the numbers of breeding birds are augmented by many transient warblers and fringillids.

Evening Grosbeaks, Purple Finches, Slate-colored Juncos, and American Tree, Harris', and Song Sparrows are usually common in winter. At the Nature Center headquarters, these and many other birds are induced to stay the winter with generously maintained feeding stations. Visitors are welcome to sit inside the Center and watch feeding activities from a glass-enclosed observation deck.

For productive bird finding elsewhere near Omaha, *see under* **Council Bluffs** *and* **Missouri Valley, Iowa.**

OSHKOSH

Crescent Lake National Wildlife Refuge

Looking north from this town on the North Platte River in west-central Nebraska, the bird finder will see the beginning of the rolling, nearly treeless Sandhills in which lies the **Crescent Lake National Wildlife Refuge** (45,818 acres). To reach the Refuge: From midtown Oshkosh at the intersection of US 26 (running east–west) and State 27 (coming from the south), drive north out of town on West 2nd Street and continue north on an unnumbered road for 28 miles. The road bisects the Refuge and passes headquarters at Gimlet Lake near the northern boundary. The road is paved for only a few miles; the remainder is sandhill trail. Hence, check on road conditions before leaving Oshkosh. Sandhill trails are best in spring and fall. In summer, the sand becomes dry and loose, and in winter, snow-covered or swampy—especially during a thaw—making travel difficult in a conventional car. (Access to the Refuge from the north is also possible by leaving State 2 just east of Lakeside on an unnumbered road—clearly marked by Refuge signs—and driving south for 28 miles.)

Though there are some blowouts and drifting sand in the Refuge, the hills are in the main excellent range country, covered with bunch grass and occasional yuccas. Scattered between the hills are many lakes and numerous potholes with growths of bulrushes and cattails, mostly along the northwest shores; all these water areas are surrounded by meadows. Several of the named lakes, including Crescent just off the Refuge, Island, Gimlet, and Goose, are passed by a paved section of the Oshkosh-Lakeside Road. Others, including Rush Lake (on private land) and Smith Lake, are bordered by the West Mail Route, which passes west from the pavement at the south end of Goose Lake. Isolated groves of trees border some of the lakes. The largest grove, near headquarters, is accessible upon request.

The Sharp-tailed Grouse is fairly common. Beginning in early June, American Avocets, Upland Sandpipers, and Long-billed Curlews nest. Other birds breeding regularly in suitable situations are the Eared, Western, and Pied-billed Grebes, Double-crested Cormorant, Great Blue and Black-crowned Night Herons, American Bittern, Canada Goose, Mallard, Gadwall, Common Pintail, Green-winged and Blue-winged Teal, American Wigeon, Northern Shoveler, Redhead, Canvasback, Lesser Scaup, Ruddy Duck, Northern Harrier, Ring-necked Pheasant, Virginia Rail, Sora, American Coot, Killdeer, Common Snipe, Willet, Wilson's

Phalarope, Forster's and Black Terns, Mourning Dove, Great Horned, Burrowing, and Short-eared Owls, Common Nighthawk, Downy Woodpecker, Eastern and Western Kingbirds, Horned Lark, Barn Swallow, Marsh Wren, American Robin, Loggerhead Shrike, Warbling Vireo, Yellow Warbler, Common Yellowthroat, Bobolink, Eastern and Western Meadowlarks, Yellow-headed Blackbird, Orchard and Baltimore Orioles, Common Grackle, American Goldfinch, Lark Bunting, and Savannah, Grasshopper, Vesper, and Lark Sparrows.

Large numbers of transient ducks, as well as American White Pelicans, Snow Geese, and Sandhill Cranes, gather on the Refuge in March, April, late September, October, and early November. Canada Geese are plentiful in late October and November.

SCOTTSBLUFF
Wildcat Hills Big Game Refuge

This city and neighboring communities in extreme western Nebraska abound with House Finches the year round. Look for them about homes and at feeding stations.

To see transient and wintering ducks in impressive numbers from October to May, drive west from the city on State 29 (20th Street) for 2.5 miles, crossing the West 20th Street Bridge over the North Platte River; then, after 200 feet, turn off north (right) onto an unmarked county road that parallels for the next 0.5 mile a warm-water channel where there are Mallards, Green-winged Teal, Common Goldeneyes, and Common Mergansers in winter and many other ducks stopping off in spring and fall. In any season, the channel and its borders of cottonwood, Russian olive, and buffaloberry are worth exploring for a wide variety of birds.

The bird finder who happens to be in Scottsbluff in early summer should drive 10 miles south on State 71 to the Wildcat Hills, a narrow strip of broken Great Plains extending eastward from the Wyoming border. Here are many steep-sided hills and buttes, together with canyons and tributary ravines. Although the vegetation is relatively sparse, owing to the general dryness of the region, there is sufficient moisture in the canyons to support ponderosa pine, with a sprinkling of juniper, and clumps of skunkbrush, snowberry, and mountain mahogany.

The **Wildcat Hills Big Game Refuge,** with entrance on the right of State 71, 12 miles south of Scottsbluff, embraces 1,003 acres of this region. About 370 acres of the Refuge is fenced to retain bison; the rest of the acreage is maintained as a recreational area. By means of trails that wind through the recreational area, the bird finder may investigate canyons and climb to high lookouts. The lack of streams and lakes limits birdlife to land species such as the Wild Turkey, American Kestrel, Poor-will, White-throated Swift, Say's Phoebe, Black-billed Magpie, Pinyon Jay, Rock Wren, Mountain Bluebird, and Lark Sparrow. Sighting a Ferruginous Hawk is always a possibility.

VALENTINE

Valentine National Wildlife Refuge | Fort Niobrara National Wildlife Refuge

South of this ranch town in north-central Nebraska, amid the Sandhills, are 36 natural lakes and numerous potholes embraced by the **Valentine National Wildlife Refuge** (71,516 acres). Although most of the lakes—characteristically long and narrow—lack outlets except at high water, a few are strongly alkaline. Nearly all are permanent, unless seasons are very dry, and support bulrushes, cattails, and other marsh vegetation. Meadowlands, with sedges and various shrubs, surround many of the lakes and merge with the extensive, drier, grass-covered uplands.

In late spring and early summer, the marshy lakes or their meadowy surroundings are nesting habitat for a wide variety of birds, notably the Eared, Western, and Pied-billed Grebes, American Bittern, Mallard, Gadwall, Common Pintail, Blue-winged Teal, Northern Shoveler, Ruddy Duck, Northern Harrier, Virginia Rail, American Coot, Black Tern, Marsh Wren, and Yellow-headed Blackbird. American Avocets nest around alkaline lakes.

On the uplands nest a few Long-billed Curlews and Upland Sandpipers along with the more common Horned Larks, Bobolinks, Dickcissels, Lark Buntings, and Grasshopper, Vesper, and

Lark Sparrows. Ring-necked Pheasants are common. Here and there are spots where the Greater Prairie Chicken and the more numerous Sharp-tailed Grouse give their courtship performances in late March, April, and early May.

In spring and fall, the lakes host vast numbers of migrating ducks including Green-winged Teal, American Wigeons, Redheads, Canvasbacks, Lesser Scaups, and Buffleheads. Their numbers peak in late March and early April, late October and early November.

The Valentine Refuge is reached from Valentine by driving south on US 83 for 17 miles, then turning right onto State Spur 16B and proceeding 13 miles to headquarters on Hackberry Lake. Beyond its junction with State Spur 16B, US 83 continues south, soon passing through part of the Refuge on its way to Thedford.

Five miles east of Valentine on State 7 is headquarters of the **Fort Niobrara National Wildlife Refuge** (19,123 acres) on the northern edge of the Sandhills. About two-thirds of the area consists of grasslands given over to ranges for bison, elk, and longhorn cattle. A prairie dog 'town' is just east of headquarters.

The same species of upland birds found in the Valentine Refuge may be observed here. In addition, there are woodland and woodland-edge birds associated with the stands of mixed deciduous trees in the 'breaks' along the Niobrara River which traverses the Refuge. Some of the birds here are the Red-tailed Hawk, Great Horned Owl, Common Nighthawk, hybrid Yellow- and Red-shafted Flickers, both Eastern and Western Kingbirds, Western Pewee, Loggerhead Shrike, Yellow-breasted Chat, Orchard Oriole, and Black-headed Grosbeak.

Nevada

BY JEAN M. LINSDALE AND DONALD H. BAEPLER

SAGE THRASHER

Ordinarily, a traveler across Nevada assumes that the desert he sees from his car covers the state and that there are no birds in the region. Even bird finders are apt to make the same assumption. Although it is true that the state contains long stretches of desert and has large areas of uncultivated land with only a small percentage of the state under irrigation, it nonetheless has a rich avifauna consisting of more than 400 species of breeding, wintering, or transient birds.

With the exception of its extreme southern part, which lies in the Mojave Desert, Nevada occupies the plateau of the Great Basin and has an area of approximately 110,540 square miles. Altitudes range from between 5,000 and 6,000 feet in the east, from less than 4,000 to 5,000 feet in the west, and sloping to the south from between 2,000 and 3,000 feet. At the extreme southern tip of the state the altitude is below 500 feet.

Nevada's topography is diverse and ruggedly beautiful, contrasting arid and semiarid lands with snow-capped mountain ranges from 50 to 100 miles long, running from north to south. On the western border, the slopes rise rapidly to the eastern crests of the Sierra Nevada.

Rivers and streams in the state typically empty into lakes without outlets or lose their water by absorption and evaporation, as they spread over the floors of the valleys. The principal rivers of the state are the Muddy and Virgin in the south, the Owyhee in the northeast, the Humboldt in the north, and the Truckee, Carson, and Walker in the west. Evaporation of ancient lakes, which covered vast areas in the interior of Nevada, has formed a series of residual and increasingly alkaline lakes, notably the Walker, Winnemucca, Pyramid, and the Carson and Humboldt sinks. Many playas (dry lakes) exist in the state which characteristically hold water only in periods of extra rain or snow. If rainy periods occur in late spring, late summer, or early fall, the playas may be thronged with transient shorebirds.

As expected of arid and semiarid lands, Nevada enjoys many days of bright sunshine, low humidity with great purity of air outside the major metropolitan areas, and wide diurnal ranges in temperature. In the northern part of the state, particularly in the mountains, temperatures range significantly below zero in the winter months; in the southern part of the state, temperatures easily reach 115 to 120 degrees F. in summer. The average annual precipitation is about 9 inches. The whole state lies within the rain shadow cast by the high Sierra Nevada on the western border, but its influence becomes less pronounced toward the east. In an average year there are 193 clear days, 87 partly cloudy days, and 85 cloudy days.

A high proportion of the birds of Nevada are migratory, obviously due to the severity of the winter climate. Although an estimated 40 per cent of the species in the state are year-round residents, the figure is usually smaller when any restricted district on the plateau is considered. Most of the birds in the mountain regions are summer residents. The few permanent residents, even though they may not leave the region, tend to move to somewhat milder areas to escape storms.

A traveler across the middle of Nevada goes over or around more than twenty separate mountain ranges. Among the high peaks that dominate their respective ranges are Mt. Rose, 10,778 feet, north of Lake Tahoe in the Carson Range (*see under* **Reno**), the only mountain in its vicinity that extends above timber line; Boundary Peak, 13,140 feet, the highest point in Nevada, on the western boundary at the north end of the Wright Mountains; Arc Dome, 11,788 feet, the highest of several peaks in the Toiyabe Mountains (*see under* **Austin**) in the center of the state; Wheeler Peak, 13,061 feet, the highest point in the Snake Mountains (*see under* **Ely**), just inside the eastern border; and Charleston Peak, 11,918 feet, in the Charleston Mountains just west of Las Vegas (*see under* **Las Vegas**), the southernmost of the high-mountain peaks.

The mountains provide islands of habitat on a desert plateau. In the higher parts of the mountains, except in the south, there are prominent stands of aspen, and still higher, stands of limber pine and Engelmann spruce. Other conifers in the mountains are in scattered small stands that harbor some birds, but the kinds of trees do not have a marked effect on the ranges of birds across the state. Toward the west the association of birds in the forests is almost like the one across the border in California. Toward the eastern border the conifers are somewhat isolated from the ones in the mountains in Utah, but the species of birds are much the same on both sides of the border.

The following birds breed regularly in association with the pines and other conifers on the higher mountains:

Blue Grouse
Yellow-bellied Sapsucker
Williamson's Sapsucker
White-headed Woodpecker
Black-backed Three-toed Woodpecker
Northern Three-toed Woodpecker
Olive-sided Flycatcher
Steller's Jay
Clark's Nutcracker
Mountain Chickadee

White-breasted Nuthatch
Red-breasted Nuthatch
Pygmy Nuthatch
Brown Creeper
American Robin
Mountain Bluebird
Townsend's Solitaire
Golden-crowned Kinglet
Ruby-crowned Kinglet
Solitary Vireo
Audubon's Warbler
Evening Grosbeak
Cassin's Finch
Pine Siskin
Oregon Junco

Meadows in the mountains usually are small and far apart. When they are flooded in spring and early summer, the abundant moisture, vegetation, and insect food make them concentration points for many kinds of birds. Among the nesting species are the Broad-tailed Hummingbird, Calliope Hummingbird, Wilson's Warbler, White-crowned Sparrow, Fox Sparrow, and Lincoln's Sparrow.

In the coarse soils of the foothills of the many mountain ranges, junipers and pinyon pines form open park-like woodlands. Here some of the characteristic nesting birds are the Scrub Jay, Pinyon Jay, Plain Titmouse, Bushtit, Blue-gray Gnatcatcher, Gray Vireo, Virginia's Warbler, and Black-throated Gray Warbler. Not all these birds are limited to this type of vegetation, and some stands of juniper-pinyon pines, especially in the southern part of the state, attract many additional kinds of birds.

Cliffs and rocky slopes are a common feature of the topography in every section of the state, in a wide range of temperature and moisture conditions. Among the birds that nest in these sites are the Red-tailed Hawk, Golden Eagle, Prairie Falcon, American Kestrel, White-throated Swift, Say's Phoebe, Violet-green Swallow, Cliff Swallow, Canyon Wren, Rock Wren, and House Finch. In vertical sand banks the Bank and Rough-winged Swallows and the Belted Kingfisher dig nesting tunnels.

Nearly all the northern valleys and the desert plateau area originally supported a sparse cover of sagebrush with a rich stand of palatable perennial grasses and weeds. But as sagebrush increased in density, the other plants almost completely disappeared—and much of the forage value was lost. Sagebrush habitat now extends across the north end of the state with broad tongues of it along the west border and into the eastern half of the state. Good examples are to be seen close to the communities of Elko, Reno, Carson City, and Minden (near Carson City).

Birds generally present over the northern sagebrush desert include the following species:

Turkey Vulture
Sage Grouse
Mourning Dove
Poor-will
Western Kingbird
Horned Lark
Northern Raven
Sage Thrasher
Loggerhead Shrike
Brewer's Blackbird
Brown-headed Cowbird
Lark Sparrow
Black-throated Sparrow
Sage Sparrow
Brewer's Sparrow

In the Mojave Desert in the extreme southern part of Nevada, creosote bush, saltbush, and burroweed are common plants over great areas, with deciduous shrubs, yuccas (including some stands of the Joshua tree), and cacti scattered throughout and with short-lived annuals abundant in the spring months. This type of vegetation extends northward in Nevada to approximately the 37th parallel (Alamo on the east and just south of Tonopah on the west). Characteristic nesting birds are the Gambel's Quail, Greater Roadrunner, Poor-will, Lesser Nighthawk, Horned Lark, Loggerhead Shrike, and Black-throated Sparrow.

In Nevada, which is so arid, the lakes, marshes, wet meadows, and rivers are naturally important for bird finding. The large deep lakes without islands are not generally productive ornithologically, but they nonetheless warrant visiting in any search for birds. Some of the lakes invite special interest because they are remnants of much larger bodies of water in the earlier history of the land. Important lakes are Pyramid, Tahoe, and Washoe (*see under* **Reno**) and Ruby (*see under* **Elko**). Man-made Lakes Mead and Mohave (*see under* **Boulder City**) have modified the avifauna in their areas. In the valleys of the northeastern part of the state there are many marshes and wet, grassy meadows.

Birds nesting in Nevada that are associated with its lakes, marshes, and wet meadows include the following:

Eared Grebe
Western Grebe
Pied-billed Grebe
American White Pelican (*Pyramid Lake*)
Double-crested Cormorant
Great Blue Heron
Green Heron
Great Egret
Snowy Egret
Black-crowned Night Heron
American Bittern
Least Bittern
White-faced Ibis
Canada Goose
Mallard
Gadwall
Common Pintail
Green-winged Teal
Cinnamon Teal
American Wigeon
Northern Shoveler
Redhead
Ruddy Duck
Northern Harrier
Sandhill Crane
Virginia Rail
Sora
American Coot
Killdeer
Common Snipe
Long-billed Curlew
Spotted Sandpiper
Willet
American Avocet
Black-necked Stilt
Wilson's Phalarope
California Gull
Forster's Tern
Caspian Tern
Black Tern
Short-eared Owl
Marsh Wren
Western Meadowlark
Yellow-headed Blackbird
Red-winged Blackbird
Brewer's Blackbird
Savannah Sparrow

The Truckee, Walker, and Carson Rivers flow eastward from the heights of the Sierra Nevada. When the snow melts in spring and early summer, these streams run full. The Humboldt River and the Colorado River (*see under* **Boulder City**) drain the north and south parts of the state. The rivers generally are lined with cottonwoods, willows, shrubs, and other deciduous growth adjacent to open country. Among the birds breeding in this riparian habitat are the following:

Mourning Dove
Yellow-billed Cuckoo
Long-eared Owl
Hairy Woodpecker
Downy Woodpecker
Western Kingbird
Traill's Flycatcher
Black-billed Magpie
House Wren
Warbling Vireo
Yellow Warbler
MacGillivray's Warbler
Bullock's Oriole
Lazuli Bunting
American Goldfinch
Song Sparrow

During migration in Nevada the main concentrations are along the western border, at the east base of the Sierra Nevada, and across southern California. Another line of travel parallels the eastern border of the state. Many birds, especially waterfowl, migrate in an east-west direction across the northern part of the state. Time of migration is determined partly by seasonal conditions at the points of arrival in the state and along the route through the state. Weather and moisture act to shorten or prolong the stay of transients and modify their local movements.

In addition to the resident birds hardy enough to remain through the winter, there are some species that come in the winter from the north or east. Most of them are never abundant. Besides rosy finches, some of the species appearing more or less regularly are the Rough-legged Hawk, Ferruginous Hawk, Mountain Plover, Northern Shrike, Common Redpoll, American Tree Sparrow, White-throated Sparrow, and Lapland Longspur.

AUSTIN
Toiyabe Mountains | Great Smoky Valley

Located in the center of the state, Austin (elevation 6,594 feet) is the site of the famous silver-mining camp of 1862. The town lies in a steep canyon on the west side of the **Toiyabe Mountains.**

In the Toiyabes are numerous small streams that start at high seeps or springs and rush down steep slopes. Some of them disappear among the rocks after running a short distance; others extend as far as the base of the range, rarely farther. Along the lower courses of the streams and on the mountain meadows are willows;

inside narrow canyons, dense stands of birches; on flats along the upper streams and on steep moist slopes, aspens; on dry slopes, sagebrush and other shrubs among thin stands of pinyon pine and mountain mahogany.

From Austin many bird-finding localities are accessible—along streams that come down from the Toiyabes and in valleys on either side of the range. Drive southeast from Austin on US 50 for 11 miles, then turn right on State 8A to Great Smoky Valley on the east side of the Toiyabes. Worthwhile side trips can be made by turning west off State 8A near Millett on marked roads up Birch Creek and Kingston Creek Canyons to campsites from which trails lead into the higher mountains. A large rock outcrop on Birch Creek Meadows attracts many birds, which nest in its crevices. South of Millett, well-marked trails beginning at the mouths of South Twin and North Twin Rivers go to the highest peaks in the Toiyabes. The trail up Arc Dome at 11,788 feet altitude passes through a variety of montane habitats.

Large numbers of migrating birds are present in May and September in **Great Smoky Valley.** The best time for nesting birds is June, although many birds have eggs in May. Favorable localities to see them are the vicinity of Millett on the west side of the alkali flats and 5 miles south of Millett. The wet meadows, abundant springs, willows and marsh plants, heavy thickets of buffaloberry, and desert plants bordering the alkali flats ensure profitable bird finding. Birds here in the nesting season include the Common Pintail, Cinnamon Teal, Turkey Vulture, Swainson's Hawk, Northern Harrier, Western Kingbird, Say's Phoebe, Horned Lark, Black-billed Magpie, Marsh Wren, Northern Mockingbird, Sage Thrasher, Loggerhead Shrike, Yellow Warbler, Common Yellowthroat, Yellow-breasted Chat, Western Meadowlark, Yellow-headed and Red-winged Blackbirds, Bullock's Oriole, Brewer's Blackbird, and Savannah, Sage, Brewer's, and Song Sparrows.

Along the eastern slope of the Toiyabes, in the canyons, in the meadows at middle altitudes, and on drier wooded or brush-covered slopes, is another combination of species: Sharp-shinned and Cooper's Hawks, Prairie Falcon, Blue and Sage Grouse, Poor-will, White-throated Swift, Broad-tailed Hummingbird, Hairy Woodpecker, Western Flycatcher, Violet-green and Cliff

Swallows, Scrub and Pinyon Jays, Clark's Nutcracker, Mountain Chickadee, Bushtit, White-breasted Nuthatch, North American Dipper, House Wren, Canyon and Rock Wrens, Hermit and Swainson's Thrushes, Mountain Bluebird, Warbling Vireo, Virginia's, Audubon's, Black-throated Gray, and MacGillivray's Warblers, Lazuli Bunting, Cassin's and House Finches, Green-tailed Towhee, Vesper Sparrow, Gray-headed Junco, and Chipping, Brewer's, Fox, and Song Sparrows.

BAKER
Lehman Caves National Monument

From this little community in eastern Nevada close to the Utah line, drive west on State 74 for 6 miles to headquarters and visitor center in **Lehman Caves National Monument** at an elevation of 6,825 feet on the eastern slope of Wheeler Peak in the Snake Mountains. The caverns are open all year.

The Monument's 640 acres of land surface support a pinyon pine and juniper woodland. Forests of pine, spruce, fir, and mountain mahogany cloak the slopes of Wheeler Peak above the Monument. Baker and Lehman Creeks run down from the mountain's east side. Off State 74 north of Monument headquarters, a road ascends to Stellar Lake at 10,750 feet; thence a trail leads to the summit of Wheeler Peak at 13,061 feet.

Among the birds to look for anytime between May and September, either in the Monument or on higher slopes, are the Northern Goshawk, Blue Grouse, Broad-tailed Hummingbird, Yellow-bellied Sapsucker, Black-backed Three-toed Woodpecker, Western Flycatcher, Steller's and Scrub Jays, Clark's Nutcracker, Mountain Chickadee, Plain Titmouse, Bushtit, White-breasted, Red-breasted, and Pygmy Nuthatches, Brown Creeper, North American Dipper, Hermit Thrush, Mountain Bluebird, Townsend's Solitaire, Golden-crowned Kinglet, Warbling Vireo, six warblers (Orange-crowned, Audubon's, Black-throated Gray, Townsend's, MacGillivray's, and Wilson's), Cassin's Finch, Red Crossbill, Green-tailed and Rufous-sided Towhees, Gray-headed Junco, and Fox Sparrow.

BOULDER CITY
Las Vegas Wash | Colorado River Valley

Once the construction headquarters for Hoover Dam, Boulder City in extreme southern Nevada is now headquarters for the Lake Mead National Recreation Area, which embraces nearly 2 million acres along the Colorado River from Grand Canyon National Monument in Arizona, past Boulder City south to Davis Dam near the southern tip of Nevada. Included in the Recreational Area are Lake Mead, east of Boulder City and formed by Hoover Dam, and Lake Mohave, south of Boulder City and formed by Davis Dam.

Annual temperatures in Boulder City range from 20 to 120 degrees F. Winter daytime temperatures vary as much as from 50 to 70 degrees. In spring and fall the days are pleasantly warm and the nights cool. The weather is warm from 1 June to 15 September, but not nearly so hot as it is nearer the Colorado River. Boulder City is an oasis for birds; during migrations many birds throng in the shrubs and trees of the town.

Las Vegas Wash enters Lake Mead about 10 miles north of Boulder City. Here look for the Gambel's Quail, Greater Roadrunner, Say's Phoebe, Verdin, Northern Raven, Cactus Wren, and Crissal Thrasher. To reach the Wash, drive east from Boulder City on US 93 and turn northwest onto State 41, which soon swings southeast along the shore of Lake Mead, then turns southwest, and eventually meets US 93.

Proceed southwest on US 93, then turn off south onto US 95. Within 5 miles of Searchlight there are, to the north and east, extensive Joshua tree 'forests' along with sagebrush and blackbrush. The Joshua trees are especially favored by the Scott's Oriole for nest sites. Other birds to look for here are the American Kestrel, Ladder-backed Woodpecker, and Loggerhead Shrike.

Nineteen miles south of Searchlight, turn off east onto State 77 and go 21 miles to Davis Dam. (For bird finding in the vicinity of the Dam, *see under* **Kingman, Arizona**.) Before reaching the Dam, turn south (right) onto State 76, an unimproved road that follows the **Colorado River Valley** south to Needles, California. Stop now and then to explore the willows, cottonwoods, and brushy thickets on bottomlands that were formerly flooded. Some of the nesting birds to expect are Black-chinned Hummingbirds, Bell's Vireos, Yellow Warblers, and Summer Tanagers. Farther away from the River toward the bottomland margins are arrowweeds and low shrubs; still farther away, on a narrow belt of land above high-water level, saltbush is prominent. On higher land and in the washes there is a greater variety of desert woody plants, some of them reaching the size of trees and providing cover and food for large bird populations. A search of these habitats in early May should yield such breeding birds as the White-winged Dove, Common Screech and Great Horned Owls, Poor-will, Lesser Nighthawk, Costa's Hummingbird, Gila Woodpecker, Vermilion Flycatcher, Phainopepla, Lucy's Warbler, Blue Grosbeak, Abert's Towhee, and Black-throated Sparrow. At this same time of year there may be swarms of transients on their way north.

CARSON CITY
Minden | Genoa

Nevada's capital city is situated in the foothills of the Sierra Nevada. The best spots for bird finding are in the vicinity of **Minden,** 15 miles south on US 395, and **Genoa** (the oldest town in the state), 12 miles south on US 395, then 3 miles west on State 57. In this area, watered by the Carson River, the Mormon pioneers established farms and planted orchards, gardens, and many shade trees.

About the many homes in early summer are House Wrens, Cedar Waxwings, Warbling Vireos, Yellow Warblers, Bullock's Orioles, Western Tanagers, Lazuli Buntings, House Finches, American Goldfinches, and Chipping Sparrows. On or about the ponds near Minden, many birds nest or stop in migration. Among them are Canada Geese, Mallards, Common Pintails, Cinnamon Teal, Willets, American Avocets, Wilson's and Northern Phalaropes, and Forster's and Black Terns. Where the water is deep and the bulrushes especially rank, Marsh Wrens and Yellow-headed and Red-winged Blackbirds nest.

ELKO
Ruby Lake National Wildlife Refuge | Harrison Pass

A must for the bird finder in northeastern Nevada is **Ruby Lake National Wildlife Refuge.** From Elko, drive southeast on the road to Lamoille for 7 miles, turn south onto State 46 and continue for 31 miles (passing through Jiggs at 26 miles), then turn off left and go 22 miles, through **Harrison Pass** (elevation 7,247 feet), following signs to Refuge headquarters.

The Refuge of 37,631 acres is a natural sump, fed by springs on the west side and by runoff from the snow on the surrounding mountains. Water areas, consisting of ponds, channels, and vast marshes in which bulrushes, pondweeds, and other aquatic plants thrive, cover approximately 12,000 acres. The rest of the Refuge is upland with native grasses, sagebrush, and rabbitbush. The strikingly rugged Ruby Mountains rise to 11,000 feet to the west of the Refuge whose elevation at headquarters is 6,012 feet.

Of special ornithological interest on the Refuge are nesting Sandhill Cranes and a transplanted group of Trumpeter Swans. Canada Geese nest as do a variety of ducks including the Mallard, Gadwall, Cinnamon Teal, Northern Shoveler, Redhead, Canvasback, Lesser Scaup, and Ruddy Duck. Among other birds nesting are Eared and Pied-billed Grebes, Snowy Egret, Black-crowned Night Heron, American Bittern, Northern Harrier, Sage Grouse, California Quail, American Coot, Long-billed Curlew, Spotted Sandpiper, Willet, Forster's and Black Terns, Short-eared Owl, Common Nighthawk, Violet-green and five other species of swallows, Marsh Wren, Western Meadowlark, and Yellow-headed and Red-winged Blackbirds. May and June are the best months for visiting the Refuge.

A stop at Harrison Pass on the way in to the Refuge may be worthwhile for finding the Lewis' Woodpecker and other woodland birds.

LAS VEGAS
Pahranagat Valley | Pahranagat National Wildlife Refuge | Charleston Mountains | Sunset Park

Several areas for good bird finding in southern Nevada may be reached from this city, the state's largest. For additional areas in southern Nevada, *see under* **Boulder City.**

A good bird-finding area north of Las Vegas is the **Pahranagat Valley,** a long desert plain divided by a narrow line of green vegetation. Agriculture, centered at Alamo, has developed rapidly, and warm springs provide open water the year round. To reach Alamo from Las Vegas, drive northeast on US 93 (combined with I 15 for the first 21 miles), then north on US 93 for 73 miles.

At Crystal Springs, 12 miles north of Alamo on US 93, there are ponds, green meadows, and trees that attract or hold a great variety of birds, both summer residents and transients.

About 10 miles south of Alamo on US 93, the **Pahranagat National Wildlife Refuge** (5,380 acres) embraces two impoundments, Upper and Lower Pahranagat Lakes, where there are such breeding birds as Double-crested Cormorants, Great Blue Herons, Mallards, Gadwalls, Common Pintails, Green-winged and Cinnamon Teal, Marsh Wrens, and blackbirds—Yellow-headed, Red-winged, and Brewer's. Bordering the Lakes are cottonwoods which in spring become an oasis for migrating warblers and other passerines.

On the desert of the Pahranagat Valley, which is below 4,500 feet, the most characteristic plant is creosote bush. Other common plants are mesquite, catclaw, peabush, and white bur sage. Breeding birds to look for here are the Swainson's Hawk, Gambel's Quail, Greater Roadrunner, Lesser Nighthawk, Northern Mockingbird, Bendire's and Sage Thrashers, Loggerhead Shrike, Scott's Oriole, and Black-throated Sparrow. The Vermilion Flycatcher and Blue Grosbeak range as far north in Nevada as the southern part of the Valley.

Above the creosote bush desert, most of the land supports sagebrush, with rabbitbrush on the lower areas, juniper, pinyon pine, and mountain mahogany on higher ground. Among the nesting birds are the Cooper's Hawk, Long-eared Owl, Poor-will, Gray Flycatcher, Pinyon Jay, Bushtit, and Sage Sparrow.

West of Las Vegas rise the **Charleston Mountains** (local name for the Spring Mountains), the largest and highest mountain mass in southern Nevada. This precipitous range, about 50 miles long and 30 miles wide, extends in a northeast-southwest direction. Although there are no meadows or lakes and only a few small springs and short streams, the higher slopes are well wooded. The highest point, Charleston Peak, is 11,918 feet, approximately 9,000 feet above the surrounding desert.

Summer headquarters of the Charleston Mountain Area is the Kyle Canyon ranger station. To reach Kyle Canyon, which is 35 miles northwest of Las Vegas, drive north on US 95 for 13 miles, then west on State 39. Lee Canyon, north of Kyle Canyon, may be reached from the latter, or by driving north from Las Vegas on US 95 for 28 miles and turning left onto State 52. Inquire at the Kyle ranger station about trails leading to higher elevations and to Charleston Peak.

The Charleston Mountains and the Sheep Mountains to the northeast are boreal islands isolated from other boreal islands by at least 100 miles of desert. Below 6,000 feet, the prominent plants are cresote bush, mesquite, and yucca, providing habitat for the Gambel's Quail, Common Screech Owl, Lesser Nighthawk, Ladder-backed Woodpecker, Western Kingbird, Cactus Wren, Le Conte's and Crissal Thrashers, Bullock's Oriole, and Black-throated Sparrow. On the higher slopes, where sagebrush, juniper, and pinyon pine are the predominant plants, the resident birds are the Scrub Jay, Plain Titmouse, Bushtit, Sage Thrasher, Rufous-sided Towhee, and Sage and Brewer's Sparrows. Summer residents are the Common Nighthawk, Gray Flycatcher, Blue-gray Gnatcatcher, Virginia's Warbler, Scott's Oriole, Black-headed Grosbeak, Lazuli Bunting, and Black-chinned Sparrow. From 8,000 to 9,000 feet, where there are ponderosa pines and white firs, and still higher where bristlecone and limber pines are dominant, nesting birds include the Northern Goshawk, Saw-whet Owl, Broad-tailed Hummingbird, Williamson's Sapsucker, Pygmy Nuthatch, Brown Creeper, Townsend's Solitaire, Ruby-crowned Kinglet, Cassin's Finch, Red Crossbill, and juncos.

In greater Las Vegas **Sunset Park** has a good variety of birds that includes the Verdin, Cactus Wren, Crissal Thrasher, Phainopepla, and Abert's Towhee. To reach the Park from 'the strip,' go east to Eastern Avenue; here turn south onto Sunset Road.

RENO
Pyramid Lake | Anaho Island National Wildlife Refuge | Carson Range | Washoe and Little Washoe Lakes

Lying along the Truckee River only 14 miles from the western border of Nevada, Reno is easily accessible to a variety of bird-finding areas ranging from barren, rocky desert and large lakes and rivers to high mountain slopes.

Pyramid Lake, 30 miles long and 5 to 12 miles wide with high mountains rising abruptly from its east and west sides, is 33 miles northeast of Reno. Drive north from Reno on State 33, which leads to the southwest shore; then turn right and follow the shore toward Nixon.

In view from the shore is the **Anaho Island National Wildlife Refuge** comprising a low, dark rocky peak of some 750 acres. The Island is notable for being the nesting site for practially all the American White Pelicans seen in summer anywhere in Nevada as the great birds wander afar on fishing expeditions. Using a spotting scope, one may clearly see the pelicans on the Island as well as other nesting birds: Double-crested Cormorants, Great Blue Herons, California Gulls, and possibly a few Caspian Terns. Scanning the Lake itself, one may also see Western Grebes, Canada Geese, Mallards, Gadwalls, Common Mergansers, and other waterbirds and waterfowl.

To see coniferous-forest birds during spring and summer on the east slope of the **Carson Range** in the Sierra Nevada, proceed to the Galena Creek Public Campground by taking US 395 south from Reno for 9 miles and turning west onto State 27. This road starts through sagebrush country from which there is a good view of Mt. Rose on the right. At the Campground there is riparian woodland along a stream, with pine forests in every direction and some areas of snowbush and manzanita. Breeding birds here are the Calliope Hummingbird, Hairy Woodpecker, Yellow-bellied Sapsucker, Steller's Jay, Mountain Chickadee, White-breasted

Nuthatch, Brown Creeper, House Wren, Townsend's Solitaire, Nashville and Yellow Warblers, and Chipping Sparrow.

Along the road above the Campground, at the 7,500-foot mark, are meadows bordered by an especially good stand of aspens and many willow clumps. Look here for Western Pewees, Warbling Vireos, Wilson's Warblers, and White-crowned, Fox, and Lincoln's Sparrows. At about 8,000 feet, beside a small stream north of Slide Mountain, are red firs, lodgepole pines, and western white pines, along with a good growth of shrubs. Search here for Hammond's Flycatchers, Audubon's Warblers, Western Tanagers, and Cassin's Finches.

By the second week of August there will be extensive post-nesting movements, indicating the start of fall migration. Some of the species then present along Galena Creek include the Poor-will, Common Nighthawk, Solitary Vireo, Orange-crowned, Black-throated Gray, Townsend's, and Hermit Warblers, Lazuli Bunting, Evening Grosbeak, Pine Siskin, Green-tailed Towhee, and Oregon Junco.

State 27 continues upward to Tahoe Meadows at 8,500 feet on the Lake Tahoe side of the pass; from here a trail leads 5 miles up Mt. Rose (10,778 feet), the only peak northeast of Lake Tahoe that reaches above timber line. Along the Mt. Rose trail, at and above timber line, in mid-July, there are American Kestrels, Prairie Falcons, Clark's Nutcrackers, Rock Wrens, and Mountain Bluebirds.

State 27 continues south down to Incline Village (6,250 feet) at the north end of Lake Tahoe. On the slopes north of the lakeshore there is a fine stand of pure coniferous forest, chiefly of white fir, Jeffrey pine, and sugar pine. Species common during summer in the vicinity of Incline Village include the Mountain Quail, Williamson's Sapsucker, White-headed and Black-backed Three-toed Woodpeckers, Hermit and Swainson's Thrushes, Ruby-crowned Kinglet, and other species aforementioned before reaching Incline Village.

South of Reno are **Washoe and Little Washoe Lakes,** connected by marshy land. To reach these bodies of water, drive south from Reno on US 395 for 18 miles, then turn left onto a road that runs along on the eastern side of the Lakes. Since the water comes from streams on the eastern slopes of the Carson Range, the amount varies greatly from year to year. Mallards, Gadwalls, Cinnamon Teal, Redheads, Ruddy Ducks, and American Coots nest commonly. When the water is low, shorebirds congregate in great numbers, some of them being American Avocets, Black-necked Stilts, and Wilson's and Northern Phalaropes. Regularly present in summer are American White Pelicans, White-faced Ibises, and Canada Geese. Back from the Lakes is a stretch of sagebrush with birds characteristic of sagebrush habitat.

New Mexico

BY DALE A. ZIMMERMAN

BURROWING OWLS

Great topographical and vegetational diversity, combined with a vast land area, have favored development of a rich and varied avifauna in New Mexico, fourth largest of the contiguous states and occupying over 121,000 square miles. Basically this region may be visualized as a high and uneven plateau, much higher in the north, its surface dissected by two major north-south drainage systems, the Pecos River and the Rio Grande. The latter roughly parallels impressive mountain ranges which rise both to the east and to the west. Significant mountain masses exist elsewhere in the state as well, their axes for the most part running north and south. Some ranges tend to merge with one another; others are widely separated by open plains and broad valleys often picturesquely dotted with isolated hills, buttes, or old volcanic cones. Toward the east, this ruggedness gradually subsides. Of the major waterways, the Rio Grande enters north-central New Mexico and flows the full length of the state. Farther east, the Pecos River extends about two-thirds of this distance. Lesser rivers are the Canadian and Cimarron in the northeast, draining into Texas and Oklahoma; and, on the other side of the state, the San Juan, San Francisco, and Gila flowing westward into Arizona.

Elevations in New Mexico range from 2,850 feet along the lower Pecos River near the Texas line to 13,161 feet atop Wheeler Peak, the Southwest's highest mountain. Between these extremes is a broad and complex array of plant associations, reflecting the exceedingly varied climatic conditions and producing complicated patterns of bird distribution. Broadly speaking, the state's major bird habitats may be recognized as follows: the *deserts* typically are dominated by cresote bush, tarbush, or mesquite in the south, and by big sage in the north. These xeric communities merge in many places with *grasslands* which typically are (or were) dominated by blue grama. Much country impressing the modern viewer as desert was grassland a century ago. Before domestic livestock appeared on the scene, true desert was comparatively scarce in New Mexico. *Chaparral*, of scrub oaks, manzanita, mountain mahogany, and skunkbush sumac, is largely confined to lower and middle elevations of the southern mountains. *Oak woodland*, sometimes with pine and reminiscent of similar Mexican plant associations, also is southern in distribution. Widespread at intermediate elevations is *pinyon-juniper woodland* dominated either by pinyon pine (Mexican pinyon near the International boundary) or by various junipers. At this woodland's lower limits, pines become scarce and junipers spread into the sagebrush, beargrass, or grassland tracts, producing ecotones of varying extent. *Ponderosa pine forest* is widely distributed above 6,500 feet merging with Douglas-fir in cool canyons and at higher elevations. Higher still, other evergreens enter the conifer association, and above 8,500 feet are tracts of lofty *spruce-fir forest* composed of Englemann spruce, white fir, or subalpine fir. Subclimax stands of aspen are numerous in these montane forests. *Alpine tundra* exists above timber line on the highest peaks, mostly in the Sangre de Cristo Range. Snow remains here through summer on the cold north slopes, and frost-free nights are rare. Of particular significance to New Mexican birds is *riparian forest* or *woodland*, characteristic of the state's watercourses. Although species composition varies, it typically includes cottonwoods and willows. Sycamores are important, though local, along streams and canyon bottoms in the southwestern part of the state.

Not surprisingly, New Mexico boasts one of the largest bird lists of any landlocked state, with its elements from the Great Plains, the Rocky Mountains, both northern and southern deserts, and the Mexican fauna of the southernmost canyons and desert mountains. Some 440 species have been reliably recorded within its borders and 270 of these breed in the state.

No regional avifauna remains static. Within recent years, New Mexico has lost her Sage Grouse, Sharp-tailed Grouse, and perhaps her Buff-breasted Flycatchers; Aplomado Falcons have all but vanished from the scene. But partially offsetting such losses are recent Mexican immigrants like the Violet-crowned Hummingbird and Thick-billed Kingbird which now nest regularly in the southwestern corner of the state, and the Buff-collared Nightjar which crosses the border at intervals. The Yellow-crowned Night Heron has become at least an occasional New Mexican visitant west to the Gila River. Olivaceous Cormorants and Cattle Egrets appear with increasing frequency and some breed in the state. More and more 'eastern' passerines are turning up and some of these doubltess will prove to be expected regularly. Indeed, Blue Jays, Common Grackles, and Indigo Buntings now nest in New Mexico. Species new to the state are seen nearly every year and this trend is likely to continue for some time.

Characterizing vast expanses of southern New Mexico are the deserts and desert grasslands, which in the Rio Grande Valley extend north nearly to Albuquerque. On the gravelly soils of the lowermost mountain slopes and foothills, cresote bush dominates large tracts, with tarbush and krameria as subdominants. On heavier soils and at somewhat higher elevations, these are replaced by grasses, ephedras, and yuccas. Cacti and ocotillos are locally conspicuous, and Apache-plume, acacia, or mesquite provide further diversity in washes and along streambeds. Typical breeeding birds of these habitats are listed below. Those marked by an asterisk also are present in winter.

Swainson's Hawk
* Scaled Quail
* Greater Roadrunner
Lesser Nighthawk
* Burrowing Owl
* Ladder-backed Woodpecker
Ash-throated Flycatcher
White-necked Raven
* Verdin
* Cactus Wren
* Northern Mockingbird
* Curve-billed Thrasher
* Crissal Thrasher
* Loggerhead Shrike
* Black-tailed Gnatcatcher
Scott's Oriole
* Black-throated Sparrow
* Cassin's Sparrow (*grassland*)

Where natural riparian habitats and agricultural areas are superimposed upon the arid landscape, as in the Rio Grande and Pecos Valleys, additional breeding species include a few waterbirds, the American Avocet, Black-necked Stilt, Western Meadowlark, Bullock's Oriole, Great-tailed Grackle, House Finch, Lesser Goldfinch, and Brown Towhee. During autumn and winter, waterbirds, fringillids, and others swarm into the diversified valley habitats. Along much of its length the Rio Grande is bordered by irrigated croplands, although extensive cottonwood groves or bosques remain in the north. More numerous are thickets of low willows, tamarisk, saltbush, mesquite, or screw bean. The best site for general bird finding in the Valley is the Bosque del Apache National Wildlife Refuge (*see under* **Socorro**). The Pecos Valley is similar to that of the Rio Grande but there extensive tamarisk growth has largely replaced the indigenous riparian vegetation. Its outstanding areas for birds are Bitter Lake National Wildlife Refuge (*see under* **Roswell**) and Lake McMillan (*see under* **Carlsbad**).

Still more rewarding to the bird finder is the Gila River Valley (*see under* **Silver City** *and* **Lordsburg**), where two-thirds of the species on the New Mexico list have been recorded. Here, during summer, dwell birds of Mexican affinities which penetrate but a short distance into the United States, among them the Lesser Black Hawk, Elf Owl, and Gila Woodpecker. The Gila River, born in the Mogollon Mountains, flows through spectacular canyons and emerges onto a mile-wide floodplain northwest of Silver City. Farther downstream it winds among low mountains and squeezes in and out of box canyons with precipitous sides. Finally, toward the Arizona line, it remains in the open, flowing through largely agricultural land. In places along its varied route, relict fringes of cottonwood, willow, and sycamore support their remarkable, though dwindling, avifauna. Xerophytic associations locally bring mesquite, cacti, and desert birds practically to the riverbanks. At intervals, a woodland of evergreen oaks, with hackberry, walnut, or sycamore, exists along the Gila River and especially in tributary canyons where an understory of encroaching junipers often indicates the direction of plant succession under present grazing practices. Most marshy sites in the Valley have given way to verdant agricultural fields. Above its confluence with Turkey Creek, the Gila is more of a mountain stream. Mexican elements gradually are left behind, and both flora and fauna assume a montane character. Typical summer birds of the lower Gila Valley from Virden to the Cliff-Gila area are listed below. Species marked with an asterisk remain through the winter, although in some cases uncommonly.

* Cooper's Hawk
Swainson's Hawk
Zone-tailed Hawk
Lesser Black Hawk
* Gambel's Quail
* Inca Dove (*Virden area*)
White-winged Dove
Yellow-billed Cuckoo
* Greater Roadrunner
* Barn Owl
* Common Screech Owl
Elf Owl
Poor-will
Lesser Nighthawk
* Red-shafted Flicker
* Gila Woodpecker
* Acorn Woodpecker
* Ladder-backed Woodpecker
Western Kingbird
Cassin's Kingbird
Wied's Crested Flycatcher
Ash-throated Flycatcher
* Black Phoebe
* Say's Phoebe
Western Pewee
* Vermilion Flycatcher
* Gray-breasted Jay
* Bridled Titmouse
* Verdin
* Bewick's Wren
* Cactus Wren
* Northern Mockingbird
* Curve-billed Thrasher
* Crissal Thrasher
* Phainopepla
Bell's Vireo
Lucy's Warbler

Hooded Oriole
Scott's Oriole
Bullock's Oriole
Bronzed Cowbird
Summer Tanager
* Northern Cardinal
Blue Grosbeak
* Abert's Towhee
* Black-throated Sparrow

Joining the permanent residents in winter are small numbers of waterfowl and many landbirds including:

Ferruginous Hawk
Sandhill Crane
Pinyon Jay (*irregular*)
Sage Thrasher
Hermit Thrush
Western Bluebird
Mountain Bluebird (*irregular*)
Townsend's Solitaire
Ruby-crowned Kinglet
Audubon's Warbler
Cassin's Finch (*irregular*)
Pine Siskin
Green-tailed Towhee (*irregular*)
Savannah Sparrow
Vesper Sparrow
Sage Sparrow
Oregon Junco
Gray-headed Junco
Brewer's Sparrow
White-crowned Sparrow
Lincoln's Sparrow

The San Juan River Valley, some 250 miles north of the Gila, also hosts an impressive array of birds. Nearly 150 species are known here in the summer. The San Juan River arises in Colorado, flows southwest into New Mexico, and extends westward almost to Arizona before looping northward again into the state of its origin. Flanking the scenic Valley is a panorama of sandstone cliffs, mesas, buttes, and rolling hills supporting pinyon-juniper woodland. Sagebrush, saltbush flats, and grasslands likewise are prominent. The riparian cottonwood forest here has an understory of big sage, New Mexico olive, and other shrubs. These woods and willow-dominated shrub associations, co-exist with numerous marshes and extensive agricultural lands. The Mexican element characteristic of the Gila Valley is largely missing from that of the San Juan. Nevertheless, several birds of southern affinities reach their northern limits here, among them the Gambel's Quail, Scaled Quail, Gray Vireo, Scott's Oriole, and Brown Towhee. The Lucy's Warbler recently has been reported. Some typically eastern species such as the Red-headed Woodpecker, Eastern Kingbird, and Common Grackle are at or near their western limits in this region. Although not known to nest here, the Least Bittern, Black Swift, and Bendire's Thrasher have been recorded in summer, and deserving of mention are the feral Rock Doves which breed in the sandstone cliffs. The San Juan, like the Gila, is an important pathway for birds during migration. Characteristic breeding species along the River and in adjacent uplands include:

Swainson's Hawk
Chukar
Yellow-billed Cuckoo
Common Screech Owl
Burrowing Owl
Poor-will
Western Kingbird
Cassin's Kingbird
Ash-throated Flycatcher
Say's Phoebe
Gray Flycatcher (*hillsides*)
Western Pewee
Scrub Jay
Black-billed Magpie
Pinyon Jay
Black-capped Chickadee
Plain Titmouse
Bushtit
Bewick's Wren
Marsh Wren
Northern Mockingbird
Sage Thrasher
Blue-gray Gnatcatcher
Virginia's Warbler
Black-throated Gray Warbler
Yellow-breasted Chat
Great-tailed Grackle
Black-headed Grosbeak
Blue Grosbeak
Lazuli Bunting
House Finch
Lesser Goldfinch
Black-throated Sparrow
Sage Sparrow

Of New Mexico's numerous mountain ranges, perhaps those most intriguing to bird finders are the northern outliers of Mexico's Sierra Madre—the Peloncillo, Animas, and San Luis Ranges. Rising in extreme southwest New Mexico (*see under* **Lordsburg**), these support localized populations of various Mexican species otherwise absent or rare in the state. The seldom-visited San Luis Range, rocky, steep, and heavily eroded, is primarily Mexican but extends across the International boundary east of the Animas Valley. Barely separated from it are the Animas Mountains, now closed to the public. However, the accessible southern Peloncillos support similar species. Despite their proximity to Mexico, these mountains are largely above 4,000 feet; spring arrives tardily in their cool canyons. Snow may fall in late April or in May, and certain of the 'Mexican' birds do not arrive until much later. The Violet-crowned Hummingbird, for example, in some years postpones its arrival until early July—just in time to meet the first southbound male Rufous Hummingbirds en route to Mexico after breeding in the Pacific Northwest. During August, ten species of hummers inhabit the southern New Mexico mountains. Characteristic breeding birds of these borderland ranges are listed below. Some are very localized. An L indicates those typical of the lower canyons, an H those largely restricted to the higher altitude forests, GC denotes birds largely or entirely confined in this region to Guadalupe Canyon. Asterisks mark species which remain through the winter.

Zone-tailed Hawk
* Montezuma Quail
* Gambel's Quail
* Wild Turkey
Band-tailed Pigeon (H)
White-winged Dove (L)
Common Ground Dove (GC)
* Greater Roadrunner
* Common Screech Owl
Flammulated Screech Owl (H)
* Northern Pygmy Owl (H)
Elf Owl (L)
* Spotted Owl (H)
Poor-will
* Acorn Woodpecker
* Ladder-backed Woodpecker (L)
* Brown-backed Woodpecker
Western Kingbird (L)
Cassin's Kingbird
Thick-billed Kingbird (GC)
Wied's Crested Flycatcher (L)
Ash-throated Flycatcher
Olivaceous Flycatcher (L)
Western Flycatcher (H)
Coues' Flycatcher (H)
Vermilion Flycatcher (L)
Northern Beardless Flycatcher (GC)
* Steller's Jay (H)
* Scrub Jay
* Gray-breasted Jay
* Mexican Chickadee (H, *mostly Animas Mts.*)
* Bridled Titmouse
* Plain Titmouse
* Verdin (L)
* Bushtit
* Pygmy Nuthatch (H)
House Wren (H)
* Bewick's Wren
* Cactus Wren (L)
Bendire's Thrasher
* Curve-billed Thrasher
* Crissal Thrasher
* Hermit Thrush (H)

- * Phainopepla
- * Hutton's Vireo
- Bell's Vireo (L)
- Gray Vireo
- Warbling Vireo (H)
- Virginia's Warbler (H)
- Lucy's Warbler (L)
- Olive Warbler (H)
- Black-throated Gray Warbler
- Grace's Warbler (H)
- Red-faced Warbler (H)
- Painted Redstart
- Hooded Oriole (L)
- Scott's Oriole (L)
- Bronzed Cowbird (L)
- Western Tanager (H)
- Hepatic Tanager (H)
- Summer Tanager (L)
- Pyrrhuloxia (L)
- Black-headed Grosbeak
- Varied Bunting (GC)
- * House Finch
- * Brown Towhee
- * Rufous-crowned Sparrow
- * Black-throated Sparrow (L)
- * Mexican Junco (H)
- * Black-chinned Sparrow

The Mexican influence on New Mexico's avifauna extends into the Mogollon Mountains (*see under* **Glenwood**), here meeting boreal elements not found to the south. The Mogollons are extensive and rugged with spectacularly deep canyons and at least seventeen peaks rising above 10,000 feet; most of the range lies within the great Gila Wilderness Area. Certain of its lowermost canyons, which drain into the San Francisco River, have streamside sycamores and cottonwoods and thus lure some birds characteristic of the Gila Valley. The Mountains' lower slopes are well clothed with juniper, oak, and pine-oak woodland, and, in drier places, chaparral. The vast open forests of ponderosa pine mingle at higher levels with white fir or Douglas-fir. Above 8,500 feet, Douglas-fir becomes common while Engelmann spruce and subalpine fir dominate the scene from 10,000 feet upward. Throughout the high country, aspens are conspicuous, and the riparian growth is dominated by narrowleaf cottonwood, Arizona alder, and blue spruce. In the following list of selected summer residents, L and H refer to birds more typical of low or high elevations respectively. Additionally, all highland species, except the Mexican Chickadee, listed for the Animas-Peloncillo Mountains also inhabit the Mogollons. Asterisks denote those present in winter, although not necessarily in the same elevational range occupied during summer.

Sharp-shinned Hawk (H)
* Cooper's Hawk (L)
* Northern Goshawk (H)
Lesser Black Hawk (L)
* Blue Grouse (H)
* Montezuma Quail (L)
Elf Owl (L)
Whip-poor-will (H)
Poor-will (L)
Acorn Woodpecker
Yellow-bellied Sapsucker (H)
* Williamson's Sapsucker (H)
* Northern Three-toed Woodpecker (H)
Cassin's Kingbird
Black Phoebe (L)
Ash-throated Flycatcher
Western Pewee
Olive-sided Flycatcher
* Scrub Jay (L)
* Gray-breasted Jay (L)
* Clark's Nutcracker (H)
* Mountain Chickadee (H)
* Red-breasted Nuthatch (H)
* Brown Creeper (H)
* Mountain Bluebird
* Western Bluebird
* Townsend's Solitaire (H)
* Golden-crowned Kinglet (H)
* Ruby-crowned Kinglet (H)
Orange-crowned Warbler (H)
Black-throated Gray Warbler
* Audubon's Warbler (H)
Grace's Warbler (H)
MacGillivray's Warbler (H)
Red-faced Warbler (H)
Painted Redstart
Summer Tanager (L)
Black-headed Grosbeak
* Red Crossbill (H)
* Evening Grosbeak (H)
Lesser Goldfinch (L)
* Pine Siskin (H)
* Brown Towhee (L)
Green-tailed Towhee (H)
* Rufous-crowned Sparrow
* Gray-headed Junco (H)

West of the Rio Grande in the southern part of the state are the Magdalena and San Mateo Mountains—scantily forested, rugged, desert ranges less attractive to bird finders than the more accessible Mogollons or Black Mountains (*see under* **Truth or Consequences**) to the southwest. Similarly, northwestern New Mexico's Zuni and Chuska Mountains, although not without avian attractions, are much less productive than the more diverse southern ranges.

The Rocky Mountains enter north-central New Mexico as the lofty San Juan and Sangre de Cristo Ranges, the latter east of the Rio Grande, towering above historic Sante Fe and Taos. The Sangre de Cristo Range is dominated by a north-south axis of high ridges and peaks, several (including Wheeler) exceeding 13,000 feet. Seven more extend above timber line, and throughout the Range there are many others almost as high. Diversifying the landscape further are lakes, ponds, and numerous swift, rocky, dipper-inhabited streams. Vegetation is varied and often distinctly zoned. Pinyon-juniper woodland and sagebrush clothe the foothills, and deciduous riparian woods prevail in the canyons. Between 7,000 and 8,500 feet, open stands of ponderosa pine dominate, often with an admixture of Douglas-fir and Gambel oak. Immediately above is the extensive spruce-fir forest, dappled with lighter green aspen groves as succession reclaims deforested patches. Higher still, where elevations much exceed 11,500 feet, the dwarfed, wind-gnarled, upper fringes of subalpine forest with bristlecone pine and subalpine fir mark the timber line. Beyond this is a true alpine tundra of grasses, sedges, and matted forbs—some characteristic of the Arctic and here reaching their southernmost distributional limits. Around Santa Fe, Taos, and in intervening country below the 7,000-foot level, the following breeding birds may be expected in suitable habitat:

Poor-will
Red-shafted Flicker
Western Kingbird
Cassin's Kingbird
Ash-throated Flycatcher
Say's Phoebe
Western Pewee
Scrub Jay
Pinyon Jay
Plain Titmouse
Bushtit
Blue-gray Gnatcatcher
Summer Tanager
Black-headed Grosbeak
Lazuli Bunting
Blue Grosbeak
House Finch
Rufous-sided Towhee
Brown Towhee
Sage Sparrow
Brewer's Sparrow
Chipping Sparrow

From about 7,000 feet to timber line, characteristic birds include those designated H in the Mogollons list (*see above*), excluding the Whip-poor-will, Coues' Flycatcher, Olive Warbler, and Red-faced Warbler. These are here replaced by the Gray Jay, Black-billed Magpie, Black-capped Chickadee, Swainson's Thrush, Pine Grosbeak, and Lincoln's Sparrow. At and above timber line, the breeding birds are the White-tailed Ptarmigan (very local), Horned

Lark, Water Pipit, Brown-capped Rosy Finch, and White-crowned Sparrow.

West of the Rio Grande, rise the San Juan and Jemez Ranges, biotically similar to the Sangre de Cristo although lacking the alpine-tundra forms. Also, the Dusky Flycatcher—a summer resident in both western Ranges—apparently is absent from the Sangre de Cristo; and Black-capped Chickadees seem not to breed in the San Juans. The Magnificent (Rivoli's) Hummingbird is known from the Jemez Mountains, where it may nest.

South of the Sangre de Cristo Range, between the Rio Grande and Pecos Valleys, a strip of mountainous country extends southward into Mexico and Texas. In this belt, near the Rio Grande, are several distinct ranges, the best known ornithologically being the Sandia (*see under* **Albuquerque**), with many of the habitats and birds found farther north. Others are rugged, often barren massifs, of limited access and of little interest to the bird finder. The Organ Mountains (*see under* **Las Cruces**) provide a partial exception. East of this highland chain, across the arid Tularosa Basin, rise higher, heavily forested ranges, among them the Sacramento (*see under* **Alamagordo**) and Guadalupe Mountains (*see under* **Carlsbad**). The summer birdlife of most of these mountains is similar to that of the Sangre de Cristo Range or the Mogollons, excluding the Mogollons' Mexican specialties, which in the main do not range west of, or even to, the Rio Grande. A feature of all New Mexican mountains, wherever wildflowers are numerous, is local concentrations of hummingbirds from July through September.

East of the Pecos, mountains disappear and the terrain becomes level to gently rolling—part of the Great Plains. Here and there are 'islands' of contrasting habitat such as the sandhills with their low shin-oaks, the wooded fringes along streams, or the pockets of greenery around isolated ranch houses. But basically this is grassland, miles and miles of it, still with herds of attractive pronghorns as well as cattle. Although man and his livestock have significantly modified the Great Plains, no bird species has entirely disappeared. Among the regular breeding species are:

Swainson's Hawk
Ferruginous Hawk
Lesser Prairie Chicken
Common Bobwhite
Scaled Quail
Mountain Plover
Long-billed Curlew
Greater Roadrunner
Burrowing Owl
Lesser Nighthawk
Western Kingbird
Scissor-tailed Flycatcher
Say's Phoebe
White-necked Raven
Western Meadowlark
Cassin's Sparrow

Bird migration in topographically diverse New Mexico is a complex phenomenon, often unpredictable and generally lacking the well-defined concentrations or 'waves' so familiar in the eastern United States. Frequently, in both spring and fall, birds appear to 'dribble through' in small numbers; at times the bird finder is scarcely aware that a migration is in progress. Major river valleys generally are the best places to seek migrants in number, but some species are found only in the highlands. Certain common breeding warblers of the southwest mountains—Grace's, Red-faced, and Painted Redstart—almost never appear in the adjacent valleys and foothills, flying over them in both spring and fall with no stopovers in the lowlands. Spring arrival of breeding birds may be affected by local conditions of topography or vegetation and doubtless by other factors. Each spring, for example, Black-headed Grosbeaks return to sites north of Silver City 7 to 10 days earlier than at the town itself only 8 miles away and some 300 feet lower. If little is known of the movements of these long-distance migrants, even less is known about some local altitudinal migrations. The Greater Roadrunner appears to move, presumably on foot, from the higher portion of its range in southwestern New Mexico down to lower elevations for the winter. On the other hand, Phainopeplas, which nest in the desert canyons and river valleys, regularly migrate upward after the breeding season to winter among the mistletoe-laden junipers. Flickers, Hairy and Downy Woodpeckers, American Robins, and other 'permanent residents' move through well-studied areas in what seem to be regular migratory patterns, yet the distances traveled are unknown. Even supposedly sedentary birds may move far at times, as demonstrated by a Curve-billed Thrasher banded in Silver City and recovered a short time later in

Chihuahua, Mexico. Migrations are often late and surprisingly prolonged. Various species of Mexican affinities do not appear in the south until late May or June whereas some northern breeding birds may be found along the Mexican border well into June. Northbound Lark Buntings may still be seen in southern New Mexico in late May, yet the first southbound arrivals have returned there by July. The Rufous Hummingbirds in the southwest mountains during early July often are viewed by uninformed bird finders as breeding birds but they too are merely 'early' migrants en route to Mexico.

New Mexico winters are more severe than those of adjacent Arizona, even in the southwestern counties, for the average elevation is considerably higher. Nevertheless, major waters like the Rio Grande host many wintering waterfowl and the plains support relatively large numbers of broad-winged (buteo) hawks, Golden Eagles, Northern Harriers, American Kestrels, and Loggerhead Shrikes from northern climes. Brushy roadsides and riparian thickets everywhere are important as wintering habitat for enormous numbers of juncos and White-crowned Sparrows. The extensive desert-grasslands, too, are well populated with Western Meadowlarks, Lark Buntings, Savannah Sparrows, Brewer's Sparrows, and longspurs. Some winters see great flocks of Western and Mountain Bluebirds in the wooded lowlands and foothills, joined at intervals by American Robins and Cedar Waxwings. There is much variation from year to year and, of course, from one nearby locality to another. The elevational differences which influence breeding distributions and migration exert an equally strong effect on the state's wintering birds.

Authorities

Bruce J. Hayward, John P. Hubbard, Charles A. Hundertmark, Dustin Huntington, J. Stokley Ligon, Alan P. Nelson, William Principe, Ralph J. Raitt, James L. Sands, James R. Travis, Daniel T. Washburn, Marian Washburn, Steve West, Marjorie Williams, Marian Zimmerman.

Reference

Revised Check-list of the Birds of New Mexico. By John P. Hubbard. New Mexico Ornithological Society Publication No. 6, 1978. Available from the author, 2016 Valley Rio, Santa Fe, NM 87501.

ALAMOGORDO
Sacramento Mountains | White Sands National Monument

On the plains at an elevation of 4,300 feet, Alamogordo is the gateway to the **Sacramento Mountains,** largest range in southeastern New Mexico. Here the forested peaks extend above 9,000 feet, and as these mountains lie within 100 miles of seven major cities, they serve as important summer recreational outlets for many Texans and New Mexicans seeking relief from the heat of the surrounding low country. For bird finding, a worthwhile site is reached by driving east from Alamogordo on US 82 for about 19 miles. At the community of High Rolls, leave the highway where a sign indicates Karr Canyon Picnic Grounds. The road, hard-packed and usually in good condition, traverses orchards, oak groves, streamside cottonwoods, junipers, and pines. Breeding birds include the Gambel's Quail, Whip-poor-will, Ladder-backed Woodpecker, Yellow-bellied Sapsucker, Steller's and Scrub Jays, Mountain Chickadee, Plain Titmouse, Bushtit, Hermit Thrush, Western Bluebird, Solitary and Warbling Vireos, Western Tanager, and Black-headed and Blue Grosbeaks. Among the presumably breeding warblers are Orange-crowned, Audubon's, Grace's, Black-throated Gray, and MacGillivray's. Red-faced Warblers appear in spring and may nest here.

Farther east along US 82, between Cloudcroft and Marshall, extensive open forests and shaded canyons harbor breeding Wild Turkeys, Montezuma Quail, Flammulated Screech and Northern Pygmy Owls, Broad-tailed Hummingbirds, Townsend's Solitaires, and Gray-headed Juncos, plus the species listed for Karr Canyon. June records of the Northern Three-toed Woodpecker and Clark's Nutcracker suggest possible local breeding of these erratic species.

The alkali flats and unique gypsum dunes of **White Sands National Monument** (229 square miles), 15 miles southwest of Alamogordo, are poor bird habitats, but 200-acre Garton Lake, near Monument headquarters is attractive to transient waterbirds. During the breeding season its shores may be frequented by Snowy Plovers, American Avocets, and Black-necked Stilts. The Lesser Nighthawk, Ladder-backed Woodpecker, Ash-throated Flycatcher, Black Phoebe, White-necked Raven, Crissal Thrasher, Verdin, and Scott's Oriole nest nearby. Winter visitants include Sage Thrashers, and Savannah, Vesper, and Brewer's Sparrows.

ALBUQUERQUE

Sandia Mountains | Juan Tabo Canyon | Capulin Springs | Sandia Crest | Isleta Marsh

New Mexico's largest city is situated on the Rio Grande near the base of the **Sandia Mountains,** which lie several miles east of the river. Rising impressively from 6,500 to 10,678 feet, the Sandias support eight major plant communities: chaparral, oak woodland, pinyon-juniper woodland, plus ponderosa pine, mixed conifer, and spruce-fir forests, with aspen groves at higher elevations and riparian woodland along streams throughout. Of 165 bird species known from the Sandia Mountains, over ninety breed in them. The permanent and summer residents are largely those of the northern mountains (*see introduction to this chapter*) although timber-line and tundra species are absent. A few southern birds—notably the Bendire's and Crissal Thrashers and Black-chinned Sparrow—range north to the Sandia foothills.

Species characteristic of the foothills are well represented in **Juan Tabo Canyon,** 5 miles north of the city. To reach it, drive north on I 25 toward Santa Fe, exit east on Montgomery Boulevard for 5 miles, turn left (north) on Tramway Boulevard and proceed past the road leading to the tram (a '4-way-stop' intersection) for 0.9 mile to Forest Road 333, which can be followed to La Luz Picnic Grounds. Beyond this point the road is undependable for vehicles, but it and various trails permit hiking from here. La Luz Trail extends to Sandia Crest, although the 6-hour climb may prove difficult for inexperienced hikers. Breeding birds include the Greater Roadrunner, Scrub and Pinyon Jays, Plain Titmouse, Curve-billed and Crissal Thrashers, Black-throated Gray Warbler, Scott's Oriole, Black-headed Grosbeak, Brown Towhee, and Black-throated Sparrow. In winter, many permanent residents of the mountains move downward to join such species as Sage Thrasher and Oregon Junco which arrive from farther north. At this season the Bohemian Waxwing and Northern Shrike appear at irregular intervals.

A 60-mile round-trip from Albuquerque to **Capulin Springs** and **Sandia Crest** provides access to higher altitude birds. Drive east on I 40 through Tijeras Pass, exit north on State 14, turn off left onto State 44 and then onto Forest Road 536, which winds a short distance to a sign (left) indicating Capulin Springs Picnic Ground. Park here and follow any of the trails which meander into the hills from the picnic area. Along these and the main road are Band-tailed Pigeons, Yellow-bellied and Williamson's Sapsuckers, Red-breasted Nuthatches, Ruby-crowned Kinglets, Hermit Thrushes, MacGillivray's and Audubon's Warblers, Green-tailed Towhees, and other coniferous-forest breeding species. Between Capulin Springs and the Crest, the road penetrates excellent forest. Where it terminates at a paved parking area, leave the car on the lowest level and walk along the road to the ski area. White-throated Swifts feed over the clearing here, and Northern Three-toed Woodpeckers have nested in the nearby woods. A winter trip, when the road is open only to vehicles with chains or snow tires, may reveal cold-weather visitants including, on rare occasions, Black and Brown-capped Rosy Finches which have been recorded at the parking lot.

For an alternate, but longer, return trip to Albuquerque during spring or summer, continue north along State 14 through pinyon-juniper woodland to State 22 at left, 17 miles north from the turn-off (State 44) to Sandia Crest. The Ladder-backed Woodpecker, Cassin's Kingbird, Scrub Jay, Plain Titmouse, Mountain Bluebird, Bewick's Wren, Northern Mockingbird, Loggerhead Shrike, House Finch, and Brown Towhee may be expected along the way. State 22, a good dirt road, is flanked by open stands of low junipers with patches of cholla cacti. The Bendire's Thrasher is locally common here, its far-carrying song much in evidence in May and early June. Near the road's junction with I 25 and US 85 (11.5 miles from State 14) are open, short-grass plains where often Mountain Plovers may be seen. Watch for them, particularly on the left, along the last 2 or 3 miles of State 22. From where the latter joins I 25, it is 35 miles to the Albuquerque junction of I 25 and I 40.

South of Albuquerque, in the Rio Grande Valley, **Isleta Marsh** supports various wetland species. From the junction of I 40 and I 25, drive south on I 25 for 13 miles and take Exit 213 to US 85; follow it south for about 6 miles until swampy ponds become evident beside the road. These extend for over a mile. Species diversity varies seasonally, but breeding birds include the Pied-billed Grebe, Snowy Egret, Black-crowned Night Heron, Least Bittern, several ducks, Common Gallinule, and Great-tailed Grackle.

CARLSBAD

Carlsbad Caverns National Park | Rattlesnake Springs | Lake McMillan | Guadalupe Mountain Range | Laguna Grande | Harroun Lake

World-famous **Carlsbad Caverns National Park** (73 square miles) lies 25 miles southwest of this southeastern New Mexico city in the Guadalupe foothills. Despite its generally arid character, the Park boasts a bird list of well over 200 species, and a visit is rewarding at any season. To reach the Park, drive southwest from Carlsbad on US 62 for about 20 miles, turning right at Whites City on State 7, which winds through Walnut Canyon, and proceeding to the entrance. Among the summer-resident birds are the Common and Lesser Nighthawks, Ash-throated Flycatcher, White-necked Raven, Scott's Oriole, and Blue Grosbeak. Poor-wills are ubiquitous in spring and summer and occasionally are in evidence on warm winter evenings. Regular permanent residents include the Harris' Hawk, Scaled Quail, Greater Roadrunner, Ladder-backed Woodpecker, Verdin, Cactus, Canyon and Rock Wrens, Curve-billed and Crissal Thrashers, Pyrrhuloxia, House Finch, Brown Towhee, and Black-throated Sparrow. These are joined in winter by the Ferruginous Hawk, Sage Thrasher, Eastern, Western, and Mountain Bluebirds, Green-tailed Towhee, Lark Bunting, Oregon and Gray-headed Juncos, plus the Savannah, Vesper, Sage, Chipping, and Brewer's Sparrows. The uncommon Black-chinned Sparrow is a summer resident throughout much of the Park west of the visitor center. Slaughter Canyon is a preferred site of the Gray Vireo, and Varied Buntings nest in Walnut Canyon. Cave Swallows may be seen feeding almost anywhere in the Park, particularly during May and June, although they are easily overlooked among the abundant Cliff Swallows. Both species are most easily observed near standing water where they come to drink or to gather mud for their nests. Over the natural amphitheater at Carlsbad Cave, where seating is provided for watching the evening bat flights, the swallows regularly fly to and from the cavern entrance.

Rattlesnake Springs, part of the Park, is best known as a 'migrant trap,' but bird finding is good at all times. To reach it from Carlsbad, drive south on US 62 and 180 to Whites City and thence beyond for about 2 more miles. Turn right (west) onto a paved road and follow the signs to the Springs, a distance of between 3 and 4 miles. Breeding species include the Scissor-tailed Flycatcher, Black Phoebe, Vermilion Flycatcher, Bell's Vireo, Hooded Oriole, Painted Bunting, Lesser Goldfinch, and Cassin's Sparrow. The Varied Bunting is occasionally seen here, particularly in June.

East of US 285 are **Lake McMillan** and Avalon Reservoir north of Carlsbad about 16 and 10 miles respectively. Both attract similar birdlife. On fields west of Lake McMillan, numbers of waterfowl and a few Sandhill Cranes feed during the winter. Grasslands and desert scrub east of the Lake support birdlife different from that in the cultivated, brushy, and wooded places nearer the water. A network of small dirt roads around the Lake provides access to several points, but following rains these should be driven with extra caution. The Lake itself attracts Common Loons, grebes, American White Pelicans, and many waterfowl in spring, late summer, and autumn. The west shoreline is especially good for herons—ten species have been recorded. The Lake's southwest end regularly attracts transient shorebirds, and when the water level is low extensive mud flats lure them by the thousands; thirty species have been seen here, including several rarites. The southward shorebird movement begins in mid-July and continues through September, with most individuals departing by mid-October. The spring migration is of briefer duration, in April and May. Transient gulls—Herring, Ring-billed, Bonaparte's, and Franklin's—and terns—mostly Forster's and Black—are numerous at times; other gulls and terns appear rarely. Thousands of waterfowl use the Lake in winter, but geese are comparatively scarce and disperse to the grainfields during the day; they are best seen along the west side. The common ducks are the Mallard, Gadwall, Common Pintail, Cinnamon Teal, American Wigeon, Northern Shoveler, Redhead, Lesser Scaup, Ruddy Duck, and Common Merganser. Sometimes there are moderate numbers of Ring-necked Ducks, Canvasbacks, Common Goldeneyes, and Buffleheads. Wood Ducks and Hooded Mergansers visit irregularly. The few wintering shorebirds are largely Killdeers, Common Snipes, Greater Yellowlegs, and Least Sandpipers. Ring-billed Gulls may be abundant at that season, typically with a few Herring Gulls among them.

Breeding landbirds in the vicinity of the Lake include (in addition to most of those listed for the National Park) Yellow-billed Cuckoos, Western Kingbirds, Scissor-tailed Flycatchers, Say's Phoebes, Yellow-breasted Chats, Western Meadowlarks, and Bullock's Orioles. During migration and in winter, Ferruginous Hawks, Short-eared Owls, and Sage Thrashers are present, as are numerous fringillids, among them the Savannah, Grasshopper, Vesper, Sage, Black-throated, Chipping, Clay-colored, Brewer's, White-crowned, Lincoln's, and Song Sparrows. Lark Buntings and Chestnut-collared Longspurs are common in open areas, and the Sprague's Pipit is a possibility.

The **Guadalupe Mountain Range** is shared by New Mexico and Texas, the highest parts being near the state line. (*See under* **Pine Springs, Texas.**) Much of New Mexico's portion is in the Lincoln National Forest west of Carlsbad Caverns National Park. Entering the Forest involves a 45-minute drive on State 137, starting from a point 12 miles northwest of Carlsbad on US 285; only the first 24 miles are hard-surfaced. The most exciting parts of the Range are seen by those willing to hike into the wilder places, but much of interest lies near the roads.

Among the 180 bird species found in and near the Guadalupes are the Zone-tailed Hawk, Montezuma Quail, Spotted Owl, Magnificent Hummingbird, Dusky Flycatcher, and other rarities. Summering species include the Scaled Quail, Band-tailed Pigeon, Acorn and Ladder-backed Woodpeckers, Pinyon Jay (irregular),

MONTEZUMA QUAIL

Mountain Chickadee, Plain Titmouse, Bushtit, Pygmy Nuthatch, Curve-billed and Crissal Thrashers, Western Bluebird, Orange-crowned, Virginia's, Audubon's, and Grace's Warblers, Scott's Oriole, Western and Hepatic Tanagers, Black-headed Grosbeak, Lesser Goldfinch, Red Crossbill, Brown Towhee, Rufous-crowned Sparrow, and Gray-headed Junco. Both Common and Lesser Nighthawks nest in the area. Similarly, both Northern and White-necked Ravens are here in summer, the Northern confined to the lowlands. Only the Northern Raven is likely to be present in winter. Canyons in these mountains are among the best places to find Black-chinned Sparrows and the elusive Gray Vireo—here fairly common from late April or early May to August. Listen for its Solitary Vireo-like songs from dry, wooded canyonsides. The poorly known winter birdlife includes Williamson's Sapsuckers, Mountain Bluebirds, Golden-crowned Kinglets, Red-breasted Nuthatches, Sage Thrashers, Townsend's Solitaires, and Evening Grosbeaks (irregular).

Southeast of Carlsbad is **Laguna Grande** (Loving Salt Lake), a series of partially inundated salt pans or playas, attractive to migrating waterbirds and shorebirds and surrounded by desert scrub. The area is reached by driving about 8 miles southeast of Carlsbad on US 285, turning east onto State 32, and proceeding 2 miles north of Loving. After 8 miles, travel east for about 1.0 mile on County 128 to the playas. **Harroun Lake,** a similar locality, is easily reached by driving 5 miles south of Loving on US 285 to Malaga. Here turn left (east) onto an unnumbered blacktop road and go 0.8 mile; then left again (north) and go 1.0 mile, after which a right turn leads about 0.6 mile to the Pecos River. Continue east across the bridge. About 1.5 miles from the River, watch for a dirt road to the right. This encircles the small Lake, providing easy access in dry weather. (All dirt roads in the vicinity should be driven with caution, and when afoot be alert for rattlesnakes.) Among the permanent-resident birds near these sites are the Harris' Hawk, Scaled Quail, Greater Roadrunner, Ladder-backed Woodpecker, Crissal and Curve-billed Thrashers, Verdin, Cactus Wren, Loggerhead Shrike, Pyrrhuloxia, and Black-throated Sparrow. Summer residents or visitants include the Yellow-billed Cuckoo, Common and Lesser Nighthawks, Bell's Vireo, Yellow-breasted Chat, Blue Grosbeak, Painted Bunting, and Lark and Cassin's Sparrows. During late fall and winter, Sandhill Cranes, numerous waterfowl, hawks, and most of the fringillids aforementioned for Lake McMillan have been recorded here.

CHAMA

Carson National Forest | Jicarilla Apache Indian Reservation | Lake Burford

This ancient town, at an elevation of 7,860 feet near the Colorado line, dates from a pueblo that may have been occupied at the time of the Spaniards' arrival in the Southwest. Today it is a popular recreational resort with an ideal summer climate, a beautiful setting, and a fine variety of birds.

West of Chama the **Carson National Forest** and **Jicarilla Apache Indian Reservation** occupy extensive tracts of sage, pinyon-juniper woodland, coniferous forest, wooded canyons, mountain lakes, and meadows. Access is via US 84, thence either on State 17 west toward Navajo Lake and Farmington, or on State 537 south through the Reservation. Of the several lakes, all on the Reservation, the most important is **Lake Burford**, New Mexico's largest natural impoundment (elevation 7,000 feet). It is 4 miles west of Tierra Amarilla on State 95 (a dirt road), which is reached by driving about 12 miles south from Chama on US 84. Summering species about the Lake include the Eared and Pied-billed Grebes, Black-crowned Night Heron, Gadwall, Common Pintail, all three teal, American Wigeon, Northern Shoveler, Redhead, Ruddy Duck, Common Merganser, Virginia Rail, Sora, Black Tern, plus Yellow-headed, Red-winged, and Brewer's Blackbirds. Occasional American Avocets and Black-necked Stilts appear. On the nearby wooded mesas are the Ash-throated Flycatcher, Say's Phoebe, Dusky Flycatcher, Western Pewee, Scrub and Pinyon Jays, Pygmy Nuthatch, Hermit Thrush, Western and Mountain Bluebirds, Solitary Vireo, Virginia's Warbler, and Black-headed Grosbeak. In the rugged mountains to the east and south, summer birds include the Northern Goshawk, Wild Turkey, Band-tailed Pigeon, Northern Pygmy Owl, White-throated Swift, Lewis' Woodpecker, Violet-green Swallow, Steller's Jay, Clark's Nutcracker, Black-billed Magpie, Northern Raven, Mountain Chickadee, Red-breasted Nuthatch, Townsend's Solitaire, Grace's Warbler, Western Tanager, Evening Grosbeak, Cassin's Finch, Red Crossbill, and Gray-headed Junco. The open ridges north and west of the Lake, with their covering of fragrant sage, are the breeding habitat of Sage Thrashers and Vesper, Sage, and Brewer's Sparrows.

DEMING

Gage | Nutt

About 33 miles west of this desert town in southwest New Mexico is the hamlet of **Gage**, on the north side of I 10. The dirt road leading north from Gage toward Whitewater offers an excellent array of desert-grassland birds during spring and early summer. Swainson's Hawks and Scaled Quail are conspicuous, the thrashers are singing, and the crepuscular, toad-like trilling of Lesser Nighthawks can be heard on all sides in mid-May and early June—the ideal time for a visit. The Cassin's Sparrow often is common in the thicker grass patches; as with the nighthawk, it can be heard at its vocal best around dawn. Other breeding species are the Greater Roadrunner, Burrowing Owl, Western Kingbird, Ash-throated Flycatcher, Bendire's Thrasher (nesting in yuccas and shrubs), Curve-billed Thrasher (in cholla cacti), Northern Mockingbird, Cactus Wren, Eastern Meadowlark, Scott's Oriole, and Black-throated Sparrow. Transient Lark Buntings linger into June and reappear in the area by early August. One may drive north along this road for about 10 miles before encountering a locked gate. (The motorist cannot, therefore, drive to Whitewater, although many maps show this as a 'through road.') Summer bird finding is worthwhile only between daybreak and midmorning, and again in the evening. Nocturnal visits may reveal Barn, Great Horned, and Burrowing Owls, plus Short-eared Owls in fall and winter.

The plains near **Nutt** offer excellent winter raptor viewing along a little traveled paved route, which permits the bird finder to journey to Silver City or Truth or Consequences via the Black Range. Two miles north of Deming, turn east from US 180 on State 26. About 27 miles from the junction, at Nutt, State 27 branches north to Hillsboro, another 34 miles. Wintering species along the route include the Copper's Hawk, Red-tailed Hawk (3 subspecies), Rough-legged Hawk, Ferruginous Hawk, Golden Eagle, Northern Harrier, American Kestrel, and Prairie Falcon. At twilight, watch for Barn, Great Horned, Burrowing, and Short-eared Owls. Short-eareds are more likely in autumn, but a few winter here, perhaps irregularly. The Swainson's Hawk is present from April to mid-September. Between November and March, Chestnut-collared Longspurs range widely over the plains, but seldom allow close approach except at earthen stock tanks where the flocks drink in dry weather. Two or three such tanks are near the road north of Nutt; an hour's wait in a car alongside one can produce satisfying views. Horned Larks, Crissal Thrashers (along brushy dry 'washes'), Western Meadowlarks, Lark Buntings, and Savannah, Vesper, and Sage Sparrows also may be expected in winter. Caution: In winter, motorists should be alert for weather conditions that might spawn blizzards on the open plains; avoid the route following heavy snowfalls. After summer rains, too, flash floods can make exceedingly dangerous the innocent-looking desert washes that intersect the road at intervals.

FARMINGTON

Jackson Lake Waterfowl Area | McGee Park | Navajo Dam | Morgan Lake | Aztec | Bloomfield | Waterflow

To the Navajos this city, 23 miles south of the Colorado line, was known as 'tqo tah'—'three waters, blending waters'—for here the San Juan, Animas, and La Plata Rivers join. Characteristic breeding birds of the San Juan Valley—typical of the riparian areas as a whole—are listed in the introduction to this chapter. Wintering species include various waterfowl, Bald Eagles, Bohemian Waxwings (irregular), Cedar Waxwings, Cassin's Finches, and American Tree Sparrows. Among the more accessible areas for these valley birds are **Jackson Lake Waterfowl Area,** reached by driving 3 miles west of Farmington on US 550, then north 5 miles on State 17, and **McGee Park** on State 17, 8 miles east of Farmington. Also worth visiting is a several-mile strip open to the public immediately below **Navajo Dam** (elevation 5,700 feet). Four miles south of Fruitland is **Morgan Lake,** part of the Four Corners Power Plant Operation. Over 50 species of water-associated birds use this 1,200-acre impoundment. April, May, September, and early October are the best months to see transients, expecially shorebirds. Common Loons, Horned and Western Grebes, Snowy Egrets, and White-faced Ibises are fairly regular transients, and Eared and Pied-billed Grebes nest on the Lake. Whistling Swans are among the numerous wintering waterfowl.

For finding birds in pinyon-juniper woodland, grassland, and sagebrush habitats, several places are recommended: (1) The Cedar Hill area 12 miles north of **Aztec** along US 550; (2) State 173 between Aztec and Navajo Dam; (3) the Angel Peak-Huerfano region 10 to 20 miles south of **Bloomfield** along State 44; and (4) the plains east of the San Juan Generating Plant at **Waterflow.** Typical summer species of these areas include the Ferruginous Hawk, Golden Eagle, Prairie Falcon, Scaled Quail, Poor-will, Hairy Woodpecker, Say's Phoebe, Scrub and Pinyon Jays, Northern Raven, Plain Titmouse, Rock Wren, Sage Thrasher, Western and Mountain Bluebirds, Blue-gray Gnatcatcher, Loggerhead Shrike, Gray Vireo (especially near Navajo Lake), Black-throated Gray Warbler, House Finch, Brown Towhee, Gray-headed Junco, and Vesper, Lark, Black-throated, Sage, Chipping, and Brewer's Sparrows.

Reference

Summer Birds of the San Juan Valley, New Mexico. By C. Gregory Schmitt. New Mexico Ornithological Society Publication No. 4. Albuquerque: McLeod Printing Co., 1977.

GLENWOOD

San Francisco River Valley | Whitewater Canyon | Mogollon Mountains | Luna | Apache Creek

This small community (elevation 4,800 feet), 63 miles northwest of Silver City on US 180, provides access to the seldom visited **San Francisco River Valley** which extends from here to Clifton, Arizona. Here the bird finder can see a choice example of riparian biotic communities of Mexican affinity, with ninety summering bird species, which are threatened almost everywhere in the Southwest. Vegetation and avifauna are similar to those along the Gila River (*see introduction to this chapter and under* **Silver City**), with nesting Lesser Black Hawks, White-winged Doves, Elf Owls, and Wied's Crested Flycatchers among the desert elements. This is the southernmost point in New Mexico where American Crows are numerous as a breeding species. They often feed alongside Northern Ravens in the fields near Glenwood. The only recommended way to see most of the River Valley—still a relatively wild and remote place—is on foot or horseback, a satisfying experience for more adventurous bird finders. Small cottonwood groves and scattered trees remain in and around Glenwood where Elf (and sometimes Northern Pygmy) Owls call at night around the motels. Lesser Black Hawks occasionally soar over the town.

Lower **Whitewater Canyon,** is reached by driving east from US 180 at the north edge of Glenwood, and following a narrow, blacktop road for about 5 miles. (A sign at the junction indicates the 'catwalk' and Whitewater Picnic Grounds.) Here Elf Owls, Black Phoebes, Western Flycatchers, Canyon Wrens, Warbling and Solitary Vireos, Painted Redstarts, and often a pair of North American Dippers nest along the rushing stream which spills through a narrow, boulder-studded canyon rendered accessible by a metal bridge (the catwalk) and a well-maintained foot trail. Birds in the general vicinity are those of the lower canyons and slopes of the Mogollon Mountains (*see introduction to this chapter*). Among the Canyon's mammals are the seldom-seen ring-tailed cat and introduced bighorn sheep.

The **Mogollon Mountains**, rising from the plateau a dozen miles east of Glenwood, support over one hundred species of breeding birds. The only major highway penetrating the range is State 78, which branches eastward from US 180 about 3.5 miles north of Glenwood. Nine miles from this junction is the ghost town of Mogollon. Portions of State 78 leading from Glenwood, and from Mogollon to Willow Creek, are steep and narrow ('one-way' in a few short sections), but perfectly safe for experienced drivers in dry weather. Deep snow can remain here into May. A visit in June, before the summer rains, provides the best opportunity to see a number of western New Mexico's highland birds. Recommended sites are: Silver Creek Divide, 6.5 miles southeast of Mogollon, and Bursum Camp, 3 miles farther. Both are in mixed conifer forest at the 9,000-foot level. Northern Goshawks, Blue Grouse, Northern Three-toed Woodpeckers, Williamson's Sapsuckers, Red-breasted Nuthatches, Clark's Nutcrackers, and several parulids breed here; finding them is largely a matter of spending sufficient time in the mountains. Although the forested slopes are steep, the lightly traveled road allows observers to walk easily and undisturbed for miles, if desired, sometimes at treetop level alongside tall firs and aspens growing below the roadway. Foot or horse trails provide access to still higher areas. Forest Trail 182 is 2 miles beyond the Divide and is signposted at a spot called Sandy Point (elevation 9,000 feet). Leading south, the trail climbs to the 10,000-foot level in 1.5 miles. It extends several miles farther to Whitewater Baldy and provides inviting scenery and a good chance to see Blue Grouse and Northern Three-toed Woodpeckers.

Ben Lilly and Willow Creek Camps, some 17 miles from Mogollon, lie 1,000 feet lower in the willow-alder-spruce belt along sparkling Willow Creek and adjacent to magnificent stands of mature ponderosa pine. Thereabouts, Western Flycatchers and Red-faced Warblers are regular breeding birds. House Wrens, Ruby-crowned Kinglets, MacGillivray's Warblers, and Gray-headed Juncos are conspicuous along the stream, and a few Green-tailed Towhees nest in the thickets. Northern Three-toed Woodpeckers reside in the adjacent forest but are difficult to find.

About 40 miles north of Glenwood along US 180 is the village of **Luna** (elevation 5,289 feet), one of the few reliable places for Lewis' Woodpeckers in southwestern New Mexico. A few usually are in evidence along the highway at the southern approach to town—on fenceposts, utility poles, or dead pine trunks. Often they feed on the ground near a prairie dog colony to the right of the road. Other summering birds are Mountain Bluebirds, Yellow-headed and Brewer's Blackbirds, and Vesper Sparrows—here close to the southern limits of their breeding ranges in New Mexico. From Luna, one may travel south and east on US 180 and State 12 through Reserve to **Apache Creek,** a distance of about 35 miles. At this crossroads, and at intervals for several miles along the highway to the east, small ponds and varying amounts of flowing water attract a few breeding Mexican Ducks and Cinnamon Teal. Watch overhead for Golden Eagles and, during spring and fall, for Ferruginous Hawks. Lewis' Woodpeckers often perch on roadside poles and fenceposts between Apache Creek and Aragon.

LAS CRUCES
Organ Mountains | Jornada del Muerto

The flatlands east of this city support birds of the creosote bush-mesquite desert and associated washes. Representative areas are easily reached by driving east from I 10 on University Avenue toward isolated Tortugas Mountain with its conspicuous letter A and the massive Organ Mountains beyond. Principal breeding birds of the desert are the Scaled Quail, Mourning Dove, Greater Roadrunner, Lesser Nighthawk, Verdin, Cactus Wren, Crissal Thrasher, Black-tailed Gnatcatcher, Loggerhead Shrike, and

Black-throated Sparrow. Along the west flank of the 9,000-foot **Organ Mountains,** near the mouths of canyons especially, greater habitat diversity supports more species, among them the Gambel's Quail, Poor-will, Ladder-backed Woodpecker, Ash-throated Flycatcher, Northern Mockingbird, Rock Wren, Pyrrhuloxia, and House Finch. (These species wander to the lower desert, especially after breeding.) In the higher canyons with hackberry, oak, and juniper, the Scrub Jay, Plain Titmouse, Bushtit, Black-headed Grosbeak, Brown Towhee, and Rufous-crowned and Black-chinned Sparrows number among the breeding birds. To investigate the upper vegetative belts of the Organ Mountains, drive to Aguirre Springs and hike from there: Take US 80 northwest of Las Cruces to a point a few miles east of San Augustin Pass. Here, turn right (south) onto the broad, 10-mile-long dirt road which leads through grasslands and juniper savannas, and finally into oak-juniper woodland. At the end, a trail leads from some campsites, providing a 4.5-mile hiking loop which penetrates the ponderosa pine belt. In summer, Whip-poor-wills, Grace's Warblers, and Hepatic Tanagers breed here. During winter, look for Hairy Woodpeckers, Townsend's Solitaires, and (where there is mistletoe) Phainopeplas which move up from the desert following breeding.

The Las Cruces region has changed greatly under man's influence. The plains above the valley once were largely short-grass prairie, but livestock grazing has produced a wholesale shift to the creosote bush-mesquite desert which predominates today. In the few places where grazing has been controlled, some grassland remains. One example is the **Jornada del Muerto,** reached by driving east from Las Cruces on US 80 toward Alamogordo. About 4 miles east of I 25 the road bends slightly to the right. Watch for a sign on the left directing motorists to the U.S.D.A. Jornada del Muerto Experimental Range. Here, continue north about 10 miles along the good dirt road to a windmill ('South Well'), meanwhile passing en route through fair grassland where in some winters Chestnut-collared Longspurs are common. During summer, Swainson's Hawks and Scott's Orioles are conspicuous birds. Cassin's Sparrows sing their flight-songs in spring, when both species of nighthawks are noticeable. Golden Eagles and Prairie Falcons are as common here as at any place in the West.

LORDSBURG

Lower Gila River Valley, Redrock | Peloncillo Mountains | Guadalupe Canyon | Geronimo Trail

The **Lower Gila River Valley** near **Redrock** is easily reached from this desert city in southwestern New Mexico. Follow US 70 northwest from I 10, pass the junction of US 180 to Silver City, and bear right (north) on State 464. The total distance is about 22 miles mostly over hard-surfaced roads. Summer birds en route include the Swainson's Hawk, Scaled Quail, Greater Roadrunner, Burrowing Owl, Lesser Nighthawk, Say's Phoebe, Cactus Wren, Scott's Oriole, Eastern Meadowlark, and (sometimes) Cassin's Sparrow. Golden Eagles and Prairie Falcons are possible at any season, but are most likely during fall and winter when Sage Thrashers, Western Meadowlarks, Lark Buntings, and Savannah, Vesper, and Brewer's Sparrows may be common. About 17.5 miles from the beginning of State 464, an often unmarked dirt road to Silver City (Forest Road 581) branches to the right. Just beyond here, State 464 descends steeply to a broad, brush-margined, usually dry streambed where the Verdin, Cactus Wren, Crissal Thrasher, and Phainopepla should be sought. A mile or so beyond, the road bends abruptly right, and near this curve Abert's Towhee is common in brushy thickets under remnant cottonwoods. Common Ground Doves are sometimes seen here. Bird finding often is best, however, north of the small settlement of Redrock itself. Beyond the gas station-post office building, drive over the hill, and turn left at a fork leading to the Gila River Bridge. A walk along the banks here in either direction usually discloses Black Phoebes and Vermilion Flycatchers. The Bell's Vireo, Lucy's Warbler, Northern Cardinal, and Abert's Towhee nest in the undergrowth, and Gila Woodpeckers call noisily from the leafy cottonwoods. Excepting the vireo and warbler, all are likely throughout the year. Overhead, during breeding season, watch for Zone-tailed and Lesser Black Hawks. After returning to the Bridge, cross to the west bank and follow the dirt road north (right) several miles to the New Mexico Game and Fish Department's 'Wildlife Area,' usually marked by a sign and high fencing. In exceptionally wet years, flooded fields and temporary marsh attract transient waterfowl, shorebirds, herons, and groups of White-faced Ibises. The water generally disappears by summer, but the numerous landbirds make a visit worthwhile at any time. Late April to mid-May is the best period for bird finding, however.

Lordsburg is a starting point for field work in the state's extreme southwest corner. The **Peloncillo Mountains** offer prime bird-finding localities, chief among which is **Guadalupe Canyon** which extends northeast from Sonora, across the corner of Arizona, and into New Mexico. Nearly 170 bird species have been recorded in the Canyon. Among the summering birds are the Zone-tailed Hawk, White-winged and Common Ground Doves, Common Screech and Elf Owls, Black-chinned, Violet-crowned, and Broad-billed Hummingbirds, Acorn, Gila, and Ladder-backed Woodpeckers, Cassin's Kingbird, Thick-billed Kingbird (in tall sycamores), Wied's Crested, Ash-throated, Olivaceous, and Northern Beardless Flycatchers, Bridled Titmouse, Phainopepla, Crissal Thrasher, Bell's Vireo, Lucy's Warbler, Hooded and Scott's Orioles, Bronzed Cowbird, and Summer Tanager. Appearing casually or occasionally are the Elegant Trogon, Buff-collared Nightjar, and Sulphur-bellied Flycatcher.

With an average elevation of 4,600 feet, Guadalupe Canyon is cold in winter and few birds remain. December and January temperatures may drop below zero; snowfalls are infrequent, but may occur as late as May. Many of the summer-resident birds do not return until mid- or late May, and some wait for another month. Winter and early spring birds in the Canyon include the aforementioned woodpeckers plus Red-shafted Flickers and sometimes Brown-backed Woodpeckers, Gray-breasted Jays, Bridled Titmice, Crissal Thrashers, American Robins, Audubon's Warblers, Northern Cardinals, and Black-chinned and White-crowned Sparrows. In some years, Evening Grosbeaks and Cassin's Finches appear.

Guadalupe Canyon is most easily reached from Douglas, Arizona (*see under* **Douglas, Arizona**), by driving 30 miles east from town, starting on 15th Street, to a fork in the wide gravel road where a sign directs motorists to the right and to the Canyon's access road. A far more scenic and ornithologically rewarding route, however, is via New Mexico's Animas Valley, Clanton Canyon, and the

Geronimo Trail. Drive west on I 10 from Lordsburg for 12 miles, turning south onto State 338 for 24 miles to Animas. Remaining on State 338, which becomes a dirt road, drive south from Animas for about 29 miles, then turn right (west) onto a smaller road, the Geronimo Trail, usually marked by a small sign indicating Clanton Canyon and Douglas. (The main road goes straight ahead to the ghost town of Cloverdale and the Gray Ranch.) Just after the turn, the Geronimo Trail traverses a small, shallow arroyo, which may have water flowing at the crossing; usually it is navigated with no problem. Mexican Ducks sometimes feed here in spring and summer; and during winter, flocks of Chestnut-collared Longspurs come to drink. At that season, Rough-legged and Ferruginous Hawks hunt over the plains where Sprague's Pipits and Baird's Sparrows should be watched for. Red-tailed and Swainson's Hawks are the most numerous summer raptors, but Zone-tailed Hawks are occasionally seen.

The Geronimo Trail intersects the Arizona state line about 13 miles beyond. After another 13 or 14 miles, watch for a sign indicating Guadalupe Canyon at a conspicuous side road to the left. At this point one is 55 miles from Animas, having journeyed through broad valleys and impressive mountain passes; pine and oak woodlands with Common Screech and Northern Pygmy Owls, Brown-backed Woodpeckers, and Gray-breasted Jays; mesquite-grasslands surveyed by Golden Eagles and Prairie Falcons; thicket-bordered desert washes and xeric mountain slopes with their intriguing plantlife and typical desert birds. En route from here, listen for Botteri's Sparrow wherever good stands of grass remain, and examine dense mesquite thickets for Varied Buntings. Scott's Orioles and Black-chinned and, sometimes, Costa's Hummingbirds feed in the scarlet ocotillo and penstemon blossoms during spring. Persons driving to Guadalupe Canyon by this route must allow four or five hours from Lordsburg; they should carry water and obtain gasoline in Animas. No other facilities of any kind exist between Animas and Douglas. Ranch houses are widely scattered, mostly far from the road, and few vehicles pass this way. Although regularly maintained, the Geronimo Trail becomes impassable by regular vehicles during and after wet weather. Usually it is in good condition during spring and early summer, when it is one of the

Southwest's most delightful roads for the bird finder. Near the Arizona-New Mexico line, a few undeveloped campsites exist in the pine-oak woods near Clanton Canyon where overnight visitors should listen for Whiskered Screech Owls—known in New Mexico only from this locality.

Guadalupe Canyon itself is rather remote, experiences high summer temperatures and, in some years, seemingly has more than its rightful share of rattlesnakes. More important, the Canyon 'road' closely follows—and in places is coincident with—a sandy, rock-strewn streambed, which quickly becomes a dangerous torrent when rain falls in the nearby mountains. During the rainy season, typically July–September, visitors should watch the clouds and be prepared to vacate the deceptively serene Canyon in haste. *Never* camp in the Canyon bottom.

PORTALES
Elida | Milnesand

In the plains about 15 miles west of the Texas state line, Portales is chiefly of interest to visiting bird finders for its nearby populations of Lesser Prairie Chickens. The display season for these birds is from late March or, usually, early April to very early May. It is best to obtain current information on specific localities from the Game and Fish Department in Roswell, as the birds shift their locations from year to year. However, one fairly reliable site, occupied for many years, lies near **Elida,** a crossroads hamlet, 24 miles southwest of Portales on US 70. To penetrate the 'chicken country,' drive south from Elida on State 440 to the end of the pavement, about 13.5 miles. Continue straight ahead on this sand road for 3.25 to 3.5 miles to the second cattle guard. Turn west here, through a gate, and watch for a windmill to the southwest. After driving about 0.5 mile, take the first faint sandy automobile trail (two simple tracks) leading south. Follow this to the windmill, from which the displaying ground is 200 to 300 yards to the northwest. A few birds sometimes may be seen on a small booming ground 50 yards inside the gate and south of the road. A second site may be reached by returning to State 440 and turning south, rather than heading north to Elida. Between 1.0 and 2 miles from here, turn left on a good dirt road (gate on the right) and drive 0.25 mile. The birds may be on either side of the road. Sometimes they can be heard from State 440 itself. Ferruginous and other hawks, Scaled Quail, Burrowing Owls, White-necked Ravens, Sage Thrashers, Western Meadowlarks, Lark Buntings, Grasshopper, Cassin's, and Brewer's Sparrows, and Chestnut-collared Longspurs are among the other species likely to be seen in the area, depending upon the season.

Caution: These sand roads are not designed for low cars; carry a shovel and be prepared to deflate the tires in case of getting mired. Also, one must be at the booming grounds *before* daybreak, and remain in a car or concealed in a blind. Walking about will prevent the birds from arriving and may interfere with their courtship. If present, they will be active at first light and will continue until the wind begins. Quiet, patient waiting may be rewarded by the sight of chickens displaying all around—possibly even on one's automobile, a memorable experience. In this featureless country an initial visit during the pre-dawn hours is unwise. Use the preceding afternoon to learn the terrain, clock distances, and determine driving time for a return trip in the morning darkness—or spend the night (it will be cold) in the vehicle beside the booming grounds.

Milnesand, 37 miles south of Portales on State 18, lies near other Lesser Prairie Chicken sites. One is reached by driving south from Milnesand for 9 miles on State 18 and branching left for 2 miles on a road leading to a State Management Area where the birds sometimes display. Displaying chickens also may be seen at times from State 18, 3 miles north of Milnesand.

ROSWELL
Bitter Lake National Wildlife Refuge | East Grand Plains | US 380 East

Approximately 15 miles east of this commercial center in east-central New Mexico (elevation 3,500 feet) is one of the Southwest's prime waterbird sanctuaries, **Bitter Lake National Wildlife Ref-**

uge, worth visiting at any time of year. It consists of two sections—a 14,000-acre tract north of US 70 (not open to public use except during a limited hunting season), and a southern tract of about 10,000 acres, 7 miles east of the junction of US 70 and US 285. There are about 15 miles of river bottomland within the Refuge, and many ponds and artificial lakes contribute to the available water. Nearly 250 bird species have been recorded on the Refuge and 44 have bred. During fall migration, about 50,000 waterfowl of over twenty species frequent these unique wetlands.

To reach Refuge headquarters, drive north from Roswell on Main Street (US 70 and 285), 0.5 mile beyond Berrendo Bridge. Turn right at the Refuge direction sign and continue for approximately 12 miles. Look for Burrowing Owls in the prairie dog colony on the right, just outside the Refuge entrance. Summer birds include the Pied-billed Grebe, Cinnamon Teal, Northern Shoveler, Swainson's Hawk, Harris' Hawk (occasional), Common Bobwhite, Scaled Quail, Snowy Plover, American Avocet, Black-necked Stilt, Little and Black Terns, Greater Roadrunner, Barn Owl, Vermilion Flycatcher (occasional), Northern Mockingbird, Western Meadowlark, and House Finch. Migration seasons bring many more species and daily lists in mid-May approach 100 species. The Wilson's Phalarope concentration (2,000 to 4,000 birds) occurs during the last week in April or first week in May; they are best seen from the dike between Units 15 and 16. Other transient shorebirds include the Black-bellied Plover, Long-billed Curlew, Willet, Greater and Lesser Yellowlegs, Long-billed Dowitcher, Marbled Godwit, and Baird's, Least, Stilt, and Western Sandpipers. American White Pelicans and Double-crested Cormorants are regular transients during spring and fall. In late September, great flocks of Sandhill Cranes arrive, increasing to between 15,000 and 35,000 birds by November; peak counts have reached 74,000. The hunting season sees a great reduction in numbers, although many remain until March. The cranes disperse during the day, returning to roost on the Refuge. The spectacular flights—one of the greatest bird displays in the country—are best witnessed at sunrise and sunset, a good vantage point being the west end of Dike 7 or the hilltop parking area nearby.

Winter bird finding on the Refuge may be exciting with grebes, herons, American Bitterns, Canada, Snow, and Ross' Geese, numerous ducks, raptors, rails, a few shorebirds, thousands of Sandhill Cranes, and the possibility of such uncommon species as the Whistling Swan, Merlin, Eastern Bluebird, Winter Wren, Northern Shrike, and American Tree Sparrow. Sedge Wrens should be sought in winter and during migration in unburned areas west and south of Unit 15. Christmas counts, which include parts of the Refuge, typically surpass 100 species.

A productive area southeast of Roswell known as **East Grand Plains** is reached by following Southeast Main Street for about 2 miles from town, then turning left (east) and proceeding for an equal distance. Most land is posted against trespassing but birds to be seen from the roadways include Swainson's Hawks (summer), Sandhill Cranes (winter), Mountain Plovers (rare breeder), Long-billed Curlews (transient), Red-headed Woodpeckers, and Scissor-tailed Flycatchers (breeding rarely). The Le Conte's Sparrow is irregular in winter along the canals in this area.

Lesser Prairie Chickens display in spring along roads branching from **US 380 East** of Roswell. About 38 miles east of the intersection of US 380 and Main Street, a narrow dirt road branches right (south) from the highway through a barbed-wire gate. Follow this for about 1.5 miles to its termination at a large bare area on which the birds display at sunrise. About 40 miles east of Roswell another road branches north from US 380 opposite a highway rest-stop, the only one between Roswell and Tatum. The chickens perform near a water tank about 5 miles from US 80. Observations must be made from an automobile (*see also under* **Portales**). As specific areas used by the chickens vary from year to year, some searching along the roads may be necessary at dawn on an April morning when the birds may be heard from some distance.

SANTA FE

Pecos River Canyon | Los Alamos | Bandelier National Monument | Jemez Mountains

New Mexico's capital city (elevation 7,000 feet) lies at the base of the Sangre de Cristo Mountains which, owing to their precipitous slopes, are best penetrated on foot or horseback. However, a productive automobile drive is up the **Pecos River Canyon** via US 85 south and east from Santa Fe for about 20 miles (2 miles beyond Glorieta), then left via a paved road to Pecos, 4 miles distant. From Pecos take State 63 north for 26 miles to the village of Cowles, beyond which lies the Pecos Wilderness Area. As the Canyon narrows, watch for North American Dippers along the River. Breeding birds along the route include the Lewis' Woodpecker, Cassin's Kingbird, Violet-green Swallow, Steller's and Scrub Jays, Black-billed Magpie, Pinyon Jay, Red-breasted Nuthatch, Mountain Bluebird, Western Tanager, House Finch, and Brown Towhee. Near Cowles, among the spruces and aspens, are Band-tailed Pigeons, Yellow-bellied Sapsuckers, Olive-sided Flycatchers, Hermit Thrushes, Townsend's Solitaires, Golden-crowned and Ruby-crowned Kinglets, Evening Grosbeaks, Red Crossbills, and Green-tailed Towhees.

Cowles is a starting point for pack trips into the roadless conifer forests and high peaks of Truchas, Pecos, Baldy, and Lake. Above the 8,500-foot level on these mountains watch for the Northern Goshawk, Blue Grouse, Spotted Owl, Williamson's Sapsucker, Northern Three-toed Woodpecker, Hammond's Flycatcher, Olive-sided Flycatcher, Swainson's Thrush (irregular), and Cassin's Finch (irregular). In the high forests on Truchas Peak, Clark's Nutcrackers and Ruby-crowned Kinglets are common, and nesting Gray Jays and Pine Grosbeaks may be expected. Water Pipits and Brown-capped Rosy Finches summer on the tundra.

Some 42 road-miles northeast of Santa Fe, above the Rio Grande Valley, is **Los Alamos** at the base of the Valle Grande, one of the world's largest volcanic craters. A representative sampling of the area's birdlife may be obtained by following State 4, heading west. From its crossing of the Rio Grande (elevation 5,500 feet), this road climbs through pinyon-juniper and ponderosa pine forests to aspens and spruces on the rim of the Valle Grande at 9,000 feet. Along this route five species of *Empidonax* flycatchers are present during summer within a distance of 25 miles, as indicated below. North of the bridge, local swampy areas along the river support Traill's (Willow) Flycatchers. Other birds include the Gambel's Quail, Yellow-billed Cuckoo, Common Screech Owl,

Western Kingbird, Marsh Wren, Gray Catbird, Warbling Vireo, Yellow Warbler, Common Yellowthroat, Yellow-breasted Chat, Summer Tanager (occasional), Black-headed Grosbeak, and Blue Grosbeak.

On the plateau above the river lies extensive pinyon-juniper forest. Exploration to the right of the road is prevented by a boundary fence of the Los Alamos Scientific Laboratory, but most species can be viewed from the roadside. In localities along the right-of-way where pinyon pines predominate, Gray Flycatchers reside from late April to September. Hepatic Tanagers nest among scattered ponderosa pines in the shallow canyon bottoms. More widespread breeding species are the Poor-will, Cassin's Kingbird, Ash-throated Flycatcher, Pinyon Jay, Plain Titmouse, Blue-gray Gnatcatcher, Black-throated Gray Warbler, and Lark Sparrow. In February, and again in early October, flocks of transient Sandhill Cranes fly high over the Valley. During winter, Townsend's Solitaires may be in full song on sunny days.

To reach one especially productive bird-finding locale, turn off State 4 at the entrance to **Bandelier National Monument** (46 square miles), 18 miles from the Rio Grande bridge. Monument headquarters lie in a cottonwood grove on the floor of Frijoles Canyon, 3 miles from the highway. From the picnic area in the cottonwoods upstream from headquarters to a point about 1.0 mile upstream, one of the most numerous summer residents is Hammond's Flycatcher. This species is in turn replaced by the Western Flycatcher farther and higher along the route. In July, southbound Rufous Hummingbirds and a few Calliope Hummingbirds join the locally breeding Broad-tailed Hummingbirds on the flower-covered slopes. Other summer birds are the White-throated Swift, Canyon and Rock Wrens (around cliffs), Virginia's Warbler (on brushy hillsides), Grace's Warbler (ponderosa pines), MacGillivray's Warbler (dense streamside bushes), Lazuli Bunting, and Lesser Goldfinch.

In the **Jemez Mountains,** Dusky Flycatchers dwell in local pockets surrounded by areas inhabited primarily by Hammond's Flycatchers. To reach such a site, turn right at the junction of State 4 and West Jemez Road, 6 miles beyond the Bandelier Monument entrance. After 1.0 mile, turn left onto the marked Valle Canyon truck road. This remains passable by automobile for 0.75 mile at which point it forks. From here, walk up the trail to the right until it levels out in open woods of ponderosa pine with Gambel oak. A little higher and to the west, where there are aspens, is a loose colony of Dusky Flycatchers—about 15 pairs in a typical summer. Wild Turkeys also frequent this area.

Beyond the West Jemez Road junction, State 4 climbs steeply through pine forest with summering Band-tailed Pigeons, Hammond's Flycatchers (widespread), Western Flycatchers (in the more mesic forest areas), Olive-sided Flycatchers, Clark's Nutcrackers, Red-breasted and Pygmy Nuthatches, Solitary Vireos, Western Tanagers, Black-headed Grosbeaks, and Green-tailed Towhees. The road levels out at the top and enters mixed woods of conifers and aspen—summer habitat of Yellow-bellied and Williamson's Sapsuckers, Townsend's Solitaires, Ruby-crowned Kinglets, Warbling Vireos, Orange-crowned and Audubon's Warblers, Evening Grosbeaks, and, sporadically, Cassin's Finches and Red Crossbills. Five miles beyond, at the viewpoint of the Valle Grande, Mountain Bluebirds breed.

SILVER CITY

Bear Mountain | Pinos Altos Mountains (Cherry Creek Canyon; Signal Peak) | Gila River Valley (Cliff, Gila, Mogollon Creek, Bill Evans Lake) | Forest Road 581

At an elevation of 6,000 feet amid pinyons and junipers, and but 30 minutes from the Mexican biota of the Gila Valley, the desert grasslands to the south, or the montane forests of the Pinos Altos Range, Silver City provides an ideal base for bird finding in southwestern New Mexico. During migration seasons, by carefully selecting routes, one may record 120 to 150 bird species in a day. Daily lists in the breeding season approach one hundred with a little effort.

Bear Mountain, northwest of Silver City, supports birds typical of pinyon-juniper woodland and chaparral. It is easily reached, in dry weather, by driving north from town on Alabama Street and crossing US 180. Several miles beyond, where the blacktop surfac-

RUFOUS-SIDED TOWHEE

ing changes to gravel, turn left at a fork just past a cattle guard. The road soon narrows and climbs into the wooded hills of the Gila National Forest. Among the regular breeding birds are the Greater Roadrunner, Poor-will (often on the road at dusk), Red-shafted Flicker, Western and Cassin's Kingbirds, Ash-throated Flycatcher, Western Pewee, Scrub Jay, Gray-breasted Jay (among oaks), Plain Titmouse, Northern Raven, Bewick's Wren, Solitary Vireo, Black-throated Gray Warbler, Scott's Oriole (near tall yuccas), Hepatic Tanager, Black-headed Grosbeak, Rufous-sided Towhee, and Chipping Sparrow. Rufous-crowned Sparrows inhabit the rimrock beyond the Allen Springs turnoff (Forest Trail 858). They respond well to 'pishing' or 'squeaking.' Less responsive Crissal Thrashers and Black-chinned Sparrows dwell shyly in dense brush near the Continental Divide, usually indicated by a sign. They are best found by following up their songs. Woodland and chaparral extend for several miles to grassy L S Mesa, where Montezuma Quail occasionally are encountered—most often in late summer and fall. Gambel's Quail also breed here.

The highly productive **Pinos Altos Mountains** are reached by following State 25 north from US 180 near the northeastern corner of Silver City's business district. Around Pinos Altos, 8 miles distant and 1,000 feet higher, birdlife is noticeably different from that around Silver City. Northern Mockingbirds and Brown Towhees become scarce, Gray-breasted Jays replace Scrub Jays, and, in summer, Grace's Warblers sing from the tall pines. Montezuma Quail, Magnificent Hummingbirds, and Painted Redstarts dwell in some of the small side canyons above the town. The Western Flycatcher and Solitary and Warbling Vireos are fairly common, and the Hutton's Vireo, although rare, is probably a regular summer resident. Most of these species, and many others, inhabit **Cherry Creek Canyon** farther along the road. Flanking the narrow creek is a wooded fringe of narrowleaf cottonwood, boxelder, velvet ash, Arizona alder, and wild cherry. Dense stands of New Mexico locust, various oaks, and lower shrubs occupy openings on the conifer-forested slopes. Red-faced Warblers and Painted Redstarts often are the most numerous parulids here; Virginia's and Grace's Warblers are fairly common as well. All nest near Cherry Creek and MacMillan Campgrounds and in some side canyons.

Throughout the area, loud descending songs of Canyon Wrens and chattering screams of White-throated Swifts echo from the vertical rock faces which tower above the pines. Northern Ravens, Red-tailed Hawks, and, more rarely, a Zone-tailed Hawk or Golden Eagle may pass overhead. Broad-tailed Hummingbirds, Violet-green Swallows, House Wrens, and Gray-headed Juncos are regularly present along the creek. Acorn Woodpeckers and Purple Martins nest in tall, dead pines, and Flammulated Screech Owls reside on the canyonsides. Spotted Owls formerly bred here but increasing disturbance has forced them to more secluded places away from the roads, although they still visit the Canyon from time to time. Northern Pygmy Owls are widespread, sometimes calling during the day—and sounding rather like the ubiquitous cliff chipmunks. The songs of Whip-poor-wills and Hermit Thrushes enliven the evening hours in spring and summer.

Three miles beyond Cherry Creek Campground, Forest Road 154 leads to the right, winding up the north side of 9,000-foot **Bignal Peak.** With care, all but the lowest cars can safely navigate this rough route unless it is muddy or snow-packed. About 2 miles from State 25, patches of aspens mark an area where Flammulated Screech Owls and, rarely, Williamson's Sapsuckers nest. (Bird finders should refrain from pounding on suspected 'owl trees.' An incubating or brooding Flammulated Screech Owl can be 'squeaked up' to its entrance hole with no disturbing effects. A breeding owl also spends much time perched just within the cavity entrance, peering out, when the nest contains young.) Wherever Douglas-fir grows alongside ponderosa pine, Olive Warblers breed in small numbers. Their young are fledged by early June, and by July they are associating with the roving bands of Pygmy Nuthatches and Audubon's Warblers which haunt these groves. They remain very late into autumn. A few Coues' Flycatchers and Orange-crowned Warblers summer here too. During winter, Red-breasted Nuthatches, Clark's Nutcrackers, Golden-crowned Kinglets, and Red Crossbills visit Signal Peak. The nuthatches and crossbills sometimes remain to breed.

About 11 miles beyond the Signal Peak road, State 25 descends to Sapello Creek, itself pleasant for bird finding away from the highway. Just beyond the crossing, State 25 divides, its left branch leading toward the Gila Cliff Dwellings National Monument, the right to man-made Roberts Lake. Either way, an influx of visitors may be expected on summer holidays and weekends, but in minutes one can leave them behind by wandering up any of the rarely visited side canyons. Birds around the Cliff Dwellings differ little from those about Cherry Creek, except for an occasional Lesser Black Hawk or Belted Kingfisher along the upper Gila River. Bald Eagles are present from November through April, and Montezuma Quail erratically visit the adjacent grassy slopes and oak-covered canyonsides. A Northern Goshawk or a flock of Wild Turkeys is always a possibility along this road.

Roberts Lake supports some transient waterbirds, including Western Grebes, but few breeding species. Pinyon Jays may be present at any season among the pinyons and junipers, although they are scarce during summer. The blacktop road, here labeled State 61, continues south along the Mimbres River Valley at an average elevation of 6,000 feet. Near San Lorenzo, it joins State 90 which turns west (right) to Santa Rita and Silver City. This loop drive from Silver City via Cherry Creek Canyon and return through the Mimbres Valley covers only 80 miles, but satisfactory coverage for bird finding can easily occupy a full day. The area is scenic and ornithologically rich, the road in places narrow and tortuous. It is unwise to hurry through this country.

The **Gila River Valley,** from Redrock upstream to the vicinity of

Cliff and Gila, constitutes one of the Southwest's finest bird-finding areas. More than 280 species have been recorded in the Valley itself, and 102 are known to nest. Others occupy the adjacent uplands. Annual early-May bird counts record 140 to 170 species in a 24-hour period, and Christmas count lists top ninety species. Although much of the Valley is privately owned, ample portions are not posted against trespassing. Some sections lie within the Gila National Forest. Visitors can see a great deal from the roads, and most landowners readily grant bird finders permission to enter their property when such is requested.

A traditionally rewarding area near **Cliff** and **Gila** begins where the road between these towns crosses the Gila River. From Silver City, drive northwest on US 180 for 29 miles, turning right on State 211 to Gila. Bear left here, continuing to the bridge. One may park along the road and investigate on foot the riverside thickets. Remnant cottonwood stands exist about a mile upstream from the bridge, and scattered clumps or single trees stand near the road. Along this stretch of the River Lesser Black Hawks are often seen. Unlike the Zone-tailed Hawk, which feeds largely over the dry hills and plains away from the River, this species is almost always in or over the riparian trees. Another worthwhile site adjoins the confluence of the Gila River and **Mogollon Creek,** 2 miles west and 7 miles north of the bridge. En route, check the roadside thickets for passerines, flooded fields for waders, and any conspicuous tree cavities for dozing Barn Owls. Migrating White-throated Swifts and Violet-green Swallows by the hundreds hawk insects over the fields and River. At the Gila National Forest boundary the blacktop road changes to gravel. Leaving the Valley's farmlands, it passes through cactus and mesquite before descending steeply to sycamore-lined Mogollon Creek (often dry), which enters from the northwest. Part of this area is privately owned and fenced, but much is public land where the bird finder may seek Wied's Crested and Ash-throated Flycatchers, Bridled Titmice, and Lucy's Warblers in the sycamore groves. Elf Owls are vociferous here on spring evenings. Where oaks interdigitate with the true riparian trees, Acorn Woodpeckers and Gray-breasted Jays breed. In summer, Band-tailed Pigeons fly down from the mountains to feast on acorns and hackberries. Bronzed Cowbirds extend upstream to this point, but they are more numerous farther south. Golden Eagles pass overhead rather often, and transient Ospreys and Prairie Falcons should be watched for. Resident Northern Ravens nest on the cliffs and in cottonwoods, summer-resident White-necked Ravens fly in flocks from nearby desert grasslands to water at the River, and American Crows are winter visitants, sometimes in large flocks. The tree squirrel here is the Arizona gray squirrel, a scarce animal in the Valley.

Two miles east of the US 180 bridge, 3.7 miles from Cliff on the main highway to Silver City, a well-marked gravel road (State 809) branches to the right and follows the river downstream. A sign at the intersection may indicate **Bill Evans Lake,** an artificial impoundment created by the Phelps-Dodge Corporation to store water from the Gila for copper-mining operations at nearby Tyrone. The Lake is leased by the New Mexico Game and Fish Department for recreation purposes, and on weekends and holidays it attracts many fishermen. Between these times it provides a resting place for such transient birds as the Common Loon, Horned, Eared, and Western Grebes, American White Pelicans, Double-creasted and Olivaceous Cormorants, numerous ducks, occasional shorebirds and gulls. A scope is essential for critical viewing. Evans Lake is 1.0 mile from State 809 on a conspicuous, short side road, branching left, 4.5 miles from US 180. En route from the highway, frequent stops should disclose many of the typical Gila Valley birds including Lesser Black Hawks. After scanning the Lake, return to State 809 and turn left. Two miles beyond, begins a section of the Gila National Forest near the boundary of which the passable road now terminates. Walking downstream along the old roadbed for 3 or 4 miles provides an opportunity to see most of the characteristic Gila Valley species. The Zone-tailed Hawk is perhaps more likely to be seen here than elsewhere in the Valley. Bell's Vireo is locally common in dense stands of seepwillow along the riverbanks. Small canyons draining into the Gila are inhabited by the Plain Titmouse, Rock Wren, Phainopepla, Scott's Oriole, and Rufous-crowned Sparrow. Bird finding here is always good. The first two weeks of May produce the most species as transients are moving through in numbers and many summer-resident birds have just returned. Lesser Black Hawks are on their breeding territories as early as late February, but Yellow-billed Cuckoos first appear in mid- or late May. Some transient passerines linger into June.

Although the lower Gila Valley near Redrock may be reached via the 22-mile paved State 464 from Lordsburg, a 'backroad' through the low Burro Mountains is much better for bird finding. Drive northwest from Silver City on US 180, turning left after 14 miles onto a hard-surfaced road leading toward the Phelps-Dodge copper mine. About 4.8 miles from US 180, a sign reading 'Redrock' indicates the beginning of **Forest Road 581** branching to the right. From here it is 24 miles to State 464. Upon reaching it, turn right; Redrock is 5 miles ahead. FR 581 is an earth road, normally well maintained but subject to choking by sand following storms. It should be avoided by vehicles with very low clearance, and driven with extra care at all times during wet seasons. It begins in cactus flats with Cactus Wrens and Curve-billed Thrashers, winds through pinyon-juniper woodland and chaparral with Scrub Jays, Plain Titmice, and Crissal Thrashers, then traverses pine and oak woods where the usual birds are Acorn Woodpeckers, Gray-breasted Jays, Bridled Titmice, Pygmy Nuthatches, and Hepatic Tanagers. It finally descends through chaparral inhabited by Crissal Thrashers and Black-chinned Sparrows. The last few miles are through sparsely vegetated plains with Bendire's Thrashers and Scott's Orioles in summer, Sage Thrashers, and Sage and Brewer's Sparrows in winter.

SOCORRO

Bosque del Apache National Wildlife Refuge | Magdalena | San Augustin Plains

The 57,191-acre **Bosque del Apache National Wildlife Refuge,** 20 miles south of Socorro, is one of New Mexico's stellar bird-finding areas. Some 5,000 of the 13,500 acres of Rio Grande bottomland in the Refuge have been modified into wildlife habitat units, and 1,500 acres are cultivated to provide food for the hordes of wintering waterfowl and Sandhill Cranes. Some tracts are flooded during autumn for waterfowl use. Apart from small areas of permanent cattail-bulrush marsh, the undeveloped bottomlands support stands of cottonwood, tamarisk, and screw bean. The remainder of the Refuge is dominated by desert vegetation with a few junipers scattered over the higher western hills.

Since establishment of the Refuge in 1939, 280 species of birds have been recorded and nearly 100 have nested here; ninety are considered permanent residents. Probably the Mexican Duck is more easily seen here than elsewhere in the state. Other New Mexican rarities are the Olivaceous Cormorant, a visitant from spring through fall, and Ross' Goose, present each winter. In October, thousands of Canada and Snow Geese begin to arrive, as do Sandhill Cranes and many ducks. Also wintering here are the Whooping Cranes hatched by foster-parent Sandhill Cranes in Idaho (*see under* **Soda Springs, Idaho**). Excluding accidentals, twenty-five species of waterfowl visit the Refuge. Their numbers peak from November to mid-January, but the September–October migration usually will disclose more species. Crane numbers peak in mid-December. By late February, numbers of all winter visitants have greatly declined, and March witnesses arrival of transient ducks and the first shorebirds. The cold-weather abundance of waterfowl, cranes, and hawks may obscure the presence of rare winter passerines such as the Northern Shrike and American Tree Sparrow.

Visitors approaching the Refuge via I 25 from the south should take the San Marcial Exit and follow the signs 10 miles to Refuge headquarters on the left. Driving south from Socorro, proceed 8 miles on I 25 and take the San Antonio Exit east for 1.0 mile to San Antonio; turn off right (south) and go along US 85 for about 7 miles to headquarters.

Cattail stands near the south part of the self-guided bottomland tour route shelter Least and American Bitterns, rails, and Marsh Wrens. Near the midpoint of the route are habitats for brush-inhabiting species such as Crissal Thrasher, Yellow-breasted Chat, Blue Grosbeak, and Brown Towhee. Typical breeding birds of the cottonwood bosques are Red-shafted Flicker, Western Pewee, Bewick's Wren, Bullock's Oriole, and Summer Tanager. Open fields along the northern edge of the tour route provide the best vantage for observing the spectacular flights of geese and cranes streaming into the Refuge on winter evenings. At sunrise begins the exodus to feeding grounds in the closed northern part of the Refuge and beyond. Many cranes fly well beyond San Antonio. A paved road branching north from the highway about 0.6 mile east of that village passes grainfields where the foraging birds often are visible at close range. Buteonine hawks (Red-tailed, Rough-legged, and Ferruginous), Northern Harriers, Greater Roadrunners, and other species may be seen to advantage here as on the Refuge.

About 27 miles west of Socorro along US 60 is **Magdalena** (elevation 6,548 feet). Near mileage marker 99, about 12 miles west of the town, a dirt road branches north (right) across a cattle guard. Follow this, turning left at the first fork, to reach open pinyon-juniper stands and, after about 2.5 miles, some stock watering tanks, which lure Pinyon Jays, Western and Mountain Bluebirds, Scott's Orioles, Hepatic Tanagers, and other woodland species during the breeding season. Farther west along US 60 lie the high, open **San Augustin Plains.** In short-grass areas east of Datil, watch for the very rare Mountain Plover which may still breed there. Bendire's and Sage Thrashers and a few Brewer's Sparrows nest in the saltbush flats where US 60 intersects the Sierra-Catron County line. This is excellent hawk country in spring and fall.

TRUTH OR CONSEQUENCES

Elephant Butte Reservoir | Percha Dam State Park | Black Range

Still known to many New Mexico residents as Hot Springs, 'T or C' is a resort town along the Rio Grande near **Elephant Butte Reservoir.** This impoundment is frequented by wintering and migrating waterbirds, but it is of little or no importance for nesting owing to its rocky shoreline and the pressure of summer recreationists. A marsh north of the main impoundment is a nesting site for Double-crested and Olivaceous Cormorants as well as Great and Snowy Egrets. However, access is controlled by the Bureau of Land Management (Box 1456, Socorro, NM 87801), and visits are not encouraged. These species are more readily seen at the Bosque del Apache National Wildlife Refuge (*see under* **Socorro**). During the colder months, winter birds on the Reservoir include the Common Loon, Eared and Western Grebes, Great Blue Heron, various waterfowl, Bald Eagle, and Ring-billed Gull. There is greater variety in spring and fall. To reach the Reservoir, take Exit 79 from I 25 on the north side of T or C and follow the signs. The park and several overlook points are about 3 miles from the Interstate. A scope is essential for satisfactory viewing.

When the Rio Grande was periodically inundated by flooding, it supported lush riparian vegetation dominated by cottonwoods and willows. Today the river is so channeled, dammed, and otherwise manhandled that few riparian groves exist. The best extant example in south-central New Mexico remains at **Percha Dam State Park** about 16 miles south of T or C. From I 25 it is reached via the Caballo Dam-Percha Dam Exit. After leaving the Interstate, drive south on US 85 for about 1.0 mile to the Park turnoff on the left. The Park itself is an open cottonwood grove with the underbrush removed for picnic tables, but beyond the fence to the south is a good bosque with dense undergrowth. This is on Bureau of Reclamation property and is open to the public. The Ladder-backed Woodpecker, Verdin, Crissal Thrasher, Phainopepla, and Pyrrhuloxia are among the resident birds. In winter, flocks of Sandhill Cranes and Ring-billed Gulls pass at dawn, and ducks of several species rest on the water. Olivaceous Cormorants occasionally stop here during migration. In summer, expect Western Kingbirds, Ash-throated Flycatchers, Yellow Warblers, Yellow-breasted Chats, Summer Tanagers, and Black-headed Grosbeaks.

The mountain forests of the nearby **Black Range** provide a contrast with lowland habitats along the Rio Grande. They are best reached via State 90, turning west from I 25 near Caballo and about 13 miles south of T or C, where signs direct motorists to Silver City. This fine blacktop road leads across creosote bush desert, climbs rapidly through wooded foothills and into the Gila National Forest. The coniferous and riparian growth, such as that near Iron Creek Campground in Gallinas Canyon, provides ideal habitat for montane birds. Olive Warblers have summered above Railroad Campground, on the right, about 0.5 mile beyond Iron Creek. Four miles farther is secluded Gallinas Campground along a small clear creek flowing through a picturesque canyon. Breeding species here are those characteristic of the Pinos Altos and Mogollon Mountains (*see the introduction to this chapter*).

North Dakota

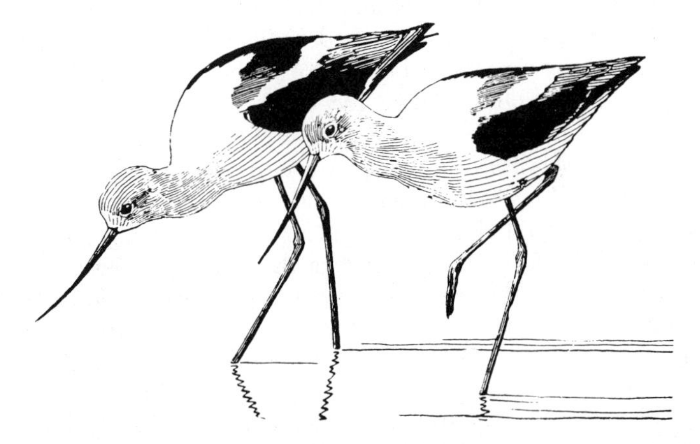

AMERICAN AVOCETS

There are numerous areas in North Dakota where the prairie dips slightly, forming shallow basins or potholes. Water collects here and becomes highly alkaline. On the shores of these basins in spring American Avocets assemble. At first there is great activity: courtship antics, during which the birds lift their wings, lower their heads, and trip along liltingly in the shallow water, pursuing and circling one another; mild contests as each pair—the mating bond once established—claims a section of shoreland for private use as nesting territory. Before long the hubbub is over, for the colony is organized and nesting is under way. If the bird finder approaches such an aggregation soon after the eggs have been laid, he is greeted by the unexpected. Instead of fleeing in panic, as is the habit of most colonial birds, the avocets hold their ground and present forthwith a series of deflection displays ('injury-feigning' performances) that are as remarkably fantastic as they are multifarious. Wing-waving and wing-trailing; careening and falling; bounding ahead on stiffened legs; taking off in erratic, zigzagging flights, and then abruptly 'crash-landing'; flopping on the ground and floundering in the water as if hopelessly maimed—all to the accompaniment of shrill staccato calls and heart-rending moans. To see this avian show, the bird finder need only visit in June one of the alkaline lakes in the Lostwood National Wildlife Refuge (*see under* **Kenmare**) or similar localities elsewhere in North Dakota.

Big skies and rolling plains. Treeless horizons with the outlines of grain elevators marking the locations of towns. Brief springs and warm summers with cool nights. Unceasing winds and ever-moving tumbleweeds. Jack rabbits bounding across wide stretches of grasslands, 'flickertails' (Richardson's ground squirrels) scurrying about or sitting 'at attention' on heavily grazed grassland, and an occasional coyote disappearing over a swell. Birds in abundance. Such are the first impressions of the visiting bird finder in North Dakota, the northernmost state that lies entirely within the interior plains.

The land surface of the state is composed primarily of three plains, which, like three great steps, rise one above the other from Minnesota westward to Montana. The lowermost, with a minimum elevation of 790 feet, is the level Red River Valley bordering the Red River of the North on the eastern boundary. Properly called the Agassiz Lake Plain, its breadth in North Dakota varies from about 40 miles near the Canadian line to 10 miles toward the South Dakota line; comparatively speaking, it is a small part of the state. Birdlife is characteristically eastern, represented by such breeding birds as the Yellow-shafted Flicker, Great Crested Flycatcher, Baltimore Oriole, Scarlet Tanager, and Indigo Bunting.

The next plain, the Drift Plain, rises gradually from the western edge of the Agassiz Lake Plain and extends westward, as gently undulating to hilly country, to meet the Missouri Plateau, whose eastern edge, the Missouri Coteau—recognized by a succession of hills and hummocks—passes diagonally through the state in a northwest-southeast direction. Due to extensive knob and kettle topography, the Missouri Coteau contains unusually large numbers of potholes that attract many aquatic breeding birds.

General elevations of the Drift Plain range from 1,500 to 1,800 feet. This large part of North Dakota is significant ornithologically because it shows a marked transition between eastern and western avifaunas. Moreover, it has several physiographic features that are attractive to particular kinds of birds. Northward, straddling the International boundary, are the Turtle Mountains (maximum elevation 2,321 feet), with timbered tracts where a few species of northern affinities nest, among them the Philadelphia Vireo. To the northwest lies the Souris River Valley, where there formerly existed some of the finest marshes on the continent. Drained many years ago for agricultural projects that proved unsuccessful, certain of them were restored by the U.S. Fish and Wildlife Service as National Wildlife Refuges (for example, *see under* **Kenmare** *and* **Towner**). Thanks to the subsequent management and protection afforded them, they are probably more attractive to wildlife now than before they were destroyed. Today they are unquestionably among the best sites in the United States for seeing grebes and ducks in exceptionally large numbers, and bird finding in their immediate vicinities—the grassy lowlands, wooded coulees (trench-like ravines), and adjacent prairie uplands—is as rewarding as anywhere in the state. Much of the Drift Plain from north to south is dotted with shallow lakes, ponds, potholes, and sloughs; owing to the naturally poor drainage of the area, very few contain running water except during wet seasons. Some have become decidedly alkaline and consequently invite American Avocet colonies and a few pairs of Piping Plovers to their shores. Nearly all these bodies of water, if they do not go dry by midsummer, have small populations of breeding waterfowl.

The Missouri Plateau, occupying the remaining half of the state, is a part of the Great Plains, with general elevations from 1,800 to 2,700 feet. Cutting through the area is the broad valley of the Missouri River, which is joined from the west by the valleys of the Little Missouri, Knife, Heart, and Cannonball Rivers. On the west side of the Missouri River, the topography of the Missouri Plateau is characterized by rolling surface and by abrupt buttes which increase in size and number toward the southwestern corner where White Butte rises to an elevation of 3,506 feet, the highest point in the state. The Plateau is further characterized by many wide valleys in which erosion has cut amazing formations, exposing layers of many-colored soils. The famous Badlands along the Little Missouri (*see under* **Belfield** *and* **Bowman**) is the best example. Here is the most spectacular scenery; here also is an area for bird finding as good as any in extreme western North Dakota. Despite the semiarid conditions, the avifauna shows a surprising variety of birds. The Golden Eagle, Red-shafted Flicker, Say's Phoebe, Rock Wren, Mountain Bluebird, Bullock's Oriole, Black-headed Grosbeak, and Lazuli Bunting are only a few of the typical species breeding regularly.

Wooded areas in North Dakota are limited to the borders of watercourses, to sheltered coulees, and to a few slopes such as those in the Turtle Mountains. Aspen, cottonwood, oak, ash, elm, birch, basswood, and boxelder are among the dominant trees; west of the Missouri River, especially in the Little Missouri Valley, Rocky Mountain cedar is common. Various shrubs—e.g., buffaloberry, wolfberry, silverberry, and wild rose—grow extensively on the edges of wooded tracts or by themselves in draws (shallow ravines).

Most of the state was originally prairie grassland, the fertile, better-watered soils of the Agassiz Lake Plain producing the tall herbaceous vegetation of the tall-grass prairie, and the less fertile, drier soils of the remainder of the state producing the shorter vegetation of the mixed-grass prairie. Now the Agassiz Lake and Drift Plains show an almost continuous succession of grainfields with intermittent pastures and haylands and, at regular intervals, a cluster of farm buildings partly surrounded by groves of trees that have been introduced as wind breaks. Because of rougher topography and lower rainfall over the Missouri Plateau, much of the land can be more profitably used for grazing than for crop production; hence the original grassland conditions have been less modified. As far as the eye can see there are vast stretches of mixed-grass prairie. Here prairie birds nest over wide areas, whereas the same species east of the Plateau are restricted mainly to small tracts of prairie and neglected fields, and other uncultivated areas.

The following birds breed more or less regularly throughout the state in the wooded areas and in the open country (prairie grasslands, pastures, fallow fields, meadow-like lowlands, woodland edges, brushy places, and dooryards):

WOODED AREAS

Black-billed Cuckoo
Yellow-shafted Flicker
Hairy Woodpecker
Downy Woodpecker
Least Flycatcher
Blue Jay
Black-capped Chickadee
White-breasted Nuthatch
Red-eyed Vireo
Warbling Vireo
American Redstart

OPEN COUNTRY

Northern Harrier
Sharp-tailed Grouse
Upland Sandpiper
Marbled Godwit (*near wetlands*)
Mourning Dove
Burrowing Owl
Short-eared Owl
Red-headed Woodpecker
Eastern Kingbird
Western Kingbird
Horned Lark
Tree Swallow (*northeastern North Dakota*)
Barn Swallow
Black-billed Magpie (*western North Dakota*)
House Wren
Gray Catbird
Brown Thrasher
Eastern Bluebird (*eastern North Dakota*)
Sprague's Pipit (*except Agassiz Lake Plain*)
Loggerhead Shrike
Yellow Warbler
Common Yellowthroat
Bobolink
Western Meadowlark
Common Grackle
American Goldfinch
Rufous-sided Towhee
Lark Bunting
Savannah Sparrow
Grasshopper Sparrow
Baird's Sparrow (*except Agassiz Lake Plain*)
Vesper Sparrow
Chipping Sparrow
Clay-colored Sparrow
Song Sparrow
Chestnut-collared Longspur

US 2 and I 94, crossing the state in an east-west direction, pass over all three plains, thus presenting a satisfying panorama of the North Dakota landscape. Both highways are remarkably straight, and there are few wayside obstructions to mar the tremendous sweep of countryside. It is, therefore, an easy matter to see many kinds of birds from the car. In late spring and summer one is almost certain to notice, either in flight or perched on poles, fence posts, and wires, Swainson's and Ferruginous Hawks, Northern Harriers, Short-eared Owls, Common Nighthawks, both Eastern and Western Kingbirds, Black-billed Magpies (western North Dakota), Loggerhead Shrikes, and Western Meadowlarks; and there is always a good chance of sighting one or two Burrowing Owls, Upland Sandpipers, and Marbled Godwits. Before the oncoming car, Horned Larks, Vesper Sparrows, and, not infrequently, Chestnut-collared Longspurs fly up from roadsides where they have been searching for food. During most years, Lark Buntings are common, sometimes abundant, along highways through the western two-thirds of the state, and occasionally Dickcissels appear in fair numbers in the eastern two-thirds of the state.

When driving over US 2 or I 94, the bird finder should leave the highway at widely separated points to investigate adjacent wooded coulees and wooded borders of streams and thus see for himself the range limits of several closely allied eastern and western birds. How far west can he find the Baltimore Oriole, Rose-breasted Grosbeak, and Indigo Bunting; and how far east their western counterparts, the Bullock's Oriole, Black-headed Grosbeak and Lazuli Bunting, respectively? In certain spots he may find both orioles, or both grosbeaks, or both buntings, in which case he is also quite likely to observe hybrids.

Waterfowl, shorebirds, and Sandhill Cranes tend to attract the greatest attention during the migration seasons because they appear in such huge numbers. Good-sized flocks of Whistling Swans are regular on the larger brackish lakes, particularly on the Missouri Coteau during fall flight. Canada and Snow Geese are especially abundant on the lakes in eastern North Dakota, ducks and shorebirds on the lakes, ponds, and sloughs throughout the state. The commonest shorebirds are the Semipalmated Plover, Common Snipe, Willet, Greater and Lesser Yellowlegs, Pectoral Sandpiper, White-rumped Sandpiper, Baird's Sandpiper, Least Sandpiper, Long-billed Dowitcher, Stilt Sandpiper, Semipalmated Sandpiper, Marbled Godwit, American Avocet, Wilson's Phalarope, and Northern Phalarope. Tens of thousands of Sandhill Cranes stop in the state while journeying to and from their Arctic breeding grounds. The largest concentrations are in south-central North Dakota (*see under* **Moffit**). Of the many passerine species that have the status of transients in North Dakota, the following are considered common to abundant in suitable environments throughout the state: Swainson's Thrush, Gray-cheeked Thrush, Water Pipit (particularly abundant in fall), Tennessee Warbler, Orange-crowned Warbler, Myrtle Warbler, Blackpoll Warbler, Slate-colored Junco, American Tree Sparrow, Harris' Sparrow, White-crowned Sparrow, White-throated Sparrow, Lincoln's Sparrow, and Lapland Longspur. In North Dakota, the main migration flights may be expected within the following dates:

Waterfowl: 5 April–5 May; 15 September–10 November
Shorebirds: 1 May–1 June; 20 July–20 September
Landbirds: 25 April–1 June: 15 August–15 October

Only the irrepressible bird finder cares to pursue his avocation in North Dakota in winter for the rewards are meager when weighed against the low, frequently sub-zero, temperatures that he must face. All the watercourses are frozen over 'for keeps,' thereby restricting his efforts entirely to landbirds. Aside from permanent residents, the species he is most likely to see are the Bohemian Waxwing, Northern Shrike, Common Redpoll, and Snow Bunting. Possibly he will note the American Tree Sparrow and Lapland Longspur, but the majority of these hardy species and others prefer the few additional degrees of warmth in the states immediately to the south.

Authorities

Robert T. and Ann M. Gammell, Robert E. Stewart.

References

Breeding Birds of North Dakota. By Robert E. Stewart. Tri-College Center for Environmental Studies, Stevens Hall, North Dakota State University, Fargo, ND 58102. 1975.

A Birder's Guide to North Dakota. By Kevin J. Zimmer. Distributed by L & P Press, Box 21604, Denver, CO 80221. 1979.

BELFIELD
Theodore Roosevelt National Park

To the footsore French trappers of bygone years, the country along the Little Missouri in the extreme west-central part of North Dakota was 'les mauvaises terres à traverser' (bad lands to travel through). However disparaging the designation, it has persisted, but in reality the Badlands of North Dakota are weirdly beautiful, not dreary or barren as might be supposed from the appellation.

The beauty of the country is due in no small measure to the extraordinary formations that mark its surface—the bizarrely shaped buttes and tablelands, their steep slopes stark and bare, showing multiple layers of variously tinted sandstone and clay capped by brick-red scoria. The beauty is considerably enhanced by diversified vegetation. Rocky Mountain cedar and shrubs such as creeping juniper, skunk sumac, buffaloberry, chokecherry, hawthorn, currant, wild rose, and serviceberry grow in cuts and draws; cottonwoods, aspens, ashes, and bur oaks along the streams and on river bottomlands. Grasses, sagebrush, rabbitbrush, and occasionally yucca and cactus grow on the floors of the wider valleys and on the tablelands. Nearly all localities having a suitable soil produce pasqueflowers, larkspurs, arnicas, and other wildflowers. In spring and early summer the blooming of the shrubs, cacti, and wildflowers creates a floral display that rivals in gaiety the blues, yellows, pinks, and reds of the earth formations.

On first entering the Badlands in the late spring and summer, the bird finder will observe a few American Kestrels, Common Nighthawks, Black-billed Magpies, Mountain Bluebirds, Loggerhead Shrikes, and perhaps one or two Turkey Vultures, Ferruginous Hawks, Golden Eagles, and Prairie Falcons. But most birds are less wide-ranging and conspicuous and must be looked for in suitable situations. Horned Larks occupy the level portions with sparse vegetation, Western Meadowlarks and Vesper Sparrows the grassy stretches, Lark Buntings the extensive sagebrush flats, Lark and Field Sparrows the slopes with scattered brush, and Say's Phoebes and Rock Wrens the vicinity of sharply cut banks of badlands or steep slopes of buttes. By far the greatest number of birds inhabit the widely spaced areas grown to bushes and trees;

the more extensive and dense the cover, the more abundant the birds. Some of the regular summer-resident species are the Gray Catbird, Brown Thrasher, Red-eyed Vireo, Yellow Warbler, Yellow-breasted Chat, Bullock's Oriole, Black-headed Grosbeak, Lazuli Bunting, and Rufous-sided Towhee, but occasionally the bird finder will meet other birds—e.g., the Least Flycatcher, Veery, Black-and-white Warbler, Ovenbird, Baltimore Oriole, Rose-breasted Grosbeak, and Indigo Bunting. Flickers, invariably common, are intermediate between the Yellow-shafted and Red-shafted forms.

Theodore Roosevelt, the outdoors-loving, conservation-minded twenty-sixth President of the United States, became so enamored of the Badlands that he chose to establish his ranch, the Elkhorn, in the immediate vicinity. Later, near the end of his life, he wrote, 'I have always said I never would have been President if it had not been for my experiences in North Dakota.' It is wholly fitting that Roosevelt's enduring contributions toward the conservation of the nation's resources should be commemorated by the **Theodore Roosevelt National Park,** which embraces country that so greatly appealed to him.

The Park consists of 110 square miles of federally owned land in three separate units: one, the *South Unit,* reached by exiting north from I 94, 16 miles west of Belfield; a second, the *North Unit,* reached by turning west off US 85, about 55 miles north of Belfield; and a third, the *Elkhorn Ranch Unit,* a small area about midway between the two, on the Little Missouri. Park headquarters is in the South Unit at Medora.

Both the North and South Units have spectacular scenery, but the North Unit is better for bird finding. It has many wooded spots and, because it is served by rough roads that are sometimes impassable in wet weather, it is little disturbed by tourists. After passing through the North Unit gate, the road winds along the base of the buttes rimming the Little Missouri to the Squaw Creek picnic grounds, in a well-forested area particularly worth investigating for birds. From here the road goes up Cedar Canyon, becoming steeper and with more sharp turns, and leads on and up to Oxbow Overlook, providing one of the most magnificent views the entire Park has to offer. Several canyons with a dense growth of cedars and valleys with sagebrush—all visible from the Point—are excellent for birds, but the bird finder will have to work his way down to them, since there are no trails.

US 85, which passes through Belfield, traverses many miles of rolling prairie both north and south of town. Where the highway crosses streams, Cliff Swallows often have colonies in the concrete culverts below. Incubation is normally under way by the first of June. An investigation of the prairie where the grasses are sufficiently dense should turn up Sprague's Pipits, Baird's Sparrows, and Chestnut-collard Longspurs. Lark Buntings are common—sometimes abundant—along and back from the shoulders of the highway. Other species to look for are the Upland Sandpiper, Burrowing Owl, Eastern and Western Kingbirds, Eastern Bluebird, and Bobolink.

BOWMAN
Southwestern Badlands

In the southwestern corner of North Dakota the **Southwestern Badlands** along the Little Missouri have associations of breeding birds unique to the state.

Where there are extensive stretches of sagebrush—largely black sage intermixed with silver sage—Sage Grouse, Lark Buntings, and Brewer's Sparrows are predominant. On the short-grass prairie of tablelands and gently sloping terrain that adjoin the Badlands are Long-billed Curlews and McCown's Longspurs along with the more common Horned Larks, Western Meadowlarks, and Chestnut-collared Longspurs. US 12, west from Bowman for 33 miles to the Montana border, traverses the Southwestern Badlands and areas where the above birds may be observed. In and around Marmarth on US 12, 27 miles from Bowman, are semi-open stands of cottonwood and shrubby thickets which are the habitat for Red-shafted Flickers, Red-headed Woodpeckers, Western Pewees, Black-billed Magpies, Orchard and Bullock's Orioles, Black-headed Grosbeaks, Lazuli Buntings, and Rufous-sided Towhees.

Here and there in the Southwestern Badlands are scattered stands of ponderosa pine that hold an association of birds which include the Sharp-shinned Hawk, Merlin, Poor-will, Red-shafted Flicker, Say's Phoebe, Black-billed Magpie, Rock Wren, Red-breasted Nuthatch, and Audubon's Warbler. A particularly fine stand covering several hundred acres with species representing the above association may be reached from Bowman by driving north for 22 miles to a T-crossroad. Here, instead of following US 85 eastward toward Amidon, turn west onto a gravel road and go 1.0 mile to Burning Coal Vein Road (marked by a sign) and follow it north and northwest for about 8 miles to the area.

JAMESTOWN

Arrowwood National Wildlife Refuge | Chase Lake National Wildlife Refuge

One of the best spots in eastern North Dakota for breeding and transient waterfowl is the **Arrowwood National Wildlife Refuge** (15,934 acres). To reach headquarters at the Refuge, drive north from Jamestown on US 52-281 for 27 miles to Edmunds, then turn off east onto an unpaved road and go 6 miles.

Extending along the James River for 16 miles, the Refuge has several lakes—notably Arrowwood (1,600 acres), Mud (470 acres), and Jim (860 acres)—and DePuy Marsh (360 acres), all formed by damming the River. The vegetation in DePuy Marsh and in marshy areas around the lakes consists principally of cord grasses, bulrushes, smartweeds, and—especially in Arrowwood Lake—an abundance of sago pondweed. A small amount of timber borders Arrowwood and Jim Lakes, stands in coulees and ravines, and forms shelterbelts. About 900 acres of the Refuge are farmed for grain, thus furnishing attractive feeding grounds for waterfowl and upland birds.

Ducks nesting commonly are the Mallard, Gadwall, Common Pintail, Blue-winged Teal, American Wigeon, Northern Shoveler, and Wood Duck. All have broods by July. Both spring and fall migrations of waterfowl bring large transient populations, but the fall population which is much heavier is well worth seeing when it reaches its peak in October. The birds begin arriving in September, and many remain until freeze-up. Included in the population are large flocks of Canada Geese, which rest on the lakes and feed in the surrounding grainfields.

During summer, great numbers of American White Pelicans commonly appear on the lakes of the Arrowwood Refuge, probably coming from Chase Lake, about 30 miles southwest, where they nest.

Chase Lake is embraced by the **Chase Lake National Wildlife Refuge** (375 acres), administered by the personnel of the Arrowwood Refuge. The pelicans colonize on two islands (4 acres and 1.5 acres, respectively), which rise above water level only slightly and have a sparse cover of grasses and weeds. The number of pelican nests varies from 2,000 to 5,000 each year; as many as 6,000 young have been counted in one year. Also nesting on the islands are a hundred or more pairs each of Double-crested Cormorants, California and Ring-billed Gulls, and Common Terns. All the species have young by late June.

To reach Chase Lake for viewing the islands, drive west from Jamestown on I 94 for 28 miles, exit north at Medina on State 30 for 9.5 miles, then turn off west and proceed 6 miles on a road that, after about 3 miles, becomes little more than a prairie trail.

While driving over the prairie country from the highway to the Lake, be alert for Ferruginous Hawks circling overhead and for Sharp-tailed Grouse near the road. Common in grasslands back from the Lake are Upland Sandpipers, Sprague's Pipits, Baird's Sparrows, and Chestnut-collared Longspurs. American Avocets have small colonies on the shore of the Lake and Le Conte's Sparrows reside in the grassy marsh at the north end of the lake.

KENMARE
Des Lacs National Wildlife Refuge | Lostwood National Wildlife Refuge

The **Des Lacs National Wildlife Refuge** (18,841 acres) in northwestern North Dakota extends for 35 miles, from the Canadian border to 7 miles south of Kenmare. Within the area are Upper, Middle, and Lower Lakes of the Rivière des Lacs, with their peripheral marshes, wooded coulees and draws, and bordering upland. Roughly, the Lakes and their marshes comprise 8,000 acres; farmlands for wheat and other small grains, 2,000 acres; prairie grassland, 7,000 acres; and tree- and brush-covered lands, 2,000 acres. Since there is no sustained flow of water through the Lakes, their size depends on the vagaries of the weather, a situation that has a profound effect on the abundance of birdlife using the Lakes for feeding and nesting purposes.

Chief among the ornithological attractions is the colony of Western Grebes which usually occupies the large marsh at the south end of Upper Lake, where bulrushes, quillreeds, and other emergent plants provide the birds with ample nesting cover, and where there are numerous channels and stretches of open water that allow sufficient space for 'water skimming'—the spectacular element in their courtship displays. The birds begin displaying almost immediately upon their arrival in late May and continue performing through most of June.

Although the Western Grebe colony is the chief ornithological attraction, this marsh and others bring an impressive aggregation of various breeding species: great numbers of Eared and Pied-billed Grebes and usually a few Red-necked Grebes; many American Bitterns and an abundance of Soras, Black Terns, and Yellow-headed Blackbirds as well as Red-winged Blackbirds; a heavy population of ducks—Mallards, Gadwalls, Common Pintails, Green-winged and Blue-winged Teal, American Wigeons, Northern Shovelers, Redheads, Canvasbacks, Lesser Scaups, and Ruddy Ducks. The wet, grassy edges of marshes invariably attract Brewer's Blackbirds and a few nesting pairs of Willets and Wilson's Phalaropes. American White Pelicans and Double-crested Cormorants do not breed on the Refuge, but are nevertheless present throughout the breeding season. Tree, Bank, and Barn Swallows are always flying low over the marshes as they forage for insects.

Elsewhere on the Refuge are many breeding birds. Species commonly inhabiting undisturbed portions of prairie grassland are the Sharp-tailed Grouse, Common Nighthawk (on bare knolls), Horned Lark, Sprague's Pipit, Western Meadowlark, Savannah Sparrow (in low grasslands), Baird's and Vesper Sparrows, and Chestnut-collared Longspur. Less common are the Upland Sandpiper, Marbled Godwit, Burrowing Owl, Bobolink (in low grasslands), and Grasshopper Sparrow. In coulees with stands of cottonwoods, aspens, and other trees, some of the species are the Common Screech and Great Horned Owls, Least Flycatcher, Black-capped Chickadee, Veery, Red-eyed Vireo, and Baltimore Oriole.

Greater in number are the species associated with shrubs—wolfberry, silverberry, hawthorn, and others—on the edges of wooded coulees, in draws, or in isolated thickets; these include the Mourning Dove, Eastern and Western Kingbirds (often perching on wires or fences), Traill's Flycatcher (near wet places), House Wren, Gray Catbird, Brown Thrasher, Cedar Waxwing, Loggerhead Shrike, Yellow Warbler, Common Yellowthroat (near wet places), Lazuli Bunting, American Goldfinch, Rufous-sided Towhee, and Clay-colored and Song Sparrows. A few Say's Phoebes frequent the vicinity of buildings, bridges, and freshly eroded banks. Common as well as conspicuous over much of the Refuge are the Swainson's Hawk (nesting usually in stunted trees, occasionally on the ground), the Northern Harrier and Short-eared Owl (nesting on the ground, generally in grassy lowlands or marshes), and the Black-billed Magpie (nesting in trees or high thickets).

Spring migration begins in March and continues until late May. One of the prominent features is the appearance of thousands of Sandhill Cranes during mid-April. Ducks reach their peak of abundance soon after the ice breaks up in mid-April; shorebirds from late April through May. The common to abundant transient warblers, namely, the Black-and-white, Tennessee, Orange-crowned, Myrtle, Blackpoll, and American Redstart, throng in wooded coulees between 25 April and 20 May. Fall migration gets under way in August with the arrival of birds from northern breeding grounds and the departure of several local summer residents. From then until freeze-up, usually in the first two weeks of November, migration proceeds steadily. Shorebirds are most abundant from late July through August; ducks during the first two weeks of October.

Headquarters of the Refuge may be reached from Kenmare, situated on the northeast shore of Middle Lake, by driving westward on a road across the north end of Middle Lake (and from which the

Western Grebe colony at the south end of Upper Lake may be viewed), then bearing left onto the entrance road. All worthwhile bird-finding sites on the Refuge are accessible by car, but directions should be obtained at headquarters before attempting to reach them.

On the hilly, prairie terrain west of Des Lacs Refuge lies the **Lostwood National Wildlife Refuge** (26,747 acres). This area is dotted with numerous potholes and several alkaline lakes, some of which have excellent marshes. Their water supply depends almost wholly on rainfall and the spring runoff. Like the Des Lacs Refuge, Lostwood has wooded coulees, draws, and thickets, but they are fewer in number and less extensive.

Refuge headquarters is on the west side of State 8, overlooking Thompson Lake, at a point 25 miles north of Stanley and 18 miles south of Bowbells. It can be reached from Kenmare, 17.5 miles distant, by driving west over an improved but unnumbered road. For route directions, inquire at headquarters of the Des Lacs Refuge.

The Lostwood Refuge has good breeding areas for the same species of waterfowl and other birds on the Des Lacs Refuge, except for Red-necked and Western Grebes, which do not nest on Lostwood.

Very near headquarters is a spot where Sharp-tailed Grouse perform their courtship 'dances' each spring from the time the snow disappears until well into June. Peak activity is usually during the last week of April. The dances, which begin in the dusk before sunrise and continue for two or three hours, may be observed at a distance of about 100 yards; if closer views are desired, the bird finder must conceal himself in a car or blind before the birds arrive.

Associated with the shallow, white-bordered alkaline lakes are two species of special interest: the Piping Plover and American Avocet. Both nest near the water's edge. One of the largest avocet colonies on the Refuge is along the shore of Dead Dog Slough. The nests are on pebbly ground among sparse grasses. By mid-June, the four or five eggs in each nest begin to hatch. Directions for reaching the Slough can be obtained at headquarters.

LISBON
Sheyenne National Grasslands

State 27 east from Lisbon in southeastern North Dakota almost immediately traverses a part of the **Sheyenne National Grasslands**—a 71,000-acre area largely of sandy soils and dunes supporting tall-grass prairie. Prevalent among the grasses are big bluestem, sand bluestem, prairie dropseed, Indian grass, and switch grass. The common breeding birds in this habitat are the Upland Sandpiper, Bobolink, Western Meadowlark, and Savannah Sparrow. Along the margins of tree thickets, the typical nesting birds include the Loggerhead Shrike, Yellow Warbler, Indigo Bunting, and four sparrows—Vesper, Lark, Field, and Song.

Always watch for Greater Prairie Chickens as the Grasslands are the last remaining stronghold for these birds in the state. Most of the population resides within a 5-mile radius of McCleod, reached from Lisbon by driving 18 miles east on State 27, then turning north onto a road and proceeding 3.5 miles. The birds begin appearing on their booming grounds in late March. Their display performances reach peak activity in late April and usually cease by mid-June.

The rich woodland along the Sheyenne River is well worth exploring for birds. It can be reached from Lisbon by driving east for 27 miles on State 27, then turning north on State 18 for 9 miles and following the road on the south side of the River westward for several miles. The trees here are primarily green ash, basswood, hackberry, and bur oak with a well-developed understory of ironwood, prickly ash, and other low growth. Both the Barred Owl and Pileated Woodpecker breed here regularly, as does the Chimney Swift which conforms to its original habit of nesting in hollow trees. Among the more common breeding birds are the Cooper's and Red-tailed Hawks, Black-billed Cuckoo, Great Crested and Least Flycatchers, Eastern Pewee, Yellow-throated, Red-eyed, and Warbling Vireos, Black-and-white Warbler, Ovenbird, American Redstart, Baltimore Oriole, Scarlet Tanager, Rose-breasted Grosbeak, and Indigo Bunting.

MOFFIT
Long Lake National Wildlife Refuge

Among the many lakes in the alkaline basins of south-central North Dakota that are particularly attractive to birds, Long Lake in the **Long Lake National Wildlife Refuge** is representative. The Lake itself at high-water level covers some 16,000 of the Refuge's 22,310 acres. The rest of the acreage embraces adjacent marshy areas, prairie grasslands, some croplands, and tree and shrub plantations.

Commonest among the nesting ducks at Long Lake are Mallards, Gadwalls, Common Pintails, and Blue-winged Teal; less common are Green-winged Teal, American Wigeons, Northern Shovelers, and Redheads. Other breeding birds associated with Long Lake or its marshy areas are Pied-billed Grebes, Black-crowned Night Herons, American Bitterns, Northern Harriers, Soras, and American Coots. American White Pelicans are present on the Lake all summer, probably coming from their colonies at Chase Lake, 30 miles to the northeast.

Lowlands bordering the marshes and dikes provide nesting sites for Spotted Sandpipers, Willets, Marbled Godwits, American Avocets, and Wilson's Phalaropes. Nesting on the grassy uplands are an assortment of birds that include the Sharp-tailed Grouse, Upland Sandpiper, Horned Lark, Sprague's Pipit, Bobolink, Western Meadowlark, Lark Bunting, Savannah, Grasshopper, and Baird's Sparrows, and Chestnut-collared Longspur.

The stellar ornithological feature at Long Lake as well as other lakes in this part of the state is the gathering of Sandhill Cranes during their fall migration. During September and October a good proportion of the tens of thousands of these big birds that pass through central North Dakota stop off, passing nights in the shallows of the Lake and foraging on the uplands in daytime.

To reach the Long Lake Refuge, exit south from I 94 on US 83 for 12.4 miles through the town of Moffit to the entrance road on the left. This passes to headquarters over a dike, offering excellent views of birds on Long Lake, especially the impressively heavy concentration of transient ducks in October and early November. At headquarters, many Cliff Swallows nest under the eaves of the utility buildings.

Near the east end of Long Lake and including a 30-mile stretch of I 94 between Steele and Medina, watch for raptors. In this part of the state, Ferruginous Hawks are exceptionally common; Red-tailed and Swainson's Hawks and Northern Harriers are quite numerous. Burrowing Owls are present in scattered groups.

ROLLA
Turtle Mountains

The **Turtle Mountains** in extreme north-central North Dakota are an island of hills and lake-filled hollows surrounded by boundless stretches of prairie. Extending for 40 miles east and west and 30 miles north and south, the area is cut across by the International boundary; thus about one-third lies in Canada. Although farms are numerous in the Turtle Mountains, much of the region is wooded, the principal trees being quaking aspen associated with balsam poplar, white birch, green ash, and bur oak. In certain places oak is abundant; ash and boxelder are sometimes common. The trees are largely second growth, having replaced the original forests that were swept by fire at one time or another. These woods-covered lands, together with marginal, shrub-covered lands and many lakes, make the Turtle Mountains one of the important wildlife areas in the state.

Ornithologically, the Turtle Mountains have the distinction of being among the few areas in the conterminous United States where the Bufflehead and Philadelphia Vireo breed. The Ruffed Grouse is a regular permanent resident, and in a few localities the Chestnut-sided Warbler, Northern Waterthrush, Mourning Warbler, and White-throated Sparrow are summer residents. Other landbirds that breed here are the Broad-winged Hawk, Ruby-throated Hummingbird, Yellow-bellied Sapsucker, Eastern Phoebe, Least Flycatcher, Veery, Red-eyed Vireo, Yellow Warbler, American Redstart, Rose-breasted Grosbeak, and Clay-colored Sparrow. Nesting on or near many of the lakes are Common Loons, all five species of grebes, and—in addition to Buffleheads—many waterfowl, including Ring-necked Ducks, which are much more common than elsewhere in the state.

The following trip through the Turtle Mountains in June or July should yield most of the characteristic birds.

From Rolla on US 281-State 5, drive northwest on State 30 for 3 miles. When State 30 bears right, continue straight ahead on an unnumbered road to St. John, then proceed west on an unnumbered and unnamed road—called in this account the East-West Road for want of another designation. After driving for 5 miles, turn north to *Lake Upsilon* (directional sign) and follow the road 1.5 miles to a picnic area on the Lake. Park the car and explore the area, walking northward on the road to its end at Wapoka Creek. Look especially for Common Loons on the Lake, for Yellow-bellied Sapsuckers and Philadelphia Vireos in deciduous trees, and, where there are spruces, for Purple Finches, Pine Siskins, and Red Crossbills.

Backtrack to the East-West Road and continue west for 7 miles to the *Brethren Camp* (marked by a sign and usually unoccupied) on the left at *School Section Lake*. Park and explore the area. The vicinity of the Camp is another good site for Philadelphia Vireos.

Again continue west on the East-West Road for 8 miles, crossing US 281-State 3 en route, then turn south to *Willow Lake* (marked by a sign at the turnoff) and proceed 1.5 miles to the Lake on the left. Park and walk along the shore. Note the small colony of Double-crested Cormorants on one of the islands, numerous pairs of Red-necked, Horned, and Eared Grebes. A careful search in the wet, grassy area along the shore may turn up Sharp-tailed Sparrows.

Where the East-West Road traverses semi-open country, expect to see both Eastern and Mountain Bluebirds and possibly Turkey Vultures. Where there are wide stretches of cleared grassy land, search for Baird's Sparrows and, in meadowy places, for Le Conte's Sparrows. Both species are regularly present but not numerous.

TOWNER

J. Clark Salyer National Wildlife Refuge

The **J. Clark Salyer National Wildlife Refuge,** one of the finest among the many great refuges administered by the U.S. Fish and

Wildlife Service, reflects the vision of the man for whom it was named. For twenty-seven years, J. Clark Salyer II (1902–66), as Chief of the Division of Wildlife Refuges, guided the development of the National Wildlife Refuge System.

Situated in the north-central part of the state, the Salyer Refuge (58,700 acres) extends from near the Canadian border southward for 75 miles along the Souris River to a point not far north of Towner on US 2. To reach Refuge headquarters, drive north from Towner on State 14 through Upham (23 miles distant) to the entrance on the right, approximately 3 miles beyond Upham.

The northern part of the Refuge consists of open water and marshes—formed by damming the Souris River at widely separated points—and adjacent wet meadows and prairie grassland. The southern part contains the wooded bottomlands of the Souris River, with numerous meander cutoffs and sloughs, and the surrounding sandhills.

The populations of nesting and transient birds associated with the marshes and their immediate vicinity are tremendous. Two outstanding features of breeding birdlife in the marshes are huge colonies of Franklin's Gulls and a large colony of Black-crowned Night Herons in which a few pairs of Green and Little Blue Herons and Cattle Egrets also nest. Other birds nesting regularly are the five species of grebes (Red-necked, Horned, Eared, Western, Pied-billed), at least eleven species of ducks (Mallard, Gadwall, Common Pintail, Green-winged and Blue-winged Teal, American Wigeon, Northern Shoveler, Redhead, Canvasback, Lesser Scaup, Ruddy Duck), and the following: American Bittern, Canada Goose, Northern Harrier, Virginia Rail, Sora, American Coot, Forster's and Black Terns, and Yellow-headed and Red-winged Blackbirds. A number of Ring-billed Gulls and Common Terns nest on islands in the marshes.

Around the marshes, in wet, grassy situations, is a variety of birds: the Willet, Wilson's Phalarope, Sedge Wren, Bobolink, Brewer's Blackbird, and Savannah, Le Conte's, and Sharp-tailed Sparrows. Always a possibility is hearing a Yellow Rail. Listen for it in spring-fed wet areas, such as quagmires with quaking surfaces that support mats of thick grasses.

NORTHERN SHOVELERS

The upland country immediately beyond the marshes is gently rolling short-grass prairie with patches of brush and tall grasses here and there. Breeding regularly in this type of environment are the Sharp-tailed Grouse, Upland Sandpiper, Marbled Godwit, Short-eared Owl, Common Nighthawk (nests on bare knolls as well as in woods), Horned Lark, Sprague's Pipit, Common Yellowthroat (in brush near wet places), Western Meadowlark, Grasshopper, Baird's, and Vesper Sparrows, Lark Sparrow (occasionally), Clay-colored Sparrow (in brush), and Chestnut-collared Longspur.

The bottomlands of the Souris River, with their cover of ashes, boxelders, and willow thickets, and the nearby sandhills, with their aspen groves and thorny bushes, provide nesting habitat for numerous species, such as the Red-tailed and Swainson's Hawks, Long-eared Owl, Traill's and Least Flycatchers, Eastern Pewee, Black-billed Magpie, House Wren, Gray Catbird, Brown Thrasher, Veery, Cedar Waxwing, Red-eyed and Warbling Vireos, Yellow Warbler, Ovenbird, American Redstart, Baltimore Oriole, Rose-breasted Grosbeak, American Goldfinch, and Song Sparrow. The Hooded Merganser nests in the wooded bottomlands, the only area in the state where it breeds regularly.

Summer-resident Eastern and Western Kingbirds and Loggerhead Shrikes may be expected on wires and other suitably exposed perches; Eastern and Say's Phoebes near buildings, bridges, and eroded riverbanks; and four species of swallows—the Tree, Bank, Rough-winged, and Barn—continually winging their way low over the marshes in search of insect food.

Waterfowl migrations in spring and fall bring enormous transient populations of geese and ducks to the Refuge. Greater White-fronted Geese appear in exceptionally large numbers, particularly during late September and early October, forming the major concentration in the state. Snow Geese are abundant in the spring (usually during mid-April), less so in the fall. Ducks, though showing up in impressive numbers in spring, with peak abundance soon after mid-April, come in far greater numbers in fall. Mallards reach peak abundance in mid-October but most other ducks congregate in largest numbers during the last week of August. Occasionally there are many Whistling Swans in the spring (mid-April).

WALHALLA
Pembina Hills

In extreme northeastern North Dakota the beautiful forested escarpments, known as the **Pembina Hills,** extend in a northwesterly direction from the community of Walhalla. The most scenic part of the Hills, as well as the most productive for viewing birds, is the Great (or Pembina) Gorge along the Pembina River into Canada. It can be reached by traveling north from Walhalla for 3.5 miles on State 32, then turning off west for 12 miles and parking near the crest of the escarpment. A wagon road continues in a west-northwest direction down the slope to the Pembina River.

The Pembina Hills resemble the lower ridges of the Appalachian Mountains and in this respect differ markedly from other parts of the state. The forests, interrupted frequently by brushy openings, are entirely deciduous, dominated by bur oak in association with quaking aspen, balsam popular, white birch, boxelder, basswood, and green ash. Some of the shrubs forming an understory and growing in openings are hazelnut, currant, serviceberry, hawthorn, chokecherry, raspberry, and smooth sumac.

Included among the principal breeding birds are the Red-tailed Hawk, Ruffed Grouse, Least Flycatcher, Eastern Pewee, Veery (unusually common), Red-eyed and Warbling Vireos, American Redstart, Baltimore Oriole, Rose-breasted Grosbeak, Rufous-sided Towhee, and Clay-colored Sparrow. Some of the other breeding birds, less common, are the Cooper's Hawk, Ruby-throated Hummingbird, White-breasted Nuthatch, Yellow-throated Vireo, Scarlet Tanager, and Indigo Bunting. Of special interest is the Orange-crowned Warbler which has been reported singing several times in late June and may nest.

Oklahoma

SCISSOR-TAILED FLYCATCHER

From early spring to late fall no living thing lends greater color or interest to the open country in Oklahoma than the handsome, graceful Scissor-tailed Flycatcher. Trim and sedate, it rests on utility and fence wires, or low trees, then takes flight, revealing the salmon pink of its underwings and flanks and spreading, as it changes course or elevation, the black and white, streamer-like feathers of its deeply forked tail. Unafraid of passers-by, courageous when protecting its nest, skilled in a variety of maneuvers—by such behavior the bird thus accentuates its unforgettable appearance.

Much of Oklahoma is open, flat to rolling prairie country, devoted to farming of one sort or another. The wide differences in the texture and composition of its soils and in its annual rainfall—

about 10 to 15 inches in the northwest, as compared with 50 to 60 in the southeast—make possible the production of nearly every type of crop raised in conterminous United States. Yet there are areas that are rough and generally unsuited to agriculture.

Of the rough areas, four are extensively mountainous and have distinctive physiographic features. First, the Ozarks, in the northeast, a westward extension of the dissected plateau of the same name in Missouri and Arkansas. These are broad flat-topped hills, or ridges, separated by the deep, often narrow, valleys with clear, cold, spring-fed streams. A few hills attain elevations as great as 1,150 feet. Characteristically the area is timbered, principally with blackjack and post oaks, black hickories, and winged elms, but about 30 per cent has been cleared for agriculture.

Second, the western chain of the Ouachita Mountains, in the southeast, the most rugged relief in the state. Rich Mountain, the highest point, is over 2,800 feet above sea level. The slopes are littered with huge rocks, and along the numerous spring-fed streams rise the sheer cliffs of canyons. Only about 15 per cent of the Ouachita area is used for farming; the rest is woodland, the finest in the state, consisting chiefly of shortleaf pines mixed with white, post, and blackjack oaks, black hickories, and black locusts. In the extreme southeast the Ouachita Mountains border upon 120 square miles of Coastal Plain, which is largely forested with loblolly and shortleaf pines.

Third, the Arbuckle Mountains, occupying a 35- by 65-mile area in south-central Oklahoma. Though they reach unspectacular heights—900 to 1,300 feet above sea level and no higher than 400 feet above the surrounding country—the area is scenically attractive, owing to the many deep canyons through which flow rapid streams with frequent and beautiful plunging waterfalls. Save in the ravines, gullies, and canyons, trees are scarce.

And fourth, the Wichita Mountains (*see under* **Lawton**) 70 miles farther west, in southwestern Oklahoma, a group of low ridges and hills somewhat widely dispersed in a 25- by 65-mile area. A few peaks exceed 2,400 feet above sea level and rise about 1,000 feet above the plains; all, regardless of elevation, are rugged, with granite outcroppings and talus boulders, steep upper slopes almost devoid of soil, and scanty tree growth.

Although the Oklahoma mountains, because of their relatively small extent and moderate elevations, hold no species of birds peculiar to them and show no altitudinal succession in birdlife, they provide opportunities for finding some birds not appearing regularly in other parts of the state. In both the Ozarks and the Ouachita Mountains, the Acadian Flycatcher, Black-and-white Warbler, Northern Parula Warbler, Louisiana Waterthrush, American Redstart, Scarlet Tanager, and Chipping Sparrow are among the breeding birds; in the Ouachita Mountains, the Pine Warbler may be expected in forests where pine is prevalent. The Arbuckle Mountains are notable for Rufous-crowned Sparrows. In the Wichita Mountains where bird finding is surprisingly productive in any season, not only do Rufous-crowned Sparrows nest, but also a host of other species, both eastern and western. It is one of the few parts of the state in which four wrens—Bewick's, Carolina, Canyon, and Rock—nest more or less side by side.

In addition to the mountain areas, another rough area warranting mention is the sandstone hills region, or Cross Timbers, its boundaries not sharply defined, which occupies a large part of east-central Oklahoma and reaches across the Texas line—where it is continuous with the East and West Cross Timbers of that state—and for a short distance across the Kansas line. Fingers of the region extend eastward between the Ozarks and the Ouachita Mountains to the Arkansas line and westward into western Oklahoma as far as Cleo Springs, Camargo, the Wichita Mountains, and other points. The roughness of the region is due to the sandstone hills and escarpments, whose tops and sides are scattered with loose sandstone boulders. Having only relatively dry, sandy soils, these areas support primarily a scrubby growth of post oaks, blackjack oaks, and black hickories, with a floor of short grasses. Between the hills, especially in the valleys of the larger streams, are level to rolling plains with rich clay soils producing tall grasses. Almost 75 per cent of the original woodland cover of the sandstone hills still remains, but the valley grasslands have lost most of their original character through agricultural uses.

Eastern and central Oklahoma, excepting the mountain areas and the sandstone hills region, is essentially tall-grass prairie in which bluestem, Indian grass, switch grass, and other tall grasses predominate, with a gradual increase westward of such short grasses as buffalo and grama. Bottomlands along the larger streams are often forested with American elms, slippery elms, hackberries, various oaks, sycamores, cottonwoods, and black willows, with extensive cutover areas grown to dense tangles of weeds and shrubs. Most of the tall-grass prairie is now a thriving agricultural land.

Western Oklahoma, excluding the Panhandle and the Wichita Mountains, is on the mixed-grass plains, level to slightly rolling, with deeply eroded channels, sometimes called ravines or canyons. Merging originally with the tall-grass prairie in the central part of the state and meeting the short-grass plains of the Great Plains in the Oklahoma and Texas Panhandles, these plains represent an area of transition. Cottonwoods and tamarisks are prevalent along the stream margins. On the floors of the deep ravines are isolated strips of forest typical of the stream bottomlands farther east, in central and eastern Oklahoma; on the slopes and terraces of the ravines are oaks and, occasionally, junipers. Here and there, notably near the larger streams, are broad, sandy areas, sometimes with dune formations, showing wide variations in cover, from none whatsoever on live dunes, to bunch grass, sagebrush, and thickets of plum and sumac. Toward the Oklahoma and Texas Panhandles, the sand areas support the so-called 'shinnery" grasslands, where shin oaks are intermixed with grasses, principally of the tall type. Usually the oaks are low, from 2 to 8 feet in height, and grow in mottes (clumps) on small hills of sand. In the extreme southwest and in a few spots farther north and east are the mesquite plains, which are part of an extensive area of that name in Texas. (For a description of the mesquite plains, see the introduction to the Texas chapter.) Nearly three-fourths of the western Oklahoma plains is under cultivation; the remainder, together with eroded spots and sandy areas, is used mainly for grazing when used at all.

Very different from the rest of the state is the Oklahoma Panhandle (*see under* **Boise City** *and* **Guymon**), which is an east-west strip of the high Great Plains that extend northward through the continent from central Texas into Canada.

An unusual spot, ornithologically as well as geologically, is the Great Salt Plains in the extreme north-central part of the state (*see under* **Cherokee**). Here, at an elevation slightly lower than the surrounding prairie, is a vast flat covering some 60 square miles. Much of it is now inundated by a reservoir, but there are nonetheless remnants still uncovered that retain the original appearance—a plain of glistening light-colored sand with a wafer-like crust of salt. Among the birds nesting in this strange environment are the Snowy Plover, American Avocet, and Little Tern.

Most of the breeding birds in the eastern third of Oklahoma are eastern in their affinities, a fact to be expected owing to the great extent of eastern forest and forest-edge habitats. Central Oklahoma, with its stretches of scrub oaks, its wooded riverbottoms, and its tree and shrub plantations about settlements, shows a westward continuation of the same avifauna, though it lacks some species that are in the forests of the Ozarks and Ouachita Mountains.

Listed below are some of the species breeding regularly in eastern and western Oklahoma (except as indicated), but not in both sections. Those marked with an asterisk range commonly into central Oklahoma.

EASTERN OKLAHOMA

Black Vulture (*mountainous areas*)
Red-shouldered Hawk
Broad-winged Hawk
Greater Prairie Chicken (*northeast only*)
* Barred Owl
Pileated Woodpecker
Acadian Flycatcher
* Eastern Pewee
* White-breasted Nuthatch
Wood Thrush
* White-eyed Vireo
Yellow-throated Vireo
Black-and-white Warbler
* Prothonotary Warbler
Northern Parula Warbler
Cerulean Warbler
Yellow-throated Warbler
Pine Warbler
Prairie Warbler
Louisiana Waterthrush (*also in Arbuckle Mountains*)
* Kentucky Warbler
American Redstart
Scarlet Tanager
Chipping Sparrow

WESTERN OKLAHOMA, EXCLUSIVE OF PANHANDLE

* Swainson's Hawk
Lesser Prairie Chicken (*northwest only*)
Long-billed Curlew
Burrowing Owl
Red-shafted Flicker

Ladder-backed Woodpecker
* Western Kingbird
White-necked Raven
Western Meadowlark
Bullock's Oriole
Rufous-crowned Sparrow
Cassin's Sparrow

A great number of species are statewide in their distribution, nesting in open country (prairie grasslands, pastures, fallow fields, meadow-like lowlands, brushy places, woodland borders, and dooryards) or in woodlands. Those species that breed regularly throughout Oklahoma, excluding the Panhandle, are as follows:

OPEN COUNTRY

Turkey Vulture
Northern Harrier (*uncommon*)
Common Bobwhite
Mourning Dove
Greater Roadrunner
Yellow-shafted Flicker
Red-headed Woodpecker
Eastern Kingbird
Scissor-tailed Flycatcher
Eastern Phoebe
Horned Lark
Barn Swallow
House Wren (*rare southward*)
Bewick's Wren
Northern Mockingbird
Gray Catbird (*rare southward*)
Brown Thrasher (*rare southward*)
Eastern Bluebird
Loggerhead Shrike
Bell's Vireo
Yellow Warbler
Common Yellowthroat
Yellow-breasted Chat
Eastern Meadowlark (*rare westward*)
Orchard Oriole
Great-tailed Grackle
Common Grackle
Northern Cardinal (*rare westward*)
Blue Grosbeak
Indigo Bunting (*rare westward*)
Painted Bunting
Dickcissel
American Goldfinch (*rare westward*)
Grasshopper Sparrow
Lark Sparrow
Field Sparrow

WOODLANDS

Mississippi Kite (*local*)
Red-tailed Hawk
Cooper's Hawk
Yellow-billed Cuckoo
Common Screech Owl
Great Horned Owl

WOODLANDS (*Cont.*)

Chuck-will's-widow (*rare westward*)
Red-bellied Woodpecker
Hairy Woodpecker
Downy Woodpecker
Great Crested Flycatcher
Blue Jay
Carolina Chickadee
Tufted Titmouse
Carolina Wren (*rare westward*)
Blue-gray Gnatcatcher
Red-eyed Vireo
Warbling Vireo
Baltimore Oriole (*chiefly northern Oklahoma*)
Summer Tanager

Spring migration in Oklahoma begins about the last week of February when the first birds appear from the south in scattered numbers. The height of landbird migration is reached during the last week of April and the first week of May, at which time the majority of northbound flycatchers, warblers, and fringillids pass through the state and most of the summer-resident species arrive. The fall migration of landbirds starts in August, when many of the summer-resident species begin leaving. Although Oklahoma is in the path of the central flyway, the birds tend to pass through the state in a widely dispersed fashion, following no particular route, Transient waterbirds, waterfowl, and shorebirds appear in appreciable numbers at the larger reservoirs. Here, frequently, the April and October flights of Franklin's Gulls are so enormous as to be a show in themselves. The many small impoundments on farms—'farm ponds'—provide resting, and sometimes feeding, places for a few ducks and shorebirds. In Oklahoma, the main migration flights may be expected within the following dates:

Waterfowl: 25 February–1 April; 20 October–15 December
Shorebirds: 15 April–15 May; 10 August–15 October
Landbirds: 25 March–5 May; 10 September–5 November

Fairly large populations of geese and ducks pass the winter on, or in the immediate vicinity of, the large reservoirs. Numerous species of northern landbirds, particularly fringillids, winter in the state. For example, in central Oklahoma, one may expect to find, in decreasing order of abundance, Harris' Sparrows, Slate-colored

Juncos, American Tree Sparrows, White-crowned Sparrows, American Goldfinches, and Song Sparrows. There are also large concentrations of American Robins, bluebirds, meadowlarks, and blackbirds. Carolina Wrens, meadowlarks, Northern Cardinals, and Harris' Sparrows enliven the winters by their frequent singing. In the southeastern part of the state several species of warblers spend the winter.

Authorities

John R. Atkin, Elizabeth Hayes, Evan V. Klett, Kenneth E. Schwindt, John Shackford, Ronald S. Sullivan, George Miksch Sutton.

References

Oklahoma Birds: Their Ecology and Distribution with Comments on the Avifauna of the Southern Great Plains. By George Miksch Sutton. Norman, University of Oklahoma Press. 1967.

Bird Finding Guide. By the Tulsa Audubon Society. 1973. Includes areas in all parts of the state except in the Panhandle. Available through the Tulsa Garden Club, 2435 South Peoria, Tulsa, OK 74114.

ALTUS
Eldorado Area

In extreme southwestern Oklahoma, reached from Altus by driving 25 miles south on State 44, is the **Eldorado Area,** comprising a remarkable combination of habitats for birds. Here are large expanses of mesquite—from dense growths to open spots—and croplands for wheat and cotton, in combination with gullies, gypsum sink holes, creeks, and riverbottoms. In the Area several southwestern birds—e.g., (Black-crested) Tufted Titmouse and Verdin—reach the northeastern limits of their breeding ranges.

From side roads leading off State 44 in the vicinity of Eldorado, bird finding can be productive. Notable year-round residents are the Scaled Quail, Greater Roadrunner, Golden-fronted and Ladder-backed Woodpeckers, (Black-crested) Tufted Titmouse, Curve-billed Thrasher, and, occasionally, the Verdin. During summer, Mississippi Kites and Scissor-tailed Flycatchers are abundant; the Ash-throated Flycatcher may be seen and the Poor-will heard calling at night. Sage Thrashers and Lark Buntings pass through in migration and may remain during milder winters.

ARNETT
Prairie Chicken Country

The area around Arnett in extreme western Oklahoma, near the Texas Panhandle, is typical shinnery grassland, with gently rolling hills and loose sandy soils covered with grasses and shrubs. This is good country for Lesser Prairie Chickens. Here the birds may be observed on their booming grounds early any morning between mid-March and mid-June, but best on an early morning in April when courtship activities reach their peak.

To find the birds on their booming grounds, leave Arnett an hour or so before dawn, driving east on US 60 for 1.0 mile, then south on US 283 for 10 miles, taking any side road along the way. Locate the grounds—they are often near roads—by listening for the birds. Having determined their direction, drive the car as close as the road will permit and wait for daylight. Stay in the car—getting out will cause the birds to disperse—and study the birds through field glasses or spotting scope. A word of caution: Most of the roads leading from US 283 are very sandy, so have a shovel handy in case the car is stalled.

After sunup, some of the birds likely to be viewed from the car besides prairie chickens are Mississippi Kites, Swainson's Hawks, White-necked Ravens, Loggerhead Shrikes, and Western Meadowlarks.

BOISE CITY
Black Mesa Country

The drive westward across the Oklahoma Panhandle on US 64 to Boise City is a 150-mile trip over high plains, where houses are few and towns small and far apart; where winds blow constantly, piling up tumbleweeds against fences and roadside embankments and sometimes lifting up clouds of fine sand; where level horizons shimmer in the heat of the summer day, frequently creating peculiar mirages. It is a journey of great sameness and yet never without interest to the bird finder (e.g., *see under* **Guymon**).

But westward beyond Boise City (elevation 4,165 feet) the level plains suddenly give away to a broken terrain of high mesas, dissected plateaus, gulches, and, occasionally, deep canyons. Unquestionably the choice area for birds is the **Black Mesa Country** in the extreme northwestern corner of the Panhandle. June, when the birds are in full song, is the best time for a visit.

To reach the Black Mesa Country from Boise City, proceed west and then north on an unnumbered road to the little town of Kenton, just south of the Black Mesa and east of the New Mexico line. The distance from Boise City is about 32 miles.

Kenton (elevation 4,349 feet) lies in the broad valley of the Cimarron River, which cuts through the northwestern Panhandle from New Mexico. Though usually dry, save for holes in quicksand, the Cimarron holds sufficient moisture to support along its banks thickets of weeds and brush and long narrow groves of such trees as cottonwoods, hackberries, and willows.

This part of the Black Mesa Country—the valley floor and the lower slopes of the nearby mesas—has a wide variety of breeding birds: *in thickets, or groves,* the Great Horned Owl, Black-chinned Hummingbird, Red-shafted Flicker, Red-headed Woodpecker, Lewis' Woodpecker (irregular), Ladder-backed Woodpecker, Cassin's Kingbird, Western Pewee, Black-billed Magpie, Brown Thrasher, Warbling Vireo, Bullock's Oriole, Blue Grosbeak, and Lesser Goldfinch; *in the sagebrush of valley floor and lower slopes,* the Scaled Quail, Greater Roadrunner, Poor-will, Curve-billed and Sage Thrashers, Western Meadowlark, Brown Towhee, and Lark, Cassin's, and Black-throated Sparrows; *around buildings,* the Barn Swallow, Bewick's Wren, Mountain Bluebird, and House Finch; *in gulches with boulders or rocky cliffs,* the Canyon and Rock Wrens and Rufous-crowned Sparrow.

The Black Mesa rises 500 feet above the valley of the Cimarron, attaining an elevation of 4,973 feet, the highest point in the state. The Mesa may be reached by driving north from Kenton on an improved gravel road for 2 miles, turning west and proceeding as far as possible in the car, then going the rest of the way on foot.

When ascending the Mesa the bird finder will climb on rocky slopes covered in part with low (4 to 5 feet high) scrubby vegetation consisting largely of hackberry mixed with live oak and juniper and also, on the north-facing slopes, with pinyon pine. Once on the Mesa's table-like top, he will enter upon a short-grass plain. During the climb it will be surprising if one or two Golden Eagles do not soar into his view. On the high slopes he will see Scrub Jays, Pinyon Jays (irregular), Plain Titmice, and Bushtits which nest among the junipers and pinyon pines.

BUFFALO
Doby Springs

In Buffalo, just east of the Panhandle, is a little 'city park' where Mississippi Kites nest. Westward from town, across open rolling prairie, leads US 64. In summer, along the way, watch for Swainson's Hawks, Greater Roadrunners, Burrowing Owls, Common Nighthawks, Horned Larks, Western Meadowlarks, Blue Grosbeaks, Dickcissels, and Lark Sparrows. Where there is sage, stop the car and listen for that most exquisite of songs, the *theeee, diddle-dit* of Cassin's Sparrow. Do not expect Scissor-tailed Flycatchers, as they are rare, but both Eastern and Western Kingbirds nest throughout the region.

About 6 miles west of Buffalo, turn north from US 64 onto a well-marked dirt road for **Doby Springs,** a choice bird-finding spot. Here, surrounded by willows and cottonwoods, is an impounded pond through which flows a clear, sandy-bottomed little stream whose water is cool and quiet enough just above the pond for watercress. Red-winged Blackbirds nest at the pond's west end, and close by in the trees nest Green Herons, Red-tailed Hawks, American Kestrels, Yellow-billed Cuckoos, Common Screech and Great Horned Owls, five species of woodpeckers, Great Crested Flycatchers, hybrid Baltimore-Bullock's Orioles, and a few Carolina Chickadees, Northern Cardinals, and Painted Buntings.

CHEROKEE
Salt Plains National Wildlife Refuge

Situated east of Cherokee in north-central Oklahoma, the **Salt Plains National Wildlife Refuge** (32,008 acres) is surrounded by level, rich farmlands producing an abundance of wheat, alfalfa, and other crops. Within the Refuge are the remarkable salt flats—now partically inundated by the Salt Plains Reservoir—and the lands immediately adjacent to them.

The Reservoir covers from 10,700 acres at 'conservation-pool level' to 29,000 acres at 'flood-control level.' Twenty-two outlying small ponds cover an additional 203 acres. The major streams entering the Reservoir are the Salt Fork (two branches) of the Arkansas River and Sand Creek from the west. A few marshy areas in the inlets and coves of the Reservoir, and in the outlying ponds, support a thin growth of bulrushes, cattails, and smartweeds, but there are no extensive marshes. West of the Reservoir is the best remnant of salt-encrusted flats.

Near the Refuge headquarters on the northeast side of the Reservoir are grass-covered sand dunes and stands of trees and shrubs, principally elm, hackberry, cottonwood, dogwood, and wild plum. In general, the Reservoir and the remaining salt flats are bordered by salt grasses and, on the adjacent higher ground that is under Refuge management, by tall-grass prairie. Some of the prairie is cultivated, producing wheat and other small grains for transient waterfowl; some of it is used by local farmers for grazing.

On the Refuge, the summer-resident bird that attracts the greatest interest is the Snowy Plover, which nests on the salt flats. Three other species nesting in the same habitat are the American Avocet, Little Tern, and Common Nighthawk. Elsewhere on the Refuge, or in the immediate vicinity, the following birds are known to nest: Mississippi Kite, American Kestrel, Common Bobwhite, Red-bellied and Red-headed Woodpeckers, Eastern and Western Kingbirds, Scissor-tailed Flycatcher, Bewick's and Carolina Wrens, Northern Mockingbird, Brown Thrasher, Eastern Bluebird, Bell's Vireo, Yellow-breasted Chat, Eastern Meadowlark, Orchard and Baltimore Orioles, Northern Cardinal, Blue Grosbeak, Indigo and Painted Buntings, Dickcissel, and Lark Sparrow.

Birds visiting the Refuge in considerable numbers during late summer are American White Pelicans and others including Great Egrets and Black-crowned Night Herons. The spring and fall flights of waterfowl take place between mid-February and the first of April and between mid-October and mid-December. Canada Geese become abundant, with peak numbers approaching 50,000. The predominant ducks in both spring and fall flights are Mallards, Gadwalls, Common Pintails, Green-winged and Blue-winged Teal, American Wigeons, Northern Shovelers, and Common Mergansers, but the following are also present: Wood Ducks, Ring-necked Ducks, Canvasbacks, Lesser Scaups, Common Goldeneyes, Buffleheads, and Ruddy Ducks.

Perhaps the most impressive ornithological spectacle on the Refuge are the enormous spring (April) and fall (October) flights of Franklin's Gulls. During winter, as many as 20,000 Canada Geese, together with a few Greater White-fronted Geese, remain in the area, resting on the Reservoir and salt flats and feeding in the nearby grainfields.

To reach Refuge headquarters from Cherokee, drive north on US 64 for 2 miles; turn right onto State 11 and go 11 miles; turn right onto State 38 and proceed 2 miles; then turn off right, following signs to headquarters. One may backtrack to State 38 and continue around the Refuge by turning right onto State 38 and driving 13 miles to Jet, then turning right onto US 64 and returning to Cherokee, a distance of 17 miles. During the trip, the Reservoir, the salt flats, and other productive habitats are in view from the highway. Bird finders desiring directions for observing certain species should make inquiries at headquarters.

CLINTON
Washita National Wildlife Refuge

Northwest of this city in west-central Oklahoma, the **Washita National Wildlife Refuge** (8,084 acres) incorporates the north end of the Foss Reservoir, which is practically surrounded by rolling, short-grass prairie and farmlands of wheat, cotton, and milo. The Reservoir receives much of its water from the Washita River which enters through the Refuge at the north end and is bordered by elms, hackberries, cottonwoods, and other woody growth. A few marshy areas line the Reservoir's coves and inlets, but there are no extensive marshes.

On the Refuge or in its immediate vicinity are good habitats for such breeding birds as the Mississippi Kite, Red-tailed Hawk, American Kestrel, Common Bobwhite, Killdeer, Mourning Dove, Yellow-billed Cuckoo, Red-bellied, Red-headed, Hairy, and Downy Woodpeckers, Eastern and Western Kingbirds, Scissor-tailed and Great Crested Flycatchers, Horned Lark, Barn and Cliff Swallows, Blue Jay, Carolina Chickadee, Tufted Titmouse, Bewick's Wren, Northern Mockingbird, Brown Thrasher, Eastern Bluebird, Blue-gray Gnatcatcher, Loggerhead Shrike, Red-winged Blackbird, Baltimore Oriole, Northern Cardinal, Dickcissel, and Lark Sparrow.

Great numbers of waterbirds and waterfowl stop off on the Refuge during their migrations. American White Pelicans, Green Herons, Snowy Egrets, and Black-crowned Night Herons appear in late summer. In spring (mid-February to early April) and fall (mid-October to mid-December) the flights of waterfowl include an abundance of Canada Geese and a few Greater White-fronted Geese and Snow Geese; large numbers of Mallards, Common Pintails, American Wigeons, and Common Mergansers, and smaller numbers of Green-winged and Blue-winged Teal, Northern Shovelers, Redheads, Ring-necked Ducks, Canvasbacks, Lesser Scaups, Buffleheads, and Ruddy Ducks.

The Refuge's most impressive ornithological feature is the gathering of Sandhill Cranes in spring and fall, particularly in the fall. During October as many as 12,000 can sometimes be seen feeding and resting in fields adjacent to the Reservoir.

To reach Refuge headquarters from Clinton, drive north on US 183 for 9 miles; turn west onto State 33 and go 17.5 miles, passing through Butler; then turn off north 1.0 mile and west 0.5 mile, following Refuge signs.

GUYMON

Optima National Wildlife Refuge | Optima Public Hunting Area

In this flat part of the Oklahoma Panhandle, some of the best spots for bird finding are the low-lying playas when water gathers in sufficient quantity to attract migrating grebes, ducks, shorebirds, Franklin's Gulls, Black Terns, Marsh Wrens, and Yellow-headed Blackbirds.

Lining the Beaver River north of Guymon and reached from US 54 and 64 are big cottonwoods and other trees where such birds nest as Mississippi Kites, Red-tailed Hawks, American Kestrels, Common Screech, Great Horned, and Long-eared Owls, several woodpecker species, American Crows, and Bewick's and House Wrens. In fall and winter, Bald Eagles often roost in the taller cottonwoods. Along the edges of these riverbottom woods, Eastern and Western Kingbirds nest, the latter in large numbers; Orchard and Bullock's Orioles are common. Blue Grosbeaks nest in shrubs and coarse weeds, sometimes well away from the woods. Since most of the bottomland woods along the Beaver River are in privately owned ranches, public access is restricted to two highway right-of-ways: One is from US 54, 5 miles northeast of Guymon, the other is from US 64, 2 miles north of town.

In open country along US 64 north and west of Guymon, look for Long-billed Curlews nesting in uncultivated areas, Burrowing Owls in prairie dog 'towns,' Barn Owls in deserted buildings, and colonies of Cliff Swallows under bridges and in cement culverts. Characteristic breeding birds of open grassland are the Horned Lark, Western Meadowlark, and three sparrows, the Grasshopper, Lark, and Cassin's. Swainson's and Ferruginous Hawks and White-necked Ravens have nests in isolated trees.

East of Guymon is the newly developed **Optima National Wildlife Refuge** (4,333 acres) on the southwest side of the Optima Reservoir, formed by the damming of the Beaver River. State 3, 15 miles east of Guymon, passes through the middle of the Refuge. Access to the Refuge is limited to dead-end roads. At Refuge headquarters in Guymon (611 East Fourth Street), inquire about the best entries for viewing birds.

The Refuge encompasses both short-grass and tall-grass vegetation, sagebrush habitat, and about 200 acres of timber, mainly cottonwood, hackberry, willow, and black locust. Among the birds breeding on the Refuge, besides many of the aforementioned species, are the Mallard, Blue-winged Teal, Northern Harrier, both the Common Bobwhite and Scaled Quail, the latter favoring the drier areas in which sagebrush is dominant, and the Spotted Sandpiper. Below the Optima Dam, east of the Refuge, nest Snowy Plovers, American Avocets, and Little Terns.

Along the edge of the Reservoir in late summer and early fall, some of the transient birds very likely to appear are the Green Heron, Great Egret, White-faced Ibis, Common Snipe, Upland Sandpiper, Greater and Lesser Yellowlegs, and Long-billed Dowitcher. By October, geese and ducks begin arriving, many remaining through the winter. Around the Reservoir in winter there are usually a few Golden as well as Bald Eagles, occasionally Rough-legged Hawks and Short-eared Owls, and sometimes a Prairie Falcon or two.

Three miles north of the Refuge on State 94, reached by turning north from State 3, is the **Optima Public Hunting Area** (3,500 acres), administered by the Oklahoma Department of Wildlife Conservation. Black-billed Magpies reside here as does a big flock of Wild Turkeys.

HINTON
Red Rock Canyon State Park

For a fine variety of small landbirds with eastern 'elements' among them, **Red Rock Canyon State Park** (310 acres) in west-central Oklahoma is a productive spot. From I 40 west of Oklahoma City, exit south on US 281 for 3 miles to Hinton, then 1.0 mile farther south into the Park.

The Park embraces one of the several canyons in this part of Oklahoma that have red sandstone walls, usually steep, with rims supporting scrub oaks and junipers. On the canyon floors, which are moist with running streams in wet seasons, stand sugar maples whose foliage turns to bright colors—red, orange, and yellow—in October. Although Red Rock Canyon Park is a popular recreational and camping area, the bird finder can avoid people by walking along the rims of the Canyon or in oak-juniper habitats covering much of the high country near the Park.

Among the regularly nesting birds in the Park are the Greater Roadrunner, Great Crested Flycatcher, Eastern Phoebe, Rough-

BLACK-CAPPED VIREO

winged and Cliff Swallows, Carolina Wren, Wood Thrush, Eastern Bluebird, Black-and-white and Prairie Warblers, Summer Tanager, American and Lesser Goldfinches, and Indigo and Painted Buntings. Other breeding birds to look or listen for: Chuck-will's-widows calling at night; Canyon Wrens and Rufous-crowned Sparrows on cliffs or rocky slopes; Black-capped Vireos in thickets on the Canyon's rims or elsewhere near the Park; Louisiana Waterthrushes near undisturbed streams; Blue Grosbeaks and Lark and Field Sparrows on the Canyon's rims.

LAWTON
Wichita Mountains Wildlife Refuge

In southwestern Oklahoma, the **Wichita Mountains Wildlife Refuge** (59,019 acres) encompasses the higher Wichitas and a wide interior valley. Included in the Refuge is Mt. Scott (2,464 feet high). About two-thirds of the area consists of hills studded with hardwoods, primarily post and blackjack oaks; the remaining third comprises prairie and 659 acres of water (17 lakes and about 50 ponds) in rocky basins, mainly behind man-made dams. The bluestem and associated grasses that thrive on the prairie form one of the finest examples of tall-grass prairie sod remaining on the continent. Extensive marshes are lacking, but the larger lakes have their marshy fringes producing cattails and pondweeds. Standing on the banks of many of the lowland streams are American elms, white ashes, spotted oaks, hackberries, black walnuts, and other woody growth requiring a moist environment.

The area that is now Wichita Mountains Refuge was proclaimed a Federal game preserve in 1905 by President Theodore Roosevelt, the primary purpose being to perpetuate suitable habitat for the bison, then in great danger of extinction. Its present name was acquired in 1935, when it was placed under the administration of the U.S. Fish and Wildlife Service. Today the Refuge has a herd of more than 600 bison. It is similarly a preserve for elk, whitetail deer, long-horned cattle, and a prairie dog 'town'—and one of the most productive bird-finding areas in southwest Oklahoma. Approximately half of the Refuge is open to the public; there are good roads (including the winding, scenic highway to the summit of Mt. Scott), well-marked foot trails, and excellent camping and picnicking facilities.

Some of the best spots for birds are in the southwestern part of the Refuge: for example, the area around headquarters and the vicinity of the Sunset Picnic Area, a short distance southwest of headquarters. Here Canyon and Rock Wrens are heard singing from April to July. Other breeding birds that may be observed here at this same time of year include the Great Blue Heron, Turkey Vulture, Mississippi Kite, Red-tailed Hawk, Common Bobwhite, Wild Turkey, Killdeer, Mourning Dove, Yellow-billed Cuckoo, Common Screech Owl, Downy Woodpecker, Western Kingbird, Scissor-tailed Flycatcher (common), Great Crested Flycatcher, Eastern Phoebe, Eastern Pewee, Barn Swallow, Carolina Chickadee, Tufted Titmouse, Bewick's, and Carolina Wrens, Northern Mockingbird, Eastern Bluebird, Red-eyed Vireo, Western Meadowlark, Red-winged Blackbird, Summer Tanager, Northern Cardinal, Painted Bunting (common), Dickcissel, and Lark, Rufous-crowned, and Chipping Sparrows. A few Greater Roadrunners are permanent residents.

About the middle of September and again about 10 February,

migrating ducks begin to appear, stopping on the larger lakes. Nearly all the species commonly using the central flyway are represented and, in addition, a few Common Goldeneyes, Buffleheads, and Hooded Mergansers. Occasionally in both spring and fall there are large flocks of American Robins and, more rarely, a few Upland Sandpipers. During open winters the Refuge has a fairly large concentration of Chestnut-collared Longspurs.

The Refuge may be reached from Lawton by driving north through Fort Sill on US 281, turning off left onto State 49 and following it past Medicine Park to a point 8 miles beyond the eastern boundary of the Refuge, then turning right onto a road leading northwest directly to headquarters. Sunset Picnic Area can be reached by a road leading southwest from headquarters.

MANGUM

Reed Area | Jaybuckle Spring

Well worth exploring for birds is the **Reed Area,** 11 miles west of Mangum on State 9 in southwestern Oklahoma. In this arid region, between the Salt Fork and Elm Fork of the Red River, the gently rolling uplands are dominated by mesquite, short grasses, and prickly pear cacti. Along the larger streams, such as Cave Creek, are dense stands of woody growth, principally elm, hackberry, willow, cottonwood, sumac, wild plum, and skunkbush.

South and west of Reed are gypsum sink holes where subterranean water erosion of soft gypsum deposits has caused overlying layers to collapse, creating wide gaping holes or caves. Some of the larger caves are used in summer by thousands of free-tailed bats for roosting and rearing young, as well as by birds for nesting. Cliff Swallows establish their colonies under the overhanging rims and Barn Owls nest on ledges near the entrances. Sometimes a pair of Great Horned Owls may be observed at an entrance or a Canyon Wren heard singing in a cave.

Reached by road 4 miles north of Reed, and not far from the Elm Fork of the Red River, is **Jaybuckle Spring,** a large natural spring bordered by huge cottonwoods with shrubby edges and bottomland woods nearby. Nesting birds here and in the adjacent open country include the Mississippi Kite, Swainson's Hawk, Scaled Quail, Greater Roadrunner, Golden-fronted and Ladder-backed Woodpeckers, Western Kingbird, Scissor-tailed Flycatcher, White-necked Raven, Western Meadowlark, Bullock's Oriole, Blue Grosbeak, and Lark and Cassin's Sparrows. In the few remaining prairie dog 'towns' near the Spring, look for Burrowing Owls on tunnel mounds.

OKLAHOMA CITY

Lake Overholser | Lake Hefner | Martin Park and Nature Center

This capital city in the central part of the state lies on the undulating plain where remnants of the blackjack and post oak woodlands meet the tall-grass prairie.

Two large reservoirs, **Lake Overholser** and **Lake Hefner,** attract sizable aggregations of waterbirds during fall, winter, and spring. Flocks of migrating American White Pelicans, which occasionally appear about the last of April and the first of October, and migrating shorebirds, particularly abundant when reservoir levels are low in late spring and late summer, are but two of the impressive aggregations to be observed. Common breeding landbirds to look for near both reservoirs include the Yellow-billed Cuckoo, Red-bellied Woodpecker, Eastern and Western Kingbirds, Scissor-tailed Flycatcher, Blue-gray Gnatcatcher, Bell's Vireo, Baltimore and Orchard Orioles, and Lark Sparrow.

Lake Overholser (1,700 acres), west of the city's center, may be reached by exiting from I 240 and proceeding westward on US 66 (Northwest 39th Expressway) for 5.1 miles, then turning off south onto Lake Overholser Drive, which passes along the Lake's east side. This road is part of a paved roadway system which completely surrounds the Lake except at the dam on the south side. There are many good vantages overlooking the Lake along this roadway.

Among the large number of transient birds attracted to Lake Overholser are Pied-billed Grebes, Double-crested Cormorants, Great Blue Herons, Canada Geese, Greater White-fronted and Snow Geese, Mallards, Common Pintails, Northern Shovelers,

Redheads, Canvasbacks, Lesser Scaups, Ruddy Ducks, Common Mergansers, American Coots, Ring-billed and Franklin's Gulls, Black Terns, and shorebirds of good variety. During the breeding season, cattails at the north end of the Lake are often suitable habitat for nesting Pied-billed Grebes, Least Bitterns, American Coots, and, occasionally, Common Gallinules and King Rails.

Worth investigating in season, while one is in the area of Lake Overholser, are: (1) A *heron colony*, 1.4 miles east of the Lake, near the intersection of Northwest 27th Street and North Wilburn Avenue, reached from Lake Overholser Drive by turning east onto Northwest 30th Street and going 1.4 miles, then proceeding south on North Wilburn Avenue for three blocks to Northwest 27th Street. Nesting here are Little Blue Herons, Cattle, Great, and Snowy Egrets, and Great-tailed Grackles. (2) A *prairie dog colony*, 1.4 miles west of the Lake, reached by traveling west on US 66 for 2.5 miles from the North Canadian River Bridge at Lake Overholser, turning south onto Sarah Road and going 0.5 mile, then turning west onto Northwest 36th Street for 0.7 mile. Besides Burrowing Owls here in any season, one is likely to see longspurs—Lapland, Smith's, and Chestnut-collared—and an occasional Ferruginous Hawk in winter.

Lake Hefner (3,500 acres), northwest of the city's center, may be reached by exiting from I 240 on Grand Boulevard north to the Northwest Expressway (1.5 miles), then turning northwest and exiting north on Portland Avenue or Meridian Avenue (0.3 mile and 1.4 miles respectively from Grand Boulevard) to the paved road which completely encircles the Lake, less than a mile in both cases. With its shallow, muddy shoreline, the Lake attracts a fine variety of transient shorebirds, among them the Killdeer, Greater and Lesser Yellowlegs, Pectoral, White-rumped, Baird's, Least, Semipalmated, and Western Sandpipers, and Wilson's Phalarope. Other shorebirds which sometimes show up are the Piping, Lesser Golden, and Black-bellied Plovers, Whimbrel, Red Knot, Buff-breasted Sandpiper, and Red and Northern Phalaropes. Particularly productive is the shoreline at either end of the dam, reached by following northward to the dam the circumferential paved roads. Dirt roads lead to the water's edge in many places, including mud flats adjacent to the dam. All the transient species of waterbirds and waterfowl common on Lake Overholser are similarly common on Lake Hefner.

Nearby **Martin Park and Nature Center** (141 acres), a wooded, city-owned tract with a small lake, has a good representation of the common landbirds in the area. To reach the tract, exit northward on North Portland Avenue, which begins at the northeast corner of Lake Hefner's circumferential road, and proceed 2 miles, then turn westward onto Memorial Road (Northwest 136th Street) and go 1.2 miles.

STILLWATER
Lake Carl Blackwell Area

The best place for birds near Stillwater is the **Lake Carl Blackwell Area,** a 21,000-acre tract administered by Oklahoma State University. This includes fine stretches of tall-grass prairie and 3,300-acre Lake Carl Blackwell, a water impoundment. Much of the prairie is dissected by timbered ravines and interspersed with patches of oak woods, resulting in a vegetative pattern that is ideal for a wide variety of birds.

Nesting species in the Area include the Mississippi Kite, Red-tailed Hawk, Northern Harrier, Mourning Dove, Chuck-will's-widow, Eastern and Western Kingbirds, Great Crested and Scissor-tailed Flycatchers, Bewick's and Carolina Wrens, Northern Mockingbird, Brown Thrasher, Bell's Vireo, Northern Cardinal, Blue Grosbeak, Painted Bunting, and Lark and Field Sparrows. Common during migration, or in winter, are Snow Geese, Franklin's Gulls, Rufous-sided Towhees, American Goldfinches, Slate-colored Juncos, American Tree and Harris' Sparrows, and Lapland Longspurs.

To reach the Area from Stillwater, drive west on State 51 for 9 miles, then approximately 2.5 miles north, to the shore of Lake Carl Blackwell. Roads and trails lead around the Lake to many parts of the Area, most of which are open to the public.

TISHOMINGO
Tishomingo National Wildlife Refuge

Part of the Washita Arm of Lake Texoma, the huge reservoir formed by the Denison Dam across the Red River that forms Oklahoma's southern boundary, is occupied by **Tishomingo National Wildlife Refuge** (16,464 acres). Roughly two-thirds of the Refuge consists of the open water of the Lake and its marshy shore. Because the water level fluctuates greatly, typical marsh conditions cannot be permanently established; nevertheless, extensive shoreline plantings of emergent aquatics, such as smartweed and wild millet, provide food attractive to ducks. The remaining third of the Refuge is upland—the river-valley terrain lying above water level. Here there are high hills, forested in places, and numerous small valleys where creeks flow into Lake Texoma. In the north-central part are flat, open areas, formerly farmlands, planted with crops suitable for waterfowl.

The principal ornithological attractions on the Tishomingo Refuge are the hordes of waterbirds, geese, and ducks that gather in late summer and fall. During August and September, Great Blue Herons and Great and Snowy Egrets are especially numerous. By September large numbers of American White Pelicans, Double-crested Cormorants, American Coots, and Franklin's Gulls begin to appear, reaching their peaks of abundance in early October. Frequently the concentrations of Franklin's Gulls are spectacular.

From September to December the following waterfowl species appear on the Refuge: Canada, Greater White-fronted, and Snow Geese, Mallard, American Black Duck, Gadwall, Common Pintail, Green-winged and Blue-winged Teal, American Wigeon, Northern Shoveler, Wood Duck, Redhead, Ring-necked Duck, Canvasback, Lesser Scaup, Common Goldeneye, Bufflehead, Ruddy Duck, and Hooded and Common Mergansers.

At peak abundance in November, the duck population usually amounts to well over 65,000 individuals, of which 40,000 are Mallards; the goose population varies from 20,000 to 30,000 individuals, most of which are Canada and Snow Geese. A great many of the waterfowl remain on the Refuge through winter.

The Tishomingo Refuge is 3 miles southeast from the town of

Tishomingo. To reach the Refuge, leave State 78 at the eastern edge of town on a gravel road that leads directly to Refuge headquarters. Most parts of the Refuge are accessible by road or trail, but first permission and directions for using them should be obtained at headquarters.

TULSA
Mohawk Park | Oxley Nature Center

The great oil refineries, skyscrapers, fine suburban homes, and tree-shaded parks of this big municipality occupy the rolling terrain on the east side of the Arkansas River in northeastern Oklahoma.

Mohawk Park, northeast of the city, surpasses all other places in or close to the city for bird finding the year round. Within this city-owned property of 2,830 acres are Lake Yahola—a reservoir of 400 acres—and the North Woods, both of which are particularly rewarding.

To reach Lake Yahola in the western section of the Park, exit north from I 244 on US 75 for approximately 3.5 miles. Exit on 36th Street North, turn left and proceed east for 0.2 mile to a traffic light on Harvard Avenue; follow Harvard north for 0.5 mile and turn right onto Mohawk Boulevard which soon passes the Lake on its south side. Enter the parking lot on the lakeside, across the Boulevard from the city water plant. The Monument at the top of the steps is a good spot for observing waterbirds in all sections of the Lake, especially in the afternoon. To reach North Woods, as well as another vantage for viewing birds on the Lake, continue on Mohawk Boulevard for 1.0 mile; then turn left onto the gravel road that leads to the Sequoyah Yacht Club on the east side of the Lake. Park here and take foot paths northward into the Woods or westward to the Lake.

From early October through March, Lake Yahola attracts large numbers of waterfowl including geese and seventeen species of ducks, many of which stay through the winter. During both spring and fall migrations, one may expect to see Common Loons, Horned and Eared Grebes, Ospreys, Franklin's and Bonaparte's

Gulls, and terns—Forster's, Little, Caspian, and Black. From 14 to 28 April and from 21 September to 15 October the Lake at times swarms with swallows. In late August through September at twilight, Chimney Swifts can be watched entering the water plant's tall chimney by the thousands for roosting. The best time for transient shorebirds is from mid-July through September during periods of high water usage. Exposed gravel bars near the Monument and the fishing pier on the east bank permit easy viewing. Plovers—Semipalmated, Piping, Lesser Golden, and Black-bellied—may be seen as well as an additional eighteen shorebird species.

The North Woods, besides its stands of various oaks, sycamore, hackberry, cottonwood, green ash, boxelder, pecan, hickory, honey locust, redbud, and persimmon, has numerous shrubby thickets of roughleaf dogwood, wild plum, green briar, and sumac. Poison ivy climbs old trees with arm-sized 'ropes,' but the paths are relatively clear of it.

Nesting in North Woods are Yellow-crowned Night Herons, Red-tailed, Red-shouldered, and Broad-winged Hawks, Common Screech, Great Horned, and Barred Owls, Yellow-shafted Flickers, and Red-bellied, Red-headed, Hairy, and Downy Woodpeckers. The Pileated Woodpecker resides throughout Mohawk Park away from well-traveled areas. Nesting tyrannids are the Eastern Kingbird, Scissor-tailed and Great Crested Flycatchers, Eastern Phoebe, and Eastern Pewee. Other breeding birds include the Carolina Chickadee, Tufted Titmouse, White-breasted Nuthatch, Carolina Wren, Gray Catbird, Brown Thrasher, Eastern Bluebird, and Blue-gray Gnatcatcher. The bubbly song of Bell's Vireos is heard from May into September in shrubbery and vines around the Lake and along sunny roadsides. Around wet areas and ponds where there are willows, Prothonotary Warblers begin nest-hunting by 17 April. Moist woodland is nesting habitat for Yellow-throated and Red-eyed Vireos and the American Redstart, the shrubby edges for the White-eyed Vireo. The Summer Tanager can usually be observed among the elms and tall oaks in the more open areas. In winter, Swamp and Song Sparrows frequent weedy patches. Field Sparrows are common the year round.

Along the dike trail east of the Yacht Club may be found wintering Rufous-sided Towhees and White-throated, Fox, and Lincoln's

HARRIS' SPARROW

Sparrows, and, with luck, a Winter Wren. Transient thrushes, including the Hermit and Swainson's, come to the forest floor; transient vireos, including the Philadelphia, and many warblers appear around the small pond. In summer, Ruby-throated Hummingbirds and Indigo Buntings perch on wires overhead, and the pond itself, rimmed by willows and buttonwoods, attracts Green Herons and both Black-crowned and Yellow-crowned Night Herons.

Returning to Mohawk Boulevard, turn left and continue 0.8 mile along the golf course to the gate of the **Oxley Nature Center** on the eastern edge of Mohawk Park. This is a limited-access sanctuary embracing 700 acres of open fields, marsh, and wood-bordered Lake Nibathe. Park the car at the gate under sycamores and listen for summer-resident Northern Parula and Yellow-throated Warblers. Drive through the gate for 0.6 mile and park near a tall black walnut tree on the right. From here one may walk south into the old tree nursery—be alert for summer-resident Kentucky Warblers and Blue Grosbeaks—or walk through the gate across the road into a wide clearing. During winter here in weeds and crabgrass, expect flocks of American Tree Sparrows; during spring and fall, Clay-colored Sparrows. In the brushy edges, Harris' Sparrows are present from October through April; almost as soon as they leave, Painted Buntings return for the summer. At the back of the clearing along Bird Creek, look for Wood Ducks at any time of the year.

Continue down the road to the Oxley Shelter and park. Walk the trails along Lake Nibathe and Coal Creek, looking for ducks, including Hooded Mergansers, in winter, herons and egrets in summer, and transient shorebirds in season. Warbling Vireos nest in the cottonwoods and elms on the Lake's north shore, as do Orchard and Baltimore Orioles. Yellow Warblers may be expected from April through September. Common Snipes are frequent through winter in the low, wet, grassy area at the east end of the north shore.

VIAN
Sequoyah National Wildlife Refuge

In east-central Oklahoma, at the confluence of the Canadian and Arkansas Rivers, lies the **Sequoyah National Wildlife Refuge.** Half of its 20,800 acreage embraces part of the Robert S. Kerr Reservoir and is consequently open water and cattail marsh. The other half consists of grasslands, upland and bottomland woods of oaks, elms, hickories, willows, and cottonwoods, and farmlands for the production of soybeans, corn, and alfalfa. Refuge headquarters is reached from I 40 by taking the Vian Exit and proceeding south for 3 miles, following signs.

Birds known to nest on the Refuge include the Double-crested Cormorant, American Anhinga, Bald Eagle, Red-headed Woodpecker, and Indigo and Painted Buntings. In all probability such species as the Wood Duck, Hooded Merganser, Red-shouldered Hawk, Barred Owl, Pileated Woodpecker, and Swainson's Warbler also nest on the Refuge.

During fall and spring, transient geese and ducks are abundant; many remain through the winter. In January and February, the number of Canada, Greater White-fronted, and Snow Geese may exceed 25,000, the number of Mallards can be as high as 250,000. Associated with the wintering waterfowl concentrations are usually twenty or more Bald Eagles, occasionally a Golden Eagle, often perched in view on dead timber.

Oregon

TUFTED PUFFIN

An unforgettable and rewarding trip for the bird finder in Oregon is along US 101, which follows the 400-mile coastline between Washington and California. For much of the way the Pacific is in view on one side of the highway and the forested slopes of the Coast Range on the other. Many times the highway cuts through or works its way around various mountain spurs that end in rocky headlands or dip into the sea, their pinnacles still exposed as rocky islets. The bolder headlands, and especially the islets that reach well above tide line, provide nesting sites for a multitude of sea-associated birds, notably the Fork-tailed Storm Petrel, Leach's Storm Petrel, Double-crested Cormorant, Brandt's Cormorant, Pelagic Cormorant, Black Oystercatcher, Western Gull, Thin-billed Murre, Pigeon Guillemot, Rhinoceros Auklet, and Tufted Puffin.

Where there are undisturbed sand beaches, the Snowy Plover is an occasional permanent resident. Immediately back from some of the beaches is a narrow strip, rarely more than a mile wide, which supports a dense shrubby growth consisting largely of salal, shot huckleberry, wax-myrtle, salmonberry, kinnikinnick, hairy manzanita, and rhododendron. The Wrentit is practically confined to this strip, except in southwestern Oregon, and a considerable list of bird species that prefer brushy habitat also reside here. In addition, there are scattered trees and extensive pure stands of lodgepole pine and Sitka spruce in which Purple Finches, Pine Siskins, and Red Crossbills are among the common birds.

For some of the best bird-finding areas along the Oregon coast, *see under* **Astoria, Bandon, Brookings, Cannon Beach, Coos Bay-North Bend, Newport,** *and* **Tillamook.**

The 96,000 square miles of Oregon are divided into sections by various mountain ranges, the highest of these being the Cascades. Several peaks exceed 10,000 feet; Mt. Hood in the extreme north reaches 11,235 feet, the highest point in the state. About 100 miles inland, the Cascades parallel the coast from the Columbia River, at the northern boundary, to the California line, thus dividing the state into a western section, comprising about one-third of the state's area, and an eastern section, which covers the remaining two-thirds.

West of the Cascades, the climate of Oregon is tempered by prevailing westerly winds from the Pacific Ocean. Winter temperatures are usually mild for so northern a latitude, and winter rainfall is heavy. The result is luxuriant forest vegetation in most areas, accompanied by birds typical of such a habitat. Most of eastern Oregon lies in the rain shadow of the Cascades so that rainfall is greatly reduced, to 20 inches or less. The climate here is more subject to influences from the interior of the continent, and low temperatures are the rule during the winter. Much of this part of the state supports only sagebrush or similarly sparse vegetation, and the birdlife consequently mirrors this difference in climate and plant cover.

The picture is interrupted in the northeast, however, by the Blue Mountains, which extend southwestward almost to the Cascades from the Rocky Mountains, with which they are contiguous. The highest and most scenic of the Blue Mountains form a group, the Wallowa Mountains, which attain elevations between 9,000 and 10,000 feet.

The mountains of western Oregon form a U. The eastern arm includes the Cascades; the western arm is formed by the lower Coast Range. The average elevation of the Coast Range is 2,000 feet, with several peaks reaching above 4,000 feet. These coastal mountains stand very close to the sea, leaving only narrow beaches, or none at all, and a few sizable bays such as Tillamook, Yaquina, and Coos. The base of the U is disproportionately thick, and is made up of hills running roughly east-west and of higher ridges. Of these cross-ranges the most impressive are the Siskiyou Mountains at the California line; their maximum elevation is reached on Mt. Ashland at 7,538 feet. The cross-ranges are separated by rivercourses, some of which—e.g., the Rogue and Umpqua—have formed valleys, now of agricultural importance. Between the arms of the U is the well-known Willamette Valley, a relatively flat agricultural area, extending 130 miles south from the Columbia River, and averaging 30 miles in width. The state's largest cities—Portland, Salem, and Eugene—lie along the Willamette River, which rises in the Cascades north of Crater Lake and flows north to join the Columbia.

The narrow coastal strip of Oregon, already described, merges into the fir-hemlock forests that cover so much of the state's western part. Typically, before the advent of lumbering, these forests were dominated by splendid forest giants—Douglas-fir and western hemlock—with a liberal sprinkling of grand fir and western red cedar, but such mature stands have been vastly depleted by clear-cutting and fire. Any recently denuded area is first covered mostly with herbaceous plants. In a few years, shrubs and deciduous trees take over and then gradually—over a long period—conifers. Traveling through the coast forest, the bird finder traverses all stages of this succession and the consequent variety of bird habitats. The following bird species breed in the forests of the

Coast Range and the higher cross-ranges and on the west slopes of the Cascades up to about 5,500 feet. Species marked with an asterisk reside mainly in deciduous growth.

Northern Goshawk
Blue Grouse
* Ruffed Grouse
* Mountain Quail
Band-tailed Pigeon
Northern Pygmy Owl
Pileated Woodpecker
Hammond's Flycatcher (*except Coast Range*)
* Dusky Flycatcher (*except Coast Range*)
Olive-sided Flycatcher
Gray Jay
Steller's Jay
Chestnut-backed Chickadee
Red-breasted Nuthatch
Brown Creeper
Winter Wren
Varied Thrush
Swainson's Thrush
Townsend's Solitaire
Golden-crowned Kinglet
* Nashville Warbler
Audubon's Warbler
Black-throated Gray Warbler
Townsend's Warbler (*above 4,000 feet*)
Hermit Warbler (*below 4,000 feet*)
* MacGillivray's Warbler
Western Tanager
* Black-headed Grosbeak
Evening Grosbeak
Purple Finch
Pine Siskin
Red Crossbill
Oregon Junco
* White-crowned Sparrow
* Fox Sparrow (*except Coast Range*)

The Willamette Valley and the other large valleys west of the Cascades have diverse habitats for birds. In addition to extensive farmlands (fields, pastures, orchards, and vineyards) are deciduous trees and shrubs growing abundantly along watercourses and roadsides, in city parks and dooryards, and adjacent foothills. Among the birds breeding regularly in the open country (farmlands and shrubby places) and in the more undisturbed wooded areas are the following:

OPEN COUNTRY

Turkey Vulture
California Quail
Mourning Dove
Barn Swallow

Scrub Jay
Bushtit
House Wren
Bewick's Wren
Western Bluebird
Orange-crowned Warbler
Yellow Warbler
MacGillivray's Warbler
Common Yellowthroat
Yellow-breasted Chat
Western Meadowlark
Lazuli Bunting
House Finch
American Goldfinch
Lesser Goldfinch
Rufous-sided Towhee
Vesper Sparrow
Chipping Sparrow
Song Sparrow

WOODED AREAS

Cooper's Hawk
Ruffed Grouse
Common Screech Owl
Red-shafted Flicker
Acorn Woodpecker
Yellow-bellied Sapsucker
Hairy Woodpecker
Downy Woodpecker
Western Kingbird (*southern valleys*)
Western Flycatcher
Western Pewee
Black-capped Chickadee
Hutton's Vireo (*mixed coniferous-deciduous woods*)
Solitary Vireo (*mainly coniferous woods*)
Warbling Vireo (*mainly deciduous woods*)
Bullock's Oriole
Black-headed Grosbeak

Eastward out of the Willamette Valley and upward into the high Cascades, heavy stands of Douglas-fir and western hemlock give way, at an altitude of 5,500 feet, to forests of lodgepole pine, western white pine, Engelmann spruce, and silver fir. These are succeeded, beginning at about 6,000 feet, by subalpine forests in which the dominant trees are the subalpine fir, mountain hemlock, and—especially at the timber line—the whitebark pine. In these coniferous forests of the high Cascades are such regularly breeding birds as the Black-backed Three-toed Woodpecker, Gray Jay, Clark's Nutcracker, Mountain Chickadee, Hermit Thrush, Mountain Bluebird, Townsend's Solitaire, Ruby-crowned Kinglet, and Cassin's Finch. The Lincoln's Sparrow is a summer resident in shrubby thickets bordering streams and wet places; the White-crowned Sparrow is a summer resident, particularly at the timber line. Above the timber line, at elevations ranging from 6,000 feet and higher, depending on slope and exposure, is an alpine region of scant vegetation—summer habitat of the Horned Lark, Water Pipit, and Gray-crowned Rosy Finch. Throughout the high Cascades are many mountain lakes, some of which attract breeding pairs of Barrow's Goldeneyes. Along swift streams coming down from the mountains nest Harlequin Ducks and North American Dippers.

On the eastern slope of the Cascades, below the forest of lodgepole pine, western white pine, Engelmann spruce, and silver fir, are park-like stands of ponderosa pine and intermittent groves of aspen, forming a wide belt in which characteristic breeding birds include the (Red-naped) Yellow-bellied Sapsucker, Williamson's Sapsucker, Hairy Woodpecker, White-headed Woodpecker, White-breasted Nuthatch, Pygmy Nuthatch, Evening Grosbeak, Cassin's Finch, and Fox Sparrow. At its lower edge the pine belt merges with the juniper woodland, of which the western juniper is the principal constituent. Breeding birds typical of this environment are the Ash-throated Flycatcher, Gray Flycatcher, Pinyon Jay, Mountain Bluebird, and Black-throated Gray Warbler. Below the juniper woodland begins the sagebrush country.

East of the Cascades, ponderosa pine clothes the slopes of mountains and foothills at similar elevations. The juniper woodland is not generally so noticeable and is sometimes lacking. In stands of ponderosa pine, where bushes form an undergrowth, the Green-tailed Towhee is a regular nesting bird. The high Blue and Wallowa Mountains in the northeast (*see under* **La Grande**) show a vertical succession of vegetation from sagebrush country to alpine region that is virtually the same as that on the eastern slope of the Cascades. Birdlife is about the same too.

The sagebrush country comprises much of eastern Oregon below the pine belt and/or juniper woodland. The most conspicuous and widespread plant is the sagebrush, but where there are alkaline soils such shrubs as greasewood and rabbitbrush thrive. Overall, the sagebrush country is high plateau, flat, semiarid, and desert-like in aspect, its monotony relieved here and there by low mountain ranges, ridges, and buttes. In the sagebrush country, the following nesting birds may be regularly expected:

Red-tailed Hawk
Swainson's Hawk
Golden Eagle
Prairie Falcon
American Kestrel
Sage Grouse (*local*)
Mourning Dove
Burrowing Owl
Poor-will
Horned Lark
Black-billed Magpie
Northern Raven
Sage Thrasher
Mountain Bluebird
Loggerhead Shrike
Lark Sparrow
Sage Sparrow
Brewer's Sparrow
Black-throated Sparrow (*local*)

The sagebrush habitat is interrupted by streams and rivers usually bordered by deciduous trees and brush. Bird species inhabiting these riparian situations are generally the same as in like situations in western Oregon. Conspicuous breeding birds in deciduous growth near streams or around ranch buildings are the Eastern Kingbird, Western Kingbird, Say's Phoebe, Bullock's Oriole, and House Finch. In northeastern Oregon, riparian situations of this type bring into the state certain bird species of eastern affinities such as the Gray Catbird, Red-eyed Vireo, and American Redstart.

Finally, in southeastern Oregon's sagebrush country, despite its being semiarid, there are nonetheless lakes, ponds, marshes, and wet meadows in widely scattered localities. All these watery environments are feeding and/or breeding grounds for great numbers and varieties of waterbirds, waterfowl, shorebirds, and marsh-loving birds. The following species are regular summer inhabitants in one watery area or another, depending on their particular preferences:

Eared Grebe
Western Grebe
Pied-billed Grebe
American White Pelican
Great Blue Heron
Great Egret
Snowy Egret
Black-crowned Night Heron

American Bittern
White-faced Ibis (*uncommon*)
Canada Goose
Mallard
Gadwall
Common Pintail
Blue-winged Teal
Northern Shoveler
Canvasback
Lesser Scaup
Ruddy Duck
Northern Harrier
Sandhill Crane
Virginia Rail
Sora
Killdeer
Common Snipe
Long-billed Curlew
Willet
American Avocet
Black-necked Stilt
Wilson's Phalarope
California Gull
Ring-billed Gull
Franklin's Gull (*uncommon*)
Forster's Tern
Caspian Tern
Black Tern
Marsh Wren
Common Yellowthroat
Yellow-headed Blackbird
Red-winged Blackbird
Tricolored Blackbird (*Klamath Basin*)
Brewer's Blackbird

Landbird migrations in Oregon are seldom spectacular and can be disappointing to any one familiar with the mass movements in eastern and middlewestern United States; but the flights of waterbirds, waterfowl, and shorebirds are always impressive in both numbers of species and numbers of individuals. In Oregon, the main migratory flights may be expected within the following dates:

Waterfowl: 10 March–20 April; 10 October–25 November
Shorebirds: 15 April–1 June; 1 August–1 October
Landbirds: 10 April–25 May; 20 August–10 October

Winters in western Oregon, being mild, hold great numbers of waterbirds, waterfowl, and landbirds. Bays and other coastal waters sheltered from the sea throng with loons, grebes, brant, ducks, and gulls. Certain landbird species summering in the mountains move down to lower elevations; others moving in from the north—e.g., the Rough-legged Hawk, Northern Shrike, and Golden-crowned Sparrow—frequently remain. Winters in eastern Oregon, although comparatively severe, nevertheless hold impressive numbers of waterbirds and waterfowl. Landbird visitants, besides the Rough-legged Hawk and Northern Shrike, include the Common Redpoll, American Tree Sparrow, and Snow Bunting.

Authorities

Kenneth C. Batchelder, Donald S. Farner, David B. Marshall, Harry B. Nehls, Fred L. Ramsey, Robert M. Storm.

Reference

Birding Oregon. By Fred L. Ramsey. Corvallis: Audubon Society of Corvallis. 1978.

ASTORIA
Fort Stevens State Park | South Jetty at the Mouth of the Columbia River

Near this city close to the mouth of the Columbia River, ornithology may be said to have had its beginnings in Oregon when the explorers Meriwether Lewis and William Clark, after their long, arduous journey down the River, arrived in October 1805. While camped for the winter, they wrote in their journal on 2 January 1806 about 'the large as well as the small or whistling swan' and other birds including 'the bald eagle.' At the prsent time from November to March, if one drives east of Astoria along the Columbia from Tongue Point to Svensen, he will see wintering flocks of Whistling Swans, small numbers of 'the large' Trumpeter Swans, and Bald Eagles. Six miles west of Astoria on US 101 is the Fort Clatsop National Historical Site on the actual spot of the Lewis and Clark encampment.

West of Astoria there are many fine areas for birds which are well worth exploring. Outstanding are **Fort Stevens State Park** on the Pacific Ocean and areas north to the **South Jetty at the Mouth of the Columbia River.** To reach the Park, drive west from Astoria on US 101 for about 8 miles, then turn off north (after passing two turnoffs north to Warrenton) onto Ridge Road; the Park is on the left.

Fort Stevens State Park (3,763 acres) extends to the ocean beach where one may drive his car on the sand in either direction.

Within the Park's 288 acres back from the beach is a large body of water, Coffenbury Lake. From the parking lot at the north end, a 2-mile trail leads around the Lake, passing through, or adjacent to, dense alder thickets on the east side, a fir-hemlock forest on the west. In the alder thickets, look especially for the permanent-resident Wrentit, which reaches here the northern limit of its range in the United States. Other birds breeding in one habitat or the other include the Blue Grouse, Ruffed Grouse, Vaux's Swift, Rufous Hummingbird, Western Flycatcher, Western Pewee, Olive-sided Flycatcher, Winter and Bewick's Wrens, Hermit and Swainson's Thrushes, five warblers—Orange-crowned, Yellow, Audubon's, Black-throated Gray, and Wilson's—Western Tanager, Black-headed Grosbeak, Purple Finch, Pine Siskin, and Red Crossbill. Townsend's and Hermit Warblers appear in small numbers during migration.

From Fort Stevens Park, continue on the road north and park near the base of the South Jetty. In the area are dunes where a few Snowy Plovers nest, Horned Larks and Savannah Sparrows reside permanently, and Lapland Lonspurs and Snow Buntings are winter visitants. There are also boggy spots where Marsh Wrens are permanent residents and Common Yellowthroats are abundant in summer.

Along the ocean beach or from the South Jetty itself (*Caution:* Do not go far out as big waves commonly crash over it), one may view fall flights of Common, Arctic, and Red-throated Loons, White-winged, Surf, and Black Scoters, Red and Northern Phalaropes, and Thin-billed Murres. In August, September, and later, many thousands of Sooty Shearwaters, often accompanied by other shearwaters such as the Pink-footed, pass by. Black-legged Kittiwakes are likely to be seen any time.

Around small ponds among the dunes as well as along the beaches on either side of the Jetty, gulls, and shorebirds frequently rest and feed during the colder months. One may expect a variety of gulls—e.g., Glaucous-winged, Western, Bonaparte's, and Heermann's. Common, Arctic, and Caspian Terns always linger in migration, with an occasional jaeger—Pomarine, Parasitic, or Long-tailed—shadowing them for a share of their fish quarry. Besides the more common transient shorebirds, such uncommon species as the Whimbrel, Wandering Tattler, Red Knot, and Pectoral and Baird's Sandpipers are regular, particularly in late summer or early fall.

BANDON
Coquille Point | Bandon Ocean State Wayside

From this community on the southwestern Oregon coast south to the California border are many offshore reefs and rocks that provide resting or nesting sites for Double-crested and Pelagic Cormorants, Black Oystercatchers, Western Gulls, Thin-billed Murres, Pigeon Guillemots, Marbled Murrelets, and Tufted Puffins. A good place from which to observe these birds is **Coquille Point.** US 101, entering Bandon from the east, makes a broad curve to the south. Three blocks south from the southern end of the curve, take the road going west to its end, 1.0 mile distant. From here, walk to the Point, where there is a Coast Guard lookout.

At **Bandon Ocean State Wayside** (15 acres), in Bandon itself adjacent to the base of the South Jetty of the Coquille River, large numbers of transient Surfbirds, Black Turnstones, and Wandering Tattlers and small numbers of transient Ruddy Turnstones and Rock Sandpipers gather at high tide and are easily observed. Follow the signs from US 101 in Bandon to this coastal spot.

BEND

This city in central Oregon is a hub of major highways radiating westward over the Cascades and eastward across sagebrush and juniper flats, often relieved by buttes and other conspicuous geologic formations. In view on the west are the snow-capped Three Sisters, towering over lesser peaks at an elevation above 10,000 feet. Suggested for the bird finder in early summer are three trips from Bend, all highly scenic but selected to show the impressive diversity of bird species within a radius of 50 miles.

1. From Bend (elevation 3,630 feet), drive northwest on US 20 for 21 miles to the little town of Sisters. As the highway traverses

open stands of juniper, watch for Pinyon Jays. Flocks frequently pass back and forth over the highway.

On arriving in Sisters, drive east on State 126 toward Redmond until the highway begins crossing juniper-sagebrush flats, habitat worth inspecting for such birds as Ash-throated and Gray Flycatchers and Brewer's Sparrows.

Backtrack to Sisters and proceed west off State 242 (McKenzie Pass Highway), which enters forests of ponderosa pine, to Cold Spring Campground, 4 miles distant. Here take gravel Forest Service Road 1424 north (unsuitable for travel in bad weather) for about 2.5 miles and turn left onto US 20 (Santiam Pass Highway) and go 1.0 mile to Indian Ford Campground. The pine forests about both Campgrounds are passed through by the gravel road, and the deciduous thickets along the creeks are ideal for bird finding. In the forests, search for the (Red-naped) Yellow-bellied and Williamson's Sapsuckers, White-headed Woodpecker, Western Pewee, Pygmy Nuthatch, Green-tailed Towhee, and Fox Sparrow; in the thickets, Traill's and Dusky Flycatchers and such warblers as the Orange-crowned, Nashville, MacGillivray's, and Wilson's. Always be alert for the sight or sounds of the Northern Pygmy and Saw-whet Owls.

From Indian Ford Campground, continue northwest and west on US 20 for about 14 miles, crossing the crest of the Cascades at Santiam Pass (elevation 4,817 feet) to Lost Lake Campground. Scan Lost Lake itself for pairs or broods of Barrow's Goldeneyes and explore the surrounding forest of Douglas-fir and other conifers for Blue Grouse, Black-backed Three-toed Woodpeckers, possibly a pair of Northern Three-toed Woodpeckers, Hermit Thrushes, Townsend's and Hermit Warblers, and, in brushy spots, Fox and Lincoln's Sparrows.

2. From Bend, drive southwest on State 46 (Cascade Lakes Highway) along the eastern slopes of the Cascades through pine forests and close to ponds, lakes, and reservoirs. Breeding birds typical of these forests are Hermit Warblers, Evening Grosbeaks, Cassin's Finches, and Oregon Juncos. One or both species of three-toed woodpeckers are possible. Where some of the larger bodies of water are bordered by marshes, expect to see Eared and Western Grebes, Forster's Terns, Wilson's Phalaropes, as well as

several species of ducks. At the Crane Prairie Reservoir, on State 46 about 50 miles from Bend, state and federal agencies maintain and protect a large breeding population of Ospreys. Some of the birds may be readily viewed with a spotting scope.

3. From Bend, drive south on US 97 for 4 miles, then turn off left on Forest Service Road 1821 which goes southward for nearly 50 miles through forests of ponderosa pine and eventually crosses sagebrush flats and rimrock country to Fort Rock State Monument and beyond. Since much of the area is quite dry in summer, birds tend to concentrate where water is available. One such place is Cabin Lake Guard Station about 40 miles from Bend. Here there is a spring that attracts so many birds—e.g., Gray and Steller's Jays, Red Crossbills, Green-tailed Towhees, juncos, and sparrows—that the Forest Service has built a stone blind near the spring for anyone wishing to photograph them coming for water.

South from Cabin Lake, the Road emerges from the forest onto sagebrush flats where breeding birds include Swainson's Hawks, Loggerhead Shrikes, Sage Thrashers, and Sage and Brewer's Sparrows. About 10 miles ahead, Fort Rock—remnant of a volcanic cone—looms some 300 feet above the flats. In its clefts nest many White-throated Swifts and its many ledge shelves provide nesting sites for Red-tailed Hawks, Golden Eagles, Prairie Falcons, Great Horned Owls, and Northern Ravens.

BROOKINGS

South Jetty | North Jetty | Harris Beach State Park

The two jetties at the mouth of the Chetco River, on the extreme southwestern Oregon coast just north of the California border, jut far enough oceanward to offer good views of waterbirds and waterfowl frequenting the coast in season. Both jetties are easily accessible by car. **South Jetty** is reached by turning west from US 101 at the south end of the Chetco River Bridge; then, within 100 yards, turning right at a sign labeled Coast Guard and proceeding to a parking area at the foot of the Jetty. **North Jetty** is reached by turning west from US 101, north of the Chetco River Bridge, and driving up a hill into Brookings; then turning left onto Alder Street and left onto Railroad Street to Del Norte Street which dead-ends at a parking area near the Jetty.

Harris Beach State Park (171 acres), on US 101, 2 miles north of Brookings, has Wrentits regularly residing in brushy habitat opposite the entrance to the campground.

BURNS
Malheur National Wildlife Refuge | Wright's Point | Steens Mountain

Long famous as one of the most important of all the federally owned refuges for its vast numbers of breeding waterbirds and waterfowl, **Malheur National Wildlife Refuge** (181,000 acres) lies in southeastern Oregon's arid Harney Basin, bounded by the Blue Mountains on the north and by Steens Mountain on the southeast, and has no outlet to the sea. The floor of the Basin (elevation about 4,100 feet), though interrupted by occasional rock-capped buttes, is generally flat and is margined by rimrock cliffs. Shallow Malheur Lake (50,000 acres in size during years of high water) is largely bulrush-cattail marsh with some open water. Harney Lake (30,000 acres during high water), a sump of the Basin, is alkaline and without vegetation except for a few spring-fed bulrush patches. The remainder of the Basin's floor, aside from its small ponds, large marshes, and grassy wet meadows, supports a plant growth typical of semidesert areas, with sagebrush on the uplands, mixed with juniper at higher elevations, and greasewood, rabbitbrush, and giant ryegrass on the alkaline flats.

Spring bird migration is at its peak in March and April, fall migration in September and October. To see nesting birds, a visit should be made during May, June, or early July. The most outstanding breeding species are probably the Trumpeter Swan and Sandhill Crane. American White Pelicans are present in the breeding season but do not always nest. Among the other breeding species regularly associated in one way or another with open water and marshes are the Eared and Western Grebes, Great and Snowy Egrets, Black-crowned Night Heron, American Bittern, White-faced Ibis, Canada Goose, Mallard, Gadwall, Common Pintail,

Blue-winged and Cinnamon Teal, American Wigeon, Northern Shoveler, Redhead, Canvasback, Lesser Scaup, Ruddy Duck, Northern Harrier, Virginia Rail, Sora, American Coot, Snowy Plover (on alkaline flats), Common Snipe, Long-billed Curlew and Willet (grassy areas near marshes), American Avocet and Black-necked Stilt (in scant grasses on alkaline flats), Wilson's Phalarope (near open water), California, Ring-billed, and Franklin's Gulls, Forster's and Black Terns, Marsh Wren, and blackbirds—Yellow-headed, Red-winged, and Brewer's.

Upland birds regularly breeding on the Refuge or in its vicinity include the Sage Grouse (uncommon), California Quail, Ring-necked Pheasant, Chukar, Great Horned and Short-eared Owls, Poor-will (uncommon), Common Nighthawk, Eastern Kingbird (uncommon), Western Kingbird, Say's Phoebe (around rock formations or buildings), Traill's Flycatcher, Tree, Barn, and Cliff Swallows, Black-billed Magpie, Northern Raven, American Robin, Mountain Bluebird, Loggerhead Shrike, Warbling Vireo, Yellow Warbler, Common Yellowthroat, Bobolink, Western Meadowlark, Bullock's Oriole, and Savannah and Song Sparrows.

The Malheur Refuge lies more than 20 miles south of Burns, a frontier town on US 20 in the northern part of the Harney Basin. To reach the Refuge from Burns, drive east for 2 miles on US 20, then turn south on State 205. For the first several miles this highway traverses a series of wet fields and meadows with a drainage ditch paralleling the road most of the way. From the car, expect to see many of the bird species mentioned above as residing on the Refuge. After about 8 miles, the highway rises to **Wright's Point,** a sagebrush-covered plateau, well worth a stop or two for finding the Sage Thrasher, Lark and Brewer's Sparrows, and possibly the Black-throated and Sage Sparrows. Look for Rock Wrens on the rough slopes. From Wright's Point, the highway drops down to the alkaline flats. Watch for Swainson's Hawks and Burrowing Owls sitting atop utility poles.

After driving south for 24 miles on State 205 and entering the Refuge, turn off east for 6 miles to headquarters, where there are maps, a bird checklist, and leaflets available. Inquire about sites for viewing such localized birds as Trumpeter Swans, Sage Grouse, Sandhill Cranes, and Bobolinks. Also obtain information about taking the self-guided, 42-mile interpretive auto route from headquarters south through the Refuge. The shrubby grounds around headquarters, shaded by cottonwoods, and—just north of headquarters—the wet meadows edging the great bulrush-cattail marsh in Malheur Lake, offer some of the best bird finding on the Refuge.

From headquarters, backtrack to State 205 and follow it south for 34 miles to Frenchglen. The highway passes along the west side of the Refuge; side roads to ponds provide excellent vantages for observing waterfowl and numerous other birds.

Southeast of the Refuge looms **Steens Mountain,** a range running northeast-southwest, its shoulders covered with sagebrush, juniper, and aspen. On its bare, rocky crest, whose maximum elevation attains 9,733 feet, are Horned Larks, Rock Wrens, Water Pipits, and the possibility of Gray-crowned and/or Black Rosy Finches, especially near snowfields. The eastern slope of the Mountain drops abruptly for about a mile to the Alvord Basin, whereas the western slope is gradual, broken up by deep canyons, and high rimrocks. To reach the crest of Steens Mountain, take the well-marked Steens Mountain Loop Road that proceeds from Frenchglen eastward for 17 miles past Page Spring and Fish Lake Campground and up the western slope. From the crest, continue on the Loop Road down the western slope another way to State 205, 10 miles south of Frenchglen. During the ascent and descent, stop frequently to listen for Canyon Wrens and to scan the sky for a soaring Golden Eagle and the rimrocks for a Prairie Falcon. Since deep winter snows close the Loop Road, it is usually passable only from mid-July through October.

CANNON BEACH

Haystack Rock | Ecola State Park | Saddle Mountain State Park

The town of Cannon Beach, in one of the more scenic sections of the northwestern Oregon coastline, is on US 101, 25 miles south of Astoria. Just offshore looms **Haystack Rock** where Double-crested, Brandt's, and Pelagic Cormorants, Black Oystercatchers, Western

Gulls, Pigeon Guillemots, and Tufted Puffins nest from April through July. Even with low-power binoculars, one can see the cormorants and gulls on their nests, and guillemots and puffins standing at the entrances to their nesting crevices or burrows.

Reached from town by a narrow, winding road north is **Ecola State Park** (1,107 acres). Here from a lookout point, one may watch many of the same species as on Haystack Rock nesting on nearby offshore rocks. During migrations or in winter, some of the birds that may be noted frequently on the water below are Horned and Western Grebes, Black Brant, Harlequin Ducks, and all three species of scoters. Thin-billed Murres and Marbled Murrelets may be seen in any season. In winter, on the rocks just above the water line, look for Surfbirds, Black Turnstones, and Rock Sandpipers.

Ecola Park has fine stands of Douglas-fir and western hemlock that hold such breeding birds as the Blue and Ruffed Grouse, Band-tailed Pigeon, Pileated Woodpecker, Varied Thrush, Red Crossbill, and Oregon Junco. But a better representation of birdlife in the coniferous forests on the west side of the Coast Range is in **Saddle Mountain State Park** (3,054 acres), reached from Cannon Beach by driving north for 5 miles on US 101, turning off east onto US 26 and going 9 miles, then turning off north onto a road leading to the entrance. Some of the breeding birds to be counted on in the vicinity of the parking area are the Vaux's Swift, Rufous Hummingbird, Gray Jay, Northern Raven, Chestnut-backed Chickadee, Winter Wren, Swainson's and Hermit Thrushes, and Hermit Warblers. If one feels like climbing, he can take an easy trail to the summit of Saddle Mountain (elevation 3,283 feet), a volcanic outcrop, for more of the same birds along the way and ultimately a superb view of the entire northwest corner of Oregon.

COOS BAY—NORTH BEND

Pony Slough | Cape Arago State Park

The Coos Bay—North Bend metropolitan area, traversed by US 101 on the southwest Oregon coast, covers much of the south side of Coos Bay, the state's largest estuary.

Pony Slough in the city of North Bend is famed on the Oregon coast for its huge concentrations of birds—grebes, herons, ducks, and shorebirds—during the migration seasons and in winter. From US 101 in North Bend, turn off west onto Virginia Avenue. Just past the Pony Village Shopping Center, turn north onto Marion Street, a gravel road running along the Slough with numerous vantages for observation. The best time to see shorebirds is at mid-tide; at low tide the birds disperse over the many channels and at high tide they disappear in the tall grasses.

On US 101 in North Bend, farther south from the exit to Virginia Avenue, watch for the well-marked exit west to Charleston and Cape Arago; take it and proceed to Charleston, 9 miles distant. After crossing the South Slough Bridge in Charleston, continue straight ahead, stopping at a series of nearly interconnecting parks, one county and three state, all highly scenic with rocky cliffs, offshore rocks, protected inlets, shell beaches, and tidal pools—and all rewarding for seeing birds. **Cape Arago State Park** (134 acres), the last in the series, is situated on a high promontory projecting 0.5 mile into the sea, thus providing fine opportunities for viewing birds in passage offshore. Some of the birds to expect in the summer or fall months are the Northern Fulmar, Sooty Shearwater, Black-legged Kittiwake, Pigeon Guillemot, Marbled Murrelet, and Tufted Puffin.

CORVALLIS

Oregon State University Campus | McDonald Forest | William L. Finley National Wildlife Refuge

Situated at the western edge of the broad central section of the Willamette Valley, Corvallis is within easy reach of several different habitat types for birds.

In the city itself, tree-shaded **Oregon State University Campus,** west of 30th Street between Jefferson Street and Western Boulevard, is often rewarding. When the elms go to seed in April and early May, tremendous flocks of Evening Grosbeaks, as well as Purple Finches and Pine Siskins, arrive to reap the harvest. Later, in fall and winter, American Robins, Varied Thrushes, and Cedar Waxwings gather to forage for berries of the hawthorns. Toward evening during fall migration, Vaux's Swifts in huge numbers swirl above certain of the large chimneys prior to descending into them for the night.

A few miles northwest of Corvallis, **McDonald Forest** (6,800 acres; owned and maintained by the School of Forestry at Oregon State University) has stands of Douglas-fir typical of the Valley's lowlands. From the city, drive north on US 99W about 4.5 miles; turn off left onto Lewisburg Road and go 1.3 miles to a fork; here take the right fork for 1.6 miles to the top of a small pass, park the car, and walk either to the right to the summit of Vineyard Mountain or to the left down into a riparian habitat. Both paths are old logging roads through deep forest and past brushy growth.

From about 10 April to 15 June is the most productive time for visiting this area. Practically all the migrating and resident warblers one encounters in western Oregon are here during the early part of May. In the tangles of thimbleberry, hazel, and young fir, look for such species as the Orange-crowned, Nashville, Yellow, MacGillivray's, and Wilson's. In the conifers, particularly in small clumps of large trees, or at the edge of larger stands, search for the Audubon's, Black-throated Gray, Townsend's, and Hermit. When thimbleberry, salmonberry, red-flowering currant, and other shrubs are in bloom, Rufous Hummingbirds become fairly numerous around them. Stop and listen at intervals for the slow, ventriloquial hooting of the Blue Grouse, given as the bird perches motionless near the top of some high fir. Other birds to look or listen for, especially in conifer areas, are the Band-tailed Pigeon, Pileated Woodpecker, (Red-breasted) Yellow-bellied Sapsucker, Olive-sided Flycatcher, Steller's Jay, Winter Wren (on the forest floor near vine tangles or exposed, entwined roots), Hutton's and Solitary Vireos, Western Tanager, Purple Finch, Pine Siskin, Red Crossbill, and Oregon Junco. Both Northern Pygmy and Saw-whet Owls are residents and, probably, the secretive Spotted Owl too.

South of Corvallis, at the southern end of the Willamette Valley, lies the 5,325-acre **William L. Finley National Wildlife Refuge,** whose habitats are in great diversity. From Corvallis, drive south on US 99W for 10 miles, then turn west at the Refuge's directional sign onto a gravel road that passes through the Refuge to headquarters on the west side.

The Refuge is situated along flood-prone Muddy Creek, bordered by heavily wooded bottomlands. Elsewhere there are lakes, marshes, wet meadows, fields, brushy stretches, different kinds of woodlots, and a natural prairie. At headquarters, obtain a map, bird checklist, and directions by car or on foot to the following areas:

Bottomlands along Muddy Creek or at the *Beaver Pond* for breeding Wood Ducks and Hooded Margansers; *oak woods with clearings* and *stands of Douglas-fir* for breeding Ruffed Grouse, Mountain Quail, Acorn and Lewis' Woodpeckers, Western Flycatcher, Western Pewee, Scrub Jay, Chestnut-backed Chickadee (possible), Red-breasted Nuthatch, Hermit and Swainson's Thrushes, Western Bluebird, Solitary and Warbling Vireos, warblers such as the Orange-crowned, Audubon's, and Black-throated Gray, and Black-headed Grosbeak; *Pigeon Springs* near the west side of Pigeon Butte where flocks of Band-tailed Pigeons gather in fall to drink; *Cabell and McFadden Marshes* for breeding American Bitterns, Virginia Rails, Soras, and Marsh Wrens, for transient shorebirds, and for transient and wintering waterfowl, notably Canada Geese, Mallards, Common Pintails, Green-winged Teal, American Wigeons, and Northern Shovelers; *natural prairie*, for White-tailed Kites, Red-tailed Hawks, and Northern Harriers the year round, for Rough-legged Hawks and Northern Shrikes in winter.

EUGENE

Autzen Stadium Area | Fern Ridge Reservoir

In this city at the southern end, or upper reaches of, the Willamette Valley, the **Autzen Stadium Area** is notable for its assortment of fringillids in the winter months. From I 5, exit west on I 105 and follow directional signs. The woodlots and open brushy spots about the Stadium, also about Alton-Baker Park just west of the Stadium and, immediately south of the Park between the jogging path and roadway, the community garden plots (dormant in winter) are especially productive. The majority of fringillids are House Finches and White-crowned, Golden-crowned, Lincoln's, and Song Sparrows, but Oregon Juncos and Chipping, White-throated, and Fox Sparrows are regularly present. Almost invariably there are other fringillids, unusual in this part of Oregon, which can be found with careful searching.

The best place for year-round bird finding is **Fern Ridge Reservoir,** with its many arms, in a hollow about 8 miles west of Eugene. In summer when the Reservoir is filled, there are fine marshes edging much of the Reservoir for breeding birds such as Pied-billed Grebes, American Bitterns, Virginia Rails, American Coots, Marsh Wrens, and Yellow-headed Blackbirds. In fall when the water level is lowered, there are mud flats attractive to herons, egrets, and shorebirds. The higher ground around the Reservoir is given over to diversified farming; hence open fields, orchards, and small woodlots are the prevalent bird habitats.

State 126, west from I 5 in Eugene, crosses the southern two arms of the Reservoir. From both sides of the highway in summer, look for colonies of Purple Martins that occupy many of the snags and for Ospreys. The swamps south of the highway are prime breeding habitat for Great Blue Herons and Wood Ducks. In fall, when the water level is down, expect to see vast numbers of shorebirds on the mud flats.

About 2 miles west of the second southern arm of the Reservoir, turn north from State 126 on Territorial Road and proceed northward. After 4.5 miles, turn east on Clear Lake Road which leads to the dam at the Reservoir's north end, meanwhile passing open water that remains deep all year. In fall and winter, rafts of ducks are in view as well as Common Loons, Western Grebes, Double-crested Cormorants, and other waterbirds. When the dam's spillways are opened in fall, gulls begin gathering and attain large numbers by winter. The most common are usually Glaucous-winged and Ring-billed, with Herring, Thayer's, California, and Mew scattered among them. On occasion this area becomes a haven for oceanic birds when gale winds or severe storms on the Pacific Coast drive them inland.

Below the dam in a park (no name) is an oak woods and brushy growth for a variety of breeding birds which may include the Common Screech Owl, Acorn and Lewis' Woodpeckers, Traill's Flycatcher, Scrub Jay, White-breasted Nuthatch, Bewick's Wren,

Warbling Vireo, MacGillivray's Warbler, Yellow-breasted Chat, Black-headed Grosbeak, Lazuli Bunting, Rufous-sided Towhee, and Song Sparrow.

HOOD RIVER

Hood River Valley | Mt. Hood

From the community of Hood River on the Columbia River, 62 miles west of Portland, exit south from I 80N on State 35 and proceed south for 45 miles to the junction with US 26. After leaving Hood River, State 35 passes for several miles through the agriculturally rich **Hood River Valley,** largely given over to fruit orchards. Birds here tend to be much the same as those in the agricultural areas of the Willamette Valley.

About 6.5 miles from Hood River, the highway winds upward through about 5 miles of oak-covered foothills, then descends into the upper Hood River Valley, where the elevation is from 1,400 to 1,900 feet. In the oak woodland, look for the Lewis' Woodpecker, Black-throated Gray Warbler, and Black-headed Grosbeak, all of which breed here.

As the distance from Hood River approaches 17 or 18 miles, the highway climbs more rapidly, entering the Douglas-fir forest, characteristic of the lower slopes of **Mt. Hood.** The continuity of this forest is interrupted by ranch clearings and natural meadows, rushing mountain streams, and logging or burned areas in various stages of reforestation. Birdlife is therefore varied and includes such species as the Blue Grouse, Band-tailed Pigeon, Pileated Woodpecker, Hammond's and Olive-sided Flycatchers, Steller's Jay, Brown Creeper, North American Dipper, Winter Wren, Varied Thrush, Townsend's Solitaire, Golden-crowned Kinglet, Hermit and MacGillivray's Warblers, Western Tanager, Evening Grosbeak, Cassin's Finch, Pine Siskin, Red Crossbill, and Oregon Junco.

Somewhat before the junction of State 35 with US 26, at an elevation on Mt. Hood of about 4,000 feet, species of smaller conifers become more and more prevalent until the forest consists almost entirely of lodgepole pine, western white pine, Englemann spruce, and grand fir. Farther up, mountain hemlock, subalpine fir, Sierra juniper, alpine mountain-ash, and other trees appear. Some of the breeding birds to look for are the Gray Jay, Clark's Nutcracker, Mountain Chickadee, Red-breasted Nuthatch, Hermit Thrush, and Ruby-crowned Kinglet.

At the junction of State 35 with US 26, turn right (west) onto US 26 and go 3 miles, then turn off right (north) and drive up the circuitous road to Timberline Lodge (elevation 6,500 feet), which, as the name implies, is near the upper level of tree growth. Numerous trails leave from the Lodge, including those leading up through the alpine region to the crest of Mt. Hood (elevation, 11,235 feet). Vegetation consists of bright-flowering annuals and perennials plus dwarfed false heathers. The special breeding bird to be alert for is the Gray-crowned Rosy Finch, often near lingering snowpacks.

The best time of the year to investigate birdlife on the slopes of Mt. Hood is June and early July, when most species are in full song.

KLAMATH FALLS

Upper Klamath Lake | Williamson River Bridge | Agency Lake

The Klamath Basin in southwest-central Oregon below the eastern slopes of the southern Cascades is noted for its great bulrush-cattail marshes. Although not as extensive as they once were, owing to diking and drainage for agricultural purposes, they are still a prime breeding area for waterbirds and waterfowl as well as a major stopping point for waterfowl during their migrations.

Upper Klamath Lake has representative marshes. At its northern end the Upper Klamath National Wildlife Refuge (14,900 acres) embraces some of the finest marshes but they are accessible only by boat. Elsewhere around the Lake and its islands are also fine marshes. Some of the birds, besides waterfowl, breeding in association with the Lake's marshes and open water, are the Eared and Western Grebes, American White Pelican, Double-crested Cormorant, Great and Snowy Egrets, Black-crowned Night

Heron, American Coot, California and Ring-billed Gulls, Forster's and Black Terns, and Yellow-headed and Red-winged Blackbirds.

A recommended trip in late spring or early summer to see birds on Upper Klamath Lake, and at other rewarding sites too, is northward from Klamath Falls on US 97 which first skirts the Lake's eastern shore. Hagelstein Park, about 10 miles north of Klamath Falls, has particularly good vantages for viewing waterbirds. During May and early June, Western Grebes often perform their fantastic courtship displays not far offshore.

Just north of Hagelstein Park a high ridge rises from the right of the highway. Watch for a Bald Eagle or Prairie Falcon. Also look for Lazuli Buntings and Green-tailed Towhees on the brushy slopes and for Rock Wrens where there are rocky outcroppings.

Continue on US 97 to Modoc Point, then turn off west onto a secondary road toward Klamath Agency. On reaching the **Williamson River Bridge,** stop and walk out on it for good views of foraging waterbirds and for swallows—Violet-green, Bank, Rough-winged, Barn, and Cliff—which nest along, or in the vicinity of, the River. Yellow and Wilson's Warblers, Bullock's Orioles, and Black-headed Grosbeaks are among the birds in the bordering trees and shrubs.

North of the Bridge the road parallels **Agency Lake,** which has several marshes teeming with birds. The marsh in view from Petric Park at the upper end of the Lake is worth close inspection because here, besides grebes, ducks, and terns, are small numbers of Tricolored Blackbirds among the abundant Yellow-headed and Red-winged Blackbirds.

After passing the junction with the road to Chiloquin, the road to Klamath Agency goes through a forest of conifers and aspens in which (Red-breasted) Yellow-bellied and Williamson's Sapsuckers, White-headed Woodpeckers, Red-breasted and Pygmy Nuthatches, Evening Grosbeaks, and Red Crossbills reside.

At Klamath Agency, take State 232 northward for about 6 miles, then turn off northwestward on State 62 for 1.0 mile to Fort Klamath. This small community is at the northern end of a broad meadow edged by a heavy coniferous forest in which there is a remarkably large population of permanent-resident Great Gray

Owls. At dusk, many of these splendid birds emerge from the forest to hunt over the meadow. The most productive vantages for watching this evening spectacle are along the 0.1-mile stretch of State 62 from State 232 to Fort Klamath.

For additional bird finding, continue north from Fort Klamath on State 62 about 6 miles to Crater Lake National Park (*see under* **Medford**).

LA GRANDE

Grand Ronde River Valley | Hilgard Junction State Park | Wallowa River Valley | Blue Mountains | Wallowa Mountains | Wallowa Lake State Park

This city in the Grande Ronde River Valley of northeastern Oregon lies mainly south of the Blue Mountains and west of the Wallowa Mountains.

The **Grand Ronde River Valley** (elevation 2,700 feet), although largely given over to farming and ranching, nonetheless retains stretches of sagebrush, rabbitbrush, and greasewood where breeding birds include California Quail, Burrowing Owls, Eastern and Western Kingbirds, Lazuli Buntings, and Savannah and Lark Sparrows. Riparian woods and thickets are nesting habitats for the Calliope Hummingbird, Gray Catbird, Veery, Red-eyed Vireo, and occasionally the American Redstart. **Hilgard Junction State Park** (233 acres) and immediate vicinity, 8 miles west of La Grande on I 80N, hold many of the species typical of the Valley.

The **Wallowa River Valley**, about the town of Enterprise (elevation 3,757 feet) on State 82, 65 miles north and east of La Grande, is a smaller version of the Grande Ronde Valley, but much more vegetated. Its bird species are about the same with the addition of Bobolinks, which nest in grassy fields and high meadows. To see a large number, drive south from Enterprise on State 82. At the point where the highway turns eastward at the south end of town, turn off south onto an unmarked road for 1.7 miles and begin watching for Bobolinks sitting on fences and chasing one another, especially in May and June when they are establishing their territories.

North and east of Enterprise rise the **Blue Mountains** with open, park-like stands of ponderosa pine and Douglas-fir on slopes up to 5,000 feet or more, where Englemann spruce and subalpine fir appear and extend up to the highest points over 8,000 feet. Most of the bird species breeding in the Blue Mountains also breed in all the higher mountain regions of Oregon. Some exceptions are the Flammulated Screech Owl, Pine Grosbeak, and White-winged Crossbill, known to breed regularly only in the Blue Mountains. State 3, running north from Enterprise into Washington, gives access to the Blue Mountains along the way. See the introduction to the Washington chapter for information about the Blue Mountains in that state.

Northeast from Enterprise, rolling prairie and rimrock lands extend to the pine-clad foothills of the Blue Mountains. To explore this area, follow road signs from Enterprise eastward and then northward to Zumwalt, a deserted town about 25 miles distant. While crossing the country, notable for its many raptors, be alert particularly for Red-tailed, Swainson's, and Ferruginous Hawks, Golden Eagles, and Prairie Falcons. In winter, flocks of Horned Larks, both Gray-crowned and Black Rosy Finches, and Snow Buntings are common sights. Continue northward from Zumwalt. After about 6 miles begin stands of ponderosa pine which are prime habitat for Flammulated Screech Owls. Pine Grosbeaks and White-winged Crossbills are residents in this general area as are Great Gray Owls, usually near meadows.

The **Wallowa Mountains,** west and south of Enterprise, are more rugged and picturesque than the Blue Mountains, having precipitous slopes and sharp crests. Several peaks reaching above 9,000 feet are snow-capped and have glaciers, which produce numerous mountain lakes. On the lower slopes the conditions of vegetation are similar to those of the Blue Mountains, but on the slopes above 6,000 feet the conditions are different. Here forests contain extensive stands of lodgepole pine, mountain hemlock, subalpine fir, and—at the timber line—whitebark pine. Above the timber line are treeless slopes with low herbaceous plants of alpine character. Among the birds in the higher forests are the Blue Grouse, probably both Black-backed and Northern Three-toed Woodpeckers, Gray Jay, Clark's Nutcracker, Townsend's Solitaire,

Ruby-crowned Kinglet, and possibly the Pine Grosbeak. Above the timber line are Water Pipits and Gray-crowned Rosy Finches, also the introduced White-tailed Ptarmigan. Throughout the Wallowas, wherever there are swift streams, Harlequin Ducks and North American Dippers are common.

Most of the Wallowa Mountains are embraced by the Eagle Cap Wilderness Area and thus inaccessible by car. To explore the Wallowas, the best procedure is to drive south from Enterprise on State 82 (passing through Joseph) for 12 miles to **Wallowa Lake State Park** (166 acres) on the edge of the Wilderness Area at an elevation of 4,427 feet. Proceed to the Park's so-called South Area where, along trails through semi-open forests of cottonwood and pine, one may see or hear birds that regularly breed on the lower slopes of the Wallowas. The species include the Mountain Chickadee, Red-breasted Nuthatch, Hermit and Varied Thrushes, Solitary Vireo, and Townsend's Warbler. To find birds on the higher slopes, one must go on foot or horseback (horses available nearby) on hiking or riding trails that start from the South Area.

LAKEVIEW

Warner Valley | Hart Mountain National Antelope Refuge | Chandler Wayside State Park | Summer Lake Wildlife Area

Northeast of this town, which is in south-central Oregon 15 miles north of the California border, lies the high **Warner Valley** (elevation 4,500 feet). On its west side is the Warner Rim, topped by broad tablelands and peaks, some reaching above 8,400 feet. On the east side is Hart Mountain, a volcanic ridge, rising abruptly with cliffs and ridges from the Valley floor to a maximum elevation of 8,065 feet. North-south through the Valley runs a chain of lakes, some of them dry beds in the fall. Lacking any water inlet, the Valley depends on the adjacent mountains for moisture. The Valley floor, consistently flat, has various habitats—meadows, marshes bordering lakes, and drier sections covered with sagebrush and greasewood. Trees are limited mainly to cottonwoods, willows, and junipers. Shrubs such as chokecherry, wild plum, and wild rose form thickets along streams which drain down from the mountains.

A wide variety of birds nest in one habitat or another in the Warner Valley as the following selected list of species will attest: Eared Grebe, Western Grebe, American White Pelican, Double-crested Cormorant, Great Egret, Snowy Egret, Black-crowned Night Heron, Cinnamon Teal, American Wigeon, Northern Shoveler, Redhead, Canvasback, Ruddy Duck, Swainson's Hawk, American Kestrel, California Quail, American Coot, Common Snipe, Willet, American Avocet, Wilson's Phalarope, California and Ring-billed Gulls, Caspian and Black Terns, Mourning Dove, Poor-will, Red-shafted Flicker, Western Kingbird, Horned Lark, Dusky Flycatcher, Marsh Wren, Sage Thrasher, Loggerhead Shrike, Yellow-breasted Chat, Yellow-headed Blackbird, American Goldfinch, Lark, Sage, and Brewer's Sparrows.

To reach the Warner Valley from Lakeview, drive north for 5 miles on US 395, then turn east onto State 140. After 9 miles, if one wishes to look for birds in higher country, he can turn off north on a secondary road (open only in summer) that leads up to Drake Peak Lookout on the Warner Rim at an elevation of 8,405 feet. In the area are coniferous forests that should yield such birds as the Calliope Hummingbird, Olive-sided Flycatcher, Steller's Jay, Hermit Thrush, Audubon's Warbler, and Fox Sparrow.

About 28 miles from US 395, State 140 reaches Adel. From here drive north into the lower Warner Valley on a secondary road that soon passes the first two of the chain of lakes, Crump and Pelican. Both have nesting colonies of American White Pelicans and Double-crested Cormorants on their small islets. Both also have many other birds on them, or in their immediate vicinities, that are typical of most lakes in the chain.

Eighteen miles from Adel the road comes to the town of Plush. From here one may continue north to explore the chain of lakes in the upper Warner Valley or go northeast on the road to the **Hart Mountain National Antelope Refuge.**

Encompassing altogether 270,000 acres, the Refuge includes part of the Warner chain of lakes and the western face of Hart Mountain, but the major portion comprises (1) the gentle, well-watered eastern slopes of Hart Mountain where, depending on elevation and exposure, there are coniferous forests, aspen groves, riparian woods, and rolling grasslands, and (2) greasewood and salt-grass flats, with intermittent lakes, that extend far eastward. So diversified are the habitats that the birds are in wide variety. True to its name, the Refuge has pronghorn 'antelopes' which roam in bands over the grasslands. At higher elevations there are mule deer and bighorn sheep.

The best time to visit the Refuge is from mid-May to November when roads are passable for conventional cars. At headquarters on the Refuge, 24 miles from Plush, obtain a bird checklist, ask for directions to the more promising habitats, and inquire about the self-guided 20-mile auto route.

Northwest from Lakeview stretches high desert country where water is scarce and most lakes are dry, with a few notable exceptions. An outstanding one for birds is Summer Lake in the **Summer Lake Wildlife Area,** maintained by the Oregon Department of Fish and Wildlife. To reach it from Lakeview, drive north on US 395 for 23 miles to Valley Falls, then northwest on State 31 for 51 miles.

On the way to Summer Lake, stop at **Chandler Wayside State Park** (85 acres) on US 395, 16 miles north of Lakeview. Here, in a forested canyon with brushy patches, are numerous breeding birds, among them Scrub Jays, Bullock's Orioles, and Green-tailed Towhees. The Park offers particularly good bird finding in spring and fall when it serves as an oasis for many migrating passerines.

At an elevation of 4,200 feet with the Winter Rim, a huge fault, rising 3,000 feet higher immediately on the west, the Summer Lake Wildlife Area embraces about 20,000 acres that include, besides open water, large fresh-water marshes, extensive alkaline flats, and about 5,500 acres of dry brushy lands. Upon reaching the Area, call in at headquarters on State 31, 1.0 mile south of the town of Summer Lake, and ask for a bird checklist and advice on roads into the Area for viewing particular species in season.

Clint Canal Road, east from headquarters into the northern part of the Area, and dike roads leading from it, make readily accessible the finest of the marshes and alkaline flats, as well as open water, where breeding birds include the Eared, Western, and Pied-billed Grebes, Black-crowned Night Heron, American Bittern, White-faced Ibis (a few), Canada Goose, Mallard, Gadwall, Common Pintail, Green-winged Teal, American Wigeon, Redhead, Ruddy Duck, Northern Harrier, Sandhill Crane, Virginia Rail, American Coot, Snowy Plover, Common Snipe, Long-billed Curlew, American Avocet, Black-necked Stilt, Wilson's Phalarope, Forster's and Black Terns, Short-eared Owl, Marsh Wren, Yellow-headed and Brewer's Blackbirds, and Savannah Sparrow. An island near the north end of Summer Lake, in view from one of the dike roads, has nesting American White Pelicans, California and Ring-billed Gulls, and Caspian Terns.

During fall and winter, huge flocks of Snow Geese, with a few Ross' Geese scattered among them, stop off to rest and feed prior to moving south into California.

MEDFORD

Roxy Ann Butte | Crater Lake National Park

In southwestern Oregon the interior valleys along the Rogue River and its tributary Bear Creek are ecologically a northern continuation of the Central Valley of California. Instead of coniferous forests on the lower slopes of the adjacent hills and mountains that one would expect in Oregon, there are oak woodlands and dense stands of woody shrubs, or chaparral, holding a mixture of bird species that is unique in the state. For a good example of this mixture, explore **Roxy Ann Butte** (elevation 3,571 feet), east of Medford and reached as follows:

From I 5 in south Medford, exit east on Barnett Road for 2 miles; turn north onto North Phoenix Road and go 0.75 mile; then turn east onto Cherry Lane and follow it to Hillcrest Road from which, in about 0.5 mile, a gravel road leads off left to the top of the Butte, about 2,000 feet above the city. During the ascent (impassable for cars in wet weather), stop to inspect the oak woods for such birds as Acorn Woodpeckers, Ash-throated Flycatchers, Scrub Jays, and Plain Titmice and the chaparral for Bushtits, Wrentits, Blue-gray Gnatcatchers, Brown Towhees, and Lark Sparrows. In the more open areas at the top, look for Western Bluebirds, Lazuli Buntings, and Lesser Goldfinches.

High in the Cascades some 80 miles northeast of Medford by

road is **Crater Lake National Park.** Even if one should fail to see birds in its 250 square miles, he would be amply rewarded by scenic grandeur. An ornithological famine is extremely unlikely, however, for its avifauna is quite representative of the southern high Cascades.

Crater Lake, with a shoreline of about 25 miles, partly fills the caldera of Mt. Mazama, destroyed by a tremendous eruption some 6,600 years ago. Of clear fresh water, without inlet or outlet and exceedingly deep (1,932 feet), the Lake reflects a rich blue of startling intensity. Rising above the Lake in sheer cliffs from 500 to 2,000 feet are the upper walls of the great caldera.

Besides Crater Lake and its surrounding cliffs, the Park has extensive coniferous forests, mountain meadows, and—at high elevations—open pumice flats. The summit of Mt. Scott (8,926 feet), the highest peak in the Park, has characteristics of an alpine tundra.

To reach the Park from Medford, take State 62 north and follow it all the way to, and into, the Park from the west side to Annie Springs. Here as the highway continues through the Park and exits south toward Klamath Falls, turn off left up to Rim Village (elevation 7,100 feet), passing headquarters along the way. State 62 and the road to Rim Village are kept open the year round. All other roads in the Park, including the entrance road from the north, are usually closed from October to June because of snow conditions.

Crater Lake Lodge, which overlooks Crater Lake from the north side of Rim Village, has conifers and gardens on its grounds. The most conspicuous bird here is probably the Clark's Nutcracker, having become accustomed to feeding on peanuts and other tidbits handed out by tourists. Not as conspicuous, but fully as forward, is the Gray Jay, which delights tourists with its fearlessness. Other birds common around the Lodge are the Rufous Hummingbird, Steller's Jay, Red-breasted Nuthatch, American Robin, Mountain Bluebird, Audubon's Warbler, Cassin's Finch, Pine Siskin, Red Crossbill (in years when cones are abundant), Oregon Junco, and Chipping Sparrow.

From Rim Village, Rim Drive, a 35-mile paved highway, completely circles the caldera clockwise and passes through subalpine forests of whitebark pine and mountain hemlock—habitat for such

GRAY-CROWNED ROSY FINCH

regularly breeding birds as the Western and Olive-sided Flycatchers, Mountain Chickadee, Townsend's Solitaire, and Western Tanager. When stopping to view the Lake, scan the sky and horizon for soaring eagles, Golden or Bald, and expect to see Northern Ravens patrolling the cliffs. At any stop above the cliffs, always watch for Gray-crowned Rosy Finches, which are reputed to nest in them.

On the east side of the caldera, a 2.5-mile trail leads east from Rim Drive to the summit of Mt. Scott through a subalpine forest. Additional species to find, if not already observed at lower elevations, include the Blue Grouse, Black-backed Three-toed Woodpecker, Hammond's Flycatcher, Brown Creeper, Rock Wren, Golden-crowned and Ruby-crowned Kinglets, and Hermit Warbler. Gray-crowned Rosy Finches are always possible, particularly about snowfields below the summit. If the bird finder's ascent is on a sunny, windy day in late August or early September, he may be rewarded at the summit by an impressive hawk flight, since Mt. Scott lies along the flyway of many migrating raptors.

North American Dippers reside along all streams, even in the higher parts of the Park, until the streams are frozen over.

In the southern part of the Park, 2.5 miles in from the exit on

State 62 to Klamath Falls, is an aspen grove with White-headed Woodpeckers and an adjacent bog with Lincoln's Sparrows. The surrounding forests of ponderosa pine have Pileated Woodpeckers and Pygmy Nuthatches.

For bird-finding opportunities along State 62 south from the Park, *see under* **Klamath Falls.**

NEWPORT

Yaquina Bay State Park | Yaquina Bay | South Jetty | Yaquina Head

Among the areas for bird finding on the central Oregon coast, one of the better is centered about Newport on US 101 north of Yaquina Bay.

Yaquina Bay State Park (32 acres), adjoining Newport and reached by turning west off US 101 just before it crosses the Yaquina Bay Bridge, has extensive brushy growth that holds a year-round population of Wrentits and attracts many warblers in migration and Golden-crowned Sparrows along with many White-crowned Sparrows in winter. The Park also has stands of conifers that bring Purple Finches, Pine Siskins, Red Crossbills, and other finches in years when cones are abundant. The main channel into the Bay, in view off the Park, is worth scanning for alcids, including Thin-billed Murres.

From the State Park, take the road east under US 101 and continue east along the upper reaches of **Yaquina Bay,** then south along the Bay to Toledo, 10 miles distant. The wharves and pilings along Newport's old bayfront, passed at the start of the trip, should be looked over for resting gulls. Pigeon Guillemots are quite common here in summer, as they nest under the wharves, in the pilings, and even under the Yaquina Bay Bridge. In winter, the Bay abounds with waterbirds and waterfowl; thus, stop frequently for views of loons, grebes, cormorants, Black Brant, and such regular duck visitants as the Common Pintail, Green-winged Teal, American Wigeon, Canvasback, Lesser Scaup, Common Goldeneye, Bufflehead, and Ruddy Duck. During migration seasons, expect to see large numbers of Bonaparte's Gulls as well as Caspian and other terns, sometimes being pursued by Parasitic

Jaegers. At Sally's Bend, about halfway to Toledo where the road swings south around the Bay, there are mud flats for a wide variety of shorebirds in season.

Returning to, and going south on, US 101, turn off west immediately onto the road marked to the Marine Science Center, but instead of following it east under the Bridge, turn off left onto the obscure South Jetty Road. Graveled and usually passable, except in bad weather, this leads, past several small rock jetties jutting into the Bay's channel, to a parking lot within a short walking distance of the **South Jetty** at the channel's entrance. In winter, around any of the jetties or resting on them at low tide, are Harlequin Ducks, Red-breasted Mergansers, and other diving ducks including scoters and an occasional Oldsquaw. Loons, Red-necked Grebes, and Black-legged Kittiwakes should be in view, especially from the South Jetty. In late summer and early fall, look for Brown Pelicans diving for fish off the South Jetty.

Three miles north of Newport on US 101, in the small community of Agate Beach, turn off west onto Lighthouse Road to scenic **Yaquina Head** with a lighthouse atop its lofty cliffs. From the parking area, walk first to the overlooks near the lighthouse. Although the rocks below the cliffs are a bit far down for ideal viewing, they will have shorebirds, among them Black Oystercatchers in any season, and transient or wintering Black Turnstones, Surfbirds, Wandering Tattlers, and (sometimes) Rock Sandpipers. On the cliff faces and rocks offshore, expect to see numerous alcids such as Pigeon Guillemots, Marbled Murrelets, Rhinoceros Auklets, and Tufted Puffins. Then walk around the south side of the lighthouse to the fence at cliff-edge, which is only a few hundred yards from an offshore rock on which there is a large nesting population of Brandt's Cormorants and Thin-billed Murres with a good representation of Western Gulls.

PORTLAND

Pittock Bird Sanctuary | Forest Park | Delta Park | Sauvie Island Wildlife Management Area

Scattered about this largest city in Oregon are many fine parks with a good variety of bird habitats. But before exploring any of them, the bird finder is advised to go first to the headquarters of the Audubon Society of Portland in the **Pittock Bird Sanctuary** for information on the most promising parks for birds at the time of his visit.

The Sanctuary itself, property of the Society, is a 32-acre tract of hilly woodland in west Portland containing fine stands of virgin Douglas-fir, a grove of old maples, and hemlocks, grand firs, cedars, yews, oaks, cascaras, alders, and cherries. A small creek meanders through the brushy-edged meadows of the southeastern section. Marked trails are maintained. Some of the birds regularly nesting are the Ruffed Grouse, Northern Pygmy and Saw-whet Owls, Pileated Woodpecker, Violet-green Swallow, Black-capped and Chestnut-backed Chickadees, Bushtit, Winter and Bewick's Wrens, American Robin, Swainson's Thrush, Audubon's Warbler, Western Tanager, Black-headed Grosbeak, Purple Finch, Rufous-sided Towhee, and White-crowned and Song Sparrows.

To reach the Sanctuary from I 5, exit west onto US 30; turn off left onto Northwest 23rd Avenue, then right onto Northwest Lovejoy Street, which becomes Cornell Road, and continue about 2 miles on Cornell Road to the entrance.

Forest Park (4,300 acres) in northwest Portland encompasses hill country largely forested with both conifers and hardwoods and is maintained in its natural state. It has few roads but many pleasant trails that wind through wooded areas to remote spots. Because of the mixed character of its forests, the Park is especially good for landbirds in the breeding season. To reach the Park from the Sanctuary, continue on Cornell Road, turn right onto Southwest 53rd Drive to the well-marked trails and parking lot on the right. For waterbirds and waterfowl, the ponds and lakes in **Delta Park** (719 acres) along both sides of I 5 in north Portland are productive in any season. They are best, however, in winter when they attract large flocks of ducks, many being American Wigeons often containing one or more Eurasian Wigeons. On outlying grassy areas, gulls gather in great numbers to rest. Glaucous-winged, California, and Mew are the most common, with Herring, Thayer's, and Ring-billed in scattered numbers and an occasional Glaucous and Western. The best way to observe these aggregations is from roads in the Park, reached by taking the Delta Park Exit from I 5.

Without question the stellar area for bird finding near Portland is Sauvie Island at the confluence of the Columbia and Willamette Rivers, 10 miles northwest of the city's center. Large, generally open and flat, with many lakes, ponds, and sloughs, the Island is a major stopping point for vast numbers of waterfowl, waterbirds, and shorebirds in migration as well as a wintering ground for great numbers of them and raptors besides. The southern part of the Island is privately owned farmland; the northern part is the state **Sauvie Island Wildlife Management Area** (13,000 acres) in three units, consisting primarily of lakes and sloughs, surrounded by lushly wooded bottomlands that are subjected to the spring floods of the Columbia River.

All three units of the Wildlife Management Area offer the best bird finding on the Island. When transient waterfowl are at their peak numbers in late October and early November, the most numerous are Common Pintails and American Wigeons with a good representation of other surface-feeding ducks and a fair representation of diving ducks. Canada Geese and Whistling Swans in impressive numbers pass the winter; Snow Geese usually appear by mid-January. In spring, and especially in fall, Sandhill Cranes stop off, sometimes a thousand or more at the peak of their migration in late October. During August and September, shorebirds throng the muddy shores and flats. Least Sandpipers, Long-billed Dowitchers, and Western Sandpipers are usually the most abundant; Semipalmated and Black-bellied Plovers, Greater and Lesser Yellowlegs, Pectoral and Baird's Sandpipers, and Northern Phalaropes are well represented. Dunlins show up later and pass the winter along with dowitchers and Sanderlings. Throughout winter the most noticeable birds of prey are Red-tailed Hawks, Northern Harriers, and Short-eared Owls; others, less noticeable but regular, include Rough-legged Hawks, Bald Eagles, American Kestrels, and Great Horned and Saw-whet Owls.

Sauvie Island may be visited profitably any time of year, except during the hunting season—from about mid-November to mid-January—when the Wildlife Management Area is closed to entry. Bird finding must then be confined to roadways through the farmlands.

To reach Sauvie Island from Portland, exit west from I 5 onto US 30. Approximately 3.5 miles beyond St. Johns Bridge, which crosses the Willamette River on the right, and 1.0 mile beyond the Portland city limits, turn off right to cross the well-marked Sauvie Island Bridge. Upon leaving it, proceed to the *Oak Island Unit* by turning sharply left onto Sauvie Island Road and going 2.5 miles, right onto Reeder Road for 1.2 miles, and then left onto Oak Island Road for 3 miles to the entrance. The open fields on the right are the main feeding and roosting area for swans, geese, and ducks in fall and winter. Along the Road, 1.0 mile beyond the entrance, is an impressive stand of white oaks edged by brushy growth, a breeding habitat for such birds as White-breasted Nuthatches, Bullock's Orioles, and Black-throated Gray Warblers. Park at the cable closing off the Road and walk north through the oaks to the tip of Oak Island and scan Sturgeon Lake to the east for waterfowl in season. Their evening flight from the fields to the Lake for the night is spectacular.

Another part of the Oak Island Unit may be reached by backtracking to Reeder Road and continuing 1.9 miles to a parking turnout on the left and climbing the dike overlooking ponds and mud flats—the best area for shorebird concentrations, also for Sandhill Cranes in September and October.

Continue on Reeder Road for 3.5 miles beyond the junction with Gillihan Road and turn left onto a gravel road into the *Eastside Unit*. The sloughs and marshes on both sides of the road are excellent for herons, Wood Ducks, rails, and blackbirds in summer, and for waterfowl of many species in winter.

The *North Unit*, reached from Sauvie Island Bridge by turning sharply left onto Sauvie Island Road and proceeding straight ahead for 8.7 miles, is largely riparian habitat with a rich variety of woodland species. After spring floods when dirt roads through the Unit are accessible, bird finding is highly rewarding.

PRINEVILLE

Ochoco National Forest | Crooked River National Grassland

Not far from this valley city at an elevation of 2,870 feet in central Oregon, one of the most scenic parts of the state, are sharply contrasting environments for birds.

North and east, largely embraced by **Ochoco National Forest,** are the westernmost foothills of the Blue Mountains with elevations reaching up to nearly 7,000 feet. For a good sampling of birdlife here, drive east from Prineville on US 26, which, in the next 11 miles, passes the south side of rimrock where Golden Eagles and Northern Ravens nest, and the north side of the Ochoco Reservoir. About 6.5 miles east of the Reservoir, as US 26 begins the northward ascent into the National Forest, turn off southeasterly on the paved road up to Ochoco ranger station, 6 miles distant, near which Ochoco and Canyon Creeks converge.

In the vicinity of the ranger station and the campground at an elevation of 4,000 feet are a park-like stand of ponderosa pine, groves of giant aspen, boggy places, and dense willow thickets lining the creeks. Breeding birds to be alert for in particular are Rufous and/or Calliope Hummingbirds about the campground, North American Dippers along the creeks, and Lincoln's Sparrows in boggy situations. Veeries are remarkably numerous and vociferous. Many of the aspens are riddled with nesting holes excavated by Red-shafted Flickers, (Red-naped) Yellow-bellied Sapsuckers, and Williamson's Sapsuckers, but often occupied by Violet-green Swallows, Mountain Chickadees, House Wrens, and Western and Mountain Bluebirds. Other breeding birds in suitable habitats of the area include the Northern Goshawk, Ruffed Grouse, White-headed Woodpecker, flycatchers (Traill's, Hammond's, Dusky, and Western), Steller's Jay, vireos (Solitary, Red-eyed, and Warbling), warblers (Orange-crowned, Nashville, Yellow, Audubon's, MacGillivray's, Yellow-breasted Chat, and Wilson's), Western Tanager, and Black-headed Grosbeak.

Northwest of Prineville, the **Crooked River National Grassland** (106,000 acres) contains extensive stretches of arid country in which grasses, juniper, and sagebrush constitute the predominant vegetation. Among the regular breeding birds are Burrowing Owls (quite localized), Poor-wills, Gray Flycatchers, Sage Thrashers, Mountain Bluebirds, Loggerhead Shrikes, Green-tailed Towhees, and Vesper, Chipping, and Brewer's Sparrows. Golden Eagles and Red-tailed and Swainson's Hawks have nesting sites on buttes and other rocky outcroppings.

For a sampling of birdlife in the National Grassland, leave Prineville on US 26 going northwest. After the first 7 miles, the highway begins traversing the Grassland. Bird finding is good almost anywhere but is more productive along roads leading off west. Some are marked to Gray Butte and Haystack Reservoir, others to particular wells or springs. During hot, dry summers, some of the springs, such as the one on the slopes of Gray Butte, attract many birds that one may approach closely as they arrive to drink.

TILLAMOOK

Tillamook Bay | Bayocean Peninsula | Cape Meares State Park | Three Arch Rocks National Wildlife Refuge | Netarts Bay | Cape Lookout State Park

Not far from this community inland from Tillamook Bay on the northwestern Oregon coast are many productive areas for birds in any season. A selection of these areas is as follows.

From US 101 in Tillamook, drive west on Third Street for 1.8 miles. After crossing the Tillamook River, turn right and follow the southern shore of **Tillamook Bay.** In winter, stop frequently to scan the waters of the Bay for large numbers of loons, grebes, and waterfowl including rafts of Black Brant. In late summer and fall, expect to see Brown Pelicans, which are regular visitants.

Where the road bends westward away from the Bay to the town of Cape Meares, turn off north onto a gravel road that runs about 0.5 mile up the **Bayocean Peninsula,** a finger of land, with mud flats and the Bay on the right, sand dunes with ponds and the ocean beach on the left. The area is ideal for watching transient shorebirds, especially at high tide when flocks gather near the road and around the ponds. Northern Ravens commonly scavenge on the beach and an occasional Peregrine Falcon shows up in fall and winter. Snowy Plovers frequent the dunes in any season and may nest. Wrentits reside permanently in dense brush flanking open areas.

Backtrack from the Bayocean Peninsula and continue on the road to Cape Meares. Here turn off west to **Cape Meares State Park** (233 acres) and park the car. The Park's rocky cliffs overlooking the Pacific provide excellent vantages for viewing nearby offshore rocks, which are packed with Thin-billed Murres and cormorants during the summer months. Black Oystercatchers commonly forage around the rocks above tide level. Just north of the parking area, Tufted Puffins have a nesting colony on grassy slopes above a small cove and can be watched resting near their burrows.

Return to the main road and continue south to the small town of Oceanside. Offshore is the **Three Arch Rocks National Wildlife Refuge,** consisting of three rugged islands totaling about 17 acres that support an enormous nesting population of Thin-billed Murres together with smaller numbers of Fork-tailed and Leach's Storm Petrels, Double-crested, Brandt's and Pelagic Cormorants, Western Gulls, and Tufted Puffins. The islands are too far offshore for satisfactory views with binoculars or spotting scope, but many of the murres and other birds which nest on them may be readily observed as they feed in waters close to the mainland.

Continue south from Oceanside on the main road to Netarts. Here turn right near the south end of town and follow the road along the entire east side of **Netarts Bay.** Being quite shallow, the Bay attracts in winter large numbers of waterbirds and waterfowl, easily seen from the car. At the same time of year there is always the likelihood of spotting a Rough-legged Hawk, Bald Eagle, Peregrine Falcon, or Merlin flying low over the Bay or along its shore.

As the main road continues south from Netarts Bay, it soon passes close to **Cape Lookout State Park** (1,946 acres), which has a rugged, narrow strip of rocky headland extending 1.5 miles straight out to sea. The very tip of the Cape, where sheer cliffs drop into the water, is about 400 feet above the high-tide mark. On the south side of the Cape is a huge colony of sea birds containing the three Oregon species of cormorants, Western Gulls, Thin-billed Murres, Pigeon Guillemots, and Tufted Puffins. In the Park and on the hills to the east are coniferous forests typical of those on the west side of the Coast Range. *See under* **Cannon Beach** for some of the birds to expect in the forests.

South Dakota

LARK BUNTING

When the eastern bird finder plans a trip west, he should arrange to drive across South Dakota on US 14 or I 90. No other crossing of the great interior plains of the United States demonstrates in so striking a manner the transition between east and west.

First, the highways traverse level, productive farmlands with freshly painted buildings sheltered by groves of trees, with neatly fenced pastures for dairy herds, with spacious fields for grain, corn, and other crops—all in all, a panorama much the same as that of Illinois, Iowa, and Minnesota. But almost imperceptibly the scene changes: the terrain becomes rolling and the farms less flourishing, the buildings shabbier and more exposed, the pastures more extensive, and the croplands more scattered.

Then, almost halfway across the state, the highways suddenly mount grassy, stone-strewn hills, descend into the wide, deep valley of the Missouri River, and ascend in sweeping curves to the high range country that marks the beginning of 'the wide open spaces.' Over the steeply undulating grasslands the roads continue their course, into draws (shallow ravines) and over ridges. A herd of cattle appears now and then. Occasionally, visible in the distance is a fence or a ranch house, but generally there is little evidence of human life. For a short distance the highways pass close to the north rim of the Big Badlands, allowing tantalizing glimpses of strange, light-colored spires and pinnacles.

Finally, the Black Hills, an outer range of the Rocky Mountains, come into view; once the highways have reached the mountains, the range country ceases almost as abruptly as it began.

The changes in birdlife along US 14 and I 90 are no less striking than the countryside transitions. From the Big Sioux River bottoms in eastern South Dakota, where Least Flycatchers, Eastern Pewees, Baltimore Orioles, Scarlet Tanagers, and Rose-breasted Grosbeaks are summer residents, the bird finder reaches by a journey of 360 miles the Western Flycatchers, Western Pewees, Bullock's Orioles, Western Tanagers, and Black-headed Grosbeaks in the Black Hills. Between these two points, some of the changes in birdlife are evident from the moving car. In June, for example, male Lark Buntings in their handsome black and white livery are usually first seen along State 14 just west of the Big Sioux River and they steadily increase in numbers until, in the range country west of the Missouri, they are common. Time after time they flush from the road shoulders or adjacent grasslands to repeat their aerial songs—ascending quickly to heights of 20 to 30 feet, bursting into a medley of tinkling notes, and spiraling down with stiff, butterfly-like wing movements.

Physiographically, as well as ornithologically, South Dakota is divided into nearly equal parts: an eastern and a western. The division roughly follows the Missouri River as it cuts down the middle of the state from North Dakota; consequently South Dakotans frequently refer to the eastern and western parts of the state as the East River and West River.

Eastern South Dakota is mostly a continuation southward of the

Drift Plain of North Dakota, with elevations decreasing from about 1,500 feet in the north to 1,200 feet in the south. The surface is flat to gently rolling except near the Minnesota boundary, where the low hills and ridges of the Coteau des Prairies arise. An abundance of lakes and sloughs characterize the topography of the northeastern corner of the state; many of these are shallow, with a rich growth of marsh vegetation (*see under* **Aberdeen** *and* **Waubay**), and provide some of the finest breeding areas for waterbirds and waterfowl in the state. Meandering southward through eastern South Dakota, and parallel to the Missouri, which they eventually join, are two prominent rivers, the James and the Big Sioux, the latter forming part of the state's eastern boundary.

Most of eastern South Dakota was originally tall-grass prairie. Willows, cottonwoods, elms, hackberries, boxelders, maples, oaks, basswoods, and ashes stood—and to some extent still do—in strips along rivers and streams and in groves around some of the lakes. In the bottomlands of the Big Sioux and Missouri Rivers in the southeastern corner of the state there are a few forests of considerable extent in which trees attain impressive height and density. Birdlife inhabiting the forest growth in eastern South Dakota is principally eastern in composition, whereas the birdlife of the open country comprises many species of western affinities.

Western South Dakota is on the Missouri Plateau, a part of the high Great Plains, whose sharply rolling surface suggests a great rumpled carpet. The extreme eastern portion of the Plateau is dissected by the Missouri River; thus a strip of Plateau, 15 to 25 miles wide, bounds the Missouri River on its eastern side. The western portion is broken by several big rivers, including the Cheyenne and the White, which empty into the Missouri. Elevations of the Missouri Plateau range from 1,400 feet along the Missouri to 3,500 feet near the western boundary. Owing to low rainfall, the soils are relatively dry, produce mainly a short-grass vegetation, and are generally unsuited to crop production except for some sorghum and winter wheat; much of western South Dakota serves, therefore, as grazing land for cattle. Various shrubs, such as wolfberry, silverberry, buffaloberry, sumac, and wild rose grow in most draws and hollows, or along streams and riverbeds, where there are also a few willows, cottonwoods, bur oaks, boxelders, and other trees.

In these environments one may expect a restricted variety of birds, which usually includes the Yellow-breasted Chat, Bullock's Oriole, Lazuli Bunting, and Rufous-sided Towhee.

The Missouri Plateau is interrupted by three distinct regions of irregular surface. The first, in the northwest, is the so-called antelope country (*see under* **Belle Fourche**), where flat-topped buttes rise to heights of 400 to 600 feet above the plains. A part of this country is noted for its 'gumbo' surface. The second region, in the south, is the remarkable Big Badlands (*see under* **Wall**), where the plains have been weirdly dissected by erosion. The third region, in the southwest, is the Black Hills, and they occupy a section of the Plateau roughly 125 miles long in a north-south direction and 60 miles wide.

The Black Hills are not hills but mountains—the highest in North America east of Denver—with a magnificence surpassing their elevation. Unless seen in silhouette, they are not 'black' but deep green, sometimes with tints of purple and blue from their forest covering. Completely surrounded by the Missouri Plateau and more than 100 miles east of the Rocky Mountains, of which they are physiographically a part, the Black Hills have an island-like separateness that gives them a distinct regional character. Ornithologically they are so distinct from South Dakota as a whole that they warrant separate consideration. The reader is, therefore, referred to a special account of the Black Hills under Rapid City.

Although the great majority of birds in South Dakota are not state-wide in their distribution—e.g., many woodland-shrub species east of the Missouri do not reside regularly west of it, and *vice versa*—the following species breed more or less regularly throughout the state in wooded areas or in open country (prairie grasslands, pastures, fallow fields, meadow-like lowlands, woodland edges, brushy places, and dooryards):

WOODED AREAS

Black-billed Cuckoo
Common Screech Owl
Yellow-shafted Flicker (*East River*)
Red-shafted Flicker (*West River*)
Hairy Woodpecker
Downy Woodpecker

Blue Jay
Black-capped Chickadee
White-breasted Nuthatch
Red-eyed Vireo
Warbling Vireo
Ovenbird
American Redstart

OPEN COUNTRY

Swainson's Hawk
Sharp-tailed Grouse
Upland Sandpiper
Mourning Dove
Red-headed Woodpecker
Eastern Kingbird
Western Kingbird
Eastern Phoebe
Say's Phoebe (*West River*)
Horned Lark
Tree Swallow
Barn Swallow
Black-billed Magpie
House Wren
Gray Catbird
Brown Thrasher
Eastern Bluebird
Mountain Bluebird (*West River*)
Loggerhead Shrike
Yellow Warbler
Common Yellowthroat
Bobolink
Western Meadowlark
Orchard Oriole (*southern South Dakota*)
Common Grackle
Northern Cardinal (*East River*)
Blue Grosbeak (*southern South Dakota*)
Indigo Bunting
Dickcissel
American Goldfinch
Rufous-sided Towhee
Lark Bunting
Savannah Sparrow
Grasshopper Sparrow
Vesper Sparrow
Chipping Sparrow
Clay-colored Sparrow
Field Sparrow
Song Sparrow
Chestnut-collared Longspur

Migration in South Dakota, which lies across the central flyway, is dominated by the spring and fall movements of waterbirds and waterfowl. In eastern South Dakota, geese steal the show as they gather in immense numbers along the valleys of the Big Sioux and James Rivers. Farther west, Sandhill Cranes are frequent along the Missouri Valley from the North Dakota line south as far as Pierre and Chamberlain, where their overland route southward into Nebraska apparently begins. Throughout eastern South Dakota, ducks and shorebirds are abundant on suitable lakes, ponds, and sloughs. Long-billed Curlews and American Avocets are conspicuous transients in western South Dakota, where they feed and rest along the shallow margins of rivers. In a very few places, such as the Lacreek National Wildlife Refuge (*see under* **Martin**), both of these splendid shorebirds remain as summer residents. A few American White Pelicans nest in the state (*see under* **Aberdeen** *and* **Martin**), but many more pass through in spring and fall, when they appear on many of the larger lakes. Among the passerine birds, the Harris' Sparrow is a common to abundant transient throughout South Dakota as are the Water Pipit, Tennessee Warbler, Orange-crowned Warbler, Myrtle Warbler, Blackpoll Warbler, White-crowned Sparrow, White-throated Sparrow, and Lincoln's Sparrow. In South Dakota the main migration flights may be expected within the following dates:

Waterfowl: 25 March–20 April; 5 October–15 November
Shorebirds: 1 May—1 June; 1 August—25 September
Landbirds: 20 April—25 May; 20 August—10 October

Owing to South Dakota's generally severe winters, most streams and other bodies of water freeze over. But where, for one reason or another, water stays open, there may be Common Goldeneyes along with a few other ducks. Bald Eagles are a familiar sight near open water below the dams on the Missouri River (for instance, *see under* **Chamberlain**). Rough-legged Hawks, Lapland Longspurs, and Snow Buntings are regular visitants in open country, Slate-colored Juncos and American Tree Sparrows in sheltered spots. In the Black Hills, lowland habitats often feature an abundance of juncos due to the permanent-resident White-winged population being swelled by an influx of many Slate-colored and a few Oregon Juncos. In the Black Hills, too, quite a few Townsend's Solitaires over-winter and, in some winters, vast numbers of Bohemian Waxwings put in an appearance.

Authorities

Leslie M. Baylor, Gilbert W. Blankespoor, Nelda Holden, Herbert Krause, Nathaniel R. Whitney, Jr.

Reference

The Birds of South Dakota: An Annotated Check List. By Nathaniel R. Whitney, Jr. and others. South Dakota Ornithologists' Union, care of the W. H. Over Museum, Vermillion, SD 57069. 1978.

ABERDEEN
Sand Lake National Wildlife Refuge

Highly rewarding for viewing waterfowl and waterbirds in the lake country of northeastern South Dakota is the **Sand Lake National Wildlife Refuge** (21,451 acres), which extends along the James River southward for about 20 miles from a point within a few miles of the North Dakota line. Two low dams across the James River within the property have created a total water surface of about 11,200 acres. Bulrushes, quillreeds, smartweeds, and other emergent plants, including small patches of cattails, grow profusely where the water is shallow, producing a dense marsh vegetation. The adjacent, relatively flat uplands are chiefly grass-covered, except where they have been cultivated to produce food for transient waterfowl, or planted with trees for shelter purposes. There is a sparse native stand of trees near the original borders of the River.

The hordes of transient waterfowl that appear on the Refuge in the spring present a remarkable spectacle. About the first of March, or soon thereafter, come the first of the Canada Geese, and during the remainder of the month their numbers increase steadily. Toward the end of March, Snow and Greater White-fronted Geese start coming. By mid-April, when all three species have reached peak abundance, the total goose population on the Refuge may exceed a half-million individuals, the majority being Snow Geese. While the goose population is building up, thousands upon thousands of ducks arrive, as do small flocks of Whistling Swans, attaining maximum numbers by mid-April. The commonest ducks in the aggregation are Mallards and Common Pintails; the next most common are usually Gadwalls, Blue-winged Teal, Northern Shovelers, Redheads, Canvasbacks, Ruddy Ducks, and Common Mergansers. The waterfowl that appear on the Refuge in the fall, from late September until late November, are the same species, but their populations, thought impressive, are not as large as in the spring.

Waterfowl, waterbirds, and other birds attracted to the Refuge for nesting are notable for their variety and abundance. Among the waterfowl are Canada Geese, Mallards, Gadwalls, Common Pintails, Blue-winged Teal, Northern Shovelers, and Ruddy Ducks. On some of the small islands in the Refuge, American White Pelicans and Double-crested Cormorants have colonies with young in them by early July. Birds nesting in the marshes include Eared, Western, and Pied-billed Grebes, Cattle and Snowy Egrets, Black-crowned Night Herons, American Bitterns, Northern Harriers, Soras, American Coots, Franklin's Gulls (in colonies), Forster's and Black Terns, Marsh Wrens, and Yellow-headed and Red-winged Blackbirds. Ring-billed Gulls and Common Terns nest on some of the small islands and Willets on the periphery of the marshes. In late summer, hundreds more American White Pelicans and Franklin's Gulls gather on the Refuge from near and far for feeding and resting.

FRANKLIN'S GULL

On the grassy uplands of the Refuge both Upland Sandpipers and Marbled Godwits are residents in late spring and summer, Ring-necked Pheasants and Gray Partridges the year round.

To reach the Refuge from Aberdeen, drive north on US 281 for 23 miles, turn east onto State 10 and go 10 miles, and turn south onto County 16 for 2.5 miles to the entrance. Inquire at headquarters about road conditions in the Refuge and good vantages for viewing birds.

BELLE FOURCHE
Antelope Country

North of the Black Hills in northwestern North Dakota, US 85 passes northward from Belle Fourche to the North Dakota line over a sparsely settled and almost desolate plain from which rise numerous buttes. Often, for lack of a better name, it is referred to as **Antelope Country** because herds of the pronghorn are common sights.

For the first 40 miles the highway traverses the 'gumbo belt,' where the clay soil, when wet, produces the 'clingingest mud in the world.' Sod is non-existent; what there is of grassy cover is extremely thin. Birdlife is consequently scarce. In draws and near creeks, however, grasses tend to attain greater density, frequently forming thick clumps. The bird finder should inspect these situations since they are likely to be breeding habitats of Brewer's Sparrows. The birds are best located by stopping the car near promising spots and listening for the song.

After the first 40 miles, the highway passes over well-grassed prairie, used principally for grazing. Common birds to watch for along the way include Lark Buntings and Chestnut-collared Longspurs. Small lakes and sloughs in sight from the highway have surprising concentrations of summer residents such as Gadwalls, Common Pintails, Ruddy Ducks, American Coots, and Wilson's Phalaropes. Adjacent to this 100-mile stretch of highway are several areas used by Sage Grouse for their April courtship performances, but it is impractical to attempt route directions to any of them.

BROOKINGS
Oakwood Lakes Area | Oakwood Lakes State Park

A good site for bird finding near Brookings in east-central South Dakota is the **Oakwood Lakes Area** comprising 3,800 state-owned acres of which 3,000 are water. There are two shallow bodies of water, Oakwood Lake and Lake Tetonkaha, surrounded by marshes supporting bulrushes, bur reeds, and quillreeds. Predominating the woody growth back from the Lakes are cottonwoods and hackberries mixed with a few hawthorns, wild plums, and chokecherries. **Oakwood Lakes State Park,** with picnic facilities and campground, occupies 255 acres in the northern part of the Area.

Migrating geese and ducks are attracted to the lakes in late March and early April, October and early November; shorebirds in May (if water is low), late August and September. In mid-May, bird finding in the Park's picnic area and campground is highly rewarding for viewing transient vireos, warblers, and other northbound small landbirds. Among summer-resident birds in the Area are the Pied-billed Grebe, Western Grebe, American Coot, Red-headed Woodpecker, Baltimore Oriole, and Yellow-headed Blackbird.

To reach the Park, drive north from Brookings on I 29 for 7 miles, exit west onto a road which, in the next 9 miles, goes around the south side of Oakwood Lake and a mile beyond, then turn north into the Park, where there is a large outdoor map showing all parts of the Oakwood Lakes Area.

CHAMBERLAIN
Big Bend Dam

Through winter the water below **Big Bend Dam** stays open, attracting Bald Eagles for their favorite food—fish. The big birds begin showing up in late fall and spend much of their time perched in the big trees along the shore.

To view the eagles, drive north on State 50 for 20 miles, west on State 34 for 6 miles to Fort Thompson, then on State 47 as far south as possible in the Big Dam Recreation Area, park the car, and scan all trees with binoculars or telescope.

MARTIN
Lacreek National Wildlife Refuge

In the extreme southwest-central part of South Dakota, on the northern margin of the Nebraska Sandhills and in the valley of the south fork of the Little White River, is **Lacreek National Wildlife Refuge** (16,147 acres), a productive spot for bird finding. Half the acreage comprises a series of pools with bordering marshes formed by impounding spring-fed sandhill streams; the other half consists of rolling short-grass prairie with scattered willow thickets and small groves of cottonwoods and other trees.

In late June and early July, Trumpeter Swans (introduced from Montana's Red Rock Lakes and well established), Canada Geese, and large numbers of ducks representing at least ten species may be viewed in the marshes with their broods. Other birds that rear young in the marshes include the Western and Pied-billed Grebes, Black-crowned Night Heron, American Bittern, Northern Harrier, Virginia Rail, Sora, American Coot, Forster's and Black Terns, Marsh Wren, Yellow-headed and Red-winged Blackbirds, and Swamp Sparrow.

A colony of American White Pelicans and Double-crested Cormorants occupies a 2-acre, treeless island in one of the pools. Hatching is usually completed in early July. The grassy edges of the pools are attractive to nesting pairs of Willets, American Avocets, and Wilson's Phalaropes.

A few Sharp-tailed Grouse reside on the surrounding stretches of prairie where it is usually possible to see a few breeding pairs of Long-billed Curlews, Upland Sandpipers, and Burrowing Owls. Lark Buntings are common summer residents in the vicinity; Cliff Swallows nest in small colonies under the eaves of Refuge buildings and under the superstructures of water-control gates.

Many thousands of Canada Geese and ducks, particularly Mallards, Gadwalls, Common Pintails, Green-winged and Blue-winged Teal, American Wigeons, Northern Shovelers, Redheads,

Canvasbacks, Lesser Scaups, Buffleheads, and Ruddy Ducks, appear on the pools during their migrations in spring (late March-early April) and fall (October–November). At about the same times of the year, migrating Sandhill Cranes show up on the Refuge in impressive numbers. A great many Mallards as well as a few Common Goldeneyes and Common Mergansers winter on the Refuge, staying in the open water below the spillways, or in spring-fed channels. Rough-legged Hawks and both Golden and Bald Eagles are observed regularly during the winter.

The Refuge may be reached from Martin by driving east on US 18 for 12 miles, turning off south onto a hard-surfaced road for 5 miles to Tuthill, and going west 1.0 to the entrance.

PIERRE

Capitol Lake | Oahe Dam | Farm Island State Park

This capital city on the Missouri River in the geographical center of the state has three notable areas for bird finding.

Capitol Lake, behind the state capitol on Capitol Avenue East, has many Canada Geese as well as Mallards, Common Goldeneyes, and Common Mergansers all winter since it is kept ice-free by a warm spring.

The water below **Oahe Dam** on the Missouri River, 6 miles north of the city, never freezes over and consequently attracts and holds in winter an impressive aggregation of waterfowl, sometimes including scoters, and gulls—Herring and Ring-billed regularly, a few Glaucous and Bonaparte's occasionally. The birds are readily viewed by driving north on State 1804 from US 14 through Pierre, then turning west onto a road which crosses the Dam. The recreation area adjacent to the Dam has marshes, brushy areas, and floodplain forest accessible by trails from which, in suitable habitats during the summer months, one way see or hear such birds as Great Blue Herons, Great Egrets, Blue-winged Teal, Northern Shovelers, American Coots, Rufous-sided Towhees, and permanent-resident Common Screech and Great Horned Owls.

Farm Island State Park, in the Missouri River 3 miles southeast of Pierre just off State 34, features an exceptionally fine floodplain forest with a fascinating mixture of eastern and western breeding birds—for example, both Eastern and Western Pewees, Scarlet Tanagers, and Black-headed Grosbeaks—besides such breeding birds as Wild Turkeys (introduced), Black-billed Cuckoos, both Red-eyed and Warbling Vireos, and Northern Cardinals. Formerly Farm Island Park's 1,800 acres offered recreational facilities but the higher water level created by Big Bend Dam forced relocation to higher ground north on the nearby mainland, in what is now Farm Island Recreational Area. Farm Island was closed to cars, but its forest, not seriously flooded, may be explored by foot trails leading from the old causeway. Thus, drive from State 34 through the Recreational Area and leave the car on the causeway, then walk past the locked gate onto the Island where trails into the forest begin.

RAPID CITY

Black Hills | Hangman's Hill | Mount Rushmore National Memorial | Harney Peak | The Needles | Jewel Cave National Monument | Spearfish Canyon

Many transcontinental tourists pass up the **Black Hills** in southwestern South Dakota as just foothills of the Rocky Mountains, never realizing that they are missing an area with superb scenery, remarkable geological formations, an abundance of wildlife, and a history as colorful and romantic as that of any other part of the west. Though tourists will doubtless continue to pass up the Black Hills—unless lured to Mt. Rushmore for a view of the great faces—may the touring bird finder know better! In this mountain range—from edge of plains to crests of ridges—there awaits an exciting avifauna.

The Black Hills, rising about 4,000 feet above the Missouri Plateau, present a varied landscape. On the eastern and southern limits of the range are hogback ridges, interrupted at intervals by water gaps where streams escape to the surrounding plains. Between the ridges, many of which have precipitous escarpments of red sandstone, and the main mass of the range lies the Red Valley, averaging about 2 miles in width. The main mass is dome-like, its eastern side dipping rather suddenly toward the Red Valley, its western side declining gently to the plains. Owing to prolonged erosion, the eastern half of the range has features that are unforgettably scenic. Deep canyons and rugged inter-canyon ridges are everywhere; in the southeastern section rise granite knobs, spires, pinnacles, and bold mountain forms, culminating in Harney Peak, the highest, with an elevation of 7,242 feet.

Because of their elevation, the Black Hills have a heavier rainfall than the surrounding plains, a fact that is reflected in their dense forest cover. A greater variety of birds is, therefore, to be expected. The predominating tree throughout the Black Hills is the ponderosa pine, often growing in pure, park-like stands; two other pines, the limber and lodgepole, are quite restricted. Mixed with pine, especially in ravines, are a few deciduous trees, principally the quaking aspen and white birch. Breeding birds commonly associated with these forests and consequently widespread in the Black Hills are the Western Flycatcher, Western Pewee, White-breasted Nuthatch, Townsend's Solitaire, Solitary Vireo, Audubon's Warbler, MacGillivray's Warbler (usually in ravines), Western Tanager, Pine Siskin, Red Crossbill (sporadic), White-winged Junco, and Chipping Sparrow. Yellow-bellied Sapsuckers are occasional where there are aspens. Along the streams at higher elevations and on cool north slopes are a few stands of white spruce. Although birds attracted to spruces include many species that are also attracted to pines, and *vice versa,* generally the Gray Jay, Red-breasted Nuthatch, Brown Creeper, Swainson's Thrush, and Golden-crowned Kinglet more commonly reside in spruces. On some of the drier ridges of the foothills—e.g., the hogback ridges on the east and south—grow mountain junipers that are habitat for Pinyon Jays.

Quite a different variety of breeding birds is found in the deciduous woods on the floors of the wider valleys and canyons and along the streams emerging upon the plains. Among the willows, cottonwoods, aspens, bur oaks, and boxelders, with their shrubby undergrowth and bordering thickets, are the Lewis' Woodpecker (uncommon), Hairy and Downy Woodpeckers, Dusky Flycatcher, Gray Catbird, Brown Thrasher, Red-eyed and Warbling Vireos, Yellow Warbler, Yellow-breasted Chat, American Redstart,

Bullock's Oriole, Black-headed Grosbeak, Indigo and Lazuli Buntings, American Goldfinch, Rufous-sided Towhee, and Song Sparrow (uncommon).

Breeding in suitable situations in many parts of the Black Hills are the following species: in steep-walled canyons or near precipitous slopes, the White-throated Swift and Rock Wren; along swift-flowing streams, the North American Dipper; in grassy stretches, weedy fields, and pastures of the Red Valley and elsewhere, the Sharp-tailed Grouse, Horned Lark, Western Meadowlark, and Grasshopper, Vesper, and Clay-colored Sparrows. Other birds likely near buildings or in view from roads through open country are the Turkey Vulture, Red-tailed and Swainson's Hawks, American Kestrel, Red-headed Woodpecker, Eastern and Western Kingbirds, Eastern and Say's Phoebes, Violet-green Swallow, Black-billed Magpie, House Wren, Eastern and Mountain Bluebirds, Loggerhead Shrike, Brewer's Blackbird, Common Grackle, and Lark Bunting.

Rapid City, lying in the gap between the hogback ridges where Rapid Creek enters the plains, is a convenient starting point to good places for viewing birds that have, coincidentally, great scenic appeal.

Bisecting Rapid City is **Hangman's Hill,** a ridge that has a few stunted pines, junipers, and oaks along the crest but is for the most part treeless, its slopes covered with grasses and scattered shrubby thickets. Skyline Drive, reached by driving west from Quincy Street and turning right, passes up Hangman's Hill to Dinosaur Park, where there are five life-sized figures of prehistoric reptiles fashioned in cement. Along the Drive, look for flickers—mostly hybrids between the Yellow-shafted and Red-shafted—and Pinyon Jays in the scrubby trees. Lark Sparrows are common nesting birds. On the slopes of the ridge, especially on the west side, below Skyline Drive, one may hear or see Western Meadowlarks and Grasshopper Sparrows.

A trip during June and July southwest of Rapid City into the most rugged part of the Black Hills provides opportunities for observing the avifauna of typical pine forests in the Black Hills. Proceed south on US 16 for 18 miles, turn left onto US 16A and go through Keystone for 5 miles, then right onto State 244 for a circuitous 2-mile climb to a parking lot in **Mount Rushmore National Memorial** (2 square miles). A few steps from the car, a height of ground affords a dramatic view of the faces of four United States Presidents—Washington, Jefferson, Lincoln, and Theodore Roosevelt—carved in heroic proportions on the granite shoulder of Mt. Rushmore (6,040 feet in elevation). White-throated Swifts nest in lofty crevices near these immense sculptures and can be seen frequently flying back and forth in front of them. Western Pewees, Solitary Vireos, Audubon's Warblers, Western Tanagers, and White-winged Juncos are some of the birds residing among the pines in the general vicinity of the parking lot and visitor center.

From the Rushmore Memorial, continue westward on State 244 which, for the next 8 miles, traverses pine forests and allows wide views of **Harney Peak** in the distance on the left; then turn south onto US 385 and travel for a short distance and left onto State 87 which, in the next 6 miles, enters Custer State Park and passes Sylvan Lake.

If one wishes to climb Harney Peak, the trail, which is very easy to follow and ascends gradually, begins at the campground on the northeastern edge of Sylvan Lake. Birds to watch or listen for along the trail include, besides those at the Rushmore Memorial, Northern Three-toed Woodpeckers, Red-breasted Nuthatches, Townsend's Solitaires, Golden-crowned Kinglets, Ovenbirds, MacGillivray's Warblers (in ravines), Pine Siskins, and, possibly, Red Crossbills. Nearing the summit, the trail leads out of the pine forest onto bare rock surfaces. (There are no timber-line or alpine conditions as on the higher peaks in the Rocky Mountains.) Birds are relatively scarce, although one is quite likely to see Turkey Vultures, Red-tailed Hawks, Common Nighthawks, White-throated Swifts, and Violet-green Swallows in flight above or below the trail. From the summit one is treated to a breath-taking view of the southern Black Hills and may observe the mountain goats that have thrived on Harney Peak since their introduction many years ago.

From Sylvan Lake, continue into Custer Park on State 87 which swings eastward and for the next 9 miles is called the Needles Highway. The source of the name is readily apparent when, 1.5 miles from Sylvan Lake, the Highway enters a tunnel in a tremendous granite promontory and emerges upon **The Needles**—gigantean knobs, pinnacles, spires, and other strangely shaped rock masses reaching skyward. A few Audubon's Warblers and Western Tanagers inhabit the scattered, somewhat stunted pines at their bases, and White-throated Swifts cut across the deep abysses between them; but in general this uncanny area is not attractive to birds.

Stay on State 87 as it goes south for 5 miles; then turn west onto US 16A and go 7 miles to Custer outside Custer Park and continue west on US 16 for 14 miles to **Jewel Cave National Monument.** Its rugged terrain of 2 square miles is forested primarily with ponderosa pines, many attaining great size. For breeding birds typical of the pine forests in the Black Hills there is probably no better place. Species such as Western Flycatchers, Western Pewees, White-breasted Nuthatches, Townsend's Solitaires, Solitary Vireos, Audubon's Warblers, Western Tanagers, and White-winged Juncos are a certainty; other species such as Black-backed Three-toed Woodpeckers, Brown Creepers, and Cassin's Finches are likely though less conspicuous.

Unquestionably the best place for bird finding in the Black Hills is **Spearfish Canyon,** far to the northwest of Rapid City. Extending for 21 miles in a north-south direction between Cheyenne Crossing and the town of Spearfish, the Canyon is comparatively open in its upper part near Cheyenne Crossing, but farther down it begins to deepen, and increasingly the light-colored sandstone and limestone walls become steeper and closer together. Clear, cool Spearfish Creek descends rapidly through the chasm and is joined about 4 miles above Spearfish by the waters of Bridal Veil Falls, which spill over the face of the east wall from a considerable height.

White-throated Swifts, North American Dippers, and Canyon and Rock Wrens are summer residents in Spearfish Canyon, but its superiority for bird finding is due in a large measure to its extensive forest growth—the spruces, sometimes in pure stands, sometimes mixed with pines, along Spearfish Creek in the upper part of the Canyon, the deciduous trees and shrub thickets along the Creek toward the southern part. Thus Spearfish Canyon attracts a wide variety of breeding birds as the following selected list attests: Saw-whet Owl, Yellow-bellied Sapsucker, both Black-

NORTH AMERICAN DIPPER

backed and Northern Three-toed Woodpeckers, Dusky and Western Flycatchers, Gray Jay, Black-capped Chickadee, White-breasted and Red-breasted Nuthatches, Brown Creeper, Swainson's Thrush, Veery, Townsend's Solitaire, Golden-crowned Kinglet, Warbling Vireo, MacGillivray's Warbler, American Redstart, Western Tanager, Black-headed Grosbeak, Lazuli Bunting, Evening Grosbeak, Purple Finch, Red Crossbill, Rufous-sided Towhee, White-winged Junco.

Spearfish Canyon may be reached from Rapid City by driving westward on State 44 (Rim Rock Drive) for 19 miles, right onto US 385 for 24 miles to Pluma, then left (west) onto US 85 through Lead for 10 miles to Cheyenne Crossing. From here, continue first on US 85 west and south along the west branch of Spearfish Creek through a delightful valley forested with spruce and pine. Along the first 5 miles of this highway, and in Hellsgate and Deadhorse Gulches whose entrances are passed on the right, bird finding is very productive.

Backtrack on US 85 to Cheyenne Crossing and go north on US

14A (Spearfish Canyon Road), which follows Spearfish Creek into the Canyon. Bird finding along the Road, which goes down the Canyon to Spearfish, is as good in one spot as another. Nearly every bridge crossed has a pair of Northern American Dippers nesting beneath it. While going through the deepest part of the gorge, stop the car now and then to listen to Canyon Wrens singing far up on the cliffs. At a spot called Savoy, 5.5 miles down from Cheyenne Crossing, walls of limestone attain heights of 1,000 feet or more. Below, in the deciduous trees and thickets near the stream, are Veeries, Warbling Vireos, Black-headed Grosbeaks, and Lazuli Buntings.

Before continuing down the Canyon past Bridal Veil Falls to Spearfish, take a side trip by turning left at Savoy onto a narrow road that follows the winding course of Little Spearfish Creek. After 1.0 mile, a short path leads from a parking lot on the left to Roughlock Falls. It is almost a certainty that a pair of dippers will have a nest behind this 30-foot drop of water and will be searching for food in the series of cascades below.

Drive 4 miles beyond Roughlock Falls to the point where the road forks. Take the right fork and, a little later, the next road left; this climbs rapidly to the crest of Cement Ridge where there is a fire-lookout just over the state line in Wyoming. Townsend's Solitaires are fairly numerous in the pine woods. The higher slopes, though treeless, are richly matted with wildflowers—balsam roots, larkspurs, lupines, chickweeds, bluebells, shooting stars, and pasqueflowers—which bloom profusely in June and July. From atop the Ridge the superlative view of the northern Black Hills is itself well worth the side trip.

SIOUX FALLS
Wall Lake | Grass Lake

In the farming country west of this large city in southeastern South Dakota are fields and pastures interrupted by marshy lakes and potholes—the most rewarding area for bird finding in the vicinity.

Drive west from Sioux Falls on State 42 for 11 miles to a crossroad; turn left (south) to **Wall Lake,** which will soon appear on the right. At the State Game Refuge on the east side of the Lake, Orchard and Baltimore Orioles are summer residents in the grove of trees and bordering shrubs. During migration, Lesser Yellowlegs and Spotted, Baird's, and Semipalmated Sandpipers often appear on the sandy shore.

Continue west on the section-line road which leads around the south side of Wall Lake. This road runs parallel to US 16 for about 3.5 miles and then crosses a blacktop road which runs north and south. Along this 3.5-mile stretch are fine examples of prairie marshes with cattails, bulrushes, and bur reeds, where, in the summer months, such birds as Pied-billed Grebes, American Bitterns, Mallards, Blue-winged Teal, Northern Shovelers, Virginia Rails, Soras, Marsh and Sedge Wrens, Yellow-headed and Red-winged Blackbirds, and Swamp Sparrows reside. In the fields and pastures between the marshes, look for the Northern Harrier, Ring-necked Pheasant, Gray Partridge, Eastern and Western Kingbirds, Brown Thrasher, Bobolink, Western Meadowlark, Dickcissel, and Grasshopper and Vesper Sparrows. Fortunate observers may obtain sightings of the Upland Sandpiper and Blue Grosbeak.

After reaching the north-south running blacktop, proceed north to State 42, cross it and continue northward for 3 miles to **Grass Lake.** At a point 0.5 mile from the highway crossing, the road passes (on the right) a shallow marsh with muddy shores and flats attractive to shorebirds in May, August, and September. Between this point and Grass Lake are large fields where thousands of Snow Geese stop to feed during their northward migration in late March. Grass Lake itself, an extensive marsh with open water, is attractive to many transient waterbirds and waterfowl. Whistling Swans in late March and American White Pelicans in the fall are two of the Lake's outstanding ornithological features. Also, the Lake is an occasional breeding site of the Western Grebe.

STAMFORD
White River Valley

I 90, between Chamberlain on the Missouri and the Black Hills, traverses a vast stretch of rolling plains. Between Murdo and Kadoka the highway is not far north of the **White River Valley** with its productive spots for bird finding. One of the best is reached by exiting at Stamford and going south on a dirt road, which, after about 7 miles, drops down into the Valley.

On the moderately high bluffs, where there are low shrubs in scattered patches, Blue Grosbeaks are surprisingly numerous. Lark and Field Sparrows occupy the same environment. Close to the White River, amid thin stands of cottonwoods, willows, boxelders, and shrubby thickets, one may locate without much difficulty such birds as Gray Catbirds, Yellow-breasted Chats, Orchard and Bullock's Orioles, Rufous-sided Towhees, and Clay-colored Sparrows.

TIMBER LAKE
Little Moreau State Recreation Area

For a good example of birdlife in wooded draws that interrupt the state's grasslands, visit **Little Moreau State Recreation Area** (320 acres) in north-central South Dakota, reached from Timber Lake on State 20, 40 miles west of Mobridge, by turning off south for 6 miles. Here in thick stands of hardwoods and shrubby thickets on slopes leading down to the Little Moreau River, and in marshy edges of lakes where the River has been impounded, are practically all the warblers breeding regularly in South Dakota outside the Black Hills—the Black-and-white, Yellow, Ovenbird, Common Yellowthroat, Yellow-breasted Chat, and American Redstart. Among other birds are the Black-billed Cuckoo, Least Flycatcher, Sedge Wren, Gray Catbird, Brown Thrasher, Wood Thrush, Black-headed Grosbeak, American Goldfinch, and Field Sparrow.

WALL
Badlands National Park

Between the Cheyenne and White Rivers east and south of the Black Hills is a vast area where the semiarid plains have been remarkably eroded, forming the Big Badlands of South Dakota. Fine soft clays, capped by a thin sod, have been deeply dissected, producing a succession of V-shaped ravines with most of their walls oddly terraced by plates of sandstone. Between closely adjacent ravines there are sharp crests; between widely separated ravines there remain, as mesas or tablelands, the grassy uncut surfaces, 500 to 600 feet above the riverbeds.

One of the most spectacular sections of the Badlands is embraced by **Badlands National Park** (410 square miles), which lies north of the White River and 60 miles east of the Black Hills. Here the plains have been cut by numerous tributaries of the White River. When they are looked down upon from the edge of the plains, or are viewed from the broad flat through which the White River wanders, the effect is startling. Between the innumerable ravines looms a veritable wilderness of sculptured formations that suggest spires, turrets, minarets, pyramids, domes, and giant tables. Nearly all have strata of different pale colors—buff, pink, gray, light green—which vary in intensity according to lighting conditions. On late, sunlit afternoons or early mornings when the sky is deep blue, they are at their brilliant best.

US 16A, proceeding south from I 90 at Wall, enters the Park after about 7 miles and soon zigzags down over the Great Wall into the strange wonderland of the Badlands. For 23 miles or so the highway winds southward and eastward, giving the traveler every opportunity to view the array of formations. After about 23 miles, US 16A ascends Cedar Pass and leaves the Park northward to join I 90 at Cactus Flat, approximately 9 miles distant. Bird finders arriving from eastern South Dakota via I 90 can exit at Cactus Flat and take US 16A through the Park in the opposite direction. In any case, whether entering the Park from Wall or Cactus Flat, bird finders should stop at the vistor center below Cedar Pass for information about interpretive services and directions to the Cliff Shelf Nature Trail.

Birdlife in Badlands Park is expectedly limited both in variety of species and in numbers of individuals, yet it is not without attractions, especially to eastern bird finders in quest of their first glimpses of such western birds as Rock Wrens, Say's Phoebes, and Mountain Bluebirds. Rock Wrens nest in the fissures of clay banks; Say's Phoebes and Mountain Bluebirds choose similar nesting sites, although in the vicinity of the visitor center the phoebes nest under the eaves of buildings and the bluebirds in bird boxes.

Other birds to look for in the Park include: Golden Eagles soaring high above the cliffs; Horned Larks and Western Meadowlarks on the flats of the White River, the tops of tablelands, and the neighboring grassy plains; Lark Sparrows on the more gradual slopes where clumps of grasses and brush manage to gain foothold; and Cliff Swallows flying about steeply cut banks where they nest in colonies under overhanging sandstone plates or sod.

WAUBAY

Waubay National Wildlife Refuge | Bitter Lake | Rush Lake | South Waubay Lake | Cormorant Island

The name of the town Waubay in northeast South Dakota comes from a word in the language of the Sioux Indians meaning 'where wild fowl build their nests.' The name is indeed appropriate, for Waubay is situated in rolling prairie dotted with many potholes and lakes. Much of the upland has the original prairie sod undisturbed by settlement, and its bodies of water, being in most cases shallow, have extensive beds of cattails, bulrushes, quillreeds, and sedges, which provide suitable cover for breeding waterbirds and waterfowl.

Waubay is 11 miles west of Webster on the south side of US 12. North of town lies the **Waubay National Wildlife Refuge** (4,651 acres), reached by turning north from US 12, 1.75 miles east of Waubay, and following Lake Road (no route number) northward for 8 miles to the southwest extremity of Enemy Swim Lake, then turning off left at a directional sign to headquarters. Inquire here about road conditions in the Refuge.

About half the Refuge is comprised of lakes and marshes that provide nesting areas for Canada Geese and for many ducks representing twelve or more species, the more common being the Mallard, Gadwall, Common Pintail, Blue-winged Teal, Northern Shoveler, Redhead, and Ruddy Duck. All have broods by late June. Certain marshes in the Refuge or around lakes near the Refuge have nesting Red-necked Grebes—close to the southernmost breeding area of the species in the conterminous United States—as well as many Western Grebes. Other birds reported breeding in the Refuge or in similar habitats nearby include Pied-billed Grebes, American Bitterns, Northern Harriers, Soras, American Coots, Franklin's Gulls, Forster's and Black Terns, Marsh Wrens, and Yellow-headed and Red-winged Blackbirds.

Besides visiting the Refuge, the bird finder should investigate especially three lakes in the Waubay area as directed below. On these bodies of water or in their peripheral marshes he will note particular species, in some instances more numerous than in the Refuge, in other instances poorly represented in the Refuge, if at all.

Bitter Lake, reached from US 12 by turning south into Waubay and continuing southward through the business section to the south edge of town, turning west after about 1.0 mile, then going south about 2 miles to the northwest shore. From here a road leads south and southeast along the west shore where there is a fine marsh.

Rush Lake, crossed at the south end by US 12, 2.5 miles west from Waubay. From the road grade are excellent views of the open water and the marsh vegetation that gives the Lake its name. Black-crowned Night Herons nest on a wooded island in the Lake 1.5 miles north of US 12. In the late spring and summer the Lake attracts a large bird population whose species composition is practically the same as in the Waubay Refuge. If the water level is low in August and September, shorebirds are numerous on the muddy shores and flats. From spring to fall a few American White Pelicans are nearly always somewhere on the Lake.

South Waubay Lake, reached from US 12, 5 miles west of Waubay by turning and going north on a road for about 3.5 miles and then turning west onto another road which leads 1.5 miles to the west shore. In view out in the Lake from here is **Cormorant Island** with a large nesting colony of Double-crested Cormorants. They have eggs by 1 June.

Texas

GREATER ROADRUNNER

Texas, true to its reputation as a land of superlatives, has the longest list of birds of any state—altogether 555 species. Since the total number of bird species in North America north of Mexico is about 825, one may say that it is possible to find in Texas three-fourths of them.

The list of Texas birds, though impressive, is to be expected, for reasons peculiar to the state itself. *In the first place,* Texas covers an immense and diversified area of some 267,000 square miles. Whether from the northern boundary of the Texas Panhandle south to the Rio Grande, or from the Sabine River in the extreme east to the westernmost corner of the state near El Paso, the distance across Texas is nearly 800 miles. The state thus brings within its borders segments of the Coastal Plain, the Great Plains that sweep north into Canada, the southwestern arid lands that characterize New Mexico and Arizona, as well as several environments that are distinctly semitropical or Mexican. From the Gulf Coast to the mountains that tower above the deserts in the far western part of the state, elevations range from sea level to over 8,000 feet. Annual rainfall varies from 50 or more inches in extreme eastern Texas to less than 10 inches around El Paso. Winter days at Amarillo on the high prairies of the Panhandle are frequently bitter, with sharp winds and freezing temperatures; those at Corpus Christi and Brownsville are typically warm and sunny. *In the second place,* Texas has vastly different and extensive habitats: fresh water and salt water; coastal islands, beaches, marshes, and prairies; lowland, upland, and montane forests; brushlands, grasslands, and deserts. *In the third place,* Texas lies in the path of a tremendous number of migrating bird species and within the winter range of many others.

The Texas coast, extending for 380 miles between Louisiana and Mexico, has a shoreline of barrier sand beaches. Many of them are long, slender islands and peninsulas that parallel the mainland and shelter narrow, intervening lagoons, usually called bays. Except for interruptions caused by such bays and by river mouths, these beaches are continuous. All the bays are normally shallow, with low islands; several such islands in Galveston Bay support productive nesting colonies of waterbirds. Frequently both the bays and their islands are fringed by grassy marshes. Of the various narrow bays, the longest is Laguna Madre, which extends south from Corpus Christi Bay for 130 miles to a point near the delta of the Rio Grande. Padre Island, reputed to be the longest island for its width in the world, separates Laguna Madre from the Gulf. Though averaging 6 miles in width, the Laguna is remarkably shallow—in a few places more than 10 feet deep, but most of it less than 2 feet. For thousands of waterfowl and other birds the Laguna is a winter haven, and certain of its islands hold enormous nesting colonies of waterbirds.

Along the Texas coast, on the beaches and islands, and in the salt and brackish marshes, the birds listed below breed more or less regularly. Species marked with an asterisk nest locally in colonies.

- * Olivaceous Cormorant
- * Great Blue Heron
- Green Heron
- * Cattle Egret
- * Reddish Egret
- * Great Egret
- * Snowy Egret
- * Louisiana Heron
- * Black-crowned Night Heron
- Yellow-crowned Night Heron (*uncommon*)
- * White-faced Ibis
- * White Ibis
- * Roseate Spoonbill
- Fulvous Whistling Duck
- Mottled Duck
- King Rail
- Clapper Rail
- Purple Gallinule
- Common Gallinule
- American Coot
- American Oystercatcher
- Snowy Plover
- Wilson's Plover
- Killdeer
- Willet
- Black-necked Stilt
- * Laughing Gull
- Gull-billed Tern
- Forster's Tern
- * Little Tern
- * Royal Tern
- * Sandwich Tern
- * Caspian Tern
- * Black Skimmer
- Marsh Wren
- Common Yellowthroat
- Red-winged Blackbird
- Boat-tailed Grackle
- Great-tailed Grackle
- Seaside Sparrow

The interior of Texas may be considered to have four major natural regions: the Coastal Plain, the High Plains Country, the Edwards Plateau, and the Trans-Pecos. All show sharp contrasts in topography and climate and, in turn, great differences in habitats for birds. It must not be inferred, however, that these contrasts exist only *between* these regions and that *within* each region the conditions are relatively uniform. So extensive is each region and so diverse its topography and climate that within each area, though to a somewhat lesser degree, there are marked differences in environmental conditions. In the following paragraphs each of the major regions is described to the extent of depicting its general limits—which are often arbitrary—and some of its more important environments, especially those of interest to the bird finder. A listing of characteristic breeding species is not attempted.

A. The *Coastal Plain* comprises all of eastern and southern Texas, limited on the west by the Balcones Escarpment of the

Edwards Plateau, a prominent topographic feature that runs from the Rio Grande near Del Rio eastward to San Antonio and northward through Austin, becoming lower and disappearing west of Waco. Farther north there is no sharp western boundary, but a line drawn northward through Mineral Wells and Nocoma to the Red River roughly marks the western limits of the true Coastal Plain. Though a low-lying country throughout, the Coastal Plain may nevertheless be divided into four general areas:

1. Inland for 30 to 75 miles from the beaches and marshes of the Gulf Coast is the flat Coast Prairie on which lie the cities of Beaumont, Port Arthur, Galveston, Houston, and Corpus Christi. The Coast Prairie is a broad strip of clayey soils and high water table, with small stretches of sandy soils interspersed. It is generally treeless; salt and marsh grasses thrive near the coast and coarse beard grasses, bull grasses, panic grasses, and others predominate farther inland. The only parts of the Coast Prairie with any extensive tree growth are the valleys of the larger rivers which support luxuriant stands of oak, pecan, elm, and cottonwood, and the Blackjack Peninsula and a few other places between Bay City and Corpus Christi which have huge mottes (clumps) of oak. (For sections of the Coast Prairie with rewarding sites for bird finding, *see under* **Port Arthur, Houston, Port Lavaca,** *and* **Rockport.**)

2. Extreme eastern Texas—north of the Coast Prairie and west approximately to a line beginning just north of Houston and passing northward to the Red River through Huntsville, Palestine, Mineola, and Clarksville—has sandy soils. Sometimes referred to as the East Texas Pineywoods (*see under* **Huntsville**), much of the area supports, or originally supported, upland forests of mixed pine and hardwoods, with oaks predominating, but river-bottom and swamp hardwoods occupy considerable territory. In the extreme southeast is the so-called Big Thicket, which contains, or did contain, an unusually rich flora, including large beeches, magnolias, oaks, and pines.

3. The remainder of eastern Texas is mainly the Blackland Prairie area, a belt of calcareous clayey soils running from San

Antonio northeast and north to the Red River. This is the most densely populated portion of the state and includes the cities of San Antonio, Austin, Waco, Dallas, and Fort Worth. North of Waco, the Prairie is split by the East Cross Timbers, a narrow strip of sandy soils covered by post and blackjack oaks. On the west at about the same latitudes, the Prairie is bordered by another strip of sandy soils, the West Cross Timbers, with similar vegetation. The West Cross Timbers extend without a topographical or vegetational break for a short distance westward outside the true Coastal Plain. (For bird finding in this general area, *see under* **Dallas** *and* **Fort Worth.**)

4. Southern Texas—the triangular part of the state south and southeast of the Edwards Plateau—is a rolling plain, rising imperceptibly westward to an elevation of almost 1,000 feet near Del Rio. Covered in wide areas with thorny shrubs, often in open stands and mixed with various grasses, yuccas, and cacti such as prickly pear and tasajillo, it receives the name Chaparral Country. The prominent shrub is mesquite; other important kinds are acacia, mimosa, huisache, buckthorn, lignum vitae, granjeno, cenizo, and whitebrush. Bottomlands of the larger stream valleys have many trees, among which are live oaks, pecans, elms, hackberries, and ashes. In the lower Rio Grande Valley, an extensive citrus-fruit and vegetable-raising industry has been developed, but there are still a few undisturbed areas where shrubs—the Texas ebony, retama or Jerusalem thorn, wild olive, anaqua, and many other semitropical or tropical species—grow luxuriantly, forming dense, almost impenetrable tangles. On the floodplain of the lower Rio Grande, trees of immense size still stand in a few places. (For bird-finding possibilities in typical Chaparral Country, *see under* **San Antonio;** in the lower Rio Grande Valley, *see under* **Harlingen.**)

B. The *High Plains Country* occupies all of the Panhandle and extends, as the Edwards Plateau, south to the Rio Grande between Del Rio and the mouth of the Pecos River; east and southeast to the Coastal Plain; and west to the Pecos River, which marks the eastern margin of the Trans-Pecos. Within the High Plains Country, excluding the Edwards Plateau, there are at least three natural divisions: the short-grass plains, the mixed-grass plains, and the mesquite plains.

1. The short-grass plains, called in Texas the Staked Plains or Llano Estacado, are a southward continuation of the Great Plains from western Kansas and the Oklahoma Panhandle. In Texas they stretch from the northernmost limits of the Texas Panhandle south to the Edwards Plateau, with which they merge, indiscernibly, south of Big Spring and Ballinger. On the west they stretch to the New Mexico line and the Pecos River Valley; on the east they are bounded by 'the Cap Rock,' an abrupt, rock-capped escarpment dropping to the mixed-grass and mesquite plains. In the Panhandle, the Staked Plains are exceedingly flat, interrupted only by a few deep valleys and by scattered, shallow depressions containing playa (temporary) lakes. A short-grass association, with buffalo and grama grasses as the principal constituents, is the natural cover. (For further information on the Staked Plains, together with an indication of the opportunities that await the bird finder in the area, *see under* **Amarillo.**)

2. In Texas the mixed-grass plains are limited to the eastern side of the Texas Panhandle between the Cap Rock of the Staked Plains and the Oklahoma boundary, but across the boundary in Oklahoma they include much of the western part of that state. (See the introduction to the Oklahoma chapter for a description of the mixed-grass plains.)

3. The mesquite plains lie in north-central Texas south of the Red River. (A small section also extends into the southwestern corner of Oklahoma, north of the Red River.) Like the mixed-grass plains, they are limited on the west by the Cap Rock of the Staked Plains. On the east they merge gradually with the West Cross Timbers of the Coastal Plain; on the south they meet the Edwards Plateau. In general, the country is level to rolling except where gullied or, at higher elevations approaching the Cap Rock, cut by deep canyons. The native vegetation is represented by an abundance of thorny shrubs, mainly mesquite, growing in

open stands that alternate with stretches of grama and other short-grass species. In certain places, especially in gullies, the shrubs grow in dense thickets. The bottomlands along the larger streams support stands of such trees as elms, oaks, hackberries, and cottonwoods; the canyons support a similar growth and, in addition, junipers. (For bird finding on the mesquite plains, *see under* **Abilene, Seymour,** *and* **Wichita Falls,** three cities situated in this type of country.)

C. The *Edwards Plateau* in south-central Texas reaches from the Staked Plains and mesquite plains to the Rio Grande between Del Rio and the mouth of the Pecos River. Its western limits are the Pecos River Valley; its eastern and southern limits are the outfacing Balcones Escarpment, which drops abruptly to the Coastal Plain. The height of the Escarpment increases gradually toward the west, from Austin, where it is 300 feet high, to Del Rio, where it reaches 1,000 feet. Altitudes of the Edwards Plateau show a steady decline southward and eastward from 2,500 feet at the border of the Staked Plains to 1,000 feet just west of Austin and 1,500 feet just north of Del Rio. West of the 100th meridian the Edwards Plateau resembles the Staked Plains, being a little-interrupted, level plain covered with short-grass sod, but farther east it has a rugged surface owing to the numerous out-flowing, many branched rivers—the San Saba, Pedernales, Blanco, Guadalupe, Nueces, and others—that have cut deeply into the underlying limestone producing such formations as terraced hills, canyons, and gullies—an area locally referred to as the Hill Country. Vegetation, in addition to grasses that cover the open plain, ranges from heavy forests of live oaks, elms, walnuts, hackberries, and pecans on the canyon floors—many are very wide—to scrub forests of Mexican junipers, Texas oaks, shin oaks, post oaks, blackjack oaks, stunted live oaks, and other trees on the higher, rocky slopes of the hills and canyons. Birdlife in the Edwards Plateau shows an intriguing mixture of the species characteristic of the several adjoining natural regions; only one species, the Golden-cheeked Warbler, is peculiar to the Plateau and rarely breeds elsewhere. (For a place on the Plateau to see the Golden-cheeked Warbler, as well as a variety of other breeding birds, *see under* **Austin.**)

D. The most rugged and scenic of the four major natural regions in Texas is the *Trans-Pecos*, which lies west of the Pecos River between the upper Rio Grande and the New Mexico line. In general, it is a region of arid to semiarid conditions, of varying elevations and discontinuously distributed habitats, of desert, grassland, and montane birds. Aside from the Toyah Basin (a plain about 50 miles wide lying along the Pecos River from New Mexico south to Terrell County) and the Stockton Plateau (geologically a part of the Edwards Plateau, and lying south and southeast of the Toyah Basin in most of Terrell and eastern Brewster Counties), the Trans-Pecos consists of bolsons (undrained basins) and intervening mountain ranges or 'roughlands.' The bolsons have a flat terrain and desert aspect: some stretches yield a thin cover of grasses, chiefly tobosa, galleta, and grama; others, open stands of such shrubs as creosote bush, catclaw, and blackbrush. On low mountain slopes and, not infrequently, on high slopes with southern exposure, similar grass associations exist, together with mesquite and various acacias, but on all high mountain slopes with northern exposure the vegetation features an extensive growth of trees.

The three highest mountain ranges of the Trans-Pecos—the Guadalupe Mountains in the extreme north (*see under* **Pine Springs**), the Davis Mountains in the central part (*see under* **Fort Davis**), and the Chisos Mountains in the extreme south (*see under* **Marathon**)—show a typical succession of plant associations from desert to mountain top. Although the avifauna of all three mountain ranges has many similarities, there are nonetheless a few species that reside in only one or sometimes two of the ranges. This is explained in part by their isolation in a vast expanse of open country. Although all three ranges show a rough grandeur, the bird finder will discover that each one has a distinctiveness of topography, color, and general character. Hence his exploration of each range will be unmatched.

Lying in the path of the central flyway (the 'flyway of the plains'), Texas is in spring and fall the scene of migratory movements of those avian hordes that breed in the enormous area between the Mississippi and the Rocky Mountains and the equally great area directly north in Canada. Texas is also a vantage point for viewing the passage of huge numbers of birds that are summer residents east and west of these areas: in eastern Canada and eastern United States; in western Canada and Alaska; in western Montana and Idaho, Wyoming, western Colorado and Utah, and New Mexico. Such birds as the Wood Thrush, Worm-eating Warbler, Golden-winged Warbler, Blue-winged Warbler, Northern Parula Warbler, Cerulean Warbler, Blackburnian Warbler, Hooded Warbler, Canada Warbler, and Louisiana Waterthrush, which breed mainly in the wooded parts of eastern North America, appear as common transients along the Texas coast and inland for 50 miles; and many waterbirds and waterfowl, which nest in the Bear River Marshes of Utah, fly in a northwest-southeast direction across the Texas Panhandle and central Texas to and from the Laguna Madre and other coastal areas southward, where they winter.

Birds moving through central and west Texas seem to follow no particular mountain ranges, ridges, valleys, or watercourses. Waterbirds, waterfowl, and shorebirds tend to 'hop skip' from one suitable body of water to another lying in the general direction of their flight. Flocks of open-country birds—e.g., Lark Buntings—pass north and south over the plains, stopping to rest and feed as the occasion demands. Flycatchers, warblers, and other tree and shrub birds gather temporarily in most forested areas along rivers, or in thickets sequestered in canyons and draws.

Along the Texas coast pass vast numbers of migrating waterbirds, shorebirds, and landbirds. The populations of certain species of small landbirds observed along the lower Texas coast from Port Lavaca south are sometimes so great as to be almost unbelievable. From Rockport, for example, have come reports in spring (late March through early May) of 500 Common Yellowthroats estimated at one time in a 100-foot row of tamarisk, of 57 Bay-breasted Warblers observed in one tree, and of 50,000 Barn Swallows counted within two hours. Along the upper Texas coast such concentrations of small landbirds occur in the spring usually during a cold front, or 'norther.' In good weather, apparently, the birds continue north inland where the upper coast swings eastward, but when a cold front comes in from the northwest, the usual direction, the birds are forced southeastward off their course and congregate on the upper coast until the weather improves.

Probably a few small landbirds reach the extreme upper Texas coast in the spring in the way that great numbers reach the Louisiana coast—from across the Gulf of Mexico. (For a description of how weather, both good and bad, affects their arrival, see the introduction to the Louisiana chapter.)

A great many shorebirds may be observed in Texas, especially along the coast, in any season except early summer, but they are most abundant in spring and fall, when their numbers are augmented by transient individuals that spend the winter farther south.

The majority of birds migrating through Texas may be observed within the following dates:

Shorebirds: 10 March–10 May; 20 August–25 October
Landbirds: 15 April–15 May; 15 September–10 November

Winter bird finding is highly rewarding in Texas, particularly along the coast, in the sheltered depths of western canyons, and in the lowlands of the Rio Grande Valley. Among Christmas bird counts taken in different parts of the state, those on the lower Texas coast almost invariably yield the highest figures—close to 200 species. The wintering populations of certain species are often as impressive as the variety. Geese, mostly Canada and Snow, on the coast and Coast Prairie have been estimated in excess of 300,000. The first flights arrive during the last two weeks of September; the heaviest flights come during the first two weeks of October; departures begin in late February and continue until May. In the Muleshoe National Wildlife Refuge (*see under* **Muleshoe**) as many as 700,000 ducks and 100,000 Sandhill Cranes have been estimated in midwinter.

Authorities

Charles H. Bender, Luther C. Goldman, Kathleen S. Harrington, James A. Lane, George A. Newman, Warren M. Pulich, Carrol Richardson, R. Dudley Ross, Kenneth Seyffert, Wayne A. Shifflett, Sally H. Spofford, James A. Tucker, John L. Tveten, Kevin L. Tveten, Roland H. Wauer, Frances Williams, Kevin J. Zimmer.

References

A Field Guide to the Birds of Texas and Adjacent States. By Roger Tory Peterson. Boston: Houghton Mifflin. 1960.

A Birder's Guide to the Rio Grande Valley of Texas. By James A. Lane. Distributed by L & P Press, Box 21604, Denver, CO 80221. 1978.

A Birder's Guide to the Texas Coast. By James A. Lane and John L. Tveten. Distributed by L & P Press, Box 21604, Denver, Co 80221. 1980.

Check-list of the Birds of Texas. Compiled by L. R. Wolfe, Warren M. Pulich, and James A. Tucker. Texas Ornithological Society. Available from ABA Sales, Box 4335, Austin, TX 78765. 1975.

ABILENE

Lake Fort Phantom Hill | Lake Abilene | Abilene State Recreation Area

About midway on I 20, which crosses the width of Texas just north of the central area, lies Abilene, situated in one of the stretches of typical mesquite plains.

Almost any road leading out of Abilene passes groves of mesquite, sometimes hung with mistletoe, and intervening grassy areas where cacti and yuccas are prevalent. Here are habitats for such regularly nesting birds as Greater Roadrunners, Western Kingbirds, Scissor-tailed Flycatchers, Verdins, Loggerhead Shrikes, Orchard and Bullock's Orioles, and Lark and Cassin's Sparrows.

Two reservoirs near Abilene are worth investigating for waterbirds, waterfowl, and shorebirds. One, **Lake Fort Phantom Hill,** is a few miles north of the city, reached by exiting north from I 20 on State 351, then going north on Farm Road 2833. East of the Lake, just off State 351, is the Water Reclamation Plant. The other reservoir, **Lake Abilene,** is about 17 miles southwest of the city, reached by exiting south from I 20 on US 83 (Winters Freeway), then going south on Farm Road 89 (Buffalo Gap Road). Both reservoirs and the Water Reclamation Plant have muddy shores when the water is receding. If, during late April, May, August, or September, there are muddy edges, shorebirds are frequently attracted to them in impressive numbers. Wilson's Phalaropes are sometimes very common, as are Greater and Lesser Yellowlegs, Pectoral, White-rumped, and Baird's Sandpipers, Long-billed Dowitchers, and others. American Avocets and Black-necked Stilts usually appear.

Near Lake Abilene, on the southeast, is the **Abilene State Recreation Area** (489 acres), embracing a diversity of woodland habitats including a streamside association of pecan, elm, and willow along Elm Creek that runs through the Area. Breeding birds residing here are represented by some species characteristic of eastern Texas, others western. Among the more common species, for example, are the Mississippi Kite, Golden-fronted Woodpecker, Blue Jay, Carolina Chickadee, and Blue-gray Gnatcatcher. Practically all habitats are readily accessible by roads and nature trails.

AMARILLO

Staked Plains | Palo Duro Canyon State Park | Lake Meredith | Buffalo Lake National Wildlife Refuge

The Texas Panhandle embraces an almost square area, roughly 165 miles from east to west and 150 miles from north to south. Amarillo, the one large city in the Panhandle, is a good focal point for the bird finder desiring to investigate this vast land.

Most of the Texas Panhandle is a segment of the short-grass Great Plains called the **Staked Plains** (3,400 to 3,800 feet elevation), flat to slightly rolling, and treeless except in dooryards, shelterbelts, and breaks along streams. The land is used principally for the grazing of livestock and for crop production. The eastern edge of the Staked Plains is marked by the Cap Rock trending north and south, some 30 to 50 miles west of the Oklahoma line. I 40 east and west from Amarillo, US 87 north and I 27 south from the same city, and side roads leading off these highways traverse country typifying the Staked Plains.

The grazing lands and cultivated areas of the Staked Plains show a paucity of breeding birds; the Horned Lark, Western Meadowlark, and Grasshopper and Cassin's Sparrow are among the few species that are at all common. Where there are shrubs and trees, however, additional breeding birds, such as the Mourning Dove, Western Kingbird, Scissor-tailed Flycatcher, Bewick's Wren, Northern Mockingbird, Bullock's Oriole, and, occasionally, the Loggerhead Shrike, may reside; in fact, there is scarcely a dooryard with trees that does not have at least one pair of Western

Kingbirds or Northern Mockingbirds. Burrowing Owls may be expected wherever there are prairie-dog 'towns.' Except during the summer months, raptors may be frequently observed along the highways, either in flight or on utility poles and fence posts. The Red-tailed, Rough-legged, and Ferruginuous Hawks, Northern Harrier, and American Kestrel are the predominating species, except in spring and fall (usually April and October), when there are immense flights of Swainson's Hawks. In May, and again in late summer and fall, migrating flocks of Lark Buntings, sometimes containing hundreds of individuals, are frequent on the Staked Plains. Big flocks of McCown's, Lapland, and Chestnut-collared Longspurs appear in winter.

Two large streams cut through the Staked Plains of the Panhandle from west to east. One, the South Canadian River, flows through a narrow, bluff-rimmed valley, 20 miles north of Amarillo; the other, a somewhat smaller stream called Prairie Dog Town Fork of the Red River, courses through an awesome gash in the Plains, known as Palo Duro Canyon, 24 miles southeast of Amarillo. Neither stream ordinarily carries much water, but both are subject to flash floods during the infrequent periods of heavy rainfall. Although birdlife is worth investigating almost anywhere along these streams, unquestionably the greatest variety of species is to be found in that part of Palo Duro Canyon within **Palo Duro Canyon State Park.**

The State Park (16,402 acres) includes Palo Duro Canyon's upper section, which is 1.0 mile wide and 600 feet deep, the walls precipitous and highly colored—mostly red combined with yellow, gray, lavender, and streaks of white. To reach the Park, drive south from Amarillo on I 27 for 16 miles to the town of Canyon; exit east on State 217 and proceed about 12 miles to the entrance. Once in the Park one may take a broad, well-graded road that passes from the rim of the Canyon to the floor, thence down the chasm for a distance of 10 miles through thickets of mesquite and juniper, crossing and re-crossing Prairie Dog Town Fork at frequent intervals. The stream itself is bordered by chinaberry, hackberry, willow, and cottonwood, and here and there are large, grassy flats.

Bird finding in the Canyon is excellent in any season, including the winter months, when the high walls, trees, and brush afford shelter for hosts of birds from near and far. Among the common winter residents are the American Robin, Mountain Bluebird (often in huge flocks), Oregon Junco, and White-crowned and Song Sparrows. Along with these species are the year-round residents including the American Kestrel, Scaled Quail, Greater Roadrunner, Great Horned Owl, Red-shafted Flicker, Golden-fronted and Ladder-backed Woodpeckers, (Black-crested) Tufted Titmouse, Bushtit, Bewick's, Canyon, and Rock Wrens, Northern Cardinal, House Finch, and Lesser Goldfinch. Beginning in May, when the nesting season gets under way, the following birds should be looked for: Mississippi Kite, Yellow-billed Cuckoo, Common Nighthawk, Belted Kingfisher, Yellow-shafted Flicker, Ash-throated Flycatcher, Cliff Swallow (large colonies), Blue-gray Gnatcatcher, Bullock's Oriole, Blue Grosbeak, Painted Bunting, and Lark and Rufous-crowned Sparrows. April, September, and October bring numerous transient warblers—Orange-crowned, Nashville, Yellow, Myrtle, Audubon's, Wilson's, and Common Yellowthroat—and fringillids—Pine Siskin, American Goldfinch, and Chipping and Clay-colored Sparrows.

Scattered across the broad, seemingly interminable expanse of the Staked Plains are a great many playa lakes, often called 'wet weather' lakes because of their temporary nature. In addition to these natural bodies of water are many water impoundments that have proved attractive to waterbirds, waterfowl, and shorebirds when the playas are dry. A good example of one of the larger impoundments is **Lake Meredith,** 20 miles long with 100 miles of shoreline, formed by damming the South Canadian River. Lying about 35 miles northeast of Amarillo, it may be reached by driving north from the city on State 136.

Some of the birds appearing regularly on the playas and impoundments in fall and spring are the Eared and Pied-billed Grebes, Great Blue Heron, Canada Goose, Mallard, Gadwall, Common Pintail, Green-winged and Blue-winged Teal, American Wigeon, Northern Shoveler, Redhead, Canvasback, Lesser Scaup, Ruddy Duck, Common Merganser, American Coot, Sandhill Crane, Killdeer, Long-billed Curlew, Greater and Lesser Yellowlegs. Long-billed Dowitcher, Baird's, Least, and Western

Sandpipers, American Avocet, Wilson's Phalarope, Ring-billed and Franklin's Gulls, and Black Tern.

Buffalo Lake National Wildlife Refuge (7,677 acres), southeast of Amarillo and reached by driving south on I 27 for 16 miles to the town of Canyon, exiting west on US 60 for 12 miles, then turning off south at Umbarger and going 2 miles, has fields and pastures that hold a good year-round sampling of bird species typical of the Staked Plains. Buffalo Lake itself, except when sometimes dry for long periods, has spectacular concentrations of waterfowl in winter and attracts shorebirds in spring and fall that are impressive both in numbers and variety.

AUSTIN
Pedernales Falls State Park

A wide variety of birds exists in the vicinity of this capital city, primarily because the Balcones Escarpment separating the Edwards Plateau from the Coastal Plain bisects the area, thus bringing into close proximity birds characteristic of each natural region. East of the city on the Coastal Plain are Ruby-throated Hummingbirds, Blue Jays, and Tufted Titmice; west of the city on the Edwards Plateau are Black-chinned Hummingbirds, Scrub Jays, and (Black-crested) Tufted Titmice. Other birds reach the eastern limits of their breeding ranges here and still others the western limits. The birds one may see during the nesting season in the few remaining undisturbed places within or just outside the city are often a mixture of eastern and western.

The stellar ornithological attraction of the Austin area is the summer-resident Golden-cheeked Warbler whose wooded habitat, almost entirely in the eastern part of the Edwards Plateau, consists of Mexican juniper mixed with Texas oak and scrub live oak standing on ridges, the rims of canyons, and the upper slopes of ravines. Among the several sites west of Austin that support optimum habitat for the Golden-cheeked Warbler, as well as for other species associated with the Edwards Plateau, is **Pedernales Falls State Park** (4,851 acres) along the Pedernales River, reached from Austin by exiting west from I 35 on US 290 (Ben White Boulevard) for 48

GOLDEN-CHEEKED WARBLER

miles to Johnson City, then turning off right onto Farm Road 2766 and going 8 miles.

The Park embraces typically rugged Hill Country with ridges, canyons, and ravines readily accessible by foot trails. Besides the Golden-cheeked Warbler, one should look especially for the summer-resident Black-capped Vireo in upland woods of oak with dense understory, Blue Grosbeak and Painted Bunting along woodland edges. Other summer residents include the Poor-will, Black-chinned Hummingbird, Ash-throated Flycatcher, Bell's Vireo, and Summer Tanager. In any season one may expect in suitable habitats the Common Ground Dove, Greater Roadrunner, Common Screech Owl, Golden-fronted Woodpecker, Scrub Jay, (Black-crested) Tufted Titmouse, Verdin, Bushtit, Canyon Wren, Lesser Goldfinch, and Lark, Rufous-crowned, and Black-throated Sparrows.

Reference

A Bird Finding and Naturalist's Guide for the Austin, Texas, Area. By Edward A. Kutac and S. Christopher Caran. Sponsored by the Travis Audubon Society and Austin Natural Science Association. Austin: The Oasis Press, 1976.

DALLAS
White Rock Lake | L. B. Houston Park Nature Area | Mountain Creek Lake | Greenhills Environmental Center

This huge industrial and commercial center on the prairie in northeast-central Texas has a number of good places for bird finding in different seasons.

Among the several bodies of water within metropolitan Dallas that are attractive to ducks, one of the best is **White Rock Lake** northeast of the city's center. From I 30, exit northeast on State 78 (Garland Road), then turn off left onto State L-12 (Buckner Boulevard) from which there are roads leading off west to Lawther Drive which winds around the Lake. Anytime between mid-October and mid-April one can view a variety of ducks that are likely to include the Mallard, Gadwall, Common Pintail, Green-winged and Blue-winged Teal, American Wigeon, Northern Shoveler, Lesser Scaup, and Ruddy Duck. Other birds to look for at the same time are Ring-billed and Bonaparte's Gulls, and the occasional Common Loon, Horned Grebe, and Eared Grebe.

Northwest of the city's center is the **L. B. Houston Park Nature Area,** operated by the Dallas Museum of Natural History, which embraces approximately 600 acres of bottomland habitat along the Elm Fork of the Trinity River. The Nature Area may be reached from I 35E (Stemmons Freeway) by exiting west on State 183 (John W. Carpenter Freeway) for 1.0 mile, bearing off right (before reaching the Texas Stadium) onto State 114 and going 0.7 mile, then turning off right (north) onto Tom Braniff Drive and going 0.5 mile to the parking lot at right. From here, near headquarters, self-guided nature trails lead through woodlands and swampy areas, along the River, and around a gravel-pit lake. Breeding birds in the Nature Area are, among many others, the Wood Duck, Red-shouldered Hawk, Red-bellied Woodpecker, Great Crested Flycatcher, Carolina Chickadee, Bewick's and Carolina Wrens, Red-eyed Vireo, Orchard Oriole, and Indigo and Painted Buntings. Ducks and a few shorebirds such as the Spotted and Solitary Sandpipers are attracted to the River and its shallows during the migration seasons. In winter, Purple Finches, American Goldfinches, Rufous-sided Towhees, and Slate-colored Juncos are common as are sparrows, particularly the Vesper, Field, Harris', White-crowned, White-throated, Fox, and Song.

Southwest of the city's center are two areas that are equally productive. One is **Mountain Creek Lake** amid mesquite and oak-clad hills. Here from May until the last of September there are fair numbers of Little Blue Herons and Great and Snowy Egrets. American Avocets may be present in May, September, and October. Black Terns are numerous during the last three weeks of May and again the last of August and September. The last half of September and the first week of October bring thousands of swallows of several species, the Barn Swallow often predominating. From mid-October to mid-November huge flocks of Franklin's Gulls make the Lake their headquarters while they forage for food in the surrounding countryside. The Ladder-backed Woodpecker is a regular inhabitant of the mesquite and oaks. Eastern Meadowlarks may be heard singing during spring in grassy places. Scissor-tailed Flycatcher nests are easy to find—sometimes in dead trees standing near the Lake.

To reach Mountain Creek Lake, exit west from I 35E on Illinois Avenue, crossing State L-12 and S-408 and continuing west on Mountain Lake Road, which turns south along the Lake, providing access to opportunities for bird finding described above.

The other area southwest of Dallas's center is the **Greenhills Environmental Center,** operated by the Greenhills Foundation, which embraces 800 acres of hilly countryside covered partly by oak-hickory woods and low shrubs. The Black-capped Vireo may be found here from mid-April to the first of September. Other birds to be looked for at the same time are both Turkey and Black Vultures, Swainson's Hawk, Blue-gray Gnatcatcher, White-eyed and Red-eyed Vireos, Summer Tanager, and Painted Bunting.

The Greenhills Center may be reached from US 67 (Marvin D. Love Freeway) by exiting west on Camp Wisdom Road, then turning south onto Clark Road for 2.1 miles to Danieldale Road. Thereafter turning west, go 0.4 mile to the entrance on the right, where there is an iron gate. Drive into headquarters for permission to explore the area and for advice on roads or trails to the most productive bird habitats.

DENISON
Hagerman National Wildlife Refuge

For a huge aggregation of waterfowl in fall, winter, and early spring, the bird finder in northeast Texas should visit the **Hagerman National Wildlife Refuge** (11,429 acres), across an arm of Lake Texoma, an impoundment of the Red River between Texas and Oklahoma. Headquarters on the Refuge may be reached as follows: From Denison on US 75, drive west on Farm Road 120 for about 6 miles, south on Farm Road 1417 for 4 miles, and then west on an unnumbered blacktop road for 6 miles, following directional signs.

Three thousand acres of the Refuge are water and marsh. Here southbound ducks begin gathering in late September, geese in early October, until by November the number of ducks and geese may exceed 100,000. About 75 per cent of the population consists of ducks, chiefly Mallards with nonetheless impressive numbers of Gadwalls, Common Pintails, Green-winged and Blue-winged Teal, American Wigeons, and Northern Shovelers, and small numbers of diving ducks. Canada Geese, and to a much less extent, Greater White-fronted and Snow Geese, make up the goose population. A great many Mallards and a few other ducks over-winter. From late February through March the Refuge's water and marsh again teem with waterfowl, stopping off temporarily during their north-bound flights.

Except in winter and early spring, the marsh commonly attracts wading birds, particularly Great Blue, Green, and Little Blue Herons, and both Great and Snowy Egrets. In May and late summer, if the water level is low, shorebirds are usually numerous. Ring-billed and Franklin's Gulls are abundant in spring and fall, Ring-billed Gulls also through winter.

The remaining 8,000 acres of the Refuge are upland, some 600 acres of which are cultivated for duck food. Elsewhere natural habitats vary from open meadows and brushy areas to woodlands of oak, pecan, ash, and some juniper. Among the more common birds breeding in one habitat or another are the Common Bob-white, Yellow-billed Cuckoo, Common Nighthawk, Red-bellied Woodpecker, Scissor-tailed Flycatcher, Blue Jay, Carolina Chickadee, Tufted Titmouse, Eastern Bluebird, Bell's Vireo, Eastern Meadowlark, Orchard Oriole, Blue Grosbeak, Indigo and Painted Buntings, Dickcissel, and Grasshopper and Lark Sparrows.

EAGLE LAKE
Attwater Prairie Chicken National Wildlife Refuge

As long ago as 1880 the Attwater's subspecies of the Greater Prairie Chicken ranged over much of the Coast Prairie from the Mississippi River in Louisiana to Corpus Christi; but since then excessive hunting and destruction of its natural habitat of tall grasses caused its gradual disappearance until by the 1960s there were estimated fewer than two thousand birds, in small groups widely scattered. Thanks to the concerted efforts of conservation agencies and the U.S. Fish and Wildlife Service, the **Attwater Prairie Chicken National Wildlife Refuge** was established in 1972 to preserve and restore some of the prairie chicken's natural habitat and to ensure the survival of at least a small population of this endangered subspecies.

Located about 75 miles inland from the Gulf of Mexico on the western fringe of the Coast Prairie, the Refuge may be reached from the community of Eagle Lake (south of I 10 between Houston and San Antonio) by driving 6 miles northeast on Farm Road 3013 to the entrance. Of the Refuge's 5,600 acres, 4,074 consist mainly of tall-grass prairie with some cropland, woodlands, and open water in potholes.

The bird finder wishing to see prairie chickens on the Refuge should plan his trip for the period from February through April when he can view the birds on their booming grounds at sunup and at dusk. After April the birds no longer appear on the grounds and stay widely dispersed. Prior to his trip the bird finder must write or call the Refuge Manager (address: Box 518, Eagle Lake, TX 77434; telephone (713) 234-3021) to arrange for his being conducted to the booming grounds, which he may observe from towers approximately 100 yards distant.

Besides prairie chickens there are numerous other birds to see on the Refuge, including such nesting species as the Fulvous Whistling Duck, Mottled Duck, and White-tailed Hawk.

WHITE-TAILED HAWK

EL PASO

Hueco Tanks State Historical Park | McKelligon Canyon | Rio Grande Levee | Fort Bliss Sewage Ponds

Texas's westernmost city (elevation 3,762 feet) is ringed by deserts, arid mountains with canyons, and the neighboring Rio Grande, making its environs attractive to a wide variety of birds. The following sites are among the best for birds but should be visited in the early morning before the day turns hot and crowds of people gather, especially on weekends.

Hueco Tanks State Historical Park (860 acres), 32 miles northeast of El Paso, embraces scrub desert, dominated by creosote bush with admixtures of grasses and yuccas. Here and there are canyons, washes, and natural cisterns or ponds formed by natural depressions in rock that hold water except in dry seasons, particularly summer. To reach the Park, drive east from the city on US 62-180 for 26 miles, then turn off north on Farm Road 2775 for 6 miles. Along this road to the Park, be alert for birds: in any season for such species as the Red-tailed Hawk, Scaled Quail, Greater Roadrunner, Cactus Wren, Northern Mockingbird, Pyrrhuloxia, and Black-throated Sparrow; in winter, for the Northern Harrier, American Kestrel, Curve-billed Thrasher, Western and Mountain Bluebirds, Loggerhead Shrike, Western Meadowlark, and sparrows that include the Chipping, Brewer's, and White-crowned. In the Park during any season, expect to see the Say's Phoebe and Crissal Thrasher around the campgrounds and the Turkey Vulture, White-throated Swift, Ladder-backed Woodpecker, Verdin, Rock and Canyon Wrens, Brown Towhee in the neighboring canyons. Black-chinned Hummingbirds, Scott's Orioles, and Blue Grosbeaks are present in summer. The ponds, except when dry, attract a few waterfowl.

In the Franklin Mountains north of El Paso are several canyons. One of the best for birds, about 7 miles north, is **McKelligon Canyon** in McKelligon Park, reached by driving north from the city on Alabama Avenue, then turning off left onto the entrance road. In the Canyon, any time of the year, are Scaled Quail, as well as Canyon and Rock Wrens, Pyrrhuloxias, Lesser Goldfinches, Brown Towhees, and Rufous-crowned and Black-throated Sparrows. Ash-throated Flycatchers reside here in summer, Black-chinned Sparrows in winter. Finding the Gambel's Quail, though uncommon, is always a possibility and sighting a Golden Eagle or two is likely.

Worthwhile for bird finding in any season is a drive along the **Rio Grande Levee** southeast of El Paso. From the Levee one may watch the river and its banks on one side for birds and on the other for birds in the paralleling ditch and the desert country or irrigated fields beyond. For a productive trip, go southeast from the city on I 10 for 18 miles, exit right at Fabens onto North Fabens Street and stay on it for about 7 miles to the river; here turn right on the Levee road and proceed. Birds observable in any season are Gambel's Quail, White-winged Doves, Greater Roadrunners, Burrowing Owls, Common Yellowthroats, and Great-tailed Grackles. After about 7 miles, begin watching the river banks, the ditch, and the irrigated fields for Mexican Ducks. After 13.6 miles there are sewage ponds on the right that attract ducks and shorebirds in season. From here one may continue on the Levee for more birds or take the next turn right and return to I 10.

The best location for waterbirds, waterfowl, and shorebirds in the El Paso area is the **Fort Bliss Sewage Ponds.** From US 54 in northeast El Paso, turn off east onto Fred Wilson Road, staying in the left lane for 1.8 miles; turn north (left) onto a road at the traffic light (a sign here says Ammunition Supply Point 1.5 miles north) and go 9 miles; turn right (east) onto a small dirt road angling off to the northeast and go 0.7 mile; then turn right to the pumping station. From here, except in wet weather, one may drive out on the dikes which overlook the Sewage Ponds.

Some of the Ponds have deep water; others are shallow with many hummocks supporting a cover of vegetation. In any season, look for the Common Gallinule; in summer, the Eared Grebe, American Avocet, Black-necked Stilt, Wilson's Phalarope, and Black Tern; in winter, the Marsh Wren, Swamp Sparrow, and most of the ducks common in the interior United States. To be expected in migration are such species as the White-faced Ibis, Semipalmated Plover, Spotted and Solitary Sandpipers, Willet, Greater and Lesser Yellowlegs, Baird's Sandpiper, Long-billed Dowitcher, Western Sandpiper, Northern Phalarope, and Franklin's Gull.

FORT DAVIS

Davis Mountains | Davis Mountains State Park | Limpia Creek | Buffalo Trail Boy Scout Ranch

North and west of Fort Davis (elevation 5,050 feet) in western Texas rise the **Davis Mountains**; the highest peak, Mt. Livermore, reaches 8,382 feet, the second highest point in the state.

Compared with the Chisos Mountains in Big Bend National Park farther south (*see under* **Marathon**), the Davis Mountains have a refreshing greenness and a less rugged, somewhat kindlier beauty—an overall impression that has some basis in fact. Despite their rocky surfaces and the canyons that cut deeply into their north slopes, they have a denser cover of vegetation. This is especially true at elevations above 5,500 feet, where there are many trees, principally oaks, pinyon pine, and alligator juniper. Ponderosa pines stand in some of the higher, wetter canyons. Among the breeding birds in these high-country woodlands are the Band-tailed Pigeon, Acorn Woodpecker (in oaks), Cassin's Kingbird, Western Pewee, (Black-eared) Bushtit, White-breasted Nuthatch, Western Bluebird, Solitary Vireo, Hepatic Tanager, Black-headed Grosbeak, and Chipping Sparrow. Violet-green Swallows may be expected. Poor-wills and Elf Owls frequently call at night. Steller's Jays and Mountain Chickadees breed on Mt. Livermore and other high peaks, all inaccessible to the public. The Montezuma Quail resides on the more open mountain slopes.

From 5,500 feet down to the scrub desert at approximately 2,500 feet are grasslands with mottes of small oaks and junipers, or thickets of mesquite, acacias, yuccas, and cacti. Breeding birds in these situations are such species as the Common Ground Dove, Inca Dove, Greater Roadrunner, Ladder-backed Woodpecker, Western Kingbird, Ash-throated Flycatcher, Traill's Flycatcher (streamside willows and cottonwoods), Vermilion Flycatcher (usually near water), Scrub Jay, (Black-crested) Tufted Titmouse, Bushtit, Bewick's and Rock Wrens, Curve-billed Thrasher, Western Bluebird, Phainopepla, Bell's Vireo, four orioles (Orchard, Hooded, Scott's, and Bullock's), Summer Tanager, Blue Grosbeak, Painted Bunting, House Finch, Lesser Goldfinch, Brown Towhee, and Lark, Rufous-crowned, Cassin's, and Black-throated Sparrows.

A strongly recommended time for bird finding in the Davis Mountains is April, May, and early June when singing is at its height and many plants are in bloom, their showy flowers always attractive to Black-chinned Hummingbirds and to the few Magnificent Hummingbirds that are appearing in the Davis Mountains with increasing frequency. Although much of the Davis Mountains area is privately owned and definitely off-limits to the public—including bird finders—there are nonetheless good opportunities to see birds along the highways and in sites open to the public. Herewith are some suggestions, all for birds in the breeding season.

Davis Mountains State Park (1,869 acres in the eastern foothills), reached from Fort Davis by driving north for 1.0 mile on State 17, then west on State 118 (known locally as Loop Road 78) for 3 miles to the entrance, has many easily observed birds—

among them, Cassin's Kingbirds, Say's Phoebes, Scrub Jays, Rock Wrens, Curve-billed Thrashers, and Brown Towhees. Other birds to look for include Western Bluebirds, Phainopeplas, Orchard Orioles, and both Cassin's and Lark Sparrows. Montezuma Quail are often seen walking across camp sites or the drives to the two scenic overlooks.

From the State Park northwestward, State 118 traverses grassland interspersed with oak-juniper woodlands, eventually passing, after 11 miles, the entrance road on the right to the McDonald Astronomical Observatory on Mt. Locke (elevation 6,828 feet)—accessible by a steep road and offering at its terminus a magnificent view of the Davis Mountains.

Beyond the entrance road to the Observatory, State 118 reaches wooded high country, well above a mile in elevation and the only area accessible to mountain birds. The Madera Canyon Picnic Ground is the best stopping point although there are frequently wide shoulders where one can stop briefly to look at roadside birds.

Thirteen miles beyond the entrance road to the McDonald Observatory and shortly after leaving Madera Canyon, turn off left from State 118 and return to Fort Davis via State 166, skirting the western side of the Davis Mountains at a lower elevation in open country. Called the Scenic Loop, the trip totals 74 miles from Fort Davis and back. Good stopping points for bird finding are at Rockpile Roadside Park and Bloys Camp Meeting Grounds.

State 17 north from Fort Davis toward Balmorhea provides opportunities for birds in desert country. Beyond its intersection with State 118 (*see above*), the highway follows **Limpia Creek** for 18 miles. Oak groves and brushy thickets along the way are worth inspecting for all the birds heretofore listed for elevations below 5,500 feet. Cliff Swallows nest under bridges and Black Phoebes may be noted from most of them. Stop and listen for Canyon Wrens singing from the walls of the narrow canyon through which the Creek and highway pass. Look for Great Horned Owls sitting in high crevices. Be alert for Lesser Black Hawks which are known to nest in cottonwoods along the Creek.

Twenty-four miles north of Fort Davis on State 17, turn off left onto the road to **Buffalo Trail Boy Scout Ranch,** 11 miles distant.

This leads through brushy country, excellent for desert birds—for example, the Verdin, Cactus Wren, Pyrrhuloxia, and Rufous-crowned and Black-throated Sparrows. Where the road fords the first creek, inspect the streamside thickets for the White-winged Dove, Bell's Vireo, all four species of orioles, and both the Varied and Painted Buntings. White-throated Swifts may be watched in flight along the cliffs overlooking the Ranch entrance. If the gate into the Ranch is open, go in and request permission to hike into Aguja Canyon. But take heed: The Canyon becomes unbearably hot after 10:00 a.m. and most of the birds stop singing.

FORT WORTH

Botanic Garden | Fort Worth Nature Center and Refuge | Eagle Mountain Fish Hatchery | Benbrook Lake

Close to this big city, which lies on the prairie between the East and West Cross Timbers of northeast-central Texas, are several areas for diversified bird finding.

Only 2 miles west of the business district is the **Botanic Garden** (40 acres), reached from I 20 by exiting north on University Drive, taking the first left, which is Botanic Garden Road in Trinity Park, and proceeding to Garden Center, a building. Park here.

In addition to woodlands, the Garden has lawns, pools, lagoons, and formal gardens. From late March to mid-May, and from late August through September, as many as twenty-five species of warblers, together with other small passerines, may be identified in migration. Among the birds nesting in the Garden, or very nearby, are the following: Yellow-billed Cuckoo, Barred Owl, Black-chinned Hummingbird, Red-bellied Woodpecker, Great Crested Flycatcher, Tufted Titmouse, Bewick's and Carolina Wrens, Summer Tanager, and Painted Bunting. Between October and April, wintering birds include the Red-shouldered Hawk, Yellow-shafted and Red-shafted Flickers, Yellow-bellied Sapsucker, Red-breasted Nuthatch, Golden-crowned and Ruby-crowned Kinglets, Orange-crowned and Myrtle Warblers, Purple Finch, Pine Siskin, Rufous-sided Towhee, and numerous sparrows such as Field, Harris', White-crowned, White-throated, Fox, Lincoln's, and Song.

Highly recommended for productive bird finding is the **Fort Worth Nature Center and Refuge** on the northwestern edge of the city. Continue north on University Drive from the Botanic Garden; turn left onto State 199 (Jacksboro Highway) and proceed northwest. After crossing Lake Worth Bridge, turn right off State 199 exactly 2 miles from the Bridge and follow directional signs to the Center and Refuge.

The Center and Refuge embrace 3,400 acres of natural area that includes a fine remnant of the West Cross Timbers—sandy uplands with extensive oak growth and intervening prairie—as well as moist lowlands with pecans, hackberries, willows, and various shrubs, and a unique lotus marsh. The Robert E. Hardwicke Interpretive Center, the hub of the Nature Center, has information on trails and seasonal programs.

The lotus marsh is accessible by boardwalk from which one may observe numerous waterbirds, shorebirds, and ducks in season. Wood Ducks are present the year round and nest in the Refuge.

For waterbirds, shorebirds, and ducks, another productive spot is the **Eagle Mountain Fish Hatchery,** operated by the Texas Parks and Wildlife Department and open to visitors, except on weekends, from 8:00 a.m. to 5:00 p.m. This is reached as follows: From the turnoff to the Nature Center and Refuge, continue northwest on State 199 for 3 miles; turn off right onto Hanger Cut Off (a sign here says Eagle Mountain Lake) and go about 0.75 mile; turn right onto Wells-Burnet Road to the dam; cross it and make a sharp turn right onto Ten Mile Bridge Road and go about 1.5 miles to the *second* Eagle Mountain Circle Drive sign (a church is on the left at the intersection); here turn left and follow the Drive about 0.5 mile to the entrance.

At the Hatchery in late summer, many wading birds, including the Little Blue and Yellow-crowned Night Herons, may be watched in the shallow water of the ponds. Probably the best time for bird finding is from mid-September to mid-November when the pools are being drained, leaving mud flats attractive to Lesser Golden and Black-bellied Plovers, Solitary Sandpipers, Lesser Yellowlegs, Pectoral Sandpipers, Stilt Sandpipers, Western Sandpipers, and other shorebirds. Also, during this period, many transient and winter-resident ducks arrive, notably Mallards,

Gadwalls, Common Pintails, Green-winged and Blue-winged Teal, American Wigeons, Northern Shovelers, and Lesser Scaups. In late October there are usually Ring-billed, Franklin's, and Bonaparte's Gulls, and Black Terns.

Benbrook Lake and vicinity southwest of Forth Worth offer a good variety of birds, particularly migrating and wintering species. From I 20, exit south on State 183 (Alta Mere Drive) for 1.0 mile to traffic circle, then continue south on US 377 for 9 miles, going through the suburb of Benbrook, to Farm Road 1187; turn left onto FR 1187 and follow directional signs to Mustang Park, a recreational area on the south side of the Lake and a good vantage for scanning the surface for birds.

A reservoir with a surface area of 3,770 acres, Benbrook Lake brings wintering and transient grebes, such as the Eared and Pied-billed, and practically the same species of shorebirds and ducks that appear at the Eagle Mountain Fish Hatchery. Wooded habitats in Mustang Park and elsewhere around the Lake where creeks enter are rewarding in spring and fall for migrating warblers and other small landbirds, in winter for Red-breasted Nuthatches, Brown Creepers, Pine Siskins, and American Goldfinches. In prairie country back from the Lake, Savannah and Vesper Sparrows are common in winter, and Lark Sparrows all year, especially in summer. Always a possibility in winter on plowed and well-grazed fields in view from roads around the Lake are Chestnut-collared and Lapland Longspurs. Also a possibility in winter is sighting a Rough-legged Hawk in flight almost anywhere above open country.

GALVESTON

Bolivar Ferry | West Galveston Island

This port city and vacation center on East Galveston Island between Galveston Bay and the Gulf of Mexico on the northeast Texas coast overlooks many salt-water bays and estuaries where Brown Pelicans, Olivaceous Cormorants, gulls, terns, and Black Skimmers may be viewed in any season; mud flats and beaches where herons, egrets, ibises, and Roseate Spoonbills may also be viewed in any season, and shorebirds commonly in winter. Two trips from the city are suggested.

1. Upon arriving in the city from Houston via I 45, proceed for a ride on the free **Bolivar Ferry** by continuing east on US 75, then left onto State 87, following the signs. For bird-finding opportunities aboard the Ferry in its 3-mile crossing of the channel (Bolivar Roads) to the Bolivar Peninsula, *see under* **Port Arthur.** Unless one wishes to continue onto the Bolivar Peninsula for more bird finding, he may park his car near the Ferry dock and ride on the Ferry over and back as many times as he desires.

2. Upon approaching the city from Houston via I 45, exit south on 61st Street (Farm Road 342), then turn off west onto Stewart Road (S Street on some maps) to **West Galveston Island** which extends southwestward for some 30 miles.

Much less settled and developed than East Galveston Island, West Galveston Island offers relatively undisturbed habitats for varieties of birds in any season. Broad grassy areas and well-grazed pastures attract in the winter months such birds as Black-bellied Plovers, Long-billed Curlews, Horned Larks, Water Pipits, and occasionally Sprague's Pipits; in the spring months, Lesser Golden Plovers, Whimbrels, Upland Sandpipers, Buff-breasted Sandpipers—and possibly an Eskimo Curlew. Mud flats in winter, or in spring and fall, are almost certain to have Semipalmated and Piping Plovers, Greater and Lesser Yellowlegs, Pectoral and Least Sandpipers, Dunlins, Long-billed Dowitchers, Semipalmated and Western Sandpipers, American Avocets, and sometimes Solitary Sandpipers, Baird's Sandpipers, Short-billed Dowitchers, Stilt Sandpipers, Marbled Godwits, and Wilson's Phalaropes. The beaches on both the east and west sides of the Island have Ruddy Turnstones and Sanderlings present in any month; Red Knots in spring and fall, less so in winter.

Wilson's Plovers nest on the more remote beaches as do a few Snowy Plovers. Ponds flanked by marshes, depending on their size and extent, hold nesting birds that may include the Least Bittern, Mottled Duck, Blue-winged Teal, King Rail (occasionally), Purple and Common Gallinules, American Coot, Willet, Black-necked Stilt, Marsh Wren, and Boat-tailed Grackle.

To see birds on West Galveston Island, the bird finder should drive southwestward on Stewart Road, stopping to explore promising habitats as they appear along the Road and along side roads—Six-Mile Road, Eight-Mile Road, Ten-Mile Road, Nottingham Ranch Road, and Eleven-Mile Road. At the termination of Stewart Road, continue southwestward on Farm Road 3005 which traverses ranchlands to the end of the Island. With their extensive grassy areas and well-grazed pastures, the ranchlands are the best sites for wintering and migrating curlews and other birds preferring upland situations. A few Sandhills Cranes can usually be viewed on the ranchlands in winter.

HARLINGEN

Lower Rio Grande Valley | Laguna Atascosa National Wildlife Refuge | Santa Ana National Wildlife Refuge | Bentsen-Rio Grande Valley State Park | Falcon Dam

For a sample of what birdlife is like in Mexico, without leaving the United States, the bird finder may go to the **Lower Rio Grande Valley.** Here he will see numerous Mexican birds that breed nowhere else in the United States.

Originally the drier, sandy soils of the Valley produced an abundance of mesquite thickets, yuccas, cacti, and grasses, and the richer soils along the resacas (old beds of rivers now cut off and usually filled with water) supported forests of jungle-like density. Remnants of these habitats for birds still exist, but they are fast being destroyed by the rapid expansion of cities and towns and by increased agricultural activity. Fortunately for the visiting bird finder there are two National Wildlife Refuges in which these habitats remain relatively undisturbed and continue to hold an excellent representation once characteristic of the entire area.

The **Laguna Atascosa National Wildlife Refuge** (45,190 acres), about 25 miles east of Harlingen, extends inland from the Laguna Madre. Although nowhere more than a few feet above sea level, the Refuge nonetheless encompasses a remarkable diversity of habitats for birds in any season: grassy salt marshes; salt flats with scattered tamarisk; mud flats; farm fields; low ridges and brief stretches of dry plain that support a chaparral association in which mesquite, clepe, granjeno, and Texas ebony comprise the principal woody growth, with various kinds of short grasses, cacti, and yuccas abundant in the more open spots. Water areas within the Refuge consist of the Laguna Atascosa (3,100 acres), Laguna de los Patos (250–300 acres), Cayo Atascosa (some 300 feet to 0.5-mile wide and 6 miles long), and several fresh-water impoundments and resacas.

Birds regularly present all year in the chaparral association include the Harris' Hawk, Plain Chachalaca, Greater Roadrunner, Pauraque, Golden-fronted and Ladder-backed Woodpeckers, Green Jay, (Black-crested) Tufted Titmouse, Verdin, Bewick's and Cactus Wrens, White-eyed Vireo, Long-billed and Curve-billed Thrashers, and Olive and Black-throated Sparrows. In summer the association is augmented by the Groove-billed Ani, Yellow-billed Cuckoo, Common and Lesser Nighthawks, White-winged Dove, Tropical Kingbird, Wied's Crested Flycatcher, Blue Grosbeak, Varied Bunting (uncommon), and Painted Bunting.

Farm fields and other grassy areas have Botteri's Sparrows residing commonly in summer and Cassin's Sparrows the year round, though more commonly in summer. In winter the farm fields are feeding areas for Canada Geese, Snow Geese, and Sandhill Cranes; along with other grassy areas, the farm fields provide cover for a few wintering Sprague's Pipits and many sparrows such as the Savannah and Vesper and sometimes the Le Conte's.

Salt marshes are habitats for nesting Clapper Rails, Willets, and Seaside Sparrows, and salt flats for nesting Wilson's Plovers, Black-necked Stilts, and American Avocets.

On most fresh-water impoundments the year round are a few Least and Pied-billed Grebes, Black-bellied Whistling Ducks, and Mottled Ducks. Beginning in October, these impoundments and other water areas, particularly the Laguna Atascosa, become winter havens for vast numbers of northern ducks, notably Common Pintails, Green-winged Teal, American Wigeons, Redheads, Canvasbacks, and Ruddy Ducks, and rafts of American Coots. At the same time, and from then on through the winter, the salt flats, mud flats, and the shore of the Laguna Madre attract large aggregations of shorebirds—Black-bellied Plovers, Long-billed Curlews, White-rumped, Baird's and Least Sandpipers, Dunlins,

Long-billed Dowitchers, Semipalmated and Western Sandpipers, and many others.

The shore of the Laguna Madre, its offshore waters, and fringing marshes are alive in any season with such waterbirds and wading birds as American White Pelicans, Double-crested Cormorants, Olivaceous Cormorants (occasionally), Great Blue and Little Blue Herons, Reddish, Great, and Snowy Egrets, Louisiana Herons, White-faced Ibises, White Ibises (occasionally), Roseate Spoonbills, Laughing Gulls, terns—Gull-billed, Forster's, Little, and Caspian—and Black Skimmers.

The Refuge may be reached from US 83 Business in Harlingen by turning off east onto Harrison Avenue and continuing east on Farm Road 106 through Rio Hondo to field headquarters in the Refuge. Register here and obtain a folder with map showing two auto tour routes and walking trails leading therefrom. The routes are open from 8:00 a.m. to sunset but may be closed during wet weather. The Bayside Tour, about 14 miles long, includes chaparral, grassy areas, fresh-water impoundments, the shore of the Laguna Madre, salt marshes, salt flats, and mud flats. The Lakeside Tour, about 12 miles long, traverses farm fields, skirts a fresh-water impoundment and the Laguna Atascosa with its extensive mud flats. Both are loop tours from headquarters. The bird finder may leave the Lakeside Tour directly for Harlingen without returning to headquarters, thus cutting down on mileage.

The **Santa Ana National Wildlife Refuge** (1,981 acres) lies southwest of Harlingen in a lowland flanking a big bend in the Rio Grande. On the tract are three lakes—Pintail (200 acres), Willow (150 acres), and Cattail (150 acres)—choked with retama, mimosa, and other plants that die off each time the Rio Grande overflows. Marshes containing duckweeds and naiads are on the periphery of all three bodies of water. Elsewhere the Refuge embraces remnants of vegetational associations that once prevailed along the Rio Grande: a dense, lowland forest of red elms festooned with Spanish moss, mixed with Texas ebonies, tepejuajes, ashes, and hackberries, and having an understory of vine and herbaceous growth; desert-like open areas with an abundance of thorny growth, typically chaparral.

Plain Chachalacas are regularly present the year round in the

GREATER KISKADEE

Refuge forests and are at their noisy best in April and May. Other regular permanent residents of special interest are the Least Grebe, White-fronted Dove, Pauraque, Greater Kiskadee, Green Jay, Long-billed Thrasher, Altamira Oriole, and Olive Sparrow. Species of special interest that are mainly summer residents include the Black-bellied Whistling Duck, Groove-billed Ani, Tropical Kingbird, and Black-headed Oriole. Besides the above specialities are such regular permanent residents as the White-tailed Kite, Harris' Hawk, Common Ground Dove, Greater Roadrunner, Golden-fronted and Ladder-backed Woodpeckers, (Black-crested) Tufted Titmouse, Cactus Wren, Curve-billed Thrasher, White-eyed Vireo, and Lark Sparrow, and such summer residents as the Purple Gallinule, White-winged Dove, Yellow-billed Cuckoo, Common and Lesser Nighthawks, Wied's Crested Flycatcher, and Bronzed Cowbird. For most species, whether permanent- or summer-resident, the breeding season is prolonged, starting in late March or early April and continuing through May and June, even into July.

Gadwalls, Common Pintails, Blue-winged and Cinnamon Teal, American Wigeons, Northern Shovelers, and other ducks from northern climes are numerous during the winter months, but by the last of April most of them have departed. Transient passerines—flycatchers, vireos, and warblers among them—are abundant during March, April, September, and October. Always impressive in late March and early April are the hundreds, sometimes thousands, of Broad-winged and Swainson's Hawks that pass over the Refuge hourly, many settling in the Refuge for the night.

To reach the Santa Ana Refuge (closed to visitors at night), drive west from Harlingen on US 83 to Alamo, south on Farm Road 907 for 7.5 miles, then left onto US 281 for 0.3 mile to the entrance road that leads over a levee and down to headquarters on the right. Here obtain a descriptive folder with map showing the 6.7-mile auto tour road, 14 miles of foot trails, and the location of blinds for photography. Inquire about any current sightings or nestings of rare or unusual birds.

The times to visit the Refuge are in the early morning or late afternoon when the birds are more active and thus more readily spotted. No bird finder starting the tour of the Refuge can miss the chachalacas and White-fronted Doves along the road, the occasional roadrunner dashing across it, but there are birds he might overlook without knowing where to look for them. Using the map in the folder provided at headquarters, he should investigate the following:

Headquarters area. Look for the rare Olive-backed Warbler and the uncommon Hooded Oriole nesting in the tall, moss-draped trees that shade the buildings; for Buff-bellied Hummingbirds visiting the pink- and red-flowering plants.

Old Spanish Cemetery. Common Ground Doves, Cactus Wrens, and Curve-billed Thrashers often nest in it. In the tall trees back of it, or across the road from it, Rose-throated Becards and Northern Beardless Flycatchers may nest.

All blinds for photography. Owing to their being baited with food at one time or another, they consistently attract both Golden-fronted and Ladder-backed Woodpeckers, Green Jays, Long-billed Thrashers, Black-headed and Altamira Orioles, and Olive Sparrows.

All three lakes. Least Grebes, Black-bellied Whistling Ducks, and gallinules are usually here as well as Greater Kiskadees on overhanging perches looking for fish. Be alert for an occasional Green Kingfisher, Fulvous Whistling Duck, and Masked Duck. Scan the trees around Cattail Lake for a Red-billed Pigeon.

When leaving the Refuge, the bird finder may turn sharp right onto the levee (crossed when he entered the Refuge) and drive upon it for about 2 miles. At first the levee overlooks Pintail Lake on the right and, later, two smaller lakes on the same side—likely sites for both whistling ducks and for Olivaceous Cormorants and American Anhingas. Be aware of the possibility of a Hook-billed Kite on high trees along the way.

Bentsen-Rio Grande Valley State Park (587 acres), upriver from the Santa Ana Refuge, embraces similar environment and bird associations. Unlike the Refuge, it is open to camping and picnicking, hence draws crowds of people, especially on weekends. Despite this the Park offers the bird finder at least two advantages. He may see Green Jays and other birds, even including Red-billed Pigeons, boldly awaiting handouts from campers and scraps of food from picnic tables. Since the Park is never closed at night, he may work along the roads for Common Screech and Elf Owls, possibly Ferruginous Pygmy Owls, and with flashlight or car headlights sometimes 'shine' the eyes of a Pauraque ahead in the road.

To reach the Park, return to US 83 and continue west to Mission. At 10th Street, bear left on US 83 Business, then turn left onto Park Road 43 and go 5 miles.

For the unusual opportunity of seeing three species of kingfishers at one time in one area of United States territory—namely, below **Falcon Dam** on the Rio Grande—proceed upriver farther by returning to US 83 and following it westward through Rio Grande City. At a point 10 miles beyond Roma-Los Saenz, turn off left onto Farm Road 2098 and go about 4 miles, passing the town of Falcon Heights, then turn left onto a paved spur of FR 2098 which leads down and below Falcon Dam to a clearing and sometime campground. Leave the car and work upriver along the edge to the spillways.

In any season, look for the little Green Kingfisher on rocks in the river or on the lower limbs of overhanging trees, the big Ringed Kingfisher higher up, often on snags. In winter, Belted Kingfishers are present and common in comparison with the other two species.

In addition to kingfishers, other rewards are Vermilion Flycatchers often hawking insects over the water; Greater Kiskadees more intent on fishing than catching insects; and in the scrubby thickets on the riverbanks, both Blue-gray and Black-tailed Gnatcatchers and Pyrrhuloxias.

HOUSTON

Hermann and Memorial Parks | San Jacinto Battleground State Historical Park | Sheldon Reservoir | Dwight D. Eisenhower Park | W. G. Jones State Forest

The area of this great city, on the Coast Prairie about 50 miles inland from the Gulf, has fresh-water and brackish ponds and marshes combined with such other environments as tidewater estuaries, large woods, and extensive grasslands. Birdlife is therefore rich both in variety of breeding and winter species and in numbers of individuals. Moreover, Houston lies in the path of spring migration, which in April and early May brings enormous numbers of transients during inclement weather.

Hermann and Memorial Parks in Houston are well worth visiting for birds. Hermann Park (545 acres), southwest of the city center, on South Main Street between Hermann Avenue and Holcombe Boulevard, has a stream, pond, and fine pine-hardwood forest, much of it in natural condition. Memorial Park, west of the city center, is considerably larger than Hermann Park and, though it includes a golf course, has a more extensive pine-hardwood forest. Both areas are excellent for the more common small landbirds, resident and transient. The pond in Hermann Park attracts herons and gallinules the year round and a few grebes and ducks in winter.

When in Houston no bird finder should neglect combining ornithology with history and visiting the **San Jacinto Battleground State Historical Park** (445 acres), southeast of the city center. This is reached by exiting east from I 610 (South Loop East Freeway) on State 225 (Pasadena Freeway), then north (left) on State 134 (Battleground Road) for 4 miles to Park Road 1836. Within the area are the San Jacinto Battleground where, in 1836, the Texans under General Sam Houston defeated the Mexican forces led by Santa Anna; the spectacular 570-foot Memorial Monument; the old battleship 'Texas,' permanently anchored and open to visitors; and a continual stream of ocean-going ships passing, within a stone's throw of land, through the Houston Ship Channel between Houston and the open sea. A natural oak woods and nearby marshes and prairie attract many birds. The more common herons and egrets may be viewed here the year round and cormorants, gulls, and terns according to season. The reflecting pool before the Monument is a refuge for ducks during the hunting season. Red-shouldered Hawks, which nest in the Park, may be seen almost any time of the year. In winter, Vermilion Flycatchers are sometimes seen in the semi-open area on the south side of the Park.

A worthwhile trip from the center of Houston is north on US 90 for 15 miles to Sheldon; here turn off left onto Sheldon-Deer Park Road and go 1.0 mile, then left again onto Garrett Road and proceed west for 3 miles, stopping now and then to scan the cypress swamp on the right for American Anhingas, Little Blue Herons, Wood Ducks, and, in summer, Prothonotary Warblers.

After 3 miles, Garrett Road crosses the **Sheldon Reservoir** in the Sheldon Reservoir Wildlife Management Area, providing a good vantage for observing some of the vast numbers of waterfowl that winter in the refuge. It is usually possible to identify fifteen or more species of ducks that may include the uncommon Cinnamon Teal and Hooded Merganser. Among other birds here in winter are Double-crested Cormorants, numerous herons and egrets, Common Gallinules, American Coots, usually Ospreys, and an occasional Bald Eagle. Purple Gallinules are present in summer. Across the Reservoir on its west side, a path leads north along the levee where there are thickets attractive in winter to sparrows such as the White-throated, Lincoln's, Swamp, and Song.

Backtrack on Garrett Road for 1.5 miles and turn north onto Aqueduct Road which, after about 2 miles, leads through huge stands of pine and oak in the **Dwight D. Eisenhower Park** at the south end of Lake Houston. Nesting birds to look for in the deep woods are the Pileated Woodpecker, Brown-headed Nuthatch, Northern Parula, Yellow-throated, and Pine Warblers, and Summer Tanager; in the swampier or more moist woods are the Swainson's, Kentucky, and Hooded Warblers and, possibly, the Louisiana Waterthrush.

For woodland birds another productive area is the **W. G. Jones State Forest** (1,725 acres), owned by Texas A. and M. University, north of Houston. To reach it, drive north on I 45 for 35 miles, then turn off west onto Farm Road 1488 and go 1.0 mile to headquarters on the right. Here obtain information about the nature trail and access roads to differently forested sites and experimental plots, some of which are largely of pines, others of hardwoods.

Special birds to find are Red-cockaded Woodpeckers by taking the road going south from opposite headquarters to open pine woods. Look or listen for the birds—they are usually in clans—associated with mature or dying pines in which they nest. Other woodpeckers to be found here or elsewhere in the Forest include the Yellow-shafted Flicker, Red-bellied Woodpecker, and Red-headed Woodpecker.

Among the bird species regularly residing all year in one part of the Forest or another, depending on the type and density of tree or shrub growth, are the Barred Owl, Brown-headed Nuthatch, Blue Jay, Carolina Chickadee, Tufted Titmouse, Carolina Wren, and Pine Warbler. Present in summer are the Chuck-will's-widow, Great Crested and Acadian Flycatchers, Eastern Pewee, White-eyed and Yellow-throated Vireos, Black-and-white and Northern Parula Warblers, Orchard Oriole, and Indigo and Painted Buntings.

HUNTSVILLE
Huntsville State Park

For birdlife typical of the East Texas Pineywoods, visit **Huntsville State Park** (2,122 acres), which is 8 miles southeast of Huntsville and may be reached by exiting from I 45 on Park Road. Nine miles of hiking and nature trails make accessible the most productive habitats for birds in the forest of mixed pine and hardwoods.

Breeding birds in the Park include the Red-shouldered Hawk, Barred Owl, Pileated, Red-headed, Hairy, and Downy Woodpeckers, Acadian Flycatcher, White-breasted and Brown-headed Nuthatches, Wood Thrush, Blue-gray Gnatcatcher, White-eyed, Yellow-throated, and Red-eyed Vireos, the Pine Warbler and eleven other species of parulids, and Chipping Sparrow. In winter there are Red-breasted Nuthatches and, in some years, Pine Siskins and Evening Grosbeaks.

MARATHON
Big Bend National Park

So named because of its location in the broad, U-shaped bend of the Rio Grande in southwest Texas, **Big Bend National Park** embraces some 1,106 square miles. Within this vast area, larger than the state of Rhode Island and far off the beaten tourist path, one regains the feeling of expansiveness and urge for adventure.

Aside from the three deep canyons carved by the Rio Grande, the chief physiographic feature of the Park is the Chisos (Ghost) Mountains whose barren, sentinel-like peaks and serrated ridges occupy about 40 square miles in the approximate center. When compared with other mountains in western United States, they are not high—the loftiest point, Mt. Emory, reaches only 7,835 feet—but they stand out above the surrounding semiarid plains that are broken only by lesser ranges, mesas, and arroyos (streambeds) that are usually dry.

Vegetation in the higher elevations of the Chisos Mountains includes oaks, big-toothed maple, pinyon pine, alligator and weeping junipers, mountain mahogany, madrona, and, in restricted localities, Douglas-fir, ponderosa pine, and Arizona cypress. Below 5,500 feet, the woodlands intergrade with a belt of grasslands, recognized by the prevalence of grama grasses and sotol. Common here is the century plant or maguey although it ranges to higher elevations. Some of the other plants intermixed are catclaw, acacias, barberries, screw bean, yuccas, and persimmon. From 3,500 feet down to 2,500, the grasslands merge with the vegetation of the scrub desert which comprises the plains over which one approaches the Chisos Mountains from the north. Conspicuous among the plants are creosote bush, lechuguilla, ocotillo, and many cacti including prickly pear and cholla. Low stands of cottonwoods and willows and deciduous thickets frequently border arroyos. Between one canyon and the next along the Rio Grande, which marks the southern boundary of the Park, are floodplains, often densely plant-covered, with reeds on the banks of the channel itself, cottonwoods, willows, tamarisk, acacias, mesquite, and buttonbush on higher ground.

The Park has a fine variety of landbirds, but no spectacular concentrations of breeding or wintering species or big waves of transients. Bird finding is nonetheless highly rewarding the year round. The first of April to July is the choicest period for breeding birds; mid-November to late February for wintering birds; mid-March through early May and August to October for migrating birds. At lower elevations and in the canyons the bird finder should plan his time in the field for the early morning or toward dusk because the heat of midday, even in winter, can be uncomfortably intense. If there is considerable moisture in spring, especially from March to May, the floral displays on the scrub desert and lower slopes of the Chisos Mountains are so colorful as to defy description.

A notable ornithological feature of Big Bend is its great variety of hummingbirds. No less than six species, the Lucifer, Black-chinned, Broad-tailed, Magnificent, Blue-throated, and Broad-billed, are summer resident, and two species, the Anna's and Rufous, are regular fall transients. Four other species have been sighted two or more times. There is no pinpointing any species to elevation or habitat as each one ranges widely. The bird finder should therefore be 'hummingbird conscious' wherever he goes, keeping his eye on plants with showy flowers. Century plants, when in bloom from April to September, are especially enticing to hummers.

The Chisos Basin at an elevation of 5,400 feet in the heart of the Chisos Mountains has the Park's only regular meal services and overnight accommodations, consisting of cottages and motel-like units open all year. For reservations, a necessity, write to National

Park Concessions, Inc, Big Bend National Park, TX 79834. The Basin has a campground and a store for food and camping supplies. Other campgrounds in the Park are at Rio Grande Village and at Panther Junction near Park headquarters. There are trailer parks with utility hookups at Rio Grande Village and Panther Junction. Gasoline may be purchased at anyone of the above places but the only service station with the usual facilities is at Panther Junction.

Although there are two approach roads to the Park, the one recommended is US 385 leading south from Marathon to the Park's northernmost boundary, 40 miles distant. Like so much of southwestern Texas where the human population is sparse, the country traversed by this highway is semiarid and mainly uninhabited. Consequently at Marathon, check on gasoline, oil, and water, for the next source of supply is well into the Park.

As US 385 approaches the Park, the Santiago Range with its highest point—flat-topped Santiago Mountain at 6,521 feet—looms ahead in the Park. Along the way, inspect the desert-like, brush-studded flats and tree-lined arroyos for such year-round birds as the Red-tailed Hawk, Swainson's Hawk (except in winter), Scaled Quail, Mourning Dove, Greater Roadrunner, Ash-throated Flycatcher, White-necked Raven, Verdin, Curve-billed Thrasher (rare in summer), Black-tailed Gnatcatcher, Pyrrhuloxia, House Finch, and Black-throated Sparrow.

On reaching Persimmon Gap at 2,971 feet, the entrance to the Park, stop at the ranger station for a map and information about the Park's facilities, regulations, and interpretive programs. Then follow the Park road for 29 miles, passing the Santiago Range on the left, crossing Tornillo Flat (more scrub desert) where the massive Chisos Mountains rise ahead, and climbing to Panther Junction (elevation 3,750 feet), the site of Park headquarters. Stop in at the administration building for road and hiker's guides, detailing information that is essential for successful exploration of the Park. Then proceed to the Chisos Basin, following the Park road west for 3 miles to Basin Junction, and from there south and up for 6 miles through Green Gulch to Panther Pass at 5,800 feet.

During the twisting climb to the Pass, the desert overlaps and gives way to grasslands and eventually the grasslands overlap and give way to woodlands of oak, pinyon pine, and juniper. As these transitions appear, park the car at pulloffs and look for such breeding birds in appropriate habitats as the Ladder-backed Woodpecker, Phainopepla, Gray Vireo, Scott's Oriole, Varied Bunting, and Cassin's and Rufous-crowned Sparrows.

Panther Pass overlooks the Chisos Basin, a cliff-ringed amphitheater into which the Park road descends about 400 feet in the course of a tortuous mile. Flat-topped Casa Grande towers above the Basin from the southeast; Toll Mountain and then Emory Peak show up farther south of it.

In and around the Basin there is much to delight the bird finder in any season. Near the lodgings and campground are Scaled Quail, Greater Roadrunners, Gray-breasted Jays, Bewick's Wrens, Northern Mockingbirds, Cactus Wrens, Black-headed Grosbeaks, Lesser Goldfinches, and the ever-abundant Brown Towhees. The surrounding cliff walls and taluses are populated by White-throated Swifts (except in winter) and Canyon Wrens. Heard at night in April, May, and June are the calls of Poor-wills and Elf Owls. Transient and wintering birds include Western Bluebirds, Orange-crowned and Audubon's Warblers, Green-tailed Towhees, Oregon and Gray-headed Juncos, and Chipping Sparrows.

From the Basin the bird finder should take the 2.6-mile Window Trail leading west, at first through grasslands and open deciduous woodlands and then down into Oak Creek Canyon to a narrow defile—the Window—in the western rimrock that gives a vista of the plains far below. A few of the many birds that may be expected in the breeding season are the Say's Phoebe, Violet-green Swallow, (Black-crested) Tufted Titmouse, Rock Wren, Crissal Thrasher, Summer Tanager, and Black-chinned Sparrow. In any season from the Window Trail or elsewhere in the Chisos Mountains one may hear or see Northern Ravens and, with luck, sight a Golden Eagle.

Lost Mine Trail, starting at Panther Pass above the Basin, goes 2 miles up the north slope of Casa Grande to a promontory high on Lost Mine Peak. Although the Trail is not particularly productive for bird finding, it is popular for providing unsurpassed views of the Park and for still another reason: It is self-guided, with a booklet, available at the trailhead, for identifying plants along the way that are characteristic of Big Bend.

No bird finder will want to miss the Colima Warbler, whose only known nesting area in the United States is centered at Boot Spring (elevation about 6,500 feet) in Boot Canyon on the eastern slope of the Chisos Mountains. This Mexican species, unafraid and easily approached, resides in its habitat of oaks, maple, pinyon pine, junipers, and Arizona cypress from about mid-April to mid-September, but is more readily located when singing early in its nesting season until mid-June. It nests on the ground under overhanging grasses or other low cover.

To reach Boot Spring, in a fine area for birds besides the Colima Warbler, take the South Rim Trail from the Basin via Laguna, then the Boot Canyon Trail, for a total distance of 5.5 miles. At Laguna, a mountain 'meadow' about 3.5 miles from the Basin, investigate the stands of oak, pinyon pine, and juniper for the Band-tailed Pigeon, Bushtit, White-breasted Nuthatch, Hutton's Vireo, Hepatic Tanager, and Rufous-sided Towhee.

About 0.5 mile above Laguna, take the Boot Canyon Trail to the left for 1.5 miles into the Canyon to a cabin used by Park personnel. A short distance below the cabin, and about midway down the Canyon, is the Spring; farther down to the mouth of the Canyon is the Boot, a strangely shaped volcanic upthrust suggesting a cowboy boot upside down.

The best spot for the Colima Warbler is from above the cabin down to the Spring. Other birds to search for in the Canyon include the Sharp-shinned Hawk, Red-shafted Flicker, Acorn Woodpecker, Blue-gray Gnatcatcher, Solitary Vireo, and the Painted Redstart—a possibility only. Whip-poor-wills, Common Screech Owls, and Flamulated Screech Owls may be heard at night.

The bird finder with ample time to spend in Big Bend will enjoy exploring several areas remote from the Chisos Basin. Two especially he must not pass up: *Santa Elena Canyon* and *Rio Grande Village*. Consult a Park map for directions from Panther Junction.

Santa Elena Canyon, through which flows the Rio Grande, is one of the awesome scenic features of the Park and conveniently entered on foot. From Panther Junction to the Canyon, a distance of 35 miles, the Park road skirts the western slopes of the Chisos Mountains and finally drops down to the Rio Grande floodplain, ending in a turn-around and parking area where the entrance to the Canyon is in view across Terlingua Creek. After walking across the Creek, which presents no problem except after a heavy rain, follow the self-guided nature trail into the Canyon. Anytime from March to October, watch for Zone-tailed Hawks, as well as Turkey Vultures, soaring overhead, be alert for a Peregrine Falcon or two, and expect to see White-throated Swifts and both Rough-winged and Cliff Swallows. In any season there will be Black Phoebes and Canyon Wrens.

The area in Big Bend for the greatest variety of birds in any season is Rio Grande Village on the Rio Grande, 22 miles by Park road southeast from Panther Junction. Here, in typical floodplain habitat already described, a 0.75-mile nature trail complements the opportunities for bird finding in and around the campground and picnic sites.

Among the summer-resident birds in the area of Rio Grande Village are the Yellow-billed Cuckoo, Groove-billed Ani (occasionally), Bell's Vireo, Yellow-breasted Chat, Orchard and Hooded Orioles, Brown-headed Cowbird, Blue Grosbeak, and Painted Bunting. White-winged Doves, Common Yellowthroats, and Northern Cardinals may be observed the year round. Brewer's Blackbirds are common in migration, and practically any species of northern parulid is possible in migration though rarely in great numbers. In winter, House and Marsh Wrens may be sighted along the nature trail and as many as twelve to fourteen species of northern fringillids may be identified by careful searching.

Reference

Birds of Big Bend National Park and Vicinity. By Roland H. Wauer. Austin: University of Texas Press, 1973.

MULESHOE
Muleshoe National Wildlife Refuge

If the bird finder has occasion to be on the Staked Plains in northwestern Texas during the winter, he should not miss visiting the **Muleshoe National Wildlife Refuge,** which has one of the greatest wintering concentrations of Sandhill Cranes on the continent. The Refuge lies on either side of State 214, 20 miles south of Muleshoe. The turnoff to headquarters, 2 miles west of the highway, is indicated by the flying-goose sign. Fire lanes and patrol roads traverse the Refuge, permitting easy access by car to all parts.

The 5,809 acres of the Refuge consist mainly of rolling to hilly plains covered by short grasses, cacti, yuccas, and tumbleweeds. Near the north and west boundaries are rocky outcroppings. The only other prominent interruptions are gullies leading into playa lakes that have no outlets and are dry, except after periods of heavy rainfall. Three of the lakes—Lower Paul's (125 acres), Lower Goose (140 acres), and Lower White (160 acres)—are the result of natural depressions in the ground; three others—Upper Paul's (18 acres), Upper Goose (140 acres), and Upper White (62 acres)—are man-made. When the lakes are receding, after being filled by rainfall, they have mud flats and muddy shores, attractive to many transient shorebirds in spring and fall when the recession occurs at such times.

A few landbirds reside on the Refuge—e.g., coveys of Scaled Quail are always present and may be observed around headquarters at almost any time; and enormous numbers of ducks winter on the lakes from November to February—as many as 700,000, mostly Mallards, Gadwalls, Common Pintails, Green-winged Teal, and American Wigeons, have been the estimate in January. But Sandhill Cranes provide the signal attraction.

The cranes usually begin appearing in late September and steadily increase in number until December. Though numerous flocks may continue on southward, as many as 100,000 individuals ordinarily remain through the winter, unless unfavorable weather conditions impel them to move farther south. By late February or early March, the cranes start journeying northward (*see under* **Kearney, Nebraska**).

The cranes usually roost at night in the shallow water near the shores of the lakes. At sunrise they depart for their feeding grounds—generally a harvested grainfield; here they stay until midmorning, when they return to their roosting grounds to rest and preen. By midafternoon they again go to the feeding grounds, returning for a second time to their roosting grounds at sunset.

The pleasure in watching these magnificent birds—they are shy and have to be observed from a distance, preferably by telescope—is always heightened by their extraordinary 'dances,' performed individually and sometimes by groups. These dances, with their ludicrous bows and vertical leaps into the air, take place on both the roosting and feeding grounds and are so frequent that the watcher need not wait long to see them.

PINE SPRINGS
Salt Flats | Guadalupe Mountains National Park

From the desert plains about 100 miles east of El Paso rise the abrupt slopes of the Guadalupe Mountains, attaining on Guadalupe Peak an elevation of 8,751 feet, the highest point in Texas. Of the Guadalupes' many breath-taking escarpments, the most impressive is the sheer, 2,000-foot face of El Capitan (elevation 8,078 feet), a familiar landmark for many generations of travelers and now probably one of the most photographed mountains features in southwestern United States.

US 62-180, leading eastward from El Paso to the Guadalupes, crosses low, treeless mountains or foothills and deserts of creosote bush, mesquite, greasewood, agave, cactus, and yucca. Within 10 miles or so of the Guadalupes, the highway crosses the **Salt Flats** (elevation 3,600 feet)—a desolate stretch of glaring white that contains, after wet weather, shallow blue-green lakes of considerable extent. If there is sufficient water in late spring, early summer, or fall, Snowy Plovers, Baird's and Western Sandpipers, Sanderlings, and other shorebirds frequently appear. Birds that undoubtedly breed in the vicinity include the Scaled Quail, Lesser Nighthawk, White-necked Raven, Northern Mockingbird, Curve-billed Thrasher, and Lark Sparrow.

Not far east of the Salt Flats, US 62-180 reaches the foothills of the Guadalupes and begins its tortuous ascent through Guadalupe Canyon to Guadalupe Pass (above 5,400 feet) and northward for 2.5 miles to Pine Springs (little more than a store and filling station) in **Guadalupe Mountains National Park.**

Created in 1966, the National Park of 77,500 acres encompasses all of the Guadalupes in Texas, which cover a triangular area. El Capitan and Guadalupe Peak on the south stand at the apex of the triangle. Other peaks, many in excess of 8,000 feet, loom northeast and northwest to the New Mexico line, the base of the triangle. Habitats within this mountain mass range from deserts on the foothills to coniferous forests on the higher peaks and ridges and the upper reaches of canyons.

At the Park's information station, on US 62-180 about 1.0 mile east of Pine Springs, obtain a map, descriptive folder, and the seasonal program of interpretive services. Inquire about current conditions for trips to the following two areas that are especially rewarding for viewing birds in late spring and early summer.

1. The Bowl at the summit of Pine Top Mountain, reached by driving west from Pine Springs to Pine Spring Campground and hiking up a fairly steep trail of 3 miles to the rim of the Bowl at 8,000 feet elevation, then down as much as 250 feet into the Bowl itself. Tucked in here is a fine 250-acre coniferous forest, consisting primarily of ponderosa pine, limber pine, and Douglas-fir with scattered stands of Gambel oak and alligator juniper.

Among the birds breeding regularly in the Bowl are Cooper's and Red-tailed Hawks, Band-tailed Pigeon, Broad-tailed Hummingbird, Red-shafted Flicker (shows hybridization with Yellow-shafted), Acorn and Hairy Woodpeckers, Western Flycatcher, Violet-green Swallow, Steller's Jay, Mountain Chickadee, White-breasted and Pygmy Nuthatches, Brown Creeper, Hermit Thrush, Western Bluebird, Solitary and Warbling Vireos, Orange-crowned, Audubon's, and Grace's Warblers, Western Tanager, Black-headed Grosbeak, Red Crossbill (irregular), Rufous-sided Towhee, Chipping Sparrow, and Gray-headed Junco. Common Nighthawks appear overhead toward evening; Flammulated Screech Owls, Great Horned Owls, Spotted Owls, and Whip-poor-wills are vociferous at night.

2. McKittrick Canyon—actually a complex of three canyons, Main, North, and South—in the Park's northeastern part, reached from the information station by shuttle van (eventually in one's own car when the road is improved) to the mouth of Main Canyon. From here, hike west, going up through Main Canyon to the junction of North and South Canyons, then up either one. South Can-

yon, whose upper reaches are 6 miles from the mouth of Main, is the more scenic with a greater diversity of birds.

All three canyons, whose floors rise in elevation from 5,000 feet at the mouth of Main to 5,500 feet, are steeply walled up to 2,000 feet. Unlike the Bowl, the canyons are much more openly forested, where forested at all, with chinquapin oaks, big-toothed maples, alligator junipers, and other tree growth. Where not forested there are agaves, yuccas, and patches of brush.

At the mouth of Main Canyon, where desert brushlands prevail, such birds as the Scaled Quail, Cactus Wren, Curve-billed and Crissal Thrashers, Pyrrhuloxias, and Black-throated Sparrows may be expected. Farther on, up the canyons, on their floors and lower slopes, breeding birds include the Poor-will, Black-chinned and Blue-throated Hummingbirds, Ladder-backed Woodpecker, Cassin's Kingbird, Ash-throated Flycatcher, Say's Phoebe, Western Pewee, Olive-sided Flycatcher (uncommon), Scrub Jay (uncommon), Bushtit, Bewick's and Canyon Wrens, Blue-gray Gnatcatcher, Gray Vireo, Virginia's Warbler, Scott's Oriole, Hepatic Tanager, Blue Grosbeak, House Finch, Lesser Goldfinch, Brown Towhee, and Rufous-crowned and Black-chinned Sparrows. High up on the canyon walls and their ridges, where there are ponderosa pines and other conifers, some of the species in the Bowl (*see above*) may also reside. White-throated Swifts nest commonly in the escarpments. Sighting a Zone-tailed Hawk and a Golden Eagle, or running onto Wild Turkeys, are strong possibilities.

For bird-finding opportunities in the Guadalupes north of Guadalupe National Park, *see under* **Carlsbad, New Mexico.**

PORT ARTHUR
Sabine Pass | High Island | Anahuac National Wildlife Refuge | Bolivar Flats | Bolivar Ferry

State 87 between Port Arthur and Galveston makes readily accessible the northeasternmost Gulf shore and coastal marshes of Texas. No bird finder can take a trip in any season along this highway without seeing the hosts of birds that frequent the marshes, mud flats, lagoons, bayous, tidewater creeks and estuaries, shipping canals, beaches, and the open water of the Gulf. When promising sites come into view, the bird finder need only stop his car and scan them.

Birds regularly nesting in the coastal marshes from late April to mid-July include the Least Bittern, Mottled Duck, King Rail, Clapper Rail, Marsh Wren, Red-winged Blackbird, Great-tailed Grackle, and Seaside Sparrow. Frequenting the edges of creeks and lagoons throughout the year are Great Blue and Little Blue Herons, Cattle, Great, and Snowy Egrets, Louisiana and Black-crowned Night Herons, and White-faced and White Ibises. From October through February, the marshes and adjacent waterways are winter havens for Canada Geese, Greater White-fronted Geese, Snow Geese, and many species of ducks. Peaks of their abundance are usually in December and January. Except in June and July, transient shorebirds galore gather on the mud flats and beaches. The Black-bellied Plover, Ruddy Turnstone, Greater and Lesser Yellowlegs, Dunlin, Long-billed Dowitcher, Western Sandpiper, Marbled Godwit, and Sanderling are just a few of the species to be expected.

Along the canals and Gulf shore, Olivaceous Cormorants, Laughing Gulls, Gull-billed, Forster's, Royal, and Caspian Terns, and Black Skimmers may be observed the year round. Other birds to be looked for in similar situations are the following: American White Pelicans (common from October through March); Double-crested Cormorants (October to April); Wood Storks (most numerous in late summer); Herring and Ring-billed Gulls (October to April); Bonaparte's Gulls (November to May); Little Terns (May to September); Sandwich Terns (April to October).

Typical sections of coastal marshes are crossed by State 87 as it passes southward from Port Arthur to the fishing village of Sabine Pass, a distance of 14 miles. Bridges over several creeks and lagoons permit good views of bordering shallows and mud flats, where wading birds and shorebirds often feed. At Sabine Pass, the bird finder should leave State 87 temporarily, continuing southward a short distance on a shell road to Sabine, another fishing village, which overlooks the entrance waters to **Sabine Pass** connecting the Gulf with Sabine Lake. This is invariably a good vantage for seeing cormorants, gulls, and terns anytime in the year, and Lesser Scaups and other diving ducks during the winter months.

From the village of Sabine Pass, State 87 goes southwestward, within a few yards of the Gulf's edge, to **High Island,** 33 miles distant. Stops can be made along the way to inspect the beach on the left and the vast expanse of Coast Prairie and intermittent marshes on the right. Once on High Island, turn right on State 124 to the small town of High Island, 2 miles inland; then turn left to the oil fields on a road that circles close to various creeks and bayous commonly visited by wading birds and other waterbirds.

High Island has, as the name implies, a somewhat higher elevation than the neighboring marshes and, moreover, supports tracts of live oaks, hackberries, and other trees together with many shrubs and vines. Because of these features, tremendous numbers of migrating passerines take refuge here in the spring—April and early May—whenever a norther sets in.

From High Island, State 124 continues north, soon crossing the Intracoastal Waterway where there are marshes on either side worth inspecting. The bird finder should follow this highway north from High Island for 8 miles, turn off left onto Farm Road 1985 for 10.3 miles, then left onto a 3-mile shell road to **Anahuac National Wildlife Refuge.** At the entrance, obtain a map showing roads through the Refuge open to public use except in wet weather.

No area on the north Texas coast offers better bird finding in any season than this Federal holding of 9,836 acres extending inland from the eastern end of East Galveston Bay. Besides vast salt and brackish marshes, the Refuge embraces fresh-water ponds, sloughs, bayous, and fields as well as a stretch of beach, all readily accessible by roads.

Shoveler Pond, of fresh water and shallow, in the northwest corner of the Refuge, attracts Roseate Spoonbills the year round, Purple Gallinules in summer, and many transient Fulvous Whistling Ducks in fall. A pair or two of Masked Ducks is always a possibility in summer. Willets are year-round residents in the vicinity of the Pond and elsewhere on the Refuge.

South from Shoveler Pond to Marsh Pond runs the West Line Road on the west side of an area called Yellow Rail Prairie in which, true to its name, that elusive species is commonly present in winter, as well as Sedge Wrens, Sprague's Pipits, and Le Conte's and Sharp-tailed Sparrows. Seaside Sparrows may be observed from the Road all year and noted singing in spring on the paralleling fence.

The Refuge has a variety of sites that draw migrating shorebirds of one species or another. In April and early May, one has good chances of seeing Lesser Golden Plovers, Whimbrels, Upland, Pectoral, White-rumped, Baird's, and Stilt Sandpipers—and even the Buff-breasted Sandpiper and Hudsonian Godwit.

Established primarily for waterfowl, the Refuge amply achieves its purpose in providing nesting areas for Mottled Ducks and serving as wintering area for many thousands of geese and many more thousands of ducks, especially Mallards, Gadwalls, Common Pintails, Green-winged and Blue-winged Teal, American Wigeons, and Northern Shovelers.

Return to State 87 and continue southward over the Bolivar Peninsula for 23 miles, turn left onto Farm Road 2612 and go 0.5 mile to the Gulf, and walk or drive on the beach to the right toward the North Jetty at the entrance to Galveston Bay. When the tide is low the **Bolivar Flats,** stretching far out from the beach to the waters of the Gulf, probably attract more shorebirds at one time than any other site on the north Texas coast. Except in early summer it is possible to see two-thirds of the North American species on a single day. If such species as the American Oystercatcher, Piping, Snowy, and Wilson's Plovers, Red Knot, American Avocet, and Black-necked Stilt were missed on the trip from Port Arthur, they are likely to be here.

Return to State 87 and complete the trip by continuing to the tip of the Bolivar Peninsula and taking the free **Bolivar Ferry** across Bolivar Roads to Galveston. During the 3-mile crossing, gulls follow the boat, often riding on the updrafts at the stern. Expect to see cormorants, wading birds, terns, and Black Skimmers in flight. Many of these birds are probably nesting on islands in Galveston Bay. In winter, scan the water for loons, grebes, and ducks. In any month, but particularly from April to September, watch overhead for a possible Magnificent Frigatebird.

PORT LAVACA
Aransas National Wildlife Refuge

On the south-central coast of Texas the **Aransas National Wildlife Refuge** embraces the well-known wintering ground of the Whooping Crane. This species, one of the largest and one of the most spectacular of all North American birds, arrives on the Refuge between mid-October and mid-December and departs between mid-March and mid-April, bound for its breeding ground in Wood Buffalo National Park, 2,500 miles distant in the Northwest Territories of Canada.

Although never abundant, the Whooping Crane once bred widely in the marshes of the Canadian prairie provinces and northern prairie states, and also wintered widely in the coastal marshes of the Gulf states and northeastern Mexico. But encroachments by man on its habitats in the last century triggered its decline in numbers, at first gradually, then alarmingly, until by 1941 it was represented by only 15 wild birds. All wintered in the Aransas Refuge where they could obtain their preferred food in coveted isolation—a requirement that the Refuge management made every effort to maintain. Thereafter, the number of Whoopers returning to the Refuge increased slowly, yet steadily. In the winter of 1979–80 the maximum count was 76 individuals.

The 54,829 acres of the Aransas Refuge take in the Blackjack Peninsula, a low-lying, gently rolling section of the Coast Prairie, indented by bays and bordered by tidal flats and salt marhes. Back from them are low dunes covered by sweet bay and oaks, bent landward by winds from the Gulf. Mottes of oak brush—tree and shrub forms of live oak and myrtleleaf oak—separated by stretches of grassy meadowland, characterize most of the interior, although stands of blackjack oak, for which the Peninsula was named, grow in scattered localities. Small bodies of fresh water dot the interior. Some are 'wet weather' ponds, others are ponds filled by overflowing wells.

The wintering ground of the Whooping Crane constitutes mainly the tidal flats and marshes on the southeast side of the Peninsula. Averaging a mile or so in width, these flats and marshes extend from Mustang Lake, a large lagoon, south for 12 miles and

WHOOPING CRANES

are interrupted by numerous tidewater creeks, long, narrow lagoons paralleling the east side, and the Intracoastal Waterway which cuts through their outermost fringes. Except for ship traffic through the Waterway, the flats and marshes are remote from most disturbances. Exploring them is prohibited during the time the Whoopers are in residence.

The Refuge is reached from State 35 between Port Lavaca and Rockport. From Port Lavaca, drive southwestward on State 35 for 20 miles to Tivoli and 2 miles beyond; turn off left onto State 113, and go 5 miles to Austwell, right onto Farm Road 774 for 0.75 mile, then left onto FR 2040 for 7 miles to headquarters in the Refuge at the base of the Blackjack Peninsula. Register here and obtain a folder with map showing the 20 miles of auto roads, including a 14-mile loop trip, and foot trails that lead through thickets and wooded areas, along the edges of flats and marshes, and across meadowy areas.

From an observation tower along the loop trip, one may with aid of a telescope sometimes see Whooping Cranes in the distance. For better viewing of these majestic birds from outside the Refuge, see the information at the end of this account about a cruise on the motor vessel 'Whooping Crane,' operated by the Sun Gun Resort Hotel.

Even though Whoopers may be observed poorly, if at all, in the Refuge, a trip on its roads and trails in winter will nevertheless be highly rewarding. Where there are tidal flats and marshes, expect to see Sandhill Cranes and other wading birds such as Great Blue Herons, Cattle, Reddish, Great, and Snowy Egrets, Louisiana Herons, White-faced and White Ibises, and Roseate Spoonbills. Scan the wider stretches of open water, or their shores, for Eared Grebes, American White Pelicans, Double-crested Cormorants, waterfowl—Canada, Greater White-fronted, and Snow Geese, Mottled Ducks, Gadwalls, Common Pintails, Green-winged and Blue-winged Teal, American Wigeons, Northern Shovelers, Redheads, Ring-necked Ducks, Lesser Scaups, and Buffleheads—Common Gallinules, American Coots, Ring-billed and Laughing Gulls, Forster's, Royal, and Caspian Terns, and Black Skimmers. Among the shorebirds well represented are the Black-bellied Plover, Common Snipe, Long-billed Curlew, Willet, Dunlin, Long-billed Dowitcher, Western Sandpiper, Sanderling, and American Avocet. Soras and Seaside Sparrows are common in the marshes; Horned Larks and Water Pipits are often in view as they forage on the flats; and Eastern Meadowlarks and Savannah and Vesper Sparrows reside in the meadowy grasslands. Northern Harriers are likely to show up almost anywhere.

Wild Turkeys inhabit wooded areas; a small number, probably a flock, appears regularly in the vicinity of headquarters. Other permanent-resident landbirds to look for when touring are the Common Bobwhite, Inca Dove, Carolina Wren, and Pyrrhuloxia. Always be alert for a White-tailed Hawk or a Crested Caracara, as well as the possibility of a White-tailed Kite. Winter birds to look for when touring include the Eastern Phoebe, Vermilion Flycatcher, Hermit Thrush, Eastern Bluebird, Blue-gray Gnatcatcher, Ruby-crowned Kinglet, Loggerhead Shrike, Orange-crowned and Myrtle Warblers, Brewer's Blackbird, American Goldfinch, and such sparrows as the Field, White-crowned, White-throated, Lincoln's, and Swamp.

When leaving the Aransas Refuge for a cruise on the 'Whooping Crane,' return to FR 774, turn left on it and go 9 miles, then left onto State 35 toward Rockport. After 14 miles, and within 9 miles of Rockport and in sight of Copano Bay, the Sun Gun Resort Hotel appears on the left. Turn in for arrangements. The Hotel operates the vessel from its marina every Wednesday, Friday, Saturday, and Sunday, usually 20 October through 10 April, starting at 1:30 p.m. In its 3.5- to 4-hour cruise the vessel crosses Aransas Bay, loops around the southern tip of the Blackjack Peninsula, and passes up the east side for good views of the Whoopers. A telescope is desirable for close views. (For further details about the cruise and for reservations, write or call the Hotel: Address, Route 1, Box 85, Rockport, TX 78382; telephone (512)729-2341.)

ROCKPORT
Liveoak Peninsula

For bird finders, Rockport has long been a mecca. This coastal town on the **Liveoak Peninsula** between Copano and Aransas Bays is so situated as to be (1) in the path of tremendous numbers of migrating birds, (2) within the wintering range of a great many species from northern North America, as well as other species from western North America, and (3) near enough to the lower Rio Grande Valley to have a few species represented from that area. Why Rockport is a mecca, when there are other similarly exciting places on the Texas coast, is due to the late Mrs. Jack (Connie) Hagar. In 1935 she and her husband moved to this town and purchased the Rockport Cottages for tourists. Beach, sand flats, and shell bars drawing great varieties of waterbirds and shorebirds were an 'extension' of their front yard, and a grove of live oaks in their backyard attracted huge aggregations of transient passerines. Before long Connie Hagar was watching birds from dawn to dusk, and soon there came reports in journals, magazines, and newspapers of her seemingly incredible observations. Bird finders started coming from near and far to see for themselves—and they are still coming. The fact that Connie Hagar recorded over 400 species of birds in the span of just fifteen years is a continuing inducement.

For the hordes of landbirds that pass through the Rockport area, there are many sites that the bird finder may reach on his own without directional information. Anywhere along the coast in March, April, October, and November he will see migrating hawks—Sharp-shinned, Red-shouldered, Broad-winged, and Swainson's among others—and falcons. Every large grove of live oaks, or big thicket of mesquite and huisache, whether in towns or on their outskirts, has throngs of transient flycatchers, thrushes, kinglets, vireos, warblers, orioles, tanagers, and fringillids from 15 April to 5 May and from 5 September to 15 October.

The following two trips will lead the bird finder to particular sites for migrating passerines and for other birds present on the Liveoak Peninsula either all year, or in one season or another.

1. Drive north from the traffic circle in Rockport on State 35 for 0.5 mile; at the first intersection, turn right onto the road to Fulton Beach and follow it north along the Beach through the town of Fulton for 7 miles. After leaving State 35 the road first skirts the Connie Hagar Wildlife Sanctuary bordering Aransas Bay, where from September through January there are usually loons, grebes, and such ducks as the Gadwall, Common Pintail, American Wigeon, Blue-winged and Green-winged Teal, Northern Shoveler, and Lesser Scaup. If it is late spring or summer, look for Wilson's Plovers on Fulton Beach where they nest. Beyond the town of Fulton and before it rejoins State 35, the road goes through a grove of live oaks at the foot of Ninemile Point. This is one of the sites where passerines swarm during their migration. Ninemile Point itself, at the tip of the Liveoak Peninsula and reached by dirt roads leading off the main road, is worth investigating for shorebirds including American Oystercatchers in any season.

When the road rejoins State 35, cross State 35 to Farm Road 1781 and follow it back toward Rockport on the west side of the Liveoak Peninsula. At the start it overlooks Copano Bay—worth scanning for loons, grebes, and ducks in season. Later it passes stands of live oaks. It also passes small ponds that may have Least Grebes, a species known to be a permanent resident on small ponds in the Rockport area for many years. After about 5 miles on FR 1781, turn left onto FR 2165 (North Pearl Street) to Rockport.

2. From the traffic circle in Rockport, proceed south on State 35 for 1.0 mile, then turn off right (west) onto FR 881 (Market Street) toward Sinton. At Peninsula Oaks, 1.7 miles distant, turn off left to

a lake for waterbirds and ducks in season and an oak grove for transient passerines. Return to FR 881 and continue west. After about 3 miles the highway passes dry meadows covered with short grasses and low catclaw growth. Cassin's Sparrows are here from April through August. After another mile the highway begins passing wet meadows and mud flats. The dry meadows previously passed together with the wet meadows and mud flats are probably the best area on or near the Liveoak Peninsula for shorebirds from February through early May. Look particularly on the dry meadows for Lesser Golden Plovers, Long-billed Curlews, Whimbrels, Upland and Buff-breasted Sandpipers, and the possibility of a Mountain Plover; on the wet meadows and mud flats for White-rumped and Baird's Sandpipers, Hudsonian Godwits, and American Avocets.

SAN ANTONIO

Edwards Plateau | Friedrich Park | Mitchell Lake | Brackenridge Park | Jack Judson Nature Trails | Chaparral Country

This city in southern Texas lies just southeast of the Edwards Plateau and north of the brush country that extends all the way south to the Rio Grande. Within or close to the city are several areas for good bird finding.

For bird finding in the **Edwards Plateau** northwest of the city, exit from Northwest Loop 410 (I 410) at Leon Valley on State 16 and proceed 9 miles to the village of Helotes. Stay on State 16 (do not go into Helotes) and turn right onto the Scenic Loop. The stream which the Loop road crosses a number of times is flanked by cliffs—breeding habitat for Canyon Wrens. Other birds to be looked for from the road in the nesting season are the Scissor-tailed Flycatcher, Eastern Phoebe, Bewick's Wren, House Finch, and Lesser Goldfinch.

Seven and a half miles from Helotes, the Scenic Loop terminates at an intersection. Here the bird finder has a choice of routes. (1) Turn left onto Beauregard Road (dirt surface). From here to Boerne, 30 miles from San Antonio where it enters I 10,

Beauregard Road traverses sparsely settled brush country where there are Vermilion Flycatchers, Scrub Jays, and (Black-crested) Tufted Titmice. (2) Go straight ahead on Boerne Stage Road to I 10 just south of Boerne. Though rarely seen from the Road, both Black-capped Vireos and Golden-cheeked Warblers are known to be summer residents in suitable habitats along the way. The birds are more readily located during their singing period from April to July.

For another area where Black-caps and Golden-cheeks nest and may be observed with greater certainty, visit **Friedrich Park** (232 acres). From Northwest Loop Road 410, exit north on I 10 for approximately 10 miles, exit on Camp Bullis Road and immediately drive back under I 10, go north on the road paralleling the west side of I 10 for 1.2 miles, then turn west at a church and go 0.5 mile to the Park entrance. Look for the Black-caps in shrubby growth on the sunny, southwest-facing slope of the ravine that cuts through the Park's central area; for the Golden-cheeks in the mixed stands of junipers and deciduous growth on the shaded, north-facing slope of the same ravine.

Mitchell Lake and vicinity south of San Antonio yield numerous birds in any season. Take I 37 south from downtown to Southeast Loop 410. Here go west on Loop 410 for 5 miles, then exit south on Moursund Boulevard for 2.2 miles to the Lake. Although the property around the Lake is fenced, permission to look for birds in the area is usually granted. If the gate just past the low-water crossing is open, feel free to drive in. Bird finding from the car is ordinarily quite productive. Many sandpipers, American Avocets, Black-necked Stilts, and Wilson's Phalaropes frequent the edge of the Lake in late summer and early fall. Late in the year and through winter the Lake has a small aggregation of waterfowl. During spring and summer, birds to look for in brushy habitats include the Verdin, Carolina and Cactus Wrens, Curve-billed Thrasher, Orchard and Bullock's Orioles, Pyrrhuloxia, Blue Grosbeak, Painted Bunting, and Black-throated Sparrow.

North of the city's center are **Brackenridge Park** and the **Jack Judson Nature Trails** which offer good bird finding.

To reach Brackenridge Park (343 acres), exit north from I 35 on Broadway (Business Route 81) and proceed north to the entrance at the Witte Museum on the left. The best place for birds is the eastern section, which is heavily wooded and has many trails and bridle paths. Besides such common birds as Golden-fronted and Ladder-backed Woodpeckers, Northern Mockingbirds, and Northern Cardinals, there are a few Long-billed Thrashers. Black-chinned Hummingbirds often nest. Brown Thrashers, Hermit Thrushes, Rufous-sided Towhees, and other northern birds winter in the area.

The Jack Judson Nature Trails north of Brackenridge Park may be reached by continuing north on Broadway and turning left onto Hildebrande Avenue. At the intersection of Devine Road (1.0 mile from Broadway where there is a high-rise apartment complex on the left), turn right and go 1.0 mile to Alamo Heights Boulevard. Turn here and follow the signs to the Alamo Heights Swimming Pool. Just beyond it the Trails begin through a wooded area. From the Trails, look for such breeding birds as Yellow-billed Cuckoos, (Black-crested) Tufted Titmice, Verdins, Carolina Wrens, White-eyed Vireos, Summer Tanagers, Painted Buntings, and most of the species to be found in the same season in Brackenridge Park.

I 35 south from San Antonio for 153 miles to Laredo on the Rio Grande crosses semiarid brush-covered lands called **Chaparral Country.** The few streams flow mainly in wet seasons, but some water is conserved in small lakes, both natural and artificial. One of the largest watercourses traversing the area from west to east, the Nueces River, is rather narrow and deep, and is lined with live oak, pecan, black walnut, ash, elm, and button willow. I 35 crosses it just south of Cotulla. Although some of the Chaparral Country is irrigated for crop production, much more of it is used for grazing livestock. All in all, it is a thinly settled area of considerable sameness and, to the average tourist bound for Mexico by way of Laredo, is a part of the trip to be hurried through; but to the bird finder it is not without interest, for it holds an abundance and variety of birds.

When one is traveling on the highway, a number of species can be seen readily from the car. In any season, Mourning Doves and White-necked Ravens are very common and conspicuous, and there is always a chance of identifying a Black Vulture, Red-tailed Hawk, Swainson's Hawk (except in winter), or Harris' Hawk. From

September to April, American Kestrels and Loggerhead Shrikes may be observed sitting on wires, posts, and utility poles, and Northern Harriers flying low over brushy terrain. Familiar sights between March and November are Scissor-tailed Flycatchers resting on exposed perches, making forays for insect food, and tending their nests poorly concealed in shrubs and low trees.

To see many other birds to advantage, the bird finder should exit from I 35 and drive along farm roads that pass brushy habitat. Regardless of the season he can observe here such species as the Scaled Quail, Common Ground Dove, Verdin, Cactus Wren, Curve-billed Thrasher, Pyrrhuloxia, and Lark and Black-throated Sparrows. Both the Common and Lesser Nighthawks are regularly present from May to October. If the road comes to a brush-filled draw where there is a pond or reservoir for cattle, he should, at the same time of year, look for the Vermilion Flycatcher and Bell's Vireo.

Near or within the limits of towns off I 35 are usually permanent-resident Inca Doves, Bewick's Wrens, and House Finches. Beginning in late April or early May, the summer residents of the tree and shrub plantations in these same localities, or of the citrus orchards, include the White-winged Dove, Black-chinned Hummingbird, Orchard and Bullock's Orioles, and Painted Bunting. During winter, Cedar Waxwings are often abundant and there are generally a few Sharp-shinned and Cooper's Hawks, House Wrens, American Robins, Ruby-crowned Kinglets, Myrtle Warblers, Lesser Goldfinches, Lark Buntings, and White-crowned Sparrows.

SEYMOUR
Mesquite Plains Country

This prosperous farming community (elevation 1,290 feet) of north-central Texas is on the rolling mesquite plains, devoted in part to wheat-raising, but chiefly to grazing. Some of the grazing area is open grassland, but much of it supports clumps of mesquite that vary in size and density. Streams are bordered by luxuriant growths of hackberry, willow, cottonwood, and other trees.

For finding nesting birds around Seymour, the best time is late May through June, preferably mid-June. Along the out-going highways, Mississippi Kites and White-necked Ravens may be readily observed, either in flight or perched on the mesquite, in which they frequently nest. Bullock's Orioles nest very commonly in the mesquite, as do Mourning Doves, Yellow-billed Cuckoos, Scissor-tailed Flycatchers, Ash-throated Flycatchers, and Northern Mockingbirds. Other birds to be expected in mesquite areas are Greater Roadrunners, Poor-wills, Common Nighthawks, Golden-fronted Woodpeckers, (Black-crested) Tufted Titmice, and Lark Sparrows.

For the best bird finding, drive north from Seymour on US 283, through ranching country, to Vernon. Turn off from time to time on country roads over grasslands studded with mesquite. In grasslands, look especially for Dickcissels and Grasshopper and Cassin's Sparrows.

UVALDE

Garner State Park | Lost Maples State Park

From Uvalde in southern Texas, on US 90 some 80 miles west of San Antonio, drive north into the Edwards Plateau on US 83 for 31 miles to **Garner State Park** on the right. Along the way during the breeding season, look for Cave Swallows on utility wires near culverts in which they and Cliff Swallows often nest.

Through Garner State Park (912 acres) winds the Frio River, flanked by pecans and oaks. Follow the entrance road around juniper-clad hills, then down to the Frio River bottom and past the concessions building to the picnic and camping areas. Park here and walk back toward the concessions building along the River, watching for a pair of Green Kingfishers. Other breeding birds to watch for include the Black Phoebe, Acadian Flycatcher, Yellow-throated Warbler, Blue Grosbeak, Painted Bunting, and Lesser Goldfinch. Yellow-throated and Red-eyed Vireos nest in trees around the picnic area. Golden-cheeked Warblers have been noted in the Park, most often from the picnic area directly north of the concessions building.

After investigating Garner State Park, continue north on US 83

GREEN KINGFISHER

for a mile, then turn east onto Farm Road 1050 and go 15 miles to its terminus at Utopia. This road traverses country frequented in summer by the Western Kingbird, Scissor-tailed Flycatcher, occasionally the Scott's Oriole, and all year by the Rufous-crowned Sparrow.

At Utopia, drive north on Farm Road 187 for 16 miles through Vanderpool to **Lost Maples State Park** on the left. Meanwhile, watch for Cave Swallows as well as Cliff Swallows near stock tanks.

The 1,280 acres of Lost Maples State Park embrace much of Sabinal Canyon in which are stands of big-toothed maple—unique in southern Texas—that attract thousands of people in fall to marvel at the foliage—brilliant in reds, oranges, and yellows. From the parking area many trails lead into the Canyon. Among the breeding birds in the maples and other trees and in riverside thickets are the Acadian Flycatcher, Western Pewee (possibly the Eastern Pewee, too), and the Yellow-throated and Red-eyed Vireos. Rock Wrens and Brown Towhees reside in suitable habitats. The Green Kingfisher should be looked for along the Sabinal River. Never take for granted what may appear to be a Turkey

Vulture—it could be a Zone-tailed Hawk which nests in Sabinal Canyon.

WICHITA FALLS
Lucy Park

This big city in north-central Texas, about 20 miles south of the Oklahoma boundary (Red River), lies along the Wichita River whose bottomlands, in contrast to the outlying rolling plains, were once heavily wooded with deciduous growth. Today in the city itself there are still woods, even though second-growth, that may hold an association of birds at least partially representative of the association that originally existed.

Lucy Park (156 acres), in a bend of the Wichita River in the northern part of the city, probably embraces the best woods. From US 277-281, exit south at the Texas Tourist Bureau, drive past it to the campground in Lucy Park, leave the car, and take the foot bridge over the River to the woods and dense thickets, where paths extend upstream and downstream for convenient exploration.

Among the birds present all year in the woods are numerous woodpeckers—e.g., Red-bellied, Golden-fronted, Red-headed, Hairy, and Downy—and such other species as the Carolina Chickadee, Tufted Titmouse, Brown Thrasher, and Northern Cardinal. In spring comes a variety of birds to nest, the stellar attraction being the Mississippi Kites. Other birds that will arrive for the summer are Swainson's Hawks, Scissor-tailed and Great Crested Flycatchers, Warbling Vireos, Indigo and Painted Buntings, and possibly a few Wood Ducks. Often abounding in April and early May are transient warblers, among them the Orange-crowned Nashville, Northern Parula, Yellow, Myrtle, Audubon's, Black-throated Green, Wilson's, and American Redstart. In winter are many sparrows including the Harris', White-crowned, and White-throated.

Utah

BY WILLIAM H. BEHLE

CALIFORNIA GULLS

There are few bird finders who have not heard of the 'miraculous' appearance of great flocks of 'sea gulls' in 1848 and subsequent years, which descended on the meager grainfields of Salt Lake Valley and destroyed the Mormon crickets that were ravaging the crops of the first white settlers. The birds were California Gulls, common today in northern Utah during the spring and summer months at parks, school grounds, and shopping centers of cities, as well as marshes and sloughs and around farms. Familiar sights in the spring are flocks of gulls in fields, 'following the plow' and eating insect larvae and worms from the furrows. Second only to the historical interest of the gulls among Utah's ornithological distinctions is the Bear River Migratory Bird Refuge (*see under* **Brigham**

City) where, in February and March, 20,000 to 25,000 or more Whistling Swans concentrate and, in spring and fall, Wilson's and Northern Phalaropes, Baird's and Western Sandpipers, and other shorebirds gather by tens of thousands. Then there are the colonies of American White Pelicans that nest with the gulls on Gunnison Island in Great Salt Lake. Pelicans are frequently seen foraging or in flight along the waterways from Utah Lake to the Bear River Refuge. These three features are localized in central-northern Utah, but ornithological opportunities exist in almost every section of the state, whose deserts, Great Salt Lake and its marshes, mountains, high plateaus, rivers, canyons, natural bridges, arches, and bizarre red-rock formations have earned Utah the name 'Center of Scenic America.'

Utah is a land of great physiographic contrasts, with alkaline desert wastes in the west, rugged mountains extending down the middle and across the northeast, bare-rock canyon country in the southeast, and numerous isolated high mountains that arise from the lowlands in western Utah and also in the southeastern portions of the state. Altitudinally the range is from about 2,800 feet along the Virgin River in the southwestern corner to the 13,528 feet of Kings Peak, Uinta Mountains, in the northeast. Less than 3 per cent of the land of the state is suitable for agriculture; most of the land thus utilized is in valleys at the base of mountains where water is available for irrigation. The western third of Utah comprises the eastern part of the Great Basin, which ranges for the most part between 4,000 and 5,000 feet elevation. Much of it is a dry wasteland, used by ranchers for grazing or by the Federal government as military sites. North-to-south mountain ranges loom up like islands in the midst of alkaline plains. Indeed, many of these mountains were once actually islands, for in Pleistocene times, about 25,000 to 50,000 years ago, Lake Bonneville covered most of the lowlands, as shown by shorelines of ancient lake levels, immense gravel bars, and wave-cut cliffs. Great Salt Lake, locked in an interior basin with no outlet, is its remnant. In this Great Basin portion of Utah one sees a northern desert shrub type of vegetation consisting essentially of sagebrush on fertile soils with shadscale and greasewood on the alkali flats. This is the nesting habitat for such birds as the Sage Grouse, Burrowing Owl, Horned Lark,

Sage Thrasher, Loggerhead Shrike, Western Meadowlark, and five sparrows, the Lark, Vesper, Black-throated, Sage, and Brewer's.

Running from north to south in central-northern Utah is the prominent Wasatch Range, extending from the Idaho border to Mount Nebo east of Nephi, and presenting on its west side a fault scarp, the Wasatch Front, which in places is 6,000 feet high. Several streams, such as Bear, Ogden, Weber, Provo, and Spanish Fork, cut through the Range, flowing from east to west into the interior sink. Nearly all the major cities of Utah, containing two-thirds of the state's population, are in the area watered by these streams at the base of the Wasatch Front. These streams are also responsible for much of the marshland of northern Utah. As the streams reach the stagnant levels around the Great Salt Lake, they form extensive marshes as at the delta of the Bear River (*see under* **Brigham City**).

South of the Wasatch Mountains and continuing down the central part of the state is a series of interconnecting ranges—the Gunnison, Pavant, Tushar, Parowan, and, curling off to the southwest, the Pine Valley Mountains. US 91 runs along the base of the mountain chain, passing through a series of small settlements wherever water and a little farmland are available. East of this mountain chain lie the high plateaus (flat-topped mountains) of central and central-southern Utah. From the southern border of the state, the Vermilion, White, and Pink Cliffs arise like great steps ascending to Markagunt Plateau, where Zion Canyon (*see under* **Springdale**) has been carved by the Virgin River, and to Paunsagunt Plateau, where Bryce Canyon (*see under* **Panguitch**) is located. South and east of Bryce Canyon National Park lies Kaiparowits Plateau; to the north are the Aquarius, Avara, Panguitch, Sevier, Fish Lake, and Wasatch Plateaus. The Sevier River in central Utah, with its two forks, rises in the high plateau country, swings north, cuts west through the mountain front between the Pavant and Gunnison Mountains, and finally turns southwest into the Sevier Desert to form the playa Sevier Lake. US 89 follows along much of the length of the Sevier River.

Southeastern Utah, part of the Colorado Plateau, is characterized by broken, red-rock country, referred to as 'the canyon lands,' typified by the Capitol Reef National Park (*see under* **Richfield**),

Arches National Park (*see under* **Moab**), and Canyonlands National Park (*see under* **Monticello**). The Green and Colorado Rivers with tributaries such as the San Juan, Fremont, and Escalante, cut deep through the rock formations. Occasional mountain masses rise up 10,000 to 13,000 feet—for example, Navajo Mountain, just north of the Arizona border, and the Henry, Abajo, and La Sal Mountains. I 70 crosses the spectacular San Rafael Desert. This is wild, isolated terrain with few birds en route.

The Book and Roan Cliffs in central-eastern Utah, at the base of which US 50 runs, mark an abrupt rise from the canyon lands of southeastern Utah to the East and West Tavaputs Plateaus and the Uintah Basin of northeastern Utah. US 40 runs across the Uintah Basin, which contains some farming land. Forming the northern border of the Basin and lying just south of the Wyoming border are the Uinta Mountains. They are unusual in that the main axis extends east and west. This range has been heavily glaciated and contains many cirques and hundreds of lakes. The dome is about 10,000 to 12,000 feet in elevation, but a great many peaks project over 13,000 feet. Much of this country is primitive, but a few roads run up to the margin of the wilderness region.

One of the distinctive features of the mountains and plateaus is the belting of vegetation. On the lower slopes and alluvial fans, between 3,200 and 7,000 feet, are distinctive forests of pinyon pine and juniper, aptly designated locally as 'pygmy forest' or 'pygmy conifers' because of the relatively low habit of the trees as compared with the taller coniferous trees found at higher elevations. Associated with this forest are sage, cliff rose, bitterbrush, serviceberry, and, at higher elevations, mountain mahogany. Certain species of breeding birds are essentially restricted to this forest: Gray Flycatcher, Pinyon Jay, Plain Titmouse, Bushtit, Gray Vireo (southwestern Utah), Black-throated Gray Warbler. Others often residing in this habitat are the Poor-will, Common Nighthawk, Say's Phoebe, Scrub Jay, Blue-gray Gnatcatcher, Rufous-sided Towhee, Chipping Sparrow, and Black-chinned Sparrow.

Above the pygmy conifers in southern and eastern Utah, approximately between 6,200 and 8,000 feet elevation, is a belt of ponderosa pine. Gambel oaks are associated with this forest, growing here and there in dense continuous stands. In addition, there are other scrub types, such as serviceberry, squaw-apple, and manzanita. Essentially, however, the ponderosa pine forest is open with little underbrush and is not rich in birdlife. A few forms nest here regularly (though not all of them exclusively): Band-tailed Pigeon, Williamson's Sapsucker, Steller's Jay, Mountain Chickadee, White-breasted Nuthatch, Pygmy Nuthatch, Western Bluebird, Audubon's Warbler, Gray-headed Junco. In central-northern Utah, where ponderosa pine is sparse but scrub oaks are extensive, common breeding birds are the introduced California Quail, the Mourning Dove, Scrub Jay, Black-billed Magpie, Virginia's Warbler, Black-headed Grosbeak, and Green-tailed and Rufous-sided Towhees.

Above the ponderosa pine and oak belt in southern and eastern Utah, and above the pygmy conifer and mountain mahogany stands in the mountains of western and central-northern Utah, is a conifer-aspen forest. The conifers are principally Colorado blue and Englemann spruces, white and subalpine firs, Douglas-fir, and bristlecone, limber, and lodgepole pines, depending on the geographic location and the elevation. At lower elevations, tongues of conifers and aspens extend down into the cooler environments along the streams. The coniferous forest is very rich in variety of birdlife, and the following species of breeding birds are to be expected in Utah wherever this type of cover prevails:

Northern Goshawk
Sharp-shinned Hawk
Cooper's Hawk
Red-tailed Hawk
Blue Grouse
Ruffed Grouse
Great Horned Owl
Broad-tailed Hummingbird
Calliope Hummingbird
Red-shafted Flicker
Yellow-bellied Sapsucker
Williamson's Sapsucker
Hairy Woodpecker
Downy Woodpecker
Northern Three-toed Woodpecker
Hammond's Flycatcher
Dusky Flycatcher
Western Flycatcher
Western Pewee
Olive-sided Flycatcher
Violet-green Swallow
Gray Jay
Steller's Jay
Clark's Nutcracker
Mountain Chickadee
Red-breasted Nuthatch
Brown Creeper

- House Wren
- American Robin
- Hermit Thrush
- Swainson's Thrush
- Townsend's Solitaire
- Golden-crowned Kinglet
- Ruby-crowned Kinglet
- Warbling Vireo
- Orange-crowned Warbler
- Audubon's Warbler
- Western Tanager
- Cassin's Finch
- Pine Grosbeak
- Pine Siskin
- Red Crossbill
- Gray-headed Junco
- White-crowned Sparrow

Cutting back through the mountain escarpments are canyons with the streams along their bottoms lined with willows and cottonwoods at the lower elevations, maples, river birch, and alder in the higher reaches. Thickets of rose and dogwood are interspersed. In this canyon habitat, lying between 5,000 and 7,000 feet elevation, one finds Song Sparrows and sometimes Soras if there are wet grassy areas associated with the willows, North American Dippers along the swifter moving streams, and a number of other breeding birds, as follows: Traill's Flycatcher, Black-capped Chickadee, Black-billed Magpie, Gray Catbird, Swainson's Thrush, Solitary Vireo, Yellow Warbler, MacGillivray's Warbler, Yellow-breasted Chat, American Redstart, Lazuli Bunting, and Fox Sparrow.

In extreme southwestern Utah along the Virgin River drainage from the mouth of Zion Canyon to the Utah-Arizona border south of St. George, as well as along the Santa Clara River and Beaver Dam Wash tributaries, a different assemblage of breeding birds is present representing an extension of the Mojave Desert avifauna. Distinctive indicator species are the following:

- Gambel's Quail
- White-winged Dove
- Greater Roadrunner
- Lesser Nighthawk
- Costa's Hummingbird
- Ladder-backed Woodpecker
- Black Phoebe
- Vermilion Flycatcher
- Verdin
- Cactus Wren
- Northern Mockingbird
- Crissal Thrasher
- Black-tailed Gnatcatcher
- Phainopepla
- Bell's Vireo
- Lucy's Warbler

Hooded Oriole
Scott's Oriole
Summer Tanager
Blue Grosbeak
Indigo Bunting
Abert's Towhee

Following the roads up the canyons to the tops of the mountains or plateaus, one can therefore pass through several vegetative belts in the space of a few miles and observe representatives of the birdlife in each situation. As far as climate, vegetation, and birdlife are concerned, such a trip is equivalent to one from the southern deserts north through the great transcontinental coniferous forests of Canada to the Arctic tundra. For example, in the St. George region and up to the mouth of Zion Canyon, semitropical conditions prevail. Cedar City is in a Great Basin-pygmy forest setting. The rim of Bryce Canyon (*see under* **Panguitch**), at about 8,000 feet, is in the ponderosa pine belt; Cedar Breaks National Monument (*see under* **Cedar City**), at 10,700 feet, is in the Canadian spruce-fir forest, and the slopes of Brian Head, extending to 11,315 feet, represent an approach to Arctic-alpine conditions. Thus, in visiting the St. George area, Zion and Bryce Canyon National Parks, Cedar Breaks, and intervening areas, one passes through five climatic situations, each with its attendant birdlife.

Breeding birds that are more regular around farmlands and outskirts of cities (in dooryards, fields, wet meadows, bushy areas, orchards, and roadsides) are the following:

Red-tailed Hawk
Swainson's Hawk
Northern Harrier
American Kestrel
California Quail
Killdeer
California Gull
Mourning Dove
Broad-tailed Hummingbird
Red-shafted Flicker
Western Kingbird
Say's Phoebe
Horned Lark
Barn Swallow
Black-billed Magpie
Northern Raven
American Robin
Yellow Warbler
Western Meadowlark
Bullock's Oriole
Brewer's Blackbird
Black-headed Grosbeak
Lazuli Bunting
House Finch
Pine Siskin
Savannah Sparrow

Lark Sparrow
Chipping Sparrow
Song Sparrow

In all sections of the state there are seasonal changes in the local avifauna as a result of migration. A great many of the waterfowl, waterbirds, wading birds, shorebirds, hawks, hummingbirds, flycatchers, swallows, wrens, thrushes, vireos, warblers, finches, and sparrows are summer residents and leave for the winter. A few other species move in from the north as winter visitants. Some of the more common of the latter are the Rough-legged Hawk, Ring-billed Gull, American Crow, Bohemian Waxwing, Cedar Waxwing, Northern Shrike, Evening Grosbeak, Gray-crowned Rosy Finch, Oregon Junco, American Tree Sparrow, and Snow Bunting. Utah is on the eastern fringe of the Pacific flyway. In the series of marshes at the base of the Wasatch Front, many kinds of transient waterfowl, waterbirds, wading birds, and shorebirds are especially abundant in spring and fall. Most seem to fly along the Sevier River in central Utah on their journeys. The Colorado-Green River system serves as another migratory lane in the eastern portion of the state. Many migrate along the Virgin River in southwestern Utah. The best time to see migrating waterfowl and shorebirds is late February to April and August to September. A few landbirds migrate through the state—for example, the Rufous Hummingbird, Townsend's Warbler, and northern races of geographically variable species. A notable feature in many parts of the state is the altitudinal migration of montane landbirds such as woodpeckers, chickadees, creepers, pipits, and rosy finches, which move down from their breeding grounds in the mountains into the lowlands of the valleys for the winter. As regards breeding birds, three avifaunas are represented in the state: the Great Basin, Rocky Mountain, and, in extreme southwestern Utah, the Mojave Desert.

References

Utah Birds: Check-list, Seasonal and Ecological Occurrence Charts and Guides to Bird Finding. By William H. Behle and Michael L. Perry. Utah Museum of Natural History, University of Utah, Salt Lake City. 1975.

Birds of Utah. By C. Lynn Hayward, Clarence Cottam, Angus M. Woodbury, Herbert H. Frost. Great Basin Naturalist Memoirs No. 1, Brigham Young University Press, Provo, Utah. 1976.

BRIGHAM CITY
Bear River Migratory Bird Refuge

Outstanding among the country's National Wildlife Refuges is the **Bear River Migratory Bird Refuge** at the mouth of the Bear River on the east side of Great Salt Lake. Established by special act of Congress in 1929, the Refuge (64,895 acres) constitutes one of the most completely developed bird refuges in the country. About 40,000 acres are under water. Headquarters is 15 miles west of Brigham City. To reach the Refuge, it is easiest to leave I 15 and drive into town along Main Street. Turn west onto Forest Street at the county court house. The surfaced road eventually follows along the Bear River to its mouth. Some birds such as Cattle Egrets, Snowy Egrets, American Avocets, American Coots, Marsh Wrens, and Yellow-headed Blackbirds will be noticed in spring and summer in small marsh areas along the road. Long-billed Curlews are common on the dry flats. Canada Geese may be seen along the River. At the Duckville Gun Club clubhouse beside the road near the entrance to the Refuge, Cliff Swallows nest under the eaves in May and gather mud near the bridge across the diversion canal. The species also nests on the buildings at Refuge headquarters. American White Pelicans, Snowy Egrets, and Black-crowned Night Herons are frequently fishing in the water near the dam across the Bear River at the Refuge headquarters.

At the administration building where visitors must register, descriptive leaflets and an annotated checklist of birds of the Refuge are available. Maps of the region, charts showing migration, pictures, mounted specimens, and sets of eggs are items of interest in a small museum. A 100-foot tower, which visitors may climb, gives a marvelous panorama of the vast marsh area.

There are approximately 40 miles of dikes in the Refuge, forming five units. Roads cap the dikes. Throughout the year visitors are allowed to drive the 12 miles around Unit 2, which is adjacent to headquarters. The dikes are constructed by scooping up the dirt from either side, the result being deep channels adjacent to the dikes. This makes it possible to see at close range from the car such diving forms as Western Grebes and Ruddy Ducks, in addition to many common species of waterfowl which spring up or

BLACK-NECKED STILTS

sneak into the nearest vegetation as the car passes. Overhead are gulls, terns, and sometimes pelicans. In shallow water are coots and numerous species of wading birds and shorebirds. A thrilling sight in May is the multiple pairs of Canada Geese convoying their broods away from the dikes, one parent in front and one behind. If lucky, the bird finder may see young Western Grebes riding on the back of a parent.

The best time to visit the Refuge is during the breeding season in May and June, when, on the drive around Unit 2, one can count on seeing about fifty summer-resident species, of which the following are the most common: Western Grebe, American White Pelican, Double-crested Cormorant, Great Blue Heron, Snowy Egret, Black-crowned Night Heron, White-faced Ibis, Canada Goose, Mallard, Gadwall, Common Pintail, Cinnamon Teal, Northern Shoveler, Redhead, Ruddy Duck, Northern Harrier, American Coot, Killdeer, Willet, American Avocet, Black-necked Stilt, Wilson's Phalarope, California and Franklin's Gulls, Forster's and

Black Terns, Short-eared Owl, Cliff and Barn Swallows, Marsh Wren, and Yellow-headed and Red-winged Blackbirds.

February is the great month for migratory waterfowl, March for marsh birds, and April for shorebirds. Not only do the summer residents listed above arrive, but thousands of transients pass through, including the Snow Goose, American Wigeon, Canvasback, Greater and Lesser Scaups, Common Goldeneye, Bufflehead, Common and Red-breasted Mergansers, Black-bellied Plover, Solitary Sandpiper, Greater and Lesser Yellowlegs, Pectoral, Baird's, and Least Sandpipers, Long-billed Dowitcher, Western Sandpiper, Marbled Godwit, Sanderling, Northern Phalarope, and Ring-billed Gull. Whistling Swans by the thousands concentrate here in February and March before moving north to their breeding grounds.

August and September mark the southward migration of northern-breeding shorebirds, as well as the exodus of many of the locally breeding species. Western Sandpipers, Wilson's Phalaropes, and Northern Phalaropes form tremendous concentrations. Numerous then are hawks and corvids such as the Red-tailed and Rough-legged Hawks, Prairie Falcon, Peregrine Falcon (occasional), Black-billed Magpie, and Northern Raven. Certain ducks—e.g., Buffleheads and Common Goldeneyes—appear even later in November. In November and December, migrating Whistling Swans are again present by the thousands.

Unless snowdrifts block the roads, the Refuge may be visited during the winter months and one may then see the following species: Great Blue Heron, Canada Goose, Mallard, Green-winged Teal, Common Merganser, Bald Eagle, Northern Harrier, and Northern Raven.

CEDAR CITY
Cedar Breaks National Monument

From I 15 in southwestern Utah, exit at Cedar City and proceed east on State 14 for 18 miles, then north on State 143 for 3 miles. The road mounts ever higher to the very edge of the Markagunt Plateau in the Pink Cliffs formation, where the huge amphitheater of **Cedar Breaks National Monument** (6,154 acres) suddenly appears in all its splendor from Point Supreme View. The elevation of the rim varies from 10,400 to 10,700 feet. Two other vantages can be reached—Sunset View and Desert View—but Point Supreme View is generally regarded as the best. Brian Head, a nearby eminence on the north, rises to an altitude of 11,315 feet; a road leads to its summit. At the rim is an information office and exhibit room. Resorts just outside the National Monument, which feature skiing, have made the whole area accessible in winter but birdlife is then sparse, being confined to a few permanent resident species.

The approach to Cedar Breaks is through ponderosa pine at lower elevations, changing to a heavy forest of Englemann spruce, subalpine fir, and bristlecone pine, with some open, park-like clearings, where one may glimpse Mountain Bluebirds, Vesper Sparrows, and occasional American Kestrels or Northern Ravens. Birds seem to be especially common in the forest and on the edges. Expect to see the following: Hairy Woodpecker, Gray and Steller's Jays, Clark's Nutcracker, Mountain Chickadee, Red-breasted and Pygmy Nuthatches, Brown Creeper, Hermit Thrush, Townsend's Solitaire, Golden-crowned and Ruby-crowned Kinglets, Audubon's Warbler, Cassin's Finch, Pine Siskin, Red Crossbill, Gray-headed Junco, and White-crowned Sparrow. The last-named species is most common on windswept slopes where there are stunted patches of spruce. Sometimes Blue Grouse show up along the forested edge of the rim. White-throated Swifts careen along the rim, in contrast to the slower-moving Violet-green Swallows. Broad-tailed Hummingbirds are common in patches of flowers along the rim and around Brian Head.

HEBER
Midway

The mountain-rimmed valley east of Salt Lake City in which this town is located has an abundance of water; hence wet meadows are common and willows and cottonwoods abound along streams. This region yields good bird finding in May and June. Birds are perhaps more easily seen at **Midway,** 3 miles west on State 113. Here are the 'Hot Pots,' limestone cones formed by hot springs. Cliff Swallows nest in the 'chimney' of the largest hot pot at a resort called The Homestead. About a mile to the northwest is Snake Creek Canyon with oaks and streamside thickets.

About 0.5 mile south of Midway is the Midway State Fish Hatchery, where sizable springs give rise to a stream that winds down through the fields to enter the Provo River at Deer Creek Reservoir near Charleston. Along the stream, in the adjacent wet fields, or in the shrubbery or cottonwoods, look for the following birds in late spring and summer: Great Blue and Black-crowned Night Herons, Cinnamon Teal, Virginia Rail, Sora, Killdeer, Spotted Sandpiper, California Gull, Broad-tailed Hummingbird, Red-shafted Flicker, Traill's Flycatcher, Western Pewee, Violet-green, Rough-winged, and Cliff Swallows, Black-billed Magpie, Black-capped Chickadee, House Wren, Gray Catbird, Yellow Warbler, Common Yellowthroat, Yellow-breasted Chat, Bobolink, Western Meadowlark, Yellow-headed, Red-winged, and Brewer's Blackbirds, Black-headed Grosbeak, American Goldfinch, and Savannah and Song Sparrows. Belted Kingfishers are usually present along the Provo River just below Deer Creek Dam. Thirty miles east of Heber, through Daniels Canyon on US 40, is the Strawberry Reservoir. Primitive roads lead down either side. Waterfowl concentrate here, especially in fall migration.

KAMAS
Mirror Lake

The high-mountain country of the west portion of the Uinta Range in northeastern Utah may be reached by driving east on State 150, 33 miles from Kamas. **Mirror Lake** is but one of seventy-five lakes within a radius of 6 miles. It lies at the base of Bald Mountain, whose summit is 11,947 feet. The pass over which the road leads is about 11,000 feet. The area is heavily forested with lodgepole pine. There are numerous United States Forest Service camps en route. Trails lead to the summit of Bald Mountain.

Breeding birds in the Mirror Lake region include the Osprey,

Blue Grouse, Hairy and Northern Three-toed Woodpeckers, Hammond's and Olive-sided Flycatchers, Gray Jay, Clark's Nutcracker, Water Pipit, Cassin's Finch, Pine Grosbeak, Black Rosy Finch, Pine Siskin, Red Crossbill, Gray-headed Junco, and White-crowned and Lincoln's Sparrows.

KANAB
Upper Reservoir | Three Lakes

Located near the Arizona line in central-southern Utah and nestling in an indentation of the Vermilion Cliffs where Kanab Creek emerges from the high plateau country to the north is Kanab, the gateway to Zion and Bryce National Parks from the Grand Canyon country on the south. It is a convenient and lovely place to stop overnight, and many bird species reside in the vicinity. As US Alternate 89 approaches Kanab from the direction of the Arizona line, it passes along a straight stretch bordered on the west (left) by a row of cottonwoods, beyond which are fields and Kanab Creek. In the cottonwoods, as summer residents, are Red-shafted Flickers, Western Pewees, Western Kingbirds, Yellow Warblers, and Bullock's Orioles. In the fields and orchards, such species as the Mourning Dove, Eastern Kingbird, Horned Lark, Northern Mockingbird, Loggerhead Shrike, Blue Grosbeak, and Lark Sparrow may be expected.

On the southern outskirts of town at the southern end of Second East Street, south of US 89 going to Page, is the **Upper Reservoir.** The Lower Reservoir is three miles farther south and difficult to reach since there are no access roads. A variety of migrating waterfowl and shorebirds stop temporarily at these reservoirs in April and May and September and October. During summer the following birds nest in the vicinity: Northern Harriers, Soras, American Coots, Killdeers, Spotted Sandpipers, Yellow Warblers, Common Yellowthroats, and Yellow-headed and Red-winged Blackbirds. A propitious bird-finding area is south of town along Kanab Creek. Here the stream has washed out a huge gorge, 75 feet deep in places and 50 to 100 yards across. Seepage water, together with water remaining after diversion, fosters considerable vegetation in the present stream bottom including some localized marshy areas. One point of access is from the rim near 300 South Street and 200 West but a more accessible portion of the wash is at the residential area known as Kanab Creek Ranchos about 1.0 mile south of town. Here a road crosses the Creek.

Birds found along Kanab Creek include the Turkey Vulture, American Kestrel, Gambel's Quail, Spotted Sandpiper, Mourning Dove, Greater Roadrunner, Say's Phoebe, Violet-green, Rough-winged, and Cliff Swallows, Northern Raven, Northern Mockingbird, Yellow Warbler, Common Yellowthroat, Western Meadowlark, Red-winged and Brewer's Blackbirds, House Finch, Lesser Goldfinch, and Brewer's Sparrow.

About 5 miles northwest of Kanab along US 89 a road goes north up Kanab Canyon making the Creek bottomland again easily accessible. About 5.5 miles north of Kanab the main highway traverses Three Lakes Canyon where **Three Lakes** are adjacent to the highway on the west. These are the only bodies of water along the road, and their vicinity is a good spot for bird finding with diverse environmental situations such as open water with fringe marshy areas and wet meadows, willow thickets, cottonwoods, cliffs, oak brush, and pygmy forest. Birds regularly here are the Common Snipe, White-throated Swift, Traill's Flycatcher, Violet-green and Cliff Swallows, Scrub Jay, Yellow Warbler, Brown-headed Cowbird, Lazuli Bunting, Rufous-sided Towhee, and Song Sparrow.

For bird finding south of Kanab, *see under* **Fredonia, Arizona.**

LOGAN
Logan Canyon

Cache Valley in northern Utah contains a small remnant of the grassland that once characterized much of the state. In this habitat east of Wellsville a small population of Sharp-tailed Grouse still survives. On a Lake Bonneville level overlooking Cache Valley is the Utah State University. East of the campus is beautiful **Logan Canyon** through which US 89 passes to the Bear Lake region. Bird finding is good nearly everywhere in the Canyon, but two specially good places for mountain birds are Tony's Grove, where Purple

Martins are known to nest, and Beaver Creek, where a skiing resort operates in winter.

MOAB

Arches National Park | Dead Horse Point State Park

Nestled in Spanish Valley at the base of red cliffs adjacent to the Colorado River in central-eastern Utah is the town of Moab—the gateway to many scenic areas including **Arches National Park** and **Dead Horse Point State Park.** The bird finder will be rewarded by working along the Colorado riverbottoms, except in spring when they may be flooded. State 128, which passes up the east side of the River from Moab, is too disturbed by traffic for the bird finder, but on the west side, reached by crossing a bridge on the western outskirts of town, he can walk upstream a short distance and probably see or hear Canyon Wrens on the cliffs, and in the tamarisk thickets, willows, and cottonwoods he should see Yellow Warblers, blackbirds, and other streamside species. The same species may be seen along the River south of Moab, where a paved State 279 leads south about 16 miles to a potash plant.

Northwest of Moab along US 163 lie the approaches to the Arches National Park (129 square miles) and Dead Horse Point State Park (46,127 acres). The entrance to the former is 5 miles northwest of Moab. At the visitor center, obtain a checklist of birds for the Park. From headquarters the Park road switchbacks up the face of a cliff to more level terrain beyond. In winter, rosy finches may be viewed along the initial stretch of the road overlooking headquarters. The area includes 83 arches and countless strange rock formations. The bird finder must walk short distances to the principal sights; in so doing, he should be alert for certain open-country birds such as Red-tailed Hawks, Say's Phoebes, and Brewer's Sparrows. The area supports sparse desert-shrub and pygmy-forest vegetation.

For the road to Dead Horse Point, turn south from US 163 about 10 miles northwest of Moab and proceed on State 313 about 30 miles. In some respects the view out over the gorge of the Colorado River equals that of the Grand Canyon in Arizona. On this trip to Dead Horse Point the bird finder may see the American Kestrel, Say's Phoebe, Horned Lark, Scrub and Pinyon Jays, Plain Titmouse, Bewick's and Canyon Wrens, Sage Thrasher, and Rufous-sided Towhee.

MONTICELLO
Canyonlands National Park

At the east base of the Abajo or Blue Mountains on US 163 in southeastern Utah lies Monticello. From here, US 163 continues on south to Blanding, Bluff, and Mexican Hat. Near Mexican Hat are the geologically astounding Goosenecks, several great bends in the San Juan River. South of Mexican Hat and the San Juan River in extreme southern Utah is Monument Valley. The sand in Monument Valley makes it necessary for the bird finder to remain on the road. This is wild and isolated country; birds are sparse and the only lodgings available beyond Mexican Hat are at Gouldings, near the Arizona line, and Kayenta, farther south in Arizona.

The approach to **Canyonlands National Park** is west from US 163 at Church Rock, about 17 miles north of Monticello over State 211. The Park, embracing 403 square miles, is by design essentially undeveloped and many areas are accessible only by four-wheeled-drive vehicles or by hiking. It is spectacular, bare, red rock country and rough desert terrain with sparse stands of juniper. The birds are about the same as in the Arches National Park (*see under* **Moab**). At Park headquarters, ask for a checklist of birds. Fifty miles west of Blanding via State 95 and 275 is the Natural Bridges National Monument (7,799 acres), also in a pinyon pine and juniper environment.

OGDEN
Snow Basin | Birch Creek Hollow | Weber Riverbottoms | Ogden Bay Waterfowl Management Area

Utah's second city in size has Ogden Canyon at its backdoor and Ogden Bay on Great Salt Lake at its front, with the Ogden and

Weber Rivers passing through town. The Canyon, on State Route 39 east of 21st Street, contains the Pineview Reservoir. East of the Reservoir the Canyon opens out into a round valley where Huntsville is located. In addition to Belted Kingfishers and North American Dippers along the Ogden River, the Reservoir attracts some waterfowl. A road to the south near the east (upper) end of the Reservoir, about 10 miles from Ogden, leads to **Snow Basin,** back of Mount Ogden at an elevation of 8,000 feet. Here a spruce-fir-aspen forest prevails, with the common species likely to be encountered, as listed for that environment in the introduction to this chapter.

Birch Creek Hollow is a productive area for viewing birds within the city. From the business district, proceed south along Washington Boulevard to 40th Street, turn left (east) one block past the Birch Creek School, then go right (south) for 0.25 mile to the north end of the Country Club Golf Course. One can then work eastward up Birch Creek toward the foothills. Near the end of the approach road are three small ponds. Along the stream and around the ponds among the cottonwoods, willows, birches, and dogwoods, look or listen for the Ring-necked Pheasant, Killdeer, Great Horned, Long-eared, and Saw-whet Owls, Downy Woodpecker, Scrub Jay, Black-billed Magpie, Black-capped Chickadee, Common Yellowthroat, Yellow-breasted Chat, Western Meadowlark, Red-winged Blackbird, Black-headed Grosbeak, Rufous-sided Towhee, and Song Sparrow. Most are permanent-resident species, their numbers augmented in fall, winter, and spring by an influx of birds such as the Northern Goshawk, Brown Creeper, Ruby-crowned Kinglet, Townsend's Solitaire, Oregon Junco, and White-crowned Sparrow.

The **Weber Riverbottoms** on the south and west sides of Ogden afford good bird finding. Drive south on Wall Avenue to 33rd Street, turn right (west) and go 0.5 mile to the railroad, and continue another 0.5 mile or so to the River. Work southward on either side of the River to the Riverdale Road or north on the west bank of the River to the viaduct on 24th Street. Here there are the same species as mentioned for Birch Creek Hollow, and, in addition, such possible species as the Pied-billed Grebe, Virginia Rail, Cooper's and Red-tailed Hawks, American Kestrel, California

Quail, Common Screech Owl, Belted Kingfisher, Red-shafted Flicker, North American Dipper, Yellow Warbler, American Redstart, Brewer's Blackbird, Lazuli Bunting, and American Goldfinch.

West of Ogden on the east shore of Great Salt Lake is the **Ogden Bay Waterfowl Management Area,** a state-operated 16,680-acre project at the mouth of the Weber River. The approach is via the town of Hooper. There are several units to the marsh with roads along the dikes. Check at headquarters to see what areas may be visited. This is a heavily used duck-hunting area in fall with populations of migrating ducks occasionally peaking at half a million. Ducks, geese, marsh birds, wading birds, and shorebirds nest in large numbers. The species are essentially the same as for the Bear River Refuge (*see under* **Brigham City**).

PANGUITCH
Bryce Canyon National Park

The bird finder in central-southern Utah must visit **Bryce Canyon National Park** for its unrivaled beauty as well as for its birds. From Panguitch, drive south on US 89 for 7 miles, then east on State 12 for 14 miles. The main feature of the Park, which extends over 56 square miles, is a huge amphitheater at the edge of the pink cliffs of the Paunsagunt Plateau, with a view from the rim of highly colored, delicately and fantastically eroded rock formations. In approaching Bryce National Park, the route passes through a ponderosa pine forest in which characteristic breeding birds include Williamson's Sapsuckers, Steller's Jays, White-breasted and Pygmy Nuthatches, Western Bluebirds, Audubon's Warblers, Western Tanagers, Red Crossbills, and Gray-headed Juncos. A short walk from the parking lot near the lodge to the rim at Sunset Point (8,000 feet) also leads through this forest.

From the rim looking out over Bryce Canyon and the Paria Valley beyond, watch for Violet-green Swallows slowly sailing along and an occasional White-throated Swift zooming by in rapid flight. A surfaced road follows the rim to the southeast for about 18 miles from the checking station, with turnouts or stops for Sunset, Inspiration, and Bryce Points, Paria View, the Natural Bridge, Rainbow Point, and Yavimpa View. Rainbow Point rises to 9,105 feet; among its spruces and firs Clark's Nutcrackers are common. A dirt-road drive of about 1.0 mile extends eastward from near the checking station to Fairyland.

Numerous trails of varying lengths radiate from Sunset Point, some leading along the rim, others into Bryce Canyon. The Navajo-Comanche Loop and return is 1.5 miles; Inspiration Point, 1.0 mile; Sunrise Point, 0.5 mile; Bryce Point, via Peek-a-boo Trail, 3.75 miles; the Campbell Canyon–Fairyland Loop and return, 6 miles. None of these is particularly good for birds. More birds may be seen by wandering around the lodge and cabins or around the campground. During the summer season, short nature walks, usually starting from Sunrise Point, are conducted daily into Bryce Canyon. Birds are best seen from mid-May through early June. A heavy mantle of snow and cold weather restrict mobility in the Park during winter months.

PROVO

Utah Lake | Provo River | Jordan River | Provo Canyon

The most spectacular peak of the Wasatch Range is Mt. Timpanogos, just southwest of which lies the city of Provo. Here the Monte L. Bean Life Science Museum on the campus of Brigham Young University features many aspects of birdlife. West of Provo is **Utah Lake,** a large body of fresh water. The **Provo River** enters Utah Lake west of the city where there is a boat harbor and state park facilities. Bird finding is good in the riverside thickets, tamarisk stands, and marshes along the lakeshore. To reach this area, drive west along Center Street from 5th West Street across the viaduct and continue for about 3 miles. The outlet of Utah Lake, the **Jordan River,** flows from the north end of the Lake north into Great Salt Lake. The area along the Jordan River near its origin is good for viewing birds. To reach this area, drive west from Lehi for 4 miles along State 73 until the highway crosses the River.

US 189 passes northeastward through **Provo Canyon** with its precipitous walls. Many birds reside here in the shrubbery along the Provo River. From Wildwood on US 189, 13 miles up from Provo, the Alpine Drive (State 92) circles back of Mt. Timpanogos into American Fork Canyon, an exceptionally beautiful drive in the fall of the year. Glacier-carved 'Timpie' looms up 11,957 feet. The Sundance Ski Area and Timpanogos Cave National Monument are points of interest on this loop road. A trail leads to the summit of Timpanogos, passing through the various vegetative belts with their characteristic birds. The start of the trail is at the Brigham Young University summer camp area back (east) of Timpanogos.

RICHFIELD

Capitol Reef National Park

Departing from US 89 either at Richfield in central Utah or 8 miles north at Sigurd, State 24 extends about 71 miles to the **Capitol Reef National Park** (398 square miles), known locally as the Wayne Wonderland. This country, in the Fremont River Basin, affords a wealth of colored rock formations, gorges, vertical walls, reefs, arches, temples, buttes, and archeological items of interest. The road to the Park ascends the Fish Lake Plateau, the summit reaching 8,410 feet, and then drops down to the Fremont River through a series of small towns—Loa, Lyman, Bicknell, Torrey—to Park headquarters in a delightful setting at the base of red cliffs in a protected little valley where formerly there was a settlement called Fruita. As the name suggests, the orchards are present along with fields of alfalfa. This introduced vegetation as well as the natural habitats offers a rich area for bird finders. Canyon and Rock Wrens are common. Other birds to be expected are the Prairie Falcon, Broad-tailed Hummingbird, Say's Phoebe, Violet-green Swallow, Scrub Jay, House Wren, Yellow Warbler, House Finch, and Rufous-sided Towhee. State 24 continues beyond the Park to Hanksville through drab desert country.

SALT LAKE CITY
City Creek Canyon | Emigration Canyon | Hogle Zoo | Snyderville | Old Mill | Brighton | Little Cottonwood Canyon | Alta | Liberty Park | Great Salt Lake | Antelope Island | Farmington Bay Waterfowl Management Area

Among the advantages of Salt Lake City are the many areas for bird finding at relatively short distances from the center of the city. In Temple Square is the famous Seagull Monument. Within 0.5 mile of this and other historical points of interest is the mouth of **City Creek Canyon,** just east of the State Capitol. At State and North Temple Streets, proceed north along Canyon Road for 0.5 mile to Memory Grove. From here, walk over trails upstream past cottonwoods, grassy hillsides, and scrub oaks. The Utah Audubon Society's field trips to this area in fall and spring usually result in lists of 30 to 35 species of birds. Bohemian Waxwings frequently appear here in winter.

Several other canyons exist on the outskirts of the city on the east whose roads are open to the public. They are from north to south: Emigration, Parley's, Mill Creek, Big Cottonwood, and Little Cottonwood. At the mouth of **Emigration Canyon** along the Wasatch Drive at the east end of Sunnyside Avenue is **Hogle Zoo,** with numerous native birds in the surroundings. Across the street is the 'This Is the Place' Monument, from which one may obtain an unsurpassed view over the tree-filled city and valley and beyond to the barren alkaline desert wasteland at the south end of Great Salt Lake and farther north to the extensive marshlands on the east side of the Lake.

The University of Utah is located on the east side of the city at the base of the mountains. The first building on the right on the lower campus at the head of Second South at University Street is the Utah Museum of Natural History, which devotes some attention to birds and where literature on bird finding is available. US 40 and I 80 traverse Parley's Canyon. About 25 miles east of Salt Lake City at **Snyderville** is an open meadow region which was an important collecting station of Robert Ridgway in 1869. The area has many birds and is known to local bird finders as the 'Ridgway

Trail,' where over ninety species of birds have been seen in the area itself and 120 there and in surrounding country. The common species are the same as those listed for Midway (*see under* **Heber**).

The **Old Mill** area at the mouth of Big Cottonwood Canyon, on State 152 about 8 miles southeast of Salt Lake City beyond Holladay, supports cottonwoods along the streams and oaks, sumac, and squawbush in drier situations; it affords fair bird finding for those species that reside in the lowlands. **Brighton,** in a mountain-surrounded amphitheater at the head of Big Cottonwood Canyon, 8,730 feet, about 28 miles from the city, at the end of State 152, is an excellent area for bird finding with dense spruce-fir-aspen forests, meadows, and bodies of water close by such as Silver Lake. There are many summer cabins here, and it is an important winter skiing area. One can expect about sixty-five species of breeding birds (*see introduction to this chapter*). In winter, Mountain Chickadees, Steller's Jays, and Clark's Nutcrackers reward the bird finder.

Alta and Snowbird lie at the end of State 210 at the head of **Little Cottonwood Canyon** at a general elevation of about 8,500 feet. **Alta,** which is 17 miles east of Sandy, or about 30 miles southeast of Salt Lake City, is an old mining camp, now a summer resort and winter skiing area. Birds are less numerous than at Brighton. However, during summer the Albion Basin about 2 miles farther up beyond Alta is lush with wildflowers that are attractive to a profusion of hummingbirds in late summer. In the cliff areas of nearby Devils Castle and Sugar Loaf Mountain, Water Pipits and Black Rosy Finches nest.

Liberty Park, at 6th East and 13th South Streets, contains the Tracy Aviary. Of the many species present most are exotic, but many native birds are represented such as ducks, wading birds, and a few species of wild landbirds are attracted by the heavy foliage.

Great Salt Lake, about 15 miles west of Salt Lake City, is situated in barren, desert surroundings. From roads along the southern end of the Lake one may get glimpses, during spring and summer, of such birds as the Red-tailed, Swainson's, and Ferruginous Hawks, Golden Eagle, Northern Harrier, Prairie Falcon (occasional), Horned Lark, Northern Raven, and Brewer's Blackbird.

At times during migration (April to May and August to September) large flocks of Eared Grebes, ducks, phalaropes, and sandpipers alight on the Lake. The nesting colonies of pelicans and gulls on small islands many miles away are closed to trespass.

Another place where the visitor may reach Great Salt Lake and visit the largest island is from the town of Syracuse, 20 miles north of Salt Lake City en route to Ogden on I 15. Great Salt Lake State Park has been established on the north end of **Antelope Island** with an approach over an 8-mile causeway from the east shore along State 127. Driving along the causeway, one may see many transient shorebirds in season. Large concentrations of Wilson's and Northern Phalaropes are likely to be present in fall. On the Island the introduced Chukars frequent rocky outcrops. Other conspicuous species of birds here are the Mourning Dove, Say's Phoebe, Horned Lark, Northern Raven, Rock Wren, Sage Thrasher, Loggerhead Shrike, Brewer's Blackbird, House Finch, and Lark and Brewer's Sparrows. The Northern Mockingbird and Long-billed Curlew are occasional in spring.

Situated on the east side of Great Salt Lake near the mouth of the Jordan River about 15 miles north of Salt Lake City and west of the town of Farmington is a state-owned fresh-water marsh and popular hunting area formerly designated as the Farmington Bay Refuge but rechristened with the more appropriate title of **Farmington Bay Waterfowl Management Area** (10,652 acres). In terms of ecology and birdlife, it is similar to the better known Bear River Refuge (*see under* **Brigham City**) but is smaller. Even so it has about 20 miles of dikes with roads on top. Finding the Area from Salt Lake City is tricky. Access is from I 15, Centerville Exit, about 12 miles north of Salt Lake City. One must get onto the frontage road that parallels I 15 running north on the east side. About 1.0 mile north of the Centerville Exit on this frontage road is the Glover's Lane Overpass. Cross over I 15 and the railroad and proceed west along Glover's Lane for 1.25 miles, make a left turn (south) and drive about 0.5 mile to the Area entrance, marked by the resident manager's house, equipment sheds, and gate. The Area is closed to the general public during the nesting season (May and June) but bird finders may gain entry by requesting permission. After July 1, cars are allowed along the first 3 miles of dikes.

As at the Bear River Refuge, large concentrations of waterfowl, waterbirds, wading birds, and shorebirds congregate here during migration, but many species are resident and may be seen in great abundance close to the dikes.

SPRINGDALE

Zion National Park | Springdale | Coalpits Wash | Kolob Reservoir

A more colorful setting than Zion Canyon in **Zion National Park** (320 square miles) can scarcely be conceived, with red, pink, and white rock formations making up the Canyon walls along with conspicuous, detached, upward-projecting monoliths, all exposed by the Virgin River as it scoured out this tremendous gorge. The Canyon extends for several miles from its mouth northward into the Markagunt Plateau. Near its mouth the valley is over a mile wide. Upstream it narrows to a mere slit about 30 feet wide, and its walls rise 1,500 feet above the visitor on the Canyon floor. The walls are nearly perpendicular with sparse talus slopes at the base in the wider portions. The Virgin River, lined with cottonwoods, winds along the floor. A few small tributary canyons have minuscule streams. Seepage water from springs in places collects in pools or in wet years drips down over the rock surface, making hanging gardens consisting of such plants as mosses and ferns, false Solomon's seal, and columbines. Except for a few swampy areas, the vegetation of the Canyon floor is mostly deciduous trees along with a few kinds of shrubs. In a few cooler pockets at higher elevations, there are isolated patches of coniferous forest—ponderosa pine, Douglas-fir, and white fir. On more shaded northerly talus slopes grow stands of Gambel oak, serviceberry, maple, and ash, forming a deciduous thicket; whereas, on exposed southerly slopes there are stands of pygmy conifers. At the Canyon mouth grows typical desert-scrub vegetation. Hence a variety of environmental situations and habitats exists in a short distance, which favors good bird finding.

Zion Canyon is open to visitors the year round. At Park headquarters near the south entrance is a visitor center and museum in which attention is given to wildlife of the Park, including birds. A checklist of birds in the Park is available.

A paved highway follows along the length of the Canyon, extending for about 7 miles beyond Park headquarters to the beginning of the narrows. Occasional, short side roads lead to points of interest. Several footpaths and trails wind along the valley floor, while others ascend to the rims or to mountain tops. All are rewarding for viewing birds, but those on the valley floor through vegetated areas yield the greatest number. Short walks through the campgrounds and picnic areas and up and down the Virgin River are especially recommended. Some of the more likely prospects for seeing early-summer birds along the trails and side roadways are the following: *In the desert areas,* look for Gambel's Quail, Greater Roadrunners, Costa's Hummingbirds, Northern Mockingbirds, Blue Grosbeaks, and Black-throated Sparrows; *along streams* for Spotted Sandpipers and North American Dippers. Canyon Wrens are common at cliffs. Red-winged Blackbirds frequent the few cattail areas. In the deciduous vegetation of the valley floor are such species as the Black-chinned and Broad-tailed Hummingbirds, Western Pewee, Solitary and Warbling Vireos, Yellow Warbler, Bullock's Oriole, Black-headed Grosbeak, Lesser Goldfinch, and Chipping Sparrow. The deciduous thickets on talus slopes attract the Ash-throated Flycatcher, Black-capped Chickadee, Bewick's Wren, Blue-gray Gnatcatcher, Yellow-breasted Chat, and Rufous-sided Towhee; whereas, the pygmy conifers are attractive to the Scrub and Pinyon Jays, Bushtit, Black-throated Gray Warbler, Black-headed Grosbeak, and Chipping Sparrow.

BLACK-THROATED GRAY WARBLER

Turkey Vultures may soar overhead, while White-throated Swifts dive past the exposed rocky areas as one climbs the rim trails. Expect to see Violet-green Swallows here, too, and Peregrine Falcons occasionally.

The most popular walk is one from the end of the highway at the Temple of Sinawava to the narrows, a distance of about 1.0 mile. This is an easy, level, all-weather trail. Nature walks, taking about two hours for the round trip, are conducted along this trail by ranger naturalists during the summer months. The shortest walk is to Weeping Rock, about 0.25 mile from the parking area, over a surfaced trail. Also originating at the Weeping Rock parking area is the strenuous, 3.5 mile East Rim Trail over steep grades and switchbacks to Observation Point at 6,508 feet. Along this route is Hidden Canyon, an exceptionally fine area for bird finding. This starts about 0.6 mile along the East Rim Trail. The trail along Hidden Canyon is 1.0 mile long. The Canyon Wren is almost always present here as are the White-throated Swift, Scrub Jay, and Mountain Chickadee. Spotted Owls have been heard here. Near the Grotto Picnic Area and Zion Lodge on the Canyon floor another trail starts westward across the Virgin River by footbridge to the Emerald Pools. The lower pool is about 1.0 mile distant, the round trip taking about two hours. The upper pool is about 1.5 miles from the bridge and requires about three hours. There are a few steep grades. Painted Redstarts have been seen in April at Emerald Pools. The last trail to note is to the Canyon Overlook, an easy 0.5-mile walk from the parking area at the upper end of the mile-long Zion-Mt. Carmel tunnel. For information on longer and more strenuous trails, make inquiry at Park headquarters.

An avian peculiarity of Zion Canyon is a two-way altitudinal migration. Certain species that nest very early on the Canyon floor subsequently move to the plateaus above. This unfortunately results in a scarcity of birds in the Canyon in late summer when the tourist season is at its height. In October a reverse altitudinal migration takes place, during which the summer contingent from below as well as high-elevation breeders drop down from the plateaus to the Canyon floor to winter. Such species as Red-breasted Nuthatches, Townsend's Solitaires, Gray-headed Juncos, and White-crowned Sparrows are usually among them. Southbound transients also appear at this time. The best months for bird finding, then, in Zion Canyon are May, June, and October.

Outside the Park's southern entrance, the area around **Springdale** is propitious for bird finding, especially a pond and marsh at the south edge of town along the Virgin River. The Lesser Black Hawk has been observed in the Springdale area on several occasions in recent years in April. **Coalpits Wash** between Rockville and Virgin on State 15 is a good site for desert birds including the Ladder-backed Woodpecker. One must park the car at the bridge and walk up the draw. If the ghost town of Grafton is included in one's itinerary, the dirt road from Rockville south across the bridge and thence west on the south side of the Virgin River, passes along dry desert stretches, limited farmland, and orchards, thus affording the opportunity of seeing distinctive species of birds, among them the Gambel's Quail, Costa's Hummingbird, Vermilion Flycatcher, Phainopepla, Summer Tanager, and Blue Grosbeak. A relatively untraveled dirt road proceeds north from Virgin about 25 miles to the **Kolob Reservoir.** From the valley floor, the route is through junipers and pinyon pines where one may expect representatives of the pygmy-forest avifauna including the Scrub Jay, Plain Titmouse, and Bewick's Wren. At higher elevations come scattered stands of ponderosa pine that are frequented by Band-tailed Pigeons, White-breasted and Pygmy Nuthatches, Grace's Warblers, and Western Tanagers. Still higher, in the spruce-fir-aspen forest, are such birds as the Blue Grouse, Yellow-bellied Sapsucker, Olive-sided Flycatcher, Steller's Jay, Mountain Chickadee, Audubon's Warbler, and Gray-headed Junco. Peregrine Falcons have been seen around the Reservoir. A side road leads east to Lava Point which offers a commanding view. The Kolob Reservoir Road continues another 21 miles north to join State 14 leading to Cedar Breaks National Monument (*see under* **Cedar City**). It is a primitive road.

Reference

Birds of Zion National Park and Vicinity. By Roland H. Wauer and Dennis L. Carter. Zion Natural History Association, Springdale, Utah. 1965.

ST. GEORGE

Santa Clara Creek Bottomlands | Bloomington | Pine Valley | Santa Clara | Snow Canyon State Park | Beaver Dam Wash

The area along the Virgin River in southwestern Utah, particularly the St. George region, is known as Utah's Dixie, because the climate is semitropical and in pioneer days cotton was grown. For the many kinds of Mojave Desert birds to be found here, see the introductory portion of this chapter. Many of them are found in Utah only in this portion of the state. Sites in town for possibly seeing certain members of this contingent are the Mormon Temple grounds, the campus of Dixie College, the grounds of Brigham Young's winter home, and the Red Hills Golf Course. In the southern portion of the latter is Watercress Springs (also called City Springs), where a canopy of willows attracts Traill's Flycatchers and transient warblers. On the southeast outskirts of St. George, down 15th East Street near the old steel bridge across the Virgin River, some of the River's bottomlands with marsh areas and ponds are easily accessible. Common Gallinules have been found here. A loop road south of the River heads east to the Washington fields, south of the town of Washington, 5 miles east of St. George. There is good bird finding en route, especially in brushy areas.

About 3 miles south of St. George, along East Tabernacle Street, lies the junction of Santa Clara Creek and the Virgin River. Along the banks of the Virgin River are thickets of tamarisk and willows. Westward from here along the **Santa Clara Creek Bottomlands** there are areas where a dense tangle of brush provides cover for many species such as the Gambel's Quail, Common Screech and Great Horned Owls, Red-shafted Flicker, Western Kingbird, Black Phoebe, Traill's and Vermilion Flycatchers, Rough-winged Swallow, Bell's Vireo, Lucy's and Yellow Warblers, Common Yellowthroat, Yellow-breasted Chat, Yellow-headed and Red-winged Blackbirds, Bullock's Oriole, Black-headed and Blue Grosbeaks, Lesser Goldfinch, Abert's Towhee, and Song Sparrow.

Still farther south, beyond a cut through which the Virgin River flows and I 15 passes, is the relatively new residential area of **Bloomington.** Bird finding is productive in the River bottomlands, along the margins of the golf course, or in adjacent fields. During migration, shorebirds and waterbirds alight along the Virgin River. In the winter months there is a concentration of Ruby-crowned Kinglets, Oregon Juncos, and White-crowned and Song Sparrows in the riverbottoms, with Loggerhead Shrikes, Water Pipits, and various species of broad-winged hawks in the more open fields.

Worthwhile is a side trip north from St. George along State 18, through Diamond Valley with its black craters contrasting with the red-rock formations, to Central, then east (right) to **Pine Valley,** and east 3 more miles to the U.S. Forest Service Campground. The fields, meadows, willow-lined streams, cottonwood groves, and ponderosa pine forests support a wide variety of birds associated with these habitats. Good hiking trails lead into the Pine Valley Mountains up through spruce-fir forests to high mountain meadows around 9,500 feet.

About 5 miles west of St. George on the Old US 91, one comes to the community of **Santa Clara** in a fruit-producing area. In between the two cities is a stretch of warm, southern desert with some dry washes where Greater Roadrunners reside. A conspicuous geological formation is a black lava flow, which in prehistoric times emerged from Snow Canyon and spread out over the red sand. On the eastern outskirts of Santa Clara a road leads south across the lava flow to the Santa Clara Creek Bottomlands with farmlands and cottonwood stands that may still serve as nesting sites for Green Herons and roosts for Turkey Vultures. The road that meanders along the bottomlands connects with one leading to Bloomington. Near the western margin of Santa Clara, close to the Jacob Hamlin House (early church leader and Indian scout of the region), is a nearby wooded and brushy portion of the Santa Clara Creek Bottomlands which affords good bird finding. The Summer Tanager was first found in the state here.

In another five miles to the west near the hamlet of Ivins is **Snow Canyon State Park** (5,688 acres) with its red cliffs, red sand, and contrasting black lava. Brushy areas of Fremont barberry are conspicuous. Many desert species of birds reside in the campground area. The road up Snow Canyon reaches the rim and here joins State 18 by which one can either continue on to Veyo, and thence to Pine Valley, or return to St. George on State 18.

For the venturesome bird finder with unlimited time and preferably in all-terrain vehicle, the **Beaver Dam Wash** in extreme southwestern Utah and extreme northern Arizona offers a challenging search for Mojave Desert species. This is remote, hot, isolated country with barrel and cholla cacti, creosote bushes, a Joshua tree forest on the west slope of the Beaver Dam Mountains, deep arroyos, and some mesquite on the floor of the Wash as well as the usual groves of cottonwoods along the stream course. The former sole access road was via US 91 west of Santa Clara over the Beaver Dam Mountains. Although one can still go this way from Santa Clara, a longer but perhaps easier way is to take I`15 out from St. George through the spectacular Virgin River Gap, exit at Littlefield, Arizona, then turn north along the old highway up the slope, proceeding about 10 miles to the Eardley Ranch turnoff on the left (west) from Castle Cliff. This is about 0.5 mile beyond (north) the road to the Joshua tree forest on the right (east). Continuing west, the unimproved road winds and loops 12 miles down and up deep arroyos and across desert slopes to the Beaver Dam Wash. Bird possibilities en route are Red-tailed Hawks, American Kestrels, Verdins, Scott's Orioles, Loggerhead Shrikes, and House Finches. Dropping down into the Wash the road reaches first the old Terry Ranch and then upstream the Lytle (Eardley) Ranch, both green oases with fields of alfalfa, orchards, and garden patches. In spring and early summer before it becomes oppressively hot, one has a better chance of finding many representatives of the Mojave Desert complement of birds here than in the St. George-Santa Clara areas.

VERNAL

Merkley Park | Uinta Mountains | Stewart Lake Waterfowl Management Area | Ashley Sloughs | Dinosaur National Monument | Ouray National Wildlife Refuge

This is the principal city of the Uinta Basin in northeastern Utah. The Natural History State Museum, on the north side of US 40, just east of the business district, features exhibits on the natural resources of the region with many ornithological items. Ten miles northwest of Vernal at the junction of Dry Fork and Ashley Creek, north of Maeser, is **Merkley Park** where there is a large grove of cottonwoods, wet meadows, willow patches, and pasturelands. In the general vicinity, look for Soras, Broad-tailed Hummingbirds, Belted Kingfishers, Black-capped Chickadees, House Wrens, Gray Catbirds, Mountain Bluebirds, Solitary and Warbling Vireos, Virginia's and MacGillivray's Warblers, American Redstarts, Rufous-sided Towhees, and Song Sparrows.

State 44 leads into the eastern section of the **Uinta Mountains** en route to Manila near the Wyoming border. The highway climbs abruptly through pygmy forest to an extensive stand of ponderosa pine and finally to a dense forest of lodgepole pine. Sites for finding such birds as the Williamson's Sapsucker, White-breasted Nuthatch, Townsend's Solitaire, Western Tanager, Cassin's Finch, and Pine Siskin are at Green Lake, in Red Canyon and near Flaming Gorge Reservoir. Ospreys nest along the Reservoir and Bald Eagles concentrate around the Reservoir in winter.

US 40 crosses the Green River at Jensen, 13 miles east of Vernal. Many waterfowl and shorebirds migrate along the River in spring and fall. A road leading south from Jensen extends 2 miles to the **Stewart Lake Waterfowl Management Area** and the **Ashley Sloughs,** where Ashley Creek approaches the River. At the Lake, its surrounding marsh and adjacent river bottomlands, one should see Sharp-shinned Hawks, American Coots, Red-shafted Flickers, Western Kingbirds, House, Bewick's, and Marsh Wrens, Sage Thrashers, Yellow Warblers, Yellow-breasted Chats, Western Meadowlarks, Yellow-headed and Red-winged Blackbirds, Bullock's Orioles, and Savannah, Lark, and Brewer's Sparrows. In fall migration, hundreds of Sandhill Cranes stop off here before proceeding southeast into Colorado.

From Jensen, State 149 leads north 7 miles to that portion of the **Dinosaur National Monument** (211,053 acres) in Utah. The Museum and exhibit area here is almost entirely devoted to the dinosaur quarry and geology. Though this is barren country with greasewood and a few scattered junipers, one may see en route to the museum and thence west to Split Mountain and the campground, such birds as Mourning Doves, Say's Phoebes, Pinyon Jays, Bewick's Wrens, Sage Thrashers, Loggerhead Shrikes, Black-throated Gray Warblers, House Finches, Green-tailed Towhees, and Lark and Brewer's Sparrows.

The **Ouray National Wildlife Refuge** (14,000 acres) is situated about 25 miles southwest of Vernal along the Green River near Ouray. Access is via State 209 and 88 leading south from US 40 about 15 miles west of Vernal. Near the junction of State 209 and 88 is Pelican Lake where many waterbirds are present, both residents and transients. At the Ouray Refuge, in addition to the usual complement of waterfowl, waterbirds, wading birds, and shorebirds, Sandhill Cranes are common in migration. Recently in summer, as many as two Whooping Cranes have been present here, hatched at Grays Lake in Idaho (*see under* **Soda Springs, Idaho**). Just south of Ouray there are two tributaries to the Green River, the White River from the east and the Duchesne River from the west. One place where Lewis' Woodpeckers are still common is the extensive cottonwood forest in this vast river bottomland area, but in spring and summer mosquitoes constitute a deterrent to bird finding.

Washington

THIN-BILLED MURRES

Probably no other state shows greater contrasts of environment than Washington. Elevations range from sea level to peaks well above 10,000 feet. Average precipitation ranges from extremes of 142 inches a year—the heaviest in the conterminous United States—in one part of the state to 6 inches in another. There are some places so densely forested as to appear tropical and others so arid in character as to be desert-like. Birdlife—for example, Thin-billed Murres, White-tailed Ptarmigan, Bushtits, Hermit Thrushes, and Sage Sparrows—responding to these extremes or the intergrading conditions that lie between them, is equally diversified.

Running through the interior of Washington from the northern to the southern boundary are the Cascade Mountains. About 100 miles wide toward the northern and southern boundaries and, like an hourglass, narrowed between to a width of 60 miles, they divide the state into western and eastern sections, the latter being slightly the larger. Western Washington, favored by prevailing moist sea winds, has a high humidity and moderate temperatures at all seasons. Cloudy weather is common. But eastern Washington, lying in the rain shadow of the Cascades, is comparatively dry and subject to extremes of heat and cold. Strong winds and clear skies are usual.

In general, the peaks and ridges of the Cascades, though exceedingly abrupt and rugged, form a relatively uniform skyline with an average level of 8,000 feet in the north, gradually descending to 5,000 feet in the south. This typical evenness is broken, however, by five lofty volcanic cones, one of which—Mt. Rainier—reaches 14,410 feet, the highest point in the state. On the eastern side of the Cascades are several long spur ranges extending in a southeastwardly direction parallel to deep, steep-sided valleys.

In western Washington, the Cascades slope down to a long, broad valley called the Puget Trough, which is continuous southward with the Willamette Valley in Oregon. The northern part of the Trough is occupied by Puget Sound; the southern part, although above sea level, rarely exceeds 500 feet elevation, except in a few hilly areas. West of Puget Sound is the Olympic Peninsula, with the Olympic Mountains rising from its central portion. Many of the Olympic ridges reach heights between 4,500 and 5,000 feet, and a few peaks extend well above 7,000. In their sharpness and steepness, the Olympics strongly resemble the Cascades. South of the Olympic Peninsula and west of the Puget Trough are the rough Willapa Hills, a northward extension of the Coast Range in Oregon and generally under 3,000 feet.

The higher peaks and ridges of both the Cascade and Olympic Mountains have alpine regions characterized by glaciers, permanent snowfields, and a vegetation of hardy shrubs and herbaceous plants. A few such regions in the Cascades support nesting pairs of White-tailed Ptarmigan, Horned Larks, Water Pipits, and Gray-crowned Rosy Finches; in like places in the Olympic Mountains, Horned Larks and rosy finches breed. Below the alpine regions exists a vertical succession of tree associations, or belts. (For a description of typical succession on the moist western slopes of the

Cascades and the slopes of the Olympic Mountains, *see under* **Tacoma** *and* **Port Angeles;** for a description of succession on the drier, eastern slopes of the Cascades, *see under* **Okanogan.**)

A peculiar feature of the vegetation on the lowlands of western Washington, which enjoys a cool, humid climate, even in summer, is the presence of coniferous forests, extending almost to the edge of the sea. Before settlement and lumbering operations, Douglas-fir, western hemlock, and western red cedar covered vast areas. Deciduous trees and shrubs grew mainly in ravines and valleys near streams, rarely on higher ground. Areas of native grasses—i.e., prairies—were of minor extent. Along the coast where the rainfall is heaviest—e.g., on the west side of the Olympic Mountains—both coniferous and deciduous growth attained remarkable density and luxuriance. While excellent samples of these original conditions may still be seen in Olympic National Park (*see under* **Port Angeles**) and offer an unusual opportunity for bird finding, most of the lowland forests of western Washington have since been replaced by second-growth woods, or by agricultural lands.

In the coniferous forests of the Cascade and Olympic Mountains and the lowlands of western Washington, the bird species listed below breed regularly unless otherwise indicated. Species marked by an asterisk reside mainly at elevations above 3,000 feet.

* Northern Goshawk
Sharp-shinned Hawk
Cooper's Hawk
Blue Grouse
* Spruce Grouse (*east slopes of Cascades*)
Band-tailed Pigeon (*west side of Cascades*)
Northern Pygmy Owl
* Saw-whet Owl
Pileated Woodpecker
Williamson's Sapsucker (*east slopes of Cascades*)
White-headed Woodpecker (*east slopes of Cascades*)
Hammond's Flycatcher
Dusky Flycatcher (*east slopes of Cascades*)
Olive-sided Flycatcher
* Gray Jay
Steller's Jay
Northern Raven
* Clark's Nutcracker (*mainly east slopes of Cascades*)
Mountain Chickadee (*mainly east slopes of Cascades*)
Chestnut-backed Chickadee (*west side of Cascades*)
White-breasted Nuthatch (*mainly eastern Washington*)
Red-breasted Nuthatch

Pygmy Nuthatch (*east slopes of Cascades*)
Brown Creeper
Winter Wren
Varied Thrush
* Hermit Thrush
Swainson's Thrush
* Townsend's Solitaire
Golden-crowned Kinglet
* Ruby-crowned Kinglet (*east slopes of Cascades*)
Solitary Vireo
Audubon's Warbler
Black-throated Gray Warbler (*mainly western Washington*)
* Townsend's Warbler
Wilson's Warbler
Western Tanager
Evening Grosbeak
Purple Finch (*west side of Cascades*)
Cassin's Finch (*east slopes of Cascades*)
Pine Siskin
Red Crossbill
Oregon Junco
White-crowned Sparrow
* Fox Sparrow (*mainly east slopes of Cascades*)
* Lincoln's Sparrow (*east slopes of Cascades*)

The Washington coast has a varied and highly scenic topography. Fronting on the Pacific in the southwest are many long, straight beaches, backed by sand dunes; reaching inland from them in several places are large bays, bordered frequently by mud flats and occasionally by salt marshes. In the northwest are high, rocky bluffs and headlands, and, seldom far offshore, numerous giant rocks and ledges boldly projecting above tide level. Cape Flattery, the northwestern tip of the Olympic Peninsula, is notable for its rugged and generally bleak appearance.

Undoubtedly the outstanding feature of the Washington coast is its great indentation, Puget Sound, connected to the sea by straits or channels that separate the state from Vancouver Island. In reality a deep, drowned valley that extends southward into the state for a hundred miles, Puget Sound is exceedingly irregular, its many bays, coves, and 'canals' forming a shoreline of 1,750 miles. Although most such inlets are flanked by precipitous banks, often of impressive height, a few, having become partly filled by silt, show mud bars and marshes along their edges. Occasionally an inlet has been cut off by sediments and forms a body of water such as Lake Washington on the east side of Seattle.

The bluffs, headlands, small islands, and rocks of the outer

Washington coast are as attractive ornithologically as they are scenically, for many of them provide the isolation that most sea-associated birds prefer when nesting. Between mid-May and mid-July, the following species breed regularly in such areas:

Fork-tailed Storm Petrel
Leach's Storm Petrel
Double-crested Cormorant
Brandt's Cormorant
Pelagic Cormorant
Black Oystercatcher
Glaucous-winged Gull (*northern coast; also Puget Sound*)
Western Gull (*mainly southwestern coast*)
Thin-billed Murre
Pigeon Guillemot
Cassin's Auklet
Rhinoceros Auklet
Tufted Puffin

For a full appreciation of the birdlife of Washington's outer coast, the bird finder should take a pelagic boat trip between late April and early October. In addition to many of the aforementioned birds that nest along the outer coast, he will see other species on the open sea that breed elsewhere. Only a few, such as Sooty Shearwaters, come close enough to shore to be viewed from points or jetties; most stay far from land. Some of the birds to expect on a trip are the Black-footed Albatross and Northern Fulmar in any season; Sooty and Pink-footed Shearwaters in spring, summer, and fall; Red and Northern Phalaropes, Pomarine and Parasitic Jaegers, Sabine's Gull, and Arctic Tern in spring and fall; and Buller's Shearwater in fall. Information about agencies that schedule pelagic trips may be obtained from the Seattle Audubon Society, 714 Joshua Green Building, Fourth Avenue and Pike Street, Seattle, WA 98101.

The northern third of eastern Washington—east of the Cascades and the Okanogan River Valley; north of the Columbia River Valley (where its course is east to west) and the Spokane River—comprises the Okanogan Highlands. A western extension of the northern Rocky Mountains in Montana and Idaho, with many elevations reaching above 6,000 feet, the Highlands consist mainly of several small ranges and other uplands trending north and south, paralleled by intervening valleys. There are no alpine regions. As on the east slopes of the adjacent Cascades (*see under* **Okanogan**), the climate here is somewhat arid, a fact that has marked effect on the vegetation. Such conifers as Douglas-fir, western larch, and lodgepole pine are replaced, except at higher elevations, by park-like stands of ponderosa pine, and occasionally by grasslands and sagelands. Breeding birds of the coniferous woods include the Williamson's Sapsucker, Steller's Jay, Pygmy Nuthatch, Audubon's Warbler, Western Tanager, Cassin's Finch, and many other species found also on the east slopes of the Cascades.

The extreme southeastern corner of Washington is occupied by the Blue Mountains, the northern fringe of the range with the same name in Oregon (*see introduction to* **Oregon** *chapter*). In Washington they attain elevations between 5,000 and 6,500 feet. On the north and west slopes and the tops of many ridges are heavy stands of Douglas-fir, grand fir, Engelmann spruce, lodgepole pine, and western larch; on the highest slopes, subalpine fir is the most common. Usually the south and east slopes are either barren or have open forests of ponderosa pine. Throughout the Blue Mountains are many deep canyons well wooded with conifers down to their bottoms, where there are cottonwoods and willows. The birdlife of the Blue Mountains (*see under* **Walla Walla**) is very much like that of the Cascades.

The remainder of eastern Washington is a semiarid plateau, most of which is below 2,000 feet in elevation and is underlain by laval rock. North of the Snake River, which crosses the southeastern part of the state to join the Columbia River, the plateau is known as the Palouse Country. Here the surface, rolling in broad, wave-like swells, contains fine-textured soils used extensively for wheat production. From the Palouse Country north and west to the Columbia, the plateau is desert-like and, at least at one point, receives an average precipitation of 9 inches or less a year. Conditions of surface and vegetation show considerable variation. Certain parts have extensive dunes, owing to shifting sands; others have a peculiar type of erosion topography called channeled scablands (*see under* **Moses Lake** *and* **Spokane**); and still others have a gently rolling surface suggesting the Great Plains east of the Rocky Mountains. In the southwest, between the Columbia and the southern Cascades, the plateau has southeastward-trending ridges, 1,000 to 3,000 feet in height, drained for the most part by the Yakima River.

Although much of the plateau has been modified for agricultural purposes, there remain a few undisturbed areas where vegetation remains more or less of the desert type. Sagebrush and bunch grass constitute the predominant growth, with admixtures of greasewood, foxtail, cheat grass, and cacti. Pines stand on ridges, cottonwoods and willows along streams, but otherwise trees are scarce. Among the birds typical of the desert areas are the following:

Swainson's Hawk (*uncommon*)
Sage Grouse (*uncommon*)
Long-billed Curlew (*uncommon*)
Burrowing Owl (*uncommon*)
Poor-will
Horned Lark
Sage Thrasher
Savannah Sparrow
Grasshopper Sparrow
Vesper Sparrow
Lark Sparrow
Sage Sparrow
Brewer's Sparrow

In widely scattered localities of eastern Washington are lakes, ponds, sloughs, potholes, and marshes, in many cases created or modified by artificial means, which attract surprisingly large nesting populations of waterbirds, waterfowl, and marsh-loving birds. The fact that these watery environments and their associated birdlife exist in an otherwise arid region contributes in no small degree to their interest.

As in western Washington, many natural areas east of the Cascades have been converted into agricultural lands. There are consequently many farmlands (fields, meadows, and pastures) where once there were sagelands or grasslands. Deciduous trees, formerly prevalent only along watercourses, now stand in dooryards and parks and comprise orchards. At the present time, throughout both western and eastern Washington (unless otherwise indicated), birds that breed regularly in open country (farmlands, brushy areas, orchards, parks, and dooryards) and deciduous woods are listed below. Species marked by an asterisk breed also in mountains at elevations above 3,000 feet.

OPEN COUNTRY

Turkey Vulture
Northern Harrier (*mainly eastern Washington*)
California Quail (*local*)
Mourning Dove
* Eastern Kingbird (*mainly eastern Washington*)
Western Kingbird (*eastern Washington*)
Say's Phoebe
* Barn Swallow
Black-billed Magpie (*eastern Washington*)
Bushtit (*mainly western Washington*)
House Wren (*mainly eastern Washington*)
Bewick's Wren
Gray Catbird (*eastern Washington*)
Western Bluebird (*mainly eastern Washington*)
* Mountain Bluebird (*mainly eastern Washington*)
Loggerhead Shrike (*eastern Washington*)
* Orange-crowned Warbler
* Yellow Warbler
* MacGillivray's Warbler
* Common Yellowthroat
Yellow-breasted Chat (*eastern Washington*)
Western Meadowlark
Lazuli Bunting (*mainly eastern Washington*)
American Goldfinch
Rufous-sided Towhee
* Chipping Sparrow
Savannah Sparrow
Vesper Sparrow (*mainly eastern Washington*)
* Song Sparrow

DECIDUOUS WOODS

Ruffed Grouse
Common Screech Owl
* Red-shafted Flicker
Yellow-bellied Sapsucker
* Hairy Woodpecker
Downy Woodpecker
* Western Flycatcher
* Western Pewee
Black-capped Chickadee
White-breasted Nuthatch (*mainly eastern Washington*)
Red-eyed Vireo
* Warbling Vireo
* Nashville Warbler (*mainly eastern Washington*)

American Redstart (*eastern Washington*)
Bullock's Oriole
Black-headed Grosbeak

Bird transients through Washington, unlike transients through the states in the same latitudes on the Atlantic coast, include very few landbird species nesting exclusively in regions to the north. Migration is notable mainly in the movements of waterfowl, waterbirds, and shorebirds. A great many geese and ducks stop off on lakes and other water areas in eastern Washington as they do west of the Cascades, in Puget Sound, Willapa Bay, and similarly sheltered coastal indentations. At ocean-front spots such as Point Chehalis (*see under* **Westport**), Sooty Shearwaters are often seen passing by in immense numbers. Transient Parasitic Jaegers and Common Terns are common to abundant in Puget Sound. Shorebirds including Black-bellied Plovers, Surfbirds, Black Turnstones, Whimbrels, and Wandering Tattlers frequent many of the beaches for feeding and resting. In Washington, the main migratory flights may be expected with the following dates:

Waterfowl: 20 March–25 April; 15 September–15 November
Shorebirds: 20 April–1 June; 1 August–15 September
Landbirds: 15 April–1 June; 15 August–1 October

Eastern Washington is within the winter ranges of a few landbird species such as the Rough-legged Hawk, Bohemian Waxwing, Common Redpoll, American Tree Sparrow, and Snow Bunting. Smaller numbers of these same species also appear in western Washington. The water areas in eastern Washington—e.g., those in the National Wildlife Refuges—hold moderately large populations of geese and ducks. Western Washington, which has relatively mild winters, can be rewarding to winter bird finders, particularly along the coast and in Puget Sound. Here are opportunities to see loons—Common, Arctic, and Red-throated; grebes—Red-necked, Horned, Eared, and Western; over twenty species of ducks; gulls—Glaucous-winged, Western, Herring, Thayer's, Mew, and Bonaparte's; a few shorebirds and alcids; and a variety of landbirds including Band-tailed Pigeons, Chestnut-backed Chickadees, Bewick's Wrens, Varied Thrushes, Cedar Waxwings, Hutton's Vireos, Purple Finches, and Song Sparrows.

Authorities

Charles M. Drabek, Glen and Wanda Hoge, Harry B. Nehls, Dennis R. Paulson, R. Dudley Ross, Peter Stettenheim, Terence R. Wahl, Wayne C. Weber.

References

Birds of Washington State. By Stanley G. Jewett and others. Seattle: University of Washington Press. 1953.

A Checklist of the Birds of Washington State with Recent Changes Annotated. By Philip W. Mattocks, Eugene S. Hunn, and Terence R. Wahl. In *Western Birds*, vol. 7 (1976), no. 2. This is largely an updating of the classic work cited above.

Washington Birds: Their Location and Identification. By Earl J. Larrison and Klaus G. Sonnenberg. Seattle Audubon Society. 1968.

Birds of Southeastern Washington. By John W. Weber and Earl J. Larrison. Moscow: University of Idaho Press, 1977.

A Guide to Bird Finding in Washington. By Terence R. Wahl and Dennis R. Paulson. Published by T. R. Wahl, 3041 Eldridge, Bellingham, WA 98225. 1977.

ANACORTES
San Juan Islands | March Point | Deception Pass State Park

Anyone who enjoys looking for birds on islands will find to his liking the rocky **San Juan Islands** in the straits between the northwestern Washington mainland and Vancouver Island. A few of the larger islands are accessible by ferry, but most must be reached by chartered boat.

About 170 of the several hundred San Juan Islands are large enough to be habitable; the others, which may be called simply rocks, either are tide-washed or have high and precipitous walls, with little level surface. Orcas Island (56 square miles) and San Juan Island (55 square miles) are the largest. Since the San Juan Islands represent a partly submerged, severely glaciated mountain range, all show a rough topography. Their shorelines, consistently irregular, are cut by many long, narrow, steep-sided inlets; their interiors have numerous low mountains or hills and intermittent valleys with basins containing lakes, marshes, and bogs. Mt. Constitution on Orcas Island attains the highest elevation (2,454 feet), and there are several other mountains on Orcas Island and others that reach above 1,000 feet.

All the larger San Juan Islands support coniferous forests, though these are much less extensive than they were formerly, owing to the development of agriculture. Douglas-fir is the principal tree, with an admixture of western red cedar, western hemlock, and grand fir. In certain areas, Pacific madrone, Oregon white oak, and bigleaf maple are a conspicuous deciduous growth. Since most of the rainfall is in winter, parts of the Islands are surprisingly arid in summer—arid enough to support one species of cactus, *Opuntia fragilis*.

A suggested route to the San Juan Islands via Anacortes, with opportunities for bird finding along the way, is suggested below. Ample time should be allowed for transportation by ferry to the larger islands and short trips by chartered boat to several of the smaller islands and rocks.

From downtown Seattle, drive north on I 5 for 60 miles to Mount Vernon; here take the Anacortes Exit and proceed west on State 536. After about 5 miles, when State 536 joins State 20 going west, continue west on State 20. Five miles west from this junction, and just beyond a high bridge over the Swinomish ship channel, turn off right (north) at a sign to **March Point.** This peninsula, easily recognized by oil refineries near its tip, is excellent for birds from late summer through May.

Immediately after turning off to March Point, park the car and scan the mud flats to the east. Hundreds of Dunlins congregate on them in winter, as do surface-feeding ducks, mainly Common Pintails. During April, May, August, and September, at low tide, transient shorebirds—particularly Semipalmated Plovers, Greater Yellowlegs, Short-billed and Long-billed Dowitchers, and Least and Western Sandpipers—are abundant. Transient warblers, sparrows, and other landbirds often throng in the woods and brush along the road.

Continue around the shoreline of March Point, keeping right at all intersections. On reaching the Point's tip, 3.5 miles from the turnoff where the road bends sharply left, park near the Shell refinery. In Padilla Bay to the east and north, thousands of Black Brant pass the winter, many of them often coming close to shore. Other wintering birds to look for include Common and Red-throated Loons, Red-necked, Horned, and Western Grebes, Double-crested Cormorants, Greater Scaups, Common Goldeneyes, Buffleheads, Oldsquaws, and White-winged and Surf Scoters. Sanderlings—and sometimes Black Turnstones—frequent the pebbly beaches.

From the tip of the Point, continue around the west side of the Point for 3.3 miles, then turn right (west) onto State 20. In 0.5 mile a spur of State 20 branches off right to Anacortes and the San Juan Islands ferry terminal, 8 miles distant. To visit **Deception Pass State Park** before going on to the San Juan Islands, keep left on State 20.

A beautiful area of 2,292 acres, Deception Pass State Park occupies the south end of Fidalgo Island and the north end of Whidbey Island. Deception Pass itself, a narrow and often turbulent channel between the Islands, is spanned by a high bridge that offers a splendid view. The Park, with its coniferous forest, largely of virgin Douglas-fir, and its lakes, beaches, and rocky shores, is ornithologically rewarding in any season.

Several attractive hiking trails lead through the coniferous forest in which, among the regularly breeding birds, are Band-tailed Pigeons, Western and Olive-sided Flycatchers, Chestnut-backed Chickadees, Red-breasted Nuthatches, Brown Creepers, Swainson's Thrushes, Golden-crowned Kinglets, Audubon's and Townsend's Warblers, Red Crossbills, and Oregon Juncos.

For viewing waterbirds and waterfowl, Rosario Beach on Fidalgo Island and the Cranberry Lake-West Beach Area on Whidbey Island provide the best opportunities.

Proceed south on State 20 for 5.2 miles from the Anacortes turn-off, passing two small lakes, Campbell and Pass; immediately beyond the second one, turn right (west) onto a paved road with a sign to the Rosario Beach Area. Follow this road for 0.8 mile, then turn left onto Rosario Beach Road (poorly marked, hence easily missed), which drops down a steep hill and comes shortly to the Rosario Beach picnic area. Park here and walk south and west out onto a small peninsula where a high bluff on its west side overlooks the Strait of Juan de Fuca. Around it in winter nearly all the waterbirds and waterfowl mentioned above for March Point are present, plus others which prefer more exposed shorelines—e.g., Brandt's and Pelagic Cormorants, Harlequin Ducks, Thin-billed Murres, Pigeon Guillemots, and Marbled Murrelets. Occasionally present are Eared Grebes, Barrow's Goldeneyes, Black Oystercatchers, and Rhinoceros Auklets.

Backtracking to State 20 at Pass Lake, turn right (south), cross the Deception Pass Bridge, and—0.8 mile from the Rosario Beach turnoff—turn off right (west) again, following the signs to Cranberry Lake and West Beach. Cranberry Lake, a fresh-water lagoon, frequently has Ring-necked Ducks, Common Goldeneyes, and Hooded and Common Mergansers in winter. The road ends in a parking lot back from West Beach at the west entrance to Deception Pass. When feeding conditions are good, the waters off West Beach are likely to teem with flocks of Arctic or Red-throated Loons and Brandt's Cormorants in winter, Heermann's Gulls in fall. Transient Common Terns and Parasitic Jaegers are frequent in early fall.

In Anacortes the bird finder may take the car ferry to several of the larger San Juan Islands, notably Orcas, Lopez, and San Juan. (The passage on the ferry can be rewarding for viewing sea birds, many of which nest on several of the smaller, uninhabited San Juan Islands and groups of offshore rocks.) On one of the islands the bird finder may hire a powerboat and pilot for viewing closely some of the offshore rocks that are occupied only by sea birds. Friday Harbor on San Juan Island is recommended as a likely place to hire both skipper and suitable craft. The bird finder must be forewarned: Since all the uninhabited islands and offshore rocks constitute the San Juan Islands National Wildlife Refuge, landing on them is prohibited unless permission is obtained in advance by writing the manager at the Willapa National Wildlife Refuge (Ilwaco, WA 98624), which manages the San Juan Islands Refuge.

Suggested below are two trips from Friday Harbor for late spring and early summer. Each trip requires about two hours each way by powerboat.

1. The first trip goes northward from Friday Harbor through the San Juan and President Channels to Bare and Skipjack Islands. On

Low Island, along the way, is a small colony of Glaucous-winged Gulls. Look for Harlequin Ducks in the kelp beds off the Island's north side, and Pelagic Cormorants and Black Turnstones resting on the rocks at the water's edge. On passing Jones Island, while continuing north, scan the treetops along the east side for an aerie of the Bald Eagle. Flattop is the next island worthy of attention. Here Glaucous-winged Gulls nest on the slopes of the south side and Pigeon Guillemots on the cliffs along the north side. This is a good island on which to look for Wandering Tattlers.

On entering President Channel, inspect the high cliffs on the east side of Point Disney, Waldron Island, for colonies of cormorants and guillemots. After leaving President Channel and proceeding north, first comes Bare Island, a turf-covered hump of rock, which is practically alive with birds. The most abundant breeding birds here are Pelagic Cormorants and Glaucous-winged Gulls, but there are also a few pairs of guillemots and Tufted Puffins. Brandt's Cormorants, Black Oystercatchers, and Black Turnstones can probably be seen on shore. Upon reaching Skyjack, lying west of Bare Island, look for a few Tufted Puffins, which have nested in burrows on a bluff back of the beach on the island's south side.

2. The second trip goes southward from Friday Harbor through San Juan Channel to the south end of Lopez Island. When passing its southeastern extremity, note Goose Island where a few pairs of Black Oystercatchers may nest. Nearly all the coves and small islands at the south end of Lopez Island are excellent for cormorants, ducks, shorebirds, gulls, and alcids. This is largely due to the area being directly exposed to the strong, food-carrying currents in the Strait of Juan de Fuca.

Probably the island that is most attractive to birds is Colville, a long, somewhat oval piece of land off the southeast corner of Lopez. Occupying it are several hundred nesting pairs of Glaucous-winged Gulls and large numbers of Pigeon Guillemots. In addition, Pelagic Cormorants nest on the cliffs and ledges, and Brandt's Cormorants have their nests on a slope along the south part of the Island. Double-crested Cormorants may be observed in the vicinity.

BELLINGHAM
Mt. Baker Area

In extreme northwestern Washington, the **Mt. Baker Area** awaits the bird finder who enjoys exploring high country. State 542 west from Bellingham proceeds up the western Cascades and the lower spurs of Mt. Baker, a massive, snowy volcanic cone rising 10,778 feet. Since the whole area traversed has many sites for camping, hiking, fishing, and skiing, numerous good roads and well-marked trails lead off the main highway. The highway ends at Artist Point on Kulshan Ridge (elevation 5,000 feet), which extends between Mt. Shuksan (elevation 9,127 feet) to the southeast and Mt. Baker to the southwest. The views during this climb and at Artist Point are among the finest in the northern Cascades.

This 58-mile drive on State 542 east from Bellingham winds past hills and farmlands and, before reaching Deming, begins following up the Nooksack River. Beyond Glacier, as the elevation increases, there are heavy forests containing mangificent stands of western red cedars, Douglas-firs, and western hemlocks. Near Austin Pass, however, the forests begin to thin out. The Heather Meadows Area, which is just below Kulshan Ridge and forms its southern wall, is a superb example of alpine meadow with its profusion of hardy wildflowers. The chief tree growth consists of scattered groups of mountain hemlocks. Some of the birds to look for in the immediate vicinity are Blue Grouse, Olive-sided Flycatchers, Townsend's Solitaires, and Fox and Lincoln's Sparrows.

From Artist Point a trail leads westward along the side of Table Mountain and then branches. Take the left branch which descends to Ptarmigan Ridge and follows it for several miles on the southeast side. White-tailed Ptarmigan, Horned Larks, Water Pipits, and Gray-crowned Rosy Finches are among the birds to expect.

BLAINE
Point Roberts

A peninsula in the extreme northwest corner of Washington, and accessible by land only through Canada, **Point Roberts** is one of the best vantages in the state for land-based viewing of sea birds. A spotting scope is essential.

To reach Point Roberts from Blaine, proceed north through Canadian customs and then north and west on Provincial 99 for 13.2 miles; exit on Provincial 10 and go west for 4.4 miles, south on Provincial 17 for 3.4 miles, then turn left (south) onto Point Roberts Road and continue for 2.8 miles to the International border. Although bird finding is good anywhere along the shoreline of Point Roberts, the two most productive spots are Lighthouse Park (administered by Whatcom County) and Lily Point, located respectively at the southwest and southeast extremities of the Point.

Lighthouse Park is reached by driving south 1.2 miles from the border on Tyee Drive (a continuation of Point Roberts Road), then turning right (west) onto Gulf Road and going 0.7 mile and finally left (south again) for 0.8 mile. Very deep water comes close to shore here at the Park, creating strong tide-rips with consequent ideal feeding conditions for waterbirds and waterfowl, which are abundant from September through May. The best time for seeing them, however, is mid-October to mid-November when schools of herring attract large mixed-species flocks, often including many Arctic Loons, Brandt's Cormorants, Heermann's Gulls, and Thin-billed Murres, sometimes a few Ancient Murrelets and Rhinoceros Auklets. The flocks are less frequent in midwinter, although at this time Yellow-billed Loons are always a strong possibility. In September and October, Parasitic Jaegers are often present, harassing Bonaparte's Gulls and Common Terns. Flocks of Black Turnstones and Sanderlings winter along the gravelly beaches and may be joined in migration periods by a Ruddy Turnstone or a Wandering Tattler. A few Pigeon Guillemots and Marbled Murrelets may be expected near shore in any season.

Lily Point, although formerly accessible by car, can be reached only by walking along the beach, preferably a low tide, but close views of many sea birds should repay the effort involved. To reach Lily Point, drive east from Lighthouse Park for 1.1 miles, where the road rejoins Tyee Drive next to a marina, then east 0.7 mile on APA Road and south 0.3 mile on South Beach Road to the South Beach Store. Here, leave the car and walk 1.5 miles east along the beach to the Point. Alternatively, drive east from the U.S. customs office for 1.4 miles to Maple Beach, then south along the shoreline to the end of the road; park the car here and walk south to the Point, again about 1.5 miles distant.

The waters of Boundary Bay, just east of Lily Point, often teem with birds: huge numbers of Western Grebes and all three scoters in winter, smaller numbers all summer; a few Harlequin Ducks all year near offshore rocks; and large numbers of Oldsquaws in winter.

Much of Point Roberts is covered with young coniferous and mixed forest, such as near Lily Point, in which typical breeding birds are the Pileated Woodpecker, Chestnut-backed Chickadee, Bushtit, Red-breasted Nuthatch, Hutton's Vireo, and—in summer—Western and Olive-sided Flycatchers, Orange-crowned Warbler, and Purple Finch.

By virtue of Point Roberts's geography—jutting south into the open waters of the Strait of Georgia and Boundary Bay—bird migration is sometimes spectacular. Huge movements of loons, Black Brant, Oldsquaws, and scoters may be seen, especially in March, April, October, and November. In early fall, southbound landbirds converge on the tip of the Point, staying over land as long as possible, before venturing over the watery expanse. From either Lighthouse Park or Lily Point during the first hours after a September or October dawn, one may see numerous flocks of American Robins, Water Pipits, blackbirds, American Goldfinches, Pine Siskins, and other passerines departing south and possibly a few Turkey Vultures too. Besides acting as a funnel for many southbound species, Point Roberts becomes a 'trap' for other migratory species, holding them up from further movements. As a result, one may find certain species, especially near the tip of the Point, that are rare in the region.

BURLINGTON
North Cascades National Park | Ross Lake National Recreational Area | Pasayten Wilderness

In north-central Washington some of the most spectacular parts of the Cascade Range are within **North Cascades National Park.** Its two units, the northern (475 square miles) and southern (315 square miles) are separated by the valley of the Skagit River which comprises the **Ross Lake National Recreational Area.** Access to the Park is by trail only, but State 20, called the North Cascades Highway east from Burlington, passes through the Recreational Area, providing fine views of the Cascades and opportunities for seeing birds on the slopes dropping down to the Skagit River.

In the Recreational Area, the vicinity of Newhalem has coniferous forests with brushy openings for a variety of breeding birds including the Blue Grouse, Pileated Woodpecker, Eastern Kingbird, Western Flycatcher, Varied and Swainson's Thrushes, Nashville and MacGillivray's Warblers, Lazuli Buntings, and Oregon Juncos. Black Swifts are regular in the area in summer.

State 20, continuing eastward from the Recreational Area, enters the **Pasayten Wilderness** on the east side of the Cascades and soon crosses Rainy and Washington Passes (4,843 and 5,483 feet elevation, respectively). Here, or from side roads and trails that lead away, some of the high-country birds to expect are the Northern Tree-toed Woodpecker, Gray Jay, Clark's Nutcracker, Mountain Chickadee, Red-breasted Nuthatch, Winter Wren, Hermit Thrush, Mountain Bluebird, Townsend's Solitaire, Ruby-crowned Kinglet, Cassin's Finch, Red Crossbill, and Fox Sparrow.

Below Washington Pass, State 20 reaches Mazama on the Methow River. For more high-country birds, cross the River from State 20 and drive northwest for about 20 miles to Hart's Pass (elevation 6,200 feet). Turn right beyond Meadows Campground on a road marked to the Slate Peak Lookout. After about 3 miles, leave the car in the parking lot and hike up to the Lookout for a breath-taking view of the eastern Cascades from an elevation of 7,500 feet. Near the parking lot and during the climb, watch for White-tailed Ptarmigan. Trails leading down from Hart's Pass and the Slate Peak area pass into coniferous woods where Spruce Grouse and Boreal Chickadees are always possibilities.

From Mazama, State 20 descends through the Methow River Valley to Twisp. Woods and their edges along the River and pine forests on the adjacent slopes are worth searching for such birds as the Ruffed Grouse, Lewis' Woodpecker, Williamson's Sapsucker, White-headed Woodpecker, Dusky Flycatcher, Steller's Jay, Pygmy Nuthatch, Solitary Vireo, and Western Tanager.

NORTHERN THREE-TOED WOODPECKER

COULEE CITY
Sun Lakes State Park | Moses Coulee

In north-central Washington the Grand Coulee, an awesome canyon carved by the Columbia River during the last Ice Age, extends 50 miles from the town of Grand Coulee—site of the Grand Coulee Dam—southward to the town of Soap Lake. The Upper Grand Coulee is now filled by man-made Banks Lake, but the Lower Grand Coulee south of Banks Lake still contains a string of natural lakes (Park, Blue, Lenore, and Soap) as well as much of the original desert vegetation.

Sun Lakes State Park (3,710 acres), just southwest of Coulee City at the northern end of Park Lake, is a convenient center for exploring the birdlife of the Grand Coulee. To reach the Park from Coulee City, drive 2 miles west on US 2, then 4 miles south on State 17.

The sides of the Lower Grand Coulee in or near the Park are imposing basaltic cliffs hundreds of feet high that provide nesting sites for Red-tailed Hawks, Prairie Falcons (a few), Great Horned Owls, White-throated Swifts, Say's Phoebes, Violet-green Swallows, Cliff Swallows (several huge colonies), and Canyon Wrens. Chukars, Poor-wills, and Rock Wrens are breeding inhabitants on the talus slopes below the cliffs.

Besides the string of large lakes, numerous small lakes and marshes dot the floor of the Lower Grand Coulee. In the Park, within 3 miles of the campground, are several small lakes and marshes where breeding birds include Pied-billed Grebes, Virginia Rails, Soras, Common Snipes, Marsh Wrens, and Yellow-headed Blackbirds. In the bordering willows and other deciduous growth, breeding birds include California Quail, Eastern Kingbirds, Gray Catbirds, Yellow Warblers, Yellow-breasted Chats, Common Yellowthroats, Bullock's Orioles, and Song Sparrows.

Twenty-one miles west of Coulee City, US 2 traverses **Moses Coulee,** another old channel of the Columbia River. Although much like Lower Grand Coulee, Moses Coulee contains fewer lakes and marshes and is thus a better area for desert or 'sagebrush' birds—e.g., Sage Thrasher, Loggerhead Shrike, and four sparrows, the Vesper, Lark, Sage, and Brewer's. For a typical habitat, leave US 2 at the bottom of the Coulee on a gravel road leading south for about a mile.

MOSES LAKE
Columbia National Wildlife Refuge

The city of Moses Lake in central Washington lies in the heart of arid country—the channeled scablands—comprising sagebrush deserts with cliff-walled buttes, potholes, and streams. Largely due to water impoundments, there are numerous lakes, often marsh-bordered and surrounded by extensively irrigated agricultural lands.

Breeding birdlife is varied. Desert species include Common Nighthawks, Horned Larks, Black-billed Magpies, Sage Thrashers, Western Meadowlarks, and both Lark and Sage Sparrows. Nesting on cliffs are Red-tailed Hawks, American Kestrels, Great Horned Owls, Say's Phoebes, Northern Ravens, Cliff Swallows, Rock Wrens, and, occasionally, Canyon Wrens. Along streams lined with willows and shrubby thickets reside such birds as Eastern

Kingbirds, Yellow-breasted Chats, Bullock's Orioles, and sometimes Ash-throated Flycatchers. Where there are suitable marshes, one may expect Northern Harriers, American Avocets, Wilson's Phalaropes, Virginia Rails, Soras, American Coots, Black Terns, Marsh Wrens, and Yellow-headed, Red-winged, and Brewer's Blackbirds. In small nesting colonies—and consequently appearing here and there on lakes, canals, and other waterways—are Black-crowned Night Herons, Ringed-billed Gulls, and Caspian Terns.

Despite extensive water impoundments, the variety of breeding waterfowl is restricted mainly to Mallards, Blue-winged Teal, and Cinnamon Teal, which are common, and to fewer numbers of Gadwalls, Common Pintails, American Wigeons, Northern Shovelers, and Ruddy Ducks. In spring and fall, transient waterfowl, including an additional variety of diving ducks, show up in great numbers; many remain through winter.

The **Columbia National Wildlife Refuge,** south of Moses Lake, embraces 28,978 acres of typical scablands in which irrigation projects have created lakes, marshes, and wetlands. The Refuge thus offers a good representation of birdlife characteristic of this wide area.

To reach the Refuge from Moses Lake, leave I 90 at the Pasco Exit and proceed south on State 17 to Othello about 25 miles distant. Here the bird finder is advised to stop at Refuge headquarters, 44 South 8th Avenue, for a map, checklist of birds, and information about the self-guided tour of the Refuge. From State 17 in Othello, take Cunningham Road west through town, then turn north on Broadway out of Othello and onto McMannaman Road. After 6 miles, turn right onto Morgan Road and follow Refuge signs.

OKANOGAN
Salmon Creek Valley

This north-central Washington community in the Okanogan River Valley makes a good base for the bird finder wishing to explore, in the late spring and summer, both the scenic east slopes of the northern Cascades and the Okanogan Highlands.

A short trip on an unnumbered road northwest from Okanogan leads up through the **Salmon Creek Valley** in the eastern Cascades and passes irrigated lands with innumerable apple orchards, which bloom in early May. Just before reaching Conconully, 17 miles from Okanogan, the road crosses a stream connecting Conconully Lake, a natural body of water on the right, with Conconully Reservoir on the left.

The country in the vicinity of Conconully (elevation 2,318 feet) is relatively rough with numerous steep-sided ridges and peaks, some attaining altitudes in excess of 7,000 feet. In addition to the orchards and other plantings introduced by man, the lower areas show, along the watercourses, cottonwoods, willows, alders, and shrubby thickets. Sagebrush predominates on the adjacent hillsides (the rangelands) just above the reach of irrigation, and is succeeded by open stands of ponderosa pine as the elevation increases. The high slopes support a wider variety of conifers, such as lodgepole pine, Douglas-fir, and western larch. Birds regularly residing during the breeding season on the agricultural and rangelands, in brushy places near the road, and in wooded stream borders include the Red-tailed and Swainson's Hawks, American Kestrel, California Quail, Poor-will, Common Nighthawk, Calliope Hummingbird, Lewis' Woodpecker, Eastern and Western Kingbirds, Say's Phoebe, Western Pewee, Violet-green and Rough-winged Swallows, Black-capped Chickadee, House Wren, Gray Catbird, Veery, Warbling Vireo, MacGillivray's Warbler, Yellow-breasted Chat, Brewer's Blackbird, and Vesper, Lark, Brewer's, and Song Sparrows.

An improved road, going northward out of Conconully and open only in summer, reaches altitudes above 5,000 feet where, in coniferous forests, one may find such breeding birds as the Blue Grouse, possibly the Spruce Grouse too, Williamson's Sapsucker, Gray Jay, Clark's Nutcracker, Mountain Chickadee, Red-breasted Nuthatch, Pygmy Nuthatch, Winter Wren, Varied Thrush, Mountain Bluebird, Townsend's Solitaire, Ruby-crowned Kinglet, Townsend's Warbler, Cassin's Finch, Red Crossbill, Oregon Junco, and Fox Sparrow. Also in the forests, but more often at lower elevations, one may find the Ruffed Grouse, Pileated Woodpecker,

Dusky Flycatcher, Steller's Jay, Solitary Vireo, and Western Tanager.

PASCO
McNary National Wildlife Refuge

In southeastern Washington, near the confluence of the Snake and Columbia Rivers, **McNary National Wildlife Refuge** occupies 3,366 acres, consisting largely of sloughs from waters backed up by McNary Dam about 20 miles downstream on the Columbia. Situated and managed as it is, the Refuge provides habitat attractive to a great many waterfowl and other birds in all seasons.

The Refuge is easily reached from Pasco by driving south on US 395. After 6 miles the highway crosses the northwestern end of the Refuge just northeast of Burbank. Here watch for signs to headquarters, which lies east of the highway. Roads through the Refuge offer good vantages for bird viewing.

Waterfowl breeding regularly on the Refuge are Canada Geese, Mallards, Gadwalls, Common Pintails, Green-winged, Blue-winged, and Cinnamon Teal, American Wigeons, and Northern Shovelers. The geese have broods as early as late April, most ducks much later. Other birds nesting in association with the sloughs and their marshy borders are Northern Harriers, American Coots, Long-billed Curlews, American Avocets, Wilson's Phalaropes, Marsh Wrens, and blackbirds—Yellow-headed, Red-winged, and Brewer's.

During fall migration the aggregations of waterfowl, which stop off to feed and rest on the Refuge, are spectacular, particularly in November. Most of the population comprises many more of the same species which breed on the Refuge, namely, Canada Geese, Mallards, Common Pintails, Green-winged Teal, and American Wigeons. The variety of waterfowl is enhanced when Whistling Swans show up, usually in October. In spring migration, during March, there is again the same variety of surface-feeding ducks, but it is increased impressively by diving ducks such as Lesser Scaups, Buffleheads, and Ruddy Ducks.

PORT ANGELES
Olympic National Park | La Push Area | Cape Flattery

The westernmost extremity of Washington, the mountainous Olympic Peninsula, is a vast wilderness penetrated by few roads and interrupted only by scattered fishing, lumbering, and farming settlements. Port Angeles, on the Strait of Juan de Fuca, is the largest city on the Peninsula and the headquarters of **Olympic National Park,** which lies to the south.

Covering about 1,407 square miles, Olympic Park encompasses most of the higher, snowy Olympic Mountains—a disorganized group of sharp ridges, crags, and peaks rather than a range—and a separate area or strip along the Pacific Coast. Elevations in the main Park extend from 300 feet to 7,965 feet at the summit of Mt. Olympus, situated almost in the center. The mountain area has been severely glaciated and even today several of the high peaks support active glaciers. Though not lofty as compared with the Cascades, the Olympic Mountains are no less majestic, their heights rising abruptly from sea level.

The dense forests of Olympic Park, containing some of the finest remaining virgin stands in the Pacific Northwest, rival its mountains as the principal wilderness feature. Douglas-fir, western hemlock, western red cedar, and Sitka spruce are the predominating species in the lower valleys up to 1,500 feet, and appear occasionally at higher altitudes in the valleys of big rivers. On the western side of the Park, these trees, deluged by an average of 142 inches of rainfall annually, attain their maximum size, and they shade an understory of mosses, lichens, and deciduous shrubs so luxuriant as to suggest a tropical rain forest. On the eastern side of the Park, which is drier, coniferous trees predominate on the slopes, but a thick deciduous growth of trees and shrubs—bigleaf maples, red alders, willows, flowering dogwoods, and others—lines the streams and covers various areas of the valley floors. Western hemlock, western red cedar, and silver fir constitute the chief forest growth of the Park from 1,500 to 3,500 feet; from that elevation to the timber line at about 5,500 feet, the forests are interspersed with mountain meadows, which, in the summer, are colored by the countless blooms of wildflowers. At elevations above timber line, the vegetation is limited to low shrubs, grasses, sedges, and a few hardy wildflowers.

Enhancing the pristine splendor of Olympic Park's mountains and forests are the many blue lakes and their glacier-fed streams with their falls and cascades. The longest stretch of water in the Park is Lake Crescent in the northwest. Ringed by conifers and with the Olympic peaks in the background, it is undoubtedly one of the most beautiful lakes in western Washington.

US 101 (Olympic Loop Highway) goes along the east, north, and west sides of the Olympic Peninsula, passes through Port Angeles, encircles Olympic Park outside its boundaries, crossing the main Park at only one point—the northwest corner along the south shore of Crescent Lake—and eventually proceeds southward through the Park's ocean strip. Spur roads lead off the highway into the main Park, via big river valleys, for short distances; at their ends begin over 600 miles of trails to some of the most remote parts of the wilderness area. A map of Olympic Park, showing its roads and trails, and information on interpretive services are available at the Port Angeles Visitor Center (2800 Hurricane Ridge Road, Port Angeles, WA 98362) and the Hoh Rain Forest Visitor Center. Both are open all year.

Of places in Olympic Park accessible by car and good for observing high-mountain birds, probably the best is Hurricane Ridge (elevation 5,000 to 6,450 feet), 17 miles south of Port Angeles. This is reached by taking State 111 south from Port Angeles to the ranger station at Heart of the Hills Entrance, then following up Hurricane Ridge Road. Once on the Ridge, either turn left to Obstruction Point, passing forests at timber line and meadows, on a winding dirt spur road for 8.4 miles, or continue to Hurricane Hill from which a trail begins 1.5 miles beyond the lodge. From the Ridge, the views of Mt. Olympus's three peaks on the south and the Strait of Juan de Fuca on the north are superlative.

Some of the birds in this high country that may be viewed between mid-June and August, from turnoffs during the ascent and along the Ridge, include the Red-tailed Hawk, Golden and Bald Eagles, American Kestrel, Blue Grouse, Black and Vaux's Swifts, Rufous Hummingbird, Horned Lark, Violet-green Swallow, Gray Jay, Northern Raven, American Robin, Mountain Bluebird,

Orange-crowned Warbler, Pine Grosbeak, Gray-crowned Rosy Finch (occasional), Red Crossbill, and Oregon Junco.

Another place for high-mountain birds in Olympic Park, also accessible by car, is Deer Park (elevation 5,400 feet), reached by driving east from Port Angeles on US 101 for 5 miles, then going south on a well-marked road for 17 miles. Deer Park has bird habitats like those on Hurricane Ridge.

Nearly all the spur roads into Olympic Park are in valleys where one may find many of the bird species regularly frequenting coniferous forests of lower slopes and the thickets and open areas of valley floors. Better results will be obtained, however, if one takes a trail into a more remote valley area where there are fewer disturbances. Strongly recommended is Elwha River Trail. This is reached by driving west and south from Port Angeles on US 101 to the ranger station at Elwha and taking the winding, single-track, dirt Elwha River Road for 5 miles to Whiskey Bend. From here the Trail starts, following the Elwha River past Elkhorn Station (12 miles from the Road) into the heart of the Park. Among the breeding birds to look for along the River, or in its immediate vicinity, are the Great Blue Heron, Harlequin Duck, Blue Grouse, Spotted Sandpiper, Band-tailed Pigeon, Common Nighthawk, Vaux's Swift, Rufous Hummingbird, Red-shafted Flicker, Western and Olive-sided Flycatchers, Violet-green, Barn and Cliff Swallows, Chestnut-backed Chickadee, North American Dipper, Winter Wren, American Robin, Varied and Swainson's Thrushes, Warbling Vireo, MacGillivray's and Wilson's Warblers, Western Tanager, Black-headed Grosbeak, Pine Siskin, American Goldfinch, and Chipping and White-crowned Sparrows.

Along the ocean strip of Olympic Park is the **La Push Area,** centered in the little fishing hamlet of La Push, just south of the mouth of the Quillayute River. La Push is 69 miles from Port Angeles, reached by driving west and south on US 101 to within a mile of the town of Forks, then turning right onto a marked road and proceeding 14 miles. Worth visiting for birds, besides La Push, is Rialto Beach on the north side of the Quillayute River's mouth, reached by returning 7 miles toward Forks, then turning left, crossing the River, and following its north bank west.

In the La Push Area is a great variety of bird habitats—

coniferous forests and their edges of deciduous trees and brush, fields, broad Rialto Beach and the mud and sand flats behind it, the water at the mouth of the Quillayute River, the jagged rocks and wooded islands offshore, and the ocean.

From vantages along the River mouth and the ocean, some of the birds to be seen, usually not in the late spring and summer, are the Arctic Loon, Red-necked Grebe, Green-winged Teal, American Wigeon, Greater Scaup, Common Goldeneye, Bufflehead, Harlequin Duck (sometimes in summer but not breeding), Common and Red-breasted Mergansers, Herring Gull, California Gull (particularly in late summer and fall), and Mew Gull. Both Double-crested and Brandt's Cormorants, Bald Eagles, Ospreys, and Glaucous-winged Gulls may be sighted in any season.

Among the shorebirds that are sometimes fairly numerous on Rialto Beach and the flats behind it in May, July, August, and September are the Semipalmated Plover, Surfbird, Black Turnstone, Red Knot (most likely in May), Pectoral Sandpiper (most likely in late summer), Least Sandpiper, Dunlin, Long-billed Dowitcher, Western Sandpiper, and Sanderling.

Inland, in the coniferous forests and their edges and in grassy areas are a great variety of breeding birds that include the Ruffed Grouse, Rufous Hummingbird, American Crow, Chestnut-backed Chickadee, Bushtit, Red-breasted Nuthatch, Varied and Swainson's Thrushes, Orange-crowned and Wilson's Warblers, Western Meadowlark, Purple Finch, and Savannah and White-crowned Sparrows.

Some of the best of Olympic Park's rain forests may be reached by continuing south on State 101 from Forks for 14 miles, then taking the Hoh River Road for 19 miles to the Hoh Rain Forest visitor center. Here are enormous, moss-draped spruces, firs, and hemlocks so tall and so close together as to shut out the sun. Along the mossy trails that begin at the ranger station and wind through this forest spectacle, a remarkable stillness prevails day and night regardless of season. Birds are relatively few. Probably the Varied Thrush is the commonest of any species. Always a possibility in the dense, higher stands of trees is the Spotted Owl.

If the bird finder wants to be sure of seeing Black Oystercatchers, he should visit **Cape Flattery** on the western tip of the

Olympic Peninsula. A wild desolate place, it is one of the few spots in Washington where these big shorebirds can be seen the year round.

The Cape is reached from Port Angeles by driving west for 5 miles on US 101, then turning off right onto State 112 and going 61 miles to Neah Bay. During much of the way, the highway runs along the Strait of Juan de Fuca. If the tide is low, stop at intervals to scan the beach and outlying water for birds. From Neah Bay, drive southwest on a marked road to the Cape. After about 5 miles, begin watching for a sign on the left, 'To the Cape,' indicating the start of the foot trail. Take this path (and carry a spotting scope), which leads to the tip of Cape Flattery—a high, lonely bluff. Without getting too close to the edge, since it overhangs dangerously, inspect the rocks below for Black Oystercatchers. From this vantage, look for Harlequin Ducks (not breeding), Northern Ravens, and many of the other species also frequenting the La Push Area.

During the late spring and summer, use the spotting scope to scan the seabird colonies offshore on Tatoosh Island. Among the species nesting are the Pelagic Cormorant, three alcids—the Thin-billed Murre, Pigeon Guillemot, and Tufted Puffin—and the Glaucous-winged Gull. Known to nest on the Island are the Fork-tailed Storm Petrel, Leach's Storm Petrel, and Cassin's Auklet.

SEATTLE

University of Washington Campus | University of Washington Arboretum | Green Lake | Puget Sound

This great metropolis lies east of Puget Sound on a wide strip of hilly land between Puget Sound and Lake Washington. Puget Sound is connected to Lake Washington by the Lake Washington Canal running through the city, first to Lake Union and then from Portage Bay to Union Bay, an arm of Lake Washington.

Undoubtedly the best bird finding in Seattle is in the north part of the city on the 582-acre **University of Washington Campus** bordering the northwest end of Union Bay (main entrance reached from I 5 by exiting east to Northeast 45th Street) and the 267-acre

University of Washington Arboretum bordering the southwest end of Union Bay, opposite the Campus (entrance reached from I 5 by exiting east on State 520 and turning off at directional signs). Two spots in these areas are especially productive.

(1) In the northwest part of the Campus are coniferous trees, Douglas-fir predominating, with deciduous trees and shrubs intermixed. Here, too, flanking Union Bay is a cattail-bulrush marsh back of which is an extensive grassy area. (2) In the Arboretum are plantations of many native and exotic trees and shrubs. A small point of land called 'Foster Island,' extending into Union Bay, is largely covered with deciduous thickets and has marshes around it.

In these two spots, where habitats are suitable, the following birds are among those to be expected: *All year,* Great Blue and Green Herons, California Quail, Band-tailed Pigeon, Common Screech Owl, Steller's Jay, Black-capped Chickadee, Chestnut-backed Chickadee (in conifers), Bushtit, Winter, Bewick's and Marsh Wrens, Golden-crowned Kinglet (in conifers), Purple Finch, Pine Siskin, Rufous-sided Towhee, Oregon Junco, and Song Sparrow. *In the nesting season* (May to August), Rufous Hummingbird, Traill's, Western, and Olive-sided Flycatchers, Violet-green and Barn Swallows, Swainson's Thrush, Solitary and Warbling Vireos, Orange-crowned and Audubon's Warblers, Black-headed Grosbeak, American Goldfinch, and White-crowned Sparrow. *In spring* (April and early May) *and fall* (September and October), Hermit Thrush (also in winter), Cedar Waxwing (also in winter), Black-throated Gray and Wilson's Warblers, and Golden-crowned Sparrow.

Although Union Bay attracts a great many waterbirds and waterfowl in fall, winter, and spring (September to April), a far better place for observing them, because it permits close views, is **Green Lake** (250 acres), a body of fresh water in the northern residential part of the city. Despite its being a municipal recreational center, with bathing beaches and playgrounds on its eastern and southern shores, this does not deter the presence of such birds as Eared Grebes, Canada Geese, Mallards (some are permanent residents and tame), Common Pintails, Green-winged Teal, American Wigeons, Northern Shovelers, Common Mergansers and a few other diving ducks, Glaucous-winged Gulls (present throughout the year), California Gulls, Ring-billed Gulls, and Bonaparte's Gulls. To reach Green Lake, exit from I 5 to North 45th Street and proceed west; turn off north onto Green Lake Way which eventually goes around the eastern side of the Lake, providing many vantages for observation.

Some of the best vantages on **Puget Sound** for seeing loons, diving ducks—Common Goldeneyes, Buffleheads, Oldsquaws, scoters, Red-breasted Mergansers—and alcids in winter are along Beach Drive southwest of the city from Alki Point south to Lincoln Park. This road passes very close to Puget Sound, permitting frequent stops to scan the open water. To reach Beach Drive, exit west from I 5 to South Spokane Street and continue west to Southwest Spokane Street, which eventually becomes Admiral Way to Alki Point.

SEAVIEW

Long Beach Peninsula | Leadbetter Point Unit of the Willapa National Wildlife Refuge | Fort Canby State Park | North Jetty at the Mouth of the Columbia River

From the southwestern coast of Washington the **Long Beach Peninsula** extends northward for over 20 miles, sheltering Willapa Bay on its eastern side and terminating in Leadbetter Point. Since 1968 the northernmost part of the Peninsula that includes the Point and nearby Grassy Island in Willpa Bay—altogether 1,433 acres—comprises the **Leadbetter Point Unit of the Willapa National Wildlife Refuge.** The Point consists largely of sand dunes, partially stabilized by grasses and small shrubs, and flanked on the bayside by a broad salt marsh of pickleweed and arrowweed with intermittent tidal flats. For bird finding on the Washington coast, Leadbetter Point is one of the most rewarding in any season.

The Snowy Plover occasionally nests in late spring and early summer on or back from the beach on the oceanside of the Point. Lapland Longspurs and sometimes a few Snow Buntings frequent the dunes in winter. But the chief ornithological attraction comes from late October through November and again in late April and early May when hordes of migrating Black Brant and Canada

Geese, as well as a good representation of Greater White-fronted Geese, stop off to feed and rest in the salt marsh and on the tidal flats; a great many remain in winter. Another attraction comes in late August and early September when vast numbers of Sooty Shearwaters pass the Point in their migration.

Transient shorebirds abound on the tidal flats in late April and early May and again from late August to mid-October; many are present in winter. Besides such regular and often common species as the Semipalmated and Black-bellied Plovers, Ruddy and Black Turnstones, Whimbrel, Greater and Lesser Yellowlegs, Pectoral, Baird's, and Least Sandpipers, Dunlin, Short-billed and Long-billed Dowitchers, Western Sandpiper, Sanderling, and Northern Phalarope are other species—e.g., the Lesser Golden Plover, Long-billed Curlew, Stilt, Semipalmated, and Buff-breasted Sandpipers, and Red and Wilson's Phalaropes—less frequently observed elsewhere on the Washington coast. Always a possibility, particularly in September and early October, is the Sharp-tailed Sandpiper.

The Leadbetter Point Unit of the Willapa Refuge may be reached from US 101 by turning off west and going toward Seaview for about 2 miles. Just before entering the town, turn north on Peninsula Road which parallels Willapa Bay for 16 miles to Oysterville. Continue past Oysterville, bearing left with the road, and turn north onto Stackpole Harbor Road and proceed for about 4 miles. This passes dense stands of conifers and impenetrable thickets—worth a stop to search for a variety of breeding birds including the Ruffed Grouse, Band-tailed Pigeon, Mourning Dove, Vaux's Swift, Rufous Hummingbird, Traill's, Western, and Olive-sided Flycatchers, Steller's Jay, Chestnut-backed Chickadee, Golden-crowned Kinglet, Orange-crowned, Yellow, and Wilson's Warblers, Black-headed Grosbeak, and Red Crossbill. After about 4 miles the Road reaches the Unit boundary. Since the Unit itself is restricted to foot travel, park the car and walk north past the salt marsh to the tip of the Point, about 2.5 miles distant.

Return to Seaview and proceed south for 2 miles to Ilwaco, then follow directional signs to **Fort Canby State Park** and thence to the **North Jetty at the Mouth of the Columbia River.** Bird species to be expected in and about the State Park and the vicinity of the

North Jetty are nearly the same as in and near Fort Stevens State Park and the vicinity of the South Jetty in Oregon (*see under* **Astoria**), but there is one greater advantage for bird finding: The North Jetty extends farther out to sea and, although rough-surfaced, can be walked to the end in safety, except in tempestuous weather. Hence one may view sea birds closely, as though he were aboard a boat.

SPOKANE
Turnbull National Wildlife Refuge

About 25 miles southwest of Spokane, the **Turnbull National Wildlife Refuge** has one of the best waterfowl nesting grounds in eastern Washington. Its 17,171 acres comprise a section of the channeled scablands where the force of ancient streams cut deep gullies down to bare lava in some places—now dry for the most part—and piled sediments high in others, leaving a succession of ponds and potholes. Within the Refuge are several bodies of water or lakes whose levels are controlled by dams and dikes and which have a total surface of about 5,000 acres. An abundance of cattails, bulrushes, and other aquatic plants, including submersed types, provides excellent cover and food for marsh-loving birds.

Waterfowl that breed commonly on the Refuge are Canada Geese, Mallards, Gadwalls, Common Pintails, Green-winged, Blue-winged, and Cinnamon Teal, American Wigeons, Redheads, Lesser Scaups, and Ruddy Ducks. The geese begin nesting in early April, the ducks in mid-April. A small breeding population of Trumpeter Swans, originating from Red Rock Lakes in Montana, has been successfully established and may be viewed on a display pond near headquarters. Other birds nesting in the marshes or their wet borders are Eared and Pied-billed Grebes, Northern Harriers, Soras, American Coots, Common Snipes, Wilson's Phalaropes, Black Terns, Marsh Wrens, Common Yellowthroats, Yellow-headed, Red-winged, and Brewer's Blackbirds, and Savannah Sparrows.

During fall migration in September and particularly in October, congregations of transient Canada Geese, Mallards, Common Pintails, American Wigeons, and American Coots are impressively large. Small flocks of Whistling Swans often appear among them.

Although the birds associated with the lakes and marshes are the Refuge's principal attraction, there is a considerable variety of landbirds to be found in the willows and cottonwoods bordering some of the lakes, in the aspen groves and open stands of ponderosa pine that cover much of the uplands, in scattered thickets of serviceberry, snowberry, and wild rose, and in the nearby farmlands. A search of these habitats will yield a variety of birds including the Red-tailed Hawk, American Kestrel, Ruffed Grouse, California Quail, Black-chinned Hummingbird, Yellow-bellied Sapsucker, Downy Woodpecker, Eastern Kingbird, Traill's Flycatcher, Western Pewee, Tree Swallow, Black-capped and Mountain Chickadees, Pygmy Nuthatch, House Wren, Western Bluebird, Warbling Vireo, Audubon's Warbler, American Redstart, Western Meadowlark, House Finch, Grasshopper and Vesper Sparrows, Oregon Junco, and Song Sparrow.

To reach the Refuge from Spokane, drive southwest on I 90 for 9 miles, exit left on State 904 for 6 miles into the business section of Cheney, then go south on the Cheney-Plaza County Road, following Refuge signs for 6 miles to headquarters. The Refuge has a 5-mile auto tour and hiking trails leading to all representative habitats.

TACOMA

Spanaway Lake Park | Mount Rainier National Park

Spanaway Lake Park (340 acres), on the right side of State 7 about 8 miles south of Tacoma, is a popular recreational area. Along its west side is Spanaway Lake, 1.5 miles long and 0.5 mile wide, bordered by woods of Douglas-fir and scrub oak and sections of open prairie. Along the Lake and elsewhere in the Park are productive spots for viewing birds in May, June, and July, provided the bird finder looks for his quarry in early morning before there is too much human activity. Such breeding birds as Lewis' Woodpeckers, Western Flycatchers, Steller's Jays, Bushtits, Brown Creepers, Hutton's and Solitary Vireos, Orange-crowned Warblers, and Pine Siskins may be observed without too much difficulty. With careful searching, one should turn up a pair or two of Hermit Warblers. Horned Larks are common on the adjacent prairie.

Southeast of Tacoma towers Mt. Rainier, the loftiest of volcanoes in conterminous United States. Though 40 miles away as the crow flies, it reaches so high—14,410 feet—that the city seems almost in its shadow. For bird finders, no less than sightseers, **Mount Rainier National Park,** which embraces this massive dome, is a must; in fact, no other area in the Pacific states offers a better opportunity for finding so many mountain birds in a glorious setting.

Mt. Rainier occupies about one-fourth of the 378 square miles of the Park. Built up by its own volcanic activity ages ago and rugged with many perilous slopes, the mountain when viewed from any direction has an aspect that belies its harshness, for its summit is capped by perennial snow and its flanks—all but the most prominent crags and ridges—are blanketed by 27 glaciers, more than the number possessed by any other single peak in the country. Most of the glaciers extend down to elevations of 4,000 feet, well below timber line.

The melting of Mr. Rainier's snow and glaciers in the warmer months gives rise to numerous streams, which, in the heat of the day, increase in volume and turbulence as they descend into the deep ravines and valleys. This water supply, supplemented by the heavy moisture that characterizes the prevailing climate of the vicinity, accounts in no small measure for the richness of vegetation below the ice and snow-covered slopes, and that, in turn, for the great abundance of birds.

The lowland forests of the Park, from the lower valleys at 1,700 feet to about 3,000 feet on Mt. Rainier, are dense and shady, their trees frequently tall and branchless for over half their height. Grand fir, Douglas-fir, western hemlock, and western red cedar are the predominating growth. Occasionally along streams is a lower deciduous growth consisting of red alder, black cottonwood, bigleaf maple, and willow.

At middle elevations, from 3,000 to 4,500 feet, the forests are still dense, but the predominant trees change to Alaska yellow cedar, mountain hemlock, silver fir, noble fir, western white pine, and lodgepole pine. Around 4,500 feet, subalpine conditions appear: the trees begin to show reduction in size; subalpine fir and mountain hemlock become the principal species, joined by a scattering of whitebark pine as the elevation increases.

Between 5,000 and 6,500 feet, depending on the exposure and conditions of mountain surface, comes the interspersion of subalpine forests and subalpine meadows, the latter being by far the most exciting places on Mt. Rainier. They offer the first open spots for awe-inspiring scenery. Birds here, though probably no more numerous than in other places in the Park, are certainly more conspicuous. And the wildflowers, in their great abundance and variety, are famous the world over. From early June to early August, they produce great carpets of gorgeous colors, which change from day to day as different species come into bloom.

Above 6,000 feet, the average elevation at timber line, is the alpine region where severe winds and chilling temperatures are common even in midsummer. Only the hardiest of wildflowers, which are no match in color for those of the subalpine meadows, can tolerate this bleak climate.

Mount Rainier National Park has five principal entrances, two on the west side and three on the east. Within the Park are some 100 miles or more of paved roads but only one runs through the Park from one side to the other; and there are numerous trails, some leading to the Park's most remote sections.

The best time of the year for finding breeding birds is from early June to mid-July. All the higher elevations that may be reached by car are usually clear of snow by mid-June, occasionally not until early July. When entering the Park, obtain a map at the entrance station and a schedule of interpretive services which include slide and film programs given in the summer months.

A fair number of bird species resides in the Park in suitable habitats at almost any elevation to the timber line. These include such breeding birds as the Sharp-shinned, Cooper's, and Red-tailed Hawks, Golden Eagle, American Kestrel, Great Horned, Spotted, and Saw-whet Owls, Vaux's Swift, Rufous Hummingbird, Red-shafted Flicker, Violet-green Swallow, Gray Jay, Chestnut-backed Chickadee, Red-breasted Nuthatch, North American Dipper, Winter Wren, American Robin, Varied, Hermit, and Swainson's

Thrushes, Golden-crowned Kinglet, Western Tanager, Evening Grosbeak, Purple Finch, Cassin's Finch, Red Crossbill, and Oregon Junco. Many other species, however, are confined to suitable habitats at one elevation or another. Three places, if investigated thoroughly, should yield most of the birds regularly breeding in the Park:

1. *Longmire Meadows* (elevation 2,760 feet) at Longmire in the southwest part of the Park, near a ranger station as well as a visitor center and campground; reached by continuing south from Spanaway Lake Park on State 7 for 32 miles, then turning east onto State 706, going 14 miles to the Park's Nisqually Entrance, and following the Park road to Longmire. The Meadows is a clearing of about 5 acres, surrounded by heavy coniferous forest, and containing old beaver ponds partly filled by cattails and bordered by a few alders. Some of the birds to look for, either in the open area or in the neighboring forest are the Ruffed Grouse, Spotted Sandpiper, Band-tailed Pigeon, Hairy and Downy Woodpeckers, Yellow-bellied Sapsucker, Black-capped Chickadee, Yellow, Audubon's, and Wilson's Warblers, Red-winged Blackbird, Black-headed Grosbeak, and Song Sparrow.

2. *Paradise Valley* (elevation 5,560) in the south part of the Park; reached by continuing north and east on the Park road from Longmire. The Valley has several hundred acres of beautiful subalpine meadows—among the most spectacular for wildflowers—and forests of subalpine fir and mountain hemlock. Birds to observe here in suitable situations are the Blue Grouse, Calliope Hummingbird, Hammond's and Olive-sided Flycatchers, Gray Jay, Northern Raven, Clark's Nutcracker, Mountain Chickadee, Mountain Bluebird, Townsend's Solitaire, Ruby-crowned Kinglet, Townsend's Warbler, and Fox and Lincoln's Sparrows. The Paradise River, cutting down through the area, is a favorite haunt for North American Dippers. White-tailed Ptarmigan, Water Pipits, and Gray-crowned Rosy Finches may be found in the vicinity of the Edith Creek Glacial Cirque, just above Paradise Inn, and on the higher slopes of the Valley.

3. *Yakima Park* (elevation 6,000–6,500 feet) at Sunrise on the east side of the Park; reached by continuing on the Park road as it winds its way for 50 miles around the southeastern part of the Park

and then northward. Yakima Park is on a long, relatively flat-topped ridge that supports wide meadows interspersed with low stands of subalpine fir and whitebark pine—conditions typical of the timber line. Most of the bird species in Paradise Valley are seen here. At higher elevations on the ridge, reached by the Burroughs Mountain Trail, are some of the best spots in Mount Rainier Park for White-tailed Ptarmigan, Horned Larks, and Water Pipits.

WALLA WALLA
Blue Mountains

This city lies west of the **Blue Mountains** in southeastern Washington, an area little explored by bird finders, in part because of poor roads, inadequately mapped and marked. The following trip, however, will provide a good sampling of birdlife.

From Walla Walla, drive northeast on US 12 for a short distance, then turn south onto an unnumbered county road that leads into Mill Creek Canyon and continues through the Canyon to Kooskooskie. In the open, ponderosa pine forests and shrubby thickets of the Canyon's floor may be found, in May and June, such breeding species as the Red-tailed Hawk, American Kestrel, Great Horned and Northern Pygmy Owls, Vaux's Swift, Red-shafted Flicker, Pileated Woodpecker, Williamson's Sapsucker, Dusky and Western Flycatchers, Steller's Jay, Brown Creeper, House Wren, Swainson's Thrush, Veery, Solitary Vireo, Orange-crowned, Audubon's, and MacGillivray's Warblers, Western Tanager, Cassin's Finch, Pine Siskin, Red Crossbill, Rufous-sided Towhee, Oregon Junco, and Chipping Sparrow. North American Dippers are present along the stream.

From Mill Creek Canyon numerous logging roads lead into higher areas. By walking a mile or two up any of these roads one may reach elevations well over 3,000 feet where Douglas-fir and white fir are the principal timber. Birds to be expected here, if not already observed at lower elevations, are the Blue Grouse, Calliope Hummingbird, Northern Three-toed Woodpecker, Hammond's and Olive-sided Flycatchers, Gray Jay, Clark's Nutcracker,

Mountain Chickadee, Red-breasted Nuthatch, Winter Wren, Varied and Hermit Thrushes, Townsend's Solitaire, Townsend's Warbler, and Evening Grosbeak.

WESTPORT

Point Chehalis

One of the most productive places for transient and wintering shorebirds on the Washington coast is **Point Chehalis** in Westhaven State Park, at the north end of a sandy peninsula that guards Grays Harbor on the south and west. From US 101, north of Raymond or south of Aberdeen, turn off west onto State 105, a loop highway. After 32 miles from Raymond or 18 miles from Aberdeen, turn off north for 4 miles through Westport to the Point.

Bird finding is probably best in May, August, and September. There is an outer beach, which is attractive to Sanderlings. From this outer beach, a great stone jetty, called Damon's Point, projects over 1.0 mile into the ocean. Surfbirds, Black Turnstones, Wandering Tattlers, and Rock Sandpipers appear along its shore, which is dangerous during storms since giant waves break over it. In late summer, Sooty Shearwaters migrate by the hundreds past the end of the jetty.

The inner beach, adjacent to backwater lagoons, is lined with mud flats where Semipalmated and Lesser Golden Plovers, Whimbrels, Least Sandpipers, Dunlins, and Long-billed Dowitchers may be viewed. The lagoons are particularly worth investigating during storms as many birds seek shelter here.

VANCOUVER, BRITISH COLUMBIA

Sea and Iona Islands | Reifel Migratory Bird Sanctuary

Near Vancouver, a metropolis of over a million people only 18 miles north of the International border, are Crested Mynas and Sharp-tailed Sandpipers, birds impossible or difficult to find anywhere south of Canada. The mynas, descendants of a few individuals introduced about 1897, are permanent residents of Vancouver and surrounding communities. The sandpipers are rare but regular fall transients, mainly between 15 September and 30 October. A good place for both species is **Sea and Iona Islands,** in the delta of the Fraser River immediately south of Vancouver. Iona Island happens to be perhaps the best area for *all* species of transient shorebirds in western Canada.

To reach Sea and Iona Islands from downtown Vancouver, drive south on Provincial 99 across the Oak Street Bridge onto Lulu Island, then take the first exit (marked 'Airport') west onto Sea Island Way. (Do *not* take the Laing Bridge going directly from Vancouver to the airport because Iona Island cannot be reached from it.) One mile west on Sea Island Way, cross the Morray Channel Bridge onto Sea Island; just past the Bridge, turn right (north) onto a road which soon swings to the left (west), becoming Grauer Road. Follow it for 2 miles and turn sharply right onto Macdonald Road; then, after 0.5 mile, turn left onto Ferguson Road. This leads due west for 1.3 miles, then bears to the right (north), crosses a causeway, and ends at the Iona Island sewage treatment plant.

Although the sewage plant is surrrounded by a high fence, bird finders are permitted inside, provided that they sign the guest book, in the plant office. Park in front of the office, sign the guest book, and inspect the four settling ponds west of the office. At least one of these ponds usually has shallow water with 'islands' of sludge and grass protruding. Look here for Sharp-tailed Sandpipers, sometimes as many as six, ordinarily among flocks of Pectoral Sandpipers. While on Iona Island, scan the mud flats on its west side for more shorebirds. The best time is at high tide when the aggregations are close to shore.

On the way to Iona Island, look for Crested Mynas on Sea Island, especially along Ferguson Road. Here and along other roadsides in Vancouver between the downtown area and the Oak Street Bridge, they often feed with European Starlings. During the breeding season, mynas are scattered widely throughout the city; hence they are seldom in great numbers in any specific area at that time. From October to April, large numbers join starlings to roost at night on tall buildings or under bridges, but the location of these roosts varies from time to time.

Another good place for Sharp-tailed Sandpipers is the **Reifel Migratory Bird Sanctuary,** west of Ladner in the Fraser Delta. To reach the Sanctuary, drive south from downtown Vancouver on Provincial 99 for 13.8 miles; south on Provincial 17 for 1.4 miles; west (right) on Provincial 10 for 1.1 miles to the small town of Ladner. Continue 2.4 miles west of Ladner on River Road, then cross a bridge over a branch of the Fraser River onto Westham Island. The Refuge entrance, 2.7 miles from the bridge, is clearly marked.

The Sanctuary features an impressive collection of pinioned waterfowl in outdoor pens—the best spot for Sharp-tailed Sandpipers as well as for other shorebirds. Several thousand Snow Geese and many thousands of ducks winter in the nearby marshes. Eurasian Wigeons are regular among the vast numbers of American Wigeons. Gyrfalcons, as well as many other raptors, are seen every winter.

VICTORIA, BRITISH COLUMBIA

Victoria International Airport | University of Victoria Campus | Clover Point | Cattle Point | Strait of Juan de Fuca

Victoria, British Columbia's capital city, on the southeastern tip of Vancouver Island, may be reached by ferry from Port Angeles, Seattle, Anacortes (via the San Juan Islands), or from Tsawwassen on the British Columbia mainland. The Port Angeles and Seattle ferries dock in downtown Victoria; the Anacortes ferry lands at Sidney, 17 miles north of Victoria, and the Tsawwassen ferry at Swartz Bay, 3 miles north of Sidney.

The Victoria area has the only nesting population of Common Skylarks in North America—except for a small satellite population on San Juan Island, Washington. Introduced shortly after 1900, skylarks prefer weedy fields not regularly cultivated. Since they do not migrate, they may be found in any season; but they are more readily observed in spring and early summer when they are flight-singing.

One of the most reliable spots for skylarks is the **Victoria International Airport,** 1.0 mile west of the ferry terminal in Sidney, or 3 miles south of Swartz Bay. Follow roads around the Airport and listen for the singing. Closer to Victoria, a good spot in summer is the **University of Victoria Campus,** where a large field has been set aside for nesting skylarks. The Campus is reached from downtown Victoria by driving 2.2 miles east on Fort Street, then 2.1 miles north on Foul Bay Road, which becomes Henderson Road just before reaching the University.

In any season, but especially in winter, a visit to **Clover Point** is a must for the bird finder in Victoria. From downtown Victoria, proceed about 0.5 mile south to Dallas Road, which follows the shoreline, then east for 0.5 mile to the parking area at Clover Point, which projects a short distance into the Strait of Juan de Fuca. Particularly at low tide, scan the rocks here for Black Oystercatchers (all year); Surfbirds and Black Turnstones (late July through early May); Rock Sandpipers (October through May); and Wandering Tattlers and Ruddy Turnstones (mainly in May and in late July through September). Harlequin Ducks, Marbled Murrelets, and Rhinoceros Auklets, regular in summer, are joined in winter by a variety of loons, grebes, cormorants, waterfowl, gulls, and alcids. To see Ancient Murrelets, which are sometimes common in winter, a better place is **Cattle Point,** reached by continuing about 5 miles east and north from Clover Point along the shoreline drive to Uplands Park in Oak Bay.

The trip by ferry to Port Angeles across the **Strait of Juan de Fuca** has good possibilities for pelagic species, especially in September and October. Some of the birds one may expect, at least occasionally, are the Northern Fulmar, Sooty Shearwater, Fork-tailed Storm Petrel, Pomarine Jaeger, Black-legged Kittiwake, Sabine's Gull, and Cassin's Auklet.

Wyoming

SAGE GROUSE

Any bird finder intent on exploring the open plains and mountain fastnesses of Wyoming must count on many miles of travel, yet if he plans a late-spring or early-summer trip that will include diversified plains and high mountain ranges that are readily accessible by car, the rewards for his efforts can be gratifying. There is much birdlife to see!

In the sagelands north of Rock Springs, Sage Grouse perform their remarkable courtship displays in an environment shared with such birds as Sage Thrashers and Brewer's Sparrows. On the grassy plains west of Laramie, where Mountain Plovers nest, the flight displays of McCown's Longspurs are a familiar and ever delightful spectacle. Probably there is no stretch of the plains not regularly frequented by the larger raptors—Swainson's Hawks,

Ferruginous Hawks, Prairie Falcons, and one or more Golden Eagles.

Despite the extent of the Wyoming plains, hardly a spot anywhere is very far from mountains that have an entirely different association of birds, for Wyoming has many mountain ranges, high and low, and they are widely scattered over the state. The higher uplifts have in common a marked ruggedness, a coniferous forest cover, and many kinds of birds such as Swainson's Thrushes, Audubon's Warblers, and Western Tanagers. Nevertheless, each has a distinct topography and tree growth and certain peculiarities of avifauna. In the Black Hills—including their western extension, the Bear Lodge Mountains—in the northeastern corner of the state, the summer-resident junco is the White-winged, but in the Big Horn Mountains in the north-central part of the state it is the Oregon, and in the Laramie and Medicine Bow Mountains in the southeast-central part of the state it is the Gray-headed. What the bird finder discovers in one Wyoming mountain range is, therefore, never exactly duplicated in another.

The Wyoming plains east of the Big Horns and the Laramie Mountains are part of the Great Plains, with average elevations between 5,000 and 6,000 feet; they incline gradually, however, from 3,100 feet, in the extreme northeast, to about 7,000 feet at the base of the Laramie Mountains in the south. Drainage, provided chiefly by the Powder, Belle Fourche, and North Platte Rivers and their tributaries, is northward or eastward into the Missouri River system. The country is best described as an undulating prairie interrupted by sandhills, buttes, and talus ridges, and by rocky bluffs and rough badlands near watercourses. Rainfall is moderate. Cottonwood, willow, boxelder, and various deciduous shrubs line most of the larger streams; juniper, mountain mahogany, and, occasionally, ponderosa pine grow in scattered stands on bluffs and ridges; but on the open expanses the principal plants are short grasses, chiefly grama and buffalo, mixed in numerous localities with sagebrush, rabbitbrush, greasewood, yucca, and other scrubby growth. Where water has been put on the land through irrigation projects, crops are produced; elsewhere the prairie is used mostly for livestock grazing.

The Wyoming plains west of the Big Horns, a segment of the

Great Plains with elevations ranging from 3,500 to 5,500 feet, drain northward into the Yellowstone River. The Wyoming plains west of the Laramie Mountains, with elevations ranging from 6,000 to 7,500 feet, drain in three different directions: north to the Yellowstone via the Big Horn River, east to the Platte River, and south to the Colorado via the Green River. Both plains are almost completely rimmed by mountain ranges. As a whole, the plains have a level to undulating surface with scattered buttes and ridges and ranges of low hills. Along the larger streams the plains have been much roughened by erosion, producing colorful badlands and rugged canyons. Owing to scant rainfall, the aspect is decidedly arid. Though there are wide grassy stretches, more generally the vegetation is of the shrubby, bunch-like, desert type, consisting of such plants as sagebrush, greasewood, and rabbitbrush, with juniper and mountain mahogany on bluffs and ridges. Willow and cottonwood, bordering the streams, are among the very few common trees. Most of the country is little used save for grazing livestock.

The distribution of birds on the Wyoming plains, ridges, bluffs, and mountain foothills varies greatly according to vegetative cover. Where the plains are grassy—e.g., the plains east of Laramie—are species that do not reside on plains dotted with sagebrush—e.g., the sagelands north of Rock Springs—and *vice versa.* Slopes and ridges with juniper associations are habitats for at least a few Poorwills and Pinyon Jays. On the slopes of the hot and dry Green River Valley, near the Colorado line in southwestern Wyoming, the juniper-pinyon pine association also attracts Gray Flycatchers, Scrub Jays, Plain Titmice, and Bewick's Wrens. Listed below are other species that breed regularly in Wyoming's open country (prairie grasslands, sagelands, rangelands, wet lowlands, brushy places, woodland edges, and dooryards) and in wooded areas (deciduous trees along streams and in plantations). Species marked with an asterisk also breed in the mountains.

OPEN COUNTRY

Turkey Vulture
Swainson's Hawk
Ferruginous Hawk
Northern Harrier

Sharp-tailed Grouse (*chiefly northeastern Wyoming*)
Sage Grouse
Mountain Plover
Long-billed Curlew
Upland Sandpiper (*eastern Wyoming*)
Mourning Dove
Burrowing Owl
Short-eared Owl
Red-headed Woodpecker (*eastern Wyoming*)
Eastern Kingbird
Western Kingbird
Say's Phoebe
* Horned Lark
* Tree Swallow
Barn Swallow
* Black-billed Magpie
* House Wren
Northern Mockingbird (*southeastern Wyoming*)
Gray Catbird
Brown Thrasher (*eastern Wyoming*)
Sage Thrasher
* Mountain Bluebird
Loggerhead Shrike
Common Yellowthroat
Yellow-breasted Chat
Bobolink (*uncommon*)
Western Meadowlark
Common Grackle (*eastern Wyoming*)
Lazuli Bunting
Dickcissel (*eastern Wyoming*)
House Finch
* American Goldfinch
Green-tailed Towhee
Rufous-sided Towhee
Lark Bunting (*eastern Wyoming*)
Savannah Sparrow
Grasshopper Sparrow (*mainly northeastern Wyoming*)
Vesper Sparrow
Sage Sparrow (*southwestern Wyoming*)
* Chipping Sparrow
Clay-colored Sparrow (*eastern Wyoming*)
Brewer's Sparrow
Song Sparrow
McCown's Longspur
Chestnut-collared Longspur (*eastern Wyoming*)

WOODED AREAS

* Cooper's Hawk
Common Screech Owl
* Red-shafted Flicker
* Hairy Woodpecker
* Downy Woodpecker
* Dusky Flycatcher
* Western Pewee
Blue Jay (*eastern Wyoming*)
* Black-capped Chickadee
* White-breasted Nuthatch
Red-eyed Vireo (*eastern Wyoming*)
Warbling Vireo
American Redstart
Bullock's Oriole
Black-headed Grosbeak

All of Wyoming's mountain ranges belong to the Rocky Mountain system, but some of them are widely separated. The Black Hills—and the Bear Lodge Mountains—rise by themselves in the northeast, and the immense Big Horns tower alone above the plains in the north-central part of the state. The Absaroka, Teton, Gros Ventre, Wind River, and Salt River Ranges in the northwest and west are, as a group, cut off eastward by extensive plains from another parallel group—the Laramie, Medicine Bow, and Sierra Madre Ranges—which extends into Colorado from the southeast-central part of the state.

The mountain ranges in Wyoming that reach elevations above 6,000 feet support coniferous forests. Plateaus—e.g., the Yellowstone in the northwest—that attain the same heights have a similar growth. Six mountain ranges—the Big Horns, Absaroka, Teton, Gros Ventre, Wind River, Medicine Bow—and several isolated peaks in Yellowstone Park reach above tree growth and are capped by an alpine region. The altitude at which the tree growth ceases and the alpine regions begin varies in a general way with latitude. Thus in the southern mountains—e.g., the Medicine Bow Mountains—the trees become stunted between 10,000 and 10,500 feet and fail to grow at all above 10,500 to 11,000 feet; in the northern mountains—e.g., the Big Horns—these points of change are respectively lower by 1,000 feet. An extensive alpine region exists in the Wind River Range, the highest and most massive of all the ranges. Above timber line here are a great many broad ridges and lofty peaks, including Gannett Peak, which reaches 13,804 feet, the highest point in the state.

The vegetation of all the higher mountain ranges shows a more or less uniform succession of tree species from the sagebrush- and juniper-covered lowest slopes to the summits. Merging with the sagebrush and juniper of the lowest slopes is a belt of ponderosa pine. Above this is lodgepole pine with quaking aspen along streams and in groves, followed at high levels by Englemann spruce, sometimes in pure stands. Mixed with these trees at all altitudes is Douglas-fir. In the subalpine belt of stunted tree growth, limber pine, subalpine fir, and whitebark pine become prominent and, at timber line, make up for the most part the characteristically twisted and one-sided 'wind timber.' Much of the alpine region is rock-strewn and barren, yet there are slopes with enough soil for a luxuriant growth of mosses, grasses, and sedges, low matted willow thickets, and countless wildflowers. On at least three ranges, the Absaroka, Teton, and Wind River, the alpine regions are of sufficient height and exposure to hold snowfields and even small glaciers in the shaded heads of valleys.

Owing to the discontinuity of mountain ranges in Wyoming, coupled with their peculiarities of exposure, vegetation, and other factors, breeding birdlife is never exactly the same from one range to another. The differences in juncos, referred to earlier, are a good example. Other examples are: the apparent absence of Steller's Jays in the Black Hills—and Bear Lodge Mountains—as well as in the Big Horns; the summer residence of Gray-crowned Rosy Finches in the Big Horns, Black Rosy Finches in the northwestern mountains, and Brown-capped Rosy Finches in the Medicine Bow Mountains. Listed below are birds breeding regularly in most of the major mountain ranges.

Northern Goshawk
Sharp-shinned Hawk
Blue Grouse
Ruffed Grouse
White-tailed Ptarmigan (*above timber line*)
Yellow-bellied Sapsucker (*generally below 9,000 feet*)
Williamson's Sapsucker (*except in Black Hills*)
Northern Three-toed Woodpecker
Western Flycatcher
Olive-sided Flycatcher
Violet-green Swallow
Gray Jay
Steller's Jay (*except in Black Hills and Big Horns*)
Northern Raven
Clark's Nutcracker (*except in Black Hills*)
Mountain Chickadee
Red-breasted Nuthatch
Pygmy Nuthatch (*below 9,000 feet*)
Brown Creeper
Hermit Thrush
Swainson's Thrush
Townsend's Solitaire
Golden-crowned Kinglet
Water Pipit (*above timber line*)
Solitary Vireo (*mainly Black Hills*)
Orange-crowned Warbler (*in deciduous thickets below 9,000 feet*)
Audubon's Warbler
Ovenbird (*mainly Black Hills*)
MacGillivray's Warbler (*in deciduous thickets below 9,000 feet*)
Wilson's Warbler (*in deciduous thickets, mainly above 9,000 feet*)

Western Tanager (*mainly below 9,000 feet*)
Cassin's Finch
Pine Grosbeak (*subalpine forest*)
Pine Siskin
Red Crossbill
White-crowned Sparrow
Fox Sparrow (*northwestern ranges*)
Lincoln's Sparrow (*in deciduous thickets*)

The two great recreational centers in the state—Yellowstone and Grand Teton National Parks (*see under* **Yellowstone National Park** *and* **Jackson**)—are hardly less outstanding for bird finding than they are for their superlative scenic attractions. Both areas hold not only a fine variety of Rocky Mountain birds and birds characteristic of sagelands but also such breeding birds as Trumpeter Swans, Sandhill Cranes, and (in Yellowstone Park) American White Pelicans.

Bird migration in Wyoming is seldom impressive as there are no main routes—no great river valleys to follow and large bodies of water to attract large congregations of waterbirds and waterfowl for rest and food during passage. Birds breeding in western Canada and Montana tend to filter through the state in a relatively dispersed manner, the small landbirds moving through the mountain valleys that have a north-south direction. The principal migratory flights through the state take place within the following dates:

Waterfowl: 10 March–20 April; 10 October–25 November
Shorebirds: 1 May–1 June; 1 August–1 October
Landbirds: 15 April–20 May; 25 August–20 October

Winter bird finding in wooded and shrubby areas along watercourses on the plains and in mountain valleys will yield, with concerted effort, between thirty-five and fifty species. Some of the species may comprise rosy finches and other birds that have moved down from the mountains, others may be Bohemian Waxwings, Northern Shrikes, Evening Grosbeaks, and American Tree Sparrows that have come from northern climes. Almost invariably a few Rough-legged Hawks and a Golden Eagle or two may be sighted in open country. Flocks of Lapland Longspurs and Snow Buntings frequently show up on the plains.

Authorities

Kenneth L. Diem, Helen Downing. Robert M. Mengel, B. C. and M. Raynes, Oliver K. Scott, William J. Wilson.

JACKSON
Grand Teton National Park | National Elk Refuge | Rendezvous Mountain

Grand Teton National Park (484 square miles), north of Jackson and south of Yellowstone National Park in western Wyoming, embraces, on the west, a series of majestic, ruggedly beautiful mountains—the most scenic section of the Teton Range—and, on the east, part of a wide valley known as Jackson Hole.

The Teton Range, close to the Idaho line, extends in a north-south direction for 40 miles. It rises abruptly on the east side, more gently on the west. The highest peak, Grand Teton, reaches 13,766 feet; numerous subordinate peaks attain elevations above 10,000 feet. Together they comprise a massive, jagged crest of granite. In many a lofty crevice, snow persists the year round, and a dozen or so glaciers occupy ravines high in the east and north slopes.

Jackson Hole, 8 to 10 miles wide, parallels the Teton Range from north to south and is so encompassed by mountains as to give the impression of a huge basin. The Teton Range walls the valley on the west side; on the north, east, and south the valley is bordered by the Absaroka and Gros Ventre Ranges. This mountainous encirclement of Jackson Hole seems accentuated by the valley floor, which is remarkably flat—a level plain except for widely scattered buttes. Elevations of the valley floor range from 6,000 feet near Jackson at the southern end to 7,000 feet at the north, but the changes in elevation are almost imperceptible to the eye. Jackson Lake, in the northern section, is a natural body of water enlarged by the damming of the Snake River which flows southward through the valley, receiving streams from the adjacent highlands along its winding course. Small lakes and beaver ponds dot the valley floor, especially on the west side of the Snake River. Some are quite shallow, with marshy edges.

The Park offers a diversity of bird habitats. Much of the valley floor consists of gravely stretches supporting sagebrush, but in spots along the Snake River and around the lakes is a coniferous-deciduous growth of lodgepole pine, blue spruce, Douglas-fir, subalpine fir, cottonwood, willow, and aspen. Fine stands of conifers (lodgepole pine, limber pine, Engelmann spruce, and subalpine fir), mixed with aspen, grow on the slopes of the Teton Range, extending up to the timber line at 10,000 to 11,000 feet, where there is also whitebark pine. Above the timber line, on the west slopes of the peaks, and in some basins on the east slopes, are true alpine meadows or tundra, matted with grasses and wildflowers. In midsummer, after the snow has disappeared, plants such as Indian paintbrush, Parry primrose, larkspur, and saxifrage bloom in profusion. The mountains to the east and south of Jackson Hole have more gradual slopes, and their forests and alpine meadows are more extensive. The isolated buttes in Jackson Hole are notably rocky, with some cliffs. Scattered brush grows on their slopes and, additionally, stands of lodgepole pine mixed with Douglas-fir and aspen on their more shaded, northern slopes.

Park headquarters is on the Snake River at Moose, reached from Jackson by driving north on US 187 for 13 miles. At the visitor center here, obtain a map of the Park and a checklist of birds, and inquire about the various naturalist services.

The Teton Park Road, which starts from the west side of US 187 at Moose and proceeds north, passes entrance drives on the left to the following big lakes: Jenny, 7 miles from Moose; String, 10 miles from Moose; Leigh, reached by a hike of about a mile from the end of the road at String; and Jackson Lake, 17 miles from Moose. Some of the birds at these lakes in the summer are Barrow's Goldeneyes, Common Goldeneyes (occasionally), Common Mergansers, and numerous swallows—Violet-green, Tree, and Barn. Several pairs of Ospreys have their aeries near the shore of Jackson Lake. Where the shores of these lakes are bordered by marsh vegetation, and, on higher ground, by brush, the Yellow Warbler, Common Yellowthroat, Brewer's Blackbird, and White-crowned and Lincoln's Sparrows are among the summer residents.

The forests and forest edges around the big lakes, and around the Sawmill Ponds about 1.5 miles south of Park headquarters,

BLACK-BILLED MAGPIE

hold birds such as the Northern Goshawk, Sharp-shinned and Cooper's Hawks, Blue and Ruffed Grouse, Northern Pygmy Owl (uncommon), Great Gray Owl, Calliope Hummingbird, Yellow-bellied and Williamson's Sapsuckers, Western Pewee, Olive-sided Flycatcher, Gray and Steller's Jays, Clark's Nutcracker, Black-capped and Mountain Chickadees, White-breasted and Red-breasted Nuthatches, Hermit and Swainson's Thrushes, Ruby-crowned Kinglet, Warbling Vireo, Audubon's Warbler, Western Tanager, Cassin's Finch, Black-headed and Pine Grosbeaks, Pine Siskin, Oregon Junco, and Chipping Sparrow.

Swainson's Hawks, Black-billed Magpies, and Mountain Bluebirds are commonly viewed from roadsides in open country. Usually conspicuous along the Snake River are Great Blue Herons, Bald Eagles, Spotted Sandpipers, Belted Kingfishers, Northern Ravens, and Bank Swallows.

For bird finding along the Snake River: (1) Leave the car at the Moose River visitor center and walk along the bank on the east side of the River. (2) Inquire at the center about arrangements for taking one of the float trips, a delightful way to see birds and enjoy marvelous scenery at the same time.

For anyone wishing to combine bird finding with mountain hiking, a trip to Lake Solitude in the high, glacial 'back country' on the east side of the Teton Range is recommended. Here conditions are typically alpine. From the west side of Jenny Lake—reached by boat or by the Lakes Trail, which begins at the boat dock on the east side of Jenny Lake and passes around the Lake—take the Cascade Canyon Trail, suitable for travel on foot or on saddle horses.

The Cascade Trail leads to the base of Hidden Falls, where Cascade Creek makes a roaring plunge from a lofty glacial bench; then it ascends sharply by a series of switchbacks through Cascade Canyon—a deep chasm in which Cascade Creek rushes downward below the Trail—to the bases of the massive peaks that divide the Canyon into the north and south forks. Here the Trail enters the north fork, and, continuing upward, comes finally to Lake Solitude (elevation 9,024 feet), the headwaters of the north fork of Cascade Creek. The distance from the Jenny Lake boat dock is 7 miles if one crosses the Lake by boat; 9.4 miles if entirely on foot.

Birds to look or listen for in the ascent to Lake Solitude are North American Dippers at Cascade Creek and at Hidden Falls and both Blue and Ruffed Grouse in the coniferous forests that penetrate the Canyon. The Hermit Thrush and White-crowned Sparrow are common along the Trail. Seeing one or more large raptors, such as a Golden Eagle, in flight over the jagged, precipitous slopes is always a possibility. Above the timber line, on the alpine meadows, are Water Pipits. Quite likely, Black Rosy Finches will be noticed searching for insects on the alpine turf or moving to and from their nesting sites in cliffs or in talus slopes.

A hike to Lake Solitude cannot ordinarily be undertaken until July, when the Trail is clear of snow and ice. Before making the trip, check at the Moose visitor center or at the Jenny Lake ranger station for current trail conditions. A permit is required for overnight trips.

Adjoining the Park on the southeast is the **National Elk Refuge** (23,754 acres), administered by the U. S. Fish and Wildlife Service. Headquarters is on the eastern outskirts of the town of Jackson, reached by driving east from the town square on Broadway

Street. The Refuge is a wintering area for elk, which arrive in late October or November, depending on the snowfall, and depart in late April and early May for mountain country far north in the Teton Wilderness and Yellowstone National Park.

Lying for the most part in the flat valley of Jackson Hole, the Refuge consists largely of marshes and grassy meadows, with cottonwoods, willows, and other trees bordering the upper reaches of Flat Creek, which flows through the Refuge, and the Gros Ventre River on the northeast boundary. Breeding birds in the vast marshes and meadows include Northern Harriers, Sandhill Cranes, Long-billed Curlews, Marsh Wrens, and Savannah and Vesper Sparrows. Rising here and there are buttes—mostly grass-covered, with cliffs and rocky crevices—that provide nesting sites for birds such as the Red-tailed Hawk, Prairie Falcon, Cliff Swallow, Northern Raven, Rock Wren, and, occasionally, the Lark Sparrow. Sage Grouse, Horned Larks, Green-tailed Towhees, and Brewer's Sparrows inhabit the wide stretches of sagebrush that surround the buttes.

On Flat Creek a few pairs of Trumpeter Swans are permanent residents along with a few pairs of Cinnamon Teal and other waterfowl—Canada Geese, Mallards, Gadwalls, Common Pintails, Green-winged Teal, and Common Mergansers—that breed on the Refuge. The swans, as well as the other waterfowl, may often be viewed from State 187 along the Refuge's west boundary just north of Jackson where the highway crosses Flat Creek. From this same vantage one or more Sandhills Cranes may sometimes be watched feeding near the Creek.

To see other birds in the Elk Refuge, one may enter it at headquarters and, after obtaining permission here, drive 4.6 miles north on a dirt road to an Izaak Walton League sign commemorating the original donation of land to the Refuge. Here the road forks. The sharp *right fork* goes from the Refuge up a series of switchbacks for 4.5 miles into the forested foothills of the Gros Ventre Mountains. Along the way, watch the sky for a Golden Eagle or a Prairie Falcon; once in the forest, look or listen for the Ruffed Grouse, Gray and Steller's Jays, Townsend's Solitaire, Pine Siskin, and Cassin's Finch. The *left fork* straight ahead, and then the next fork to the right, goes from the Refuge for 2.5 miles to upper Flat Creek, lined here with cottonwoods and willow thickets. For the next three miles one may continue during dry weather in a conventional car along the Creek, stopping to search for a variety of birds including the Dusky Flycatcher, MacGillivray's and Wilson's Warblers, and Lazuli Bunting.

Before leaving Jackson, stop at the Wyoming State Visitor Center at the north end of town for a map of the Jackson Hole area. The Center itself has an elevated tower overlooking a marshy area of the Elk Refuge that attracts large numbers of Yellow-headed and Brewer's Blackbirds and sometimes teems with insect-foraging Tree and Barn Swallows.

A brushy stream bottom south of Jackson offers good prospects for Lazuli Buntings. Drive south from Jackson on US 187 for 8 miles, then turn left onto a dirt road that follows Game Creek for about 3 miles. Stop the car anywhere along the Creek, since one spot is as good as another. Besides looking for the buntings, watch for North American Dippers along the edge of the stream.

In Teton Village, reached from Jackson by driving west on State 22 for 4 miles across the Snake River and north on State 390 for 8 miles, is the Jackson Hole Aerial Tram that carries summer passengers to the summit of **Rendezvous Mountain.** Here, 4,139 feet above the valley floor and over 10,000 feet above sea level, one may take in a sweeping view of Jackson Hole and the mountains to the east and south, investigate tundra flora, and find Water Pipits and—with luck—Black Rosy Finches. Look particularly for the finches foraging for insects on or near the permanent snowfields, in view from the road about a mile down from its start at the summit.

LARAMIE

Laramie Plains | Snowy Range of the Medicine Bow Mountains | Hutton Lake National Wildlife Refuge

The **Laramie Plains** immediately northwest of this southeastern Wyoming city is good country for Mountain Plovers from mid-May through early September. For a particularly good site, drive north from the city on US 30 for about 2 miles, turn off west onto Howell Road which crosses the Laramie River and soon takes a

McCOWN'S LONGSPUR

northwestward course. Two miles after crossing the River, begin looking for the plovers—usually inconspicuous from either side of the Road.

A drive west from Laramie on State 130 crosses 30 miles of the Laramie Plains with elevation ranging from 7,000 to 7,500 feet. The best time to make the trip is between 15 May and 15 July, when this grassy, undulating country is brilliantly colored with wildflowers and when Horned Larks and McCown's Longspurs, both abundant birds, are repeating their flight songs. Along the way be prepared for sighting a Swainson's Hawk, Ferruginous Hawk, Golden Eagle, or Prairie Falcon. If stops are made to investigate the alkaline lakes that dot the Plains along the way, expect to see Snowy Egrets, Northern Harriers, Wilson's Phalaropes, California Gulls, and various ducks.

Beyond the Laramie Plains the highway comes to the foot of the **Snowy Range of the Medicine Bow Mountains** at Centennial (elevation 8,000 feet) and then proceeds to cross the Range via Snowy Range Pass (highest point, 10,800 feet). Bird finding along this scenic highway and at points to be reached from side trips on foot will yield a fine variety of birds typical of the coniferous forest and alpine meadows of the Rocky Mountains. The best time is from 20 June to 20 July, when the birds are in full song.

Lodgepole pine forests, interspersed with aspen groves, predominate in the Snowy Range from 8,000 to 9,000 feet elevation; above them are forests of spruce and fir that, from 10,500 feet to the timber line at 11,500 feet, become stunted, forming a typical subalpine forest. State 130 at its highest point passes through a part of this forest. Extending above the timber line to the summits, the highest of which is Medicine Bow Peak at 12,005 feet, are extensive alpine meadows.

Some of the characteristic breeding birds in the forests of the Snowy Range are the Northern Goshawk, Blue Grouse, Gray Jay, Clark's Nutcracker, Mountain Chickadee, Hermit and Swainson's Thrushes, Mountain Bluebird, Townsend's Solitaire, Ruby-crowned Kinglet, Cassin's Finch, Pine Grosbeak, and Gray-headed Junco. White-crowned Sparrows breed regularly at timber line. Brown-capped Rosy Finches and a few White-tailed Ptarmigan reside above timber line. In order to see them, turn right onto a secondary road that branches off north from State 130, 8.7 miles from Centennial at a point just beyond Brooklyn Lodge. This road terminates at Brooklyn Lake. From here take a 2-mile foot trail to the top of Brooklyn Ridge. Failing to find the birds at this site, climb Medicine Bow Peak, where the finches are a certainty around the summit and the ptarmigan are more likely than at lower elevations.

During the trip across the Snowy Range, look for Horned Larks and Water Pipits on Libby Flats, which State 130 traverses after the turnoff to Brooklyn Lodge.

On the Laramie Plains southwest of Laramie, the **Hutton Lake National Wildlife Refuge** (1,970 acres) enjoys a park-like entity surrounded by mountains. At an elevation of 7,150 feet, the Refuge embraces 560 acres of marshes and open water consisting mainly of five small lakes, arranged in the shape of a half-moon, that have been developed from natural sumps supplied by Sand Creek which meanders through a small western portion of the Refuge. Meadowlands, greasewood-dominated alkali flats, and grassy rangelands constitute the remaining 1,410 acres.

The ornithological feature of the Refuge is the remarkable concentration of birds attracted by the marshes and lakes. Among the species represented in the nesting season are at least ten ducks—

Mallard, Gadwall, Common Pintail, Green-winged and Blue-winged Teal, American Wigeon, Northern Shoveler, Redhead, Canvasback, and Ruddy Duck—as well as Eared and Pied-billed Grebes, Snowy Egret, Black-crowned Night Heron, American Bittern, Northern Harrier, Virginia Rail, Sora, American Coot, Common Snipe, American Avocet, Wilson's Phalarope, California Gull, Forster's and Black Terns, and Yellow-headed Blackbird.

To reach the Refuge from Laramie, drive southwest on State 230 for about 12 miles, turn off south onto County 37 for 7 miles, then turn northeast onto County 34 for 4.5 miles to the entrance.

ROCK SPRINGS

Some of the largest populations of Sage Grouse in Wyoming exist in the upper Green River Valley in the southwestern part of the state. Here the habitat is semiarid sagebrush prairie and foothills (elevations from 6,000 to 8,000 feet), scarred by streambeds, sometimes dry, and by variously colored badlands consisting of valleys and benchlands, and of buttes, mesas, ridges, and bluffs, often with precipitous slopes. The grouse may be observed commonly in the spring, when they assemble on permanently established strutting or display grounds.

Each year there are many strutting grounds in the vicinity of Eden and Farson on US 187, 36 and 40 miles north from Rock Springs. As a rule the strutting grounds are bare spots, either on exposed ridges, knolls, and small buttes, or on the flat sagebrush areas. The maximum daily strutting activity takes place on these grounds for an hour or so beginning soon after daybreak, from late March until early June. Display activities may be watched at close range from a car. From June to October, Sage Grouse frequently visit alfalfa fields to feed.

The Sage Grouse shares this sagebrush country with a number of other breeding species. The most common is probably the Horned Lark; others include the Sage Thrasher and Brewer's Sparrow. Regularly seen in flight over the sagebrush prairie are the Red-tailed Hawk, Swainson's and Ferruginous Hawks, Northern Harrier, and Prairie Falcon.

SHERIDAN
Big Horn Mountains

The bird finder headed west in northern Wyoming will encounter the **Big Horn Mountains** as the first great mountain mass in his path. Perhaps he has recently passed through the picturesque Black Hills of South Dakota, where gentle slopes and forests of ponderosa pine have given him his initial experience with the west, and where the bird fauna was a striking mixture of eastern and western species. Beyond the Black Hills a long haul across open plains—antelope country—on I 90, or on US 16, toward Sheridan brings him within view of the massive, tumbled peaks of this eastern outpost of the Rocky Mountains.

The Big Horns should be traversed, rather than bypassed, by the bird finder. For the newcomer to the west they offer a cross section of true Rocky Mountain scenery and bird fauna, with relatively few characteristic species missing; for anyone, the pleasant scenic and climatic relief from the hot summer plains will more than repay for the extra hours and miles.

On the lower slopes of these mountains the bare, upsweeping plains give way, at 4,000 to 5,000 feet, to steep, rocky inclines grown up with sagebrush and spotted with juniper and scrubby pines. These slopes are the homes of innumerable whitetail and mule deer, chipmunks, cottontail rabbits, snowshoe hares, of permanent-resident Gray Partridges and Black-billed Magpies, of summer-resident Common Nighthawks, Rock Wrens, Green-tailed and Rufous-sided Towhees, Lark Sparrows, and other species. Particularly good bird-finding sites at these levels are afforded by canyons containing streams bordered by deciduous thickets and clumps of cottonwood and willow. Here such birds nest as the Black-capped Chickadee, White-breasted Nuthatch, Gray Catbird, Orange-crowned Warbler, Common Yellowthroat, Yellow-breasted Chat, Bullock's Oriole, Black-headed Grosbeak, and Lazuli Bunting.

On the higher slopes, irregularly distributed according to soil and moisture, is the forest of lodgepole pine and Engelmann spruce (the two predominant tree species), Douglas-fir, ponderosa pine, and subalpine fir, which comprises the million-acre Bighorn

National Forest. These areas of Rocky Mountain forest are rich in birdlife. Characteristic breeding species are the Blue Grouse, Western Pewee, Gray Jay, Clark's Nutcracker, Mountain Chickadee, Red-breasted Nuthatch, Brown Creeper, Hermit Thrush, Townsend's Solitaire, Audubon's Warbler, Western Tanager, Cassin's Finch, Pine Siskin, and Oregon Junco. On the edges of forest that has been clear-cut or burned, Northern Three-toed Woodpeckers are present but always hard to find. In shrubby areas, MacGillivray's and Wilson's Warblers, White-crowned, Lincoln's, and Song Sparrows are common, and North American Dippers reside along streams.

The broad tops of the mountains, particularly in the northern parts of the range, are tablelands of subalpine meadows; decorated with stark patches of fir and littered with boulders, the meadows are very beautiful. In places the mountains rise to rocky crags well above timber line. Horned Larks, Water Pipits, and Vesper Sparrows are typical summer residents of these high meadows, which are riddled with the holes of mountain marmots. To the south the elevations are higher and the country more rugged, culminating in mighty Cloud Peak (elevation 13,175 feet), and embraced by the Cloud Peak Primitive Area, which is accessible only by pack trip. Both the Gray-crowned and Black Rosy Finches are reported in the Area during early summer.

A selection of routes for a summer trip over the Big Horns is presented below. In advance of the trip, acquire a visitors' map of Bighorn National Forest from the Forest Service Office (address: Box 2046, Sheridan, WY 82801) and a brochure listing resorts, lodges, and recreational facilities in the Big Horns from the Wyoming Travel Commission (address: 1-25 at Etchepare Circle, Cheyenne, WY 82002).

1. Drive north from Sheridan on I 90 for 18 miles, then exit west at Ranchester on US 14 toward Dayton, 12 miles distant. Upon nearing Dayton, turn right before crossing the Tongue River on a gravel road leading 4 miles to Tongue River Canyon. In the steep walls of this narrow gorge, White-throated Swifts and Violet-green Swallows nest, as do Canyon Wrens, although hearing them sing is difficult due to the sound of rushing water below. Rufous-sided Towhees are numerous in brushy ravines; Ovenbirds, Amer-

ican Redstarts, Veeries, and three vireos—Solitary, Red-eyed, Warbling—dwell in suitably wooded habitat.

Backtrack on US 14 and continue for 27 miles to Burgess Junction in the Big Horns, making a spectacular climb from 3,900 to 7,900 feet. Within 15 miles after leaving Dayton, stop at any pullout and investigate for Green-tailed Towhees. At Burgess Junction the highway divides, offering a choice of routes.

(a) *US 14A west for 59 miles to Lovell.* This route is a better way to see wilderness country from a good paved road (narrow in places) with relatively little traffic. Elk and mule deer may be seen frequently, particularly at dusk. Ferruginous Hawks, Golden Eagles, Prairie Falcons, and Northern Ravens show up regularly about the many rugged cliffs and peaks close by the road. There are numerous beaver-dammed streams, and Yellow-bellied Sapsuckers, Gray Jays, Hermit Thrushes, and Oregon Juncos in adjacent woods.

About 3 miles from Burgess Junction the highway goes along the south side of the North Tongue River, bordered by boggy areas and willow thickets where Savannah Sparrows nest along with Wilson's Warblers and White-crowned and Lincoln's Sparrows.

Near Little Bald Mountain, about 15 miles from Burgess Junction, the highway reaches 9,700 feet elevation. From the scenic overlook here, search for Water Pipits in the talus above and the snowfields below. Medicine Mountain, north of the highway, is accessible by a 3-mile drive and well worth the time. Along the way one is likely to see many White-throated Swifts executing truly astonishing flight performances over the knife-edge ridge above the timber line.

After passing the Medicine Mountain turnoff, the highway descends, providing some of the finest views in the Big Horns, through Five Springs Canyon to the plains. *Caution:* Approximately the last 10 miles of the highway down to the plains is narrow, steep, and winding, thus not recommended for inexperienced drivers or for cars pulling trailers. A good stop on the way down is the Five Springs Campground in coniferous-forest surroundings with deciduous growth along Five Springs Creek, which runs through it. A nature trail winds up from the Campground for less than a mile to a point where the Creek drops in spectacular falls before reaching the Campground.

(b) *US 14 south from Burgess Junction for 49 miles to Greybull.* This route attains 8,950 feet at Granite Pass, then gradually descends, emerging from the mountains through scenic Shell Canyon to the plains. About 5 miles south from Burgess Junction, where a road turns off right from US 14 to Owen Creek Campground, there is excellent habitat for Gray Jays, Golden-crowned Kinglets, Mountain Bluebirds, Wilson's Warblers, Cassin's Finches, Oregon Juncos, White-crowned and Lincoln's Sparrows, and other high-country birds. A few yards farther south on US 14, turn off left to Tie Flume and Dead Swede Campgrounds, both of which lie along streams near burned and logged areas where Northern Three-toed Woodpeckers are a strong possibility. Continuing south on US 14, expect to see elk and mule deer. At Granite Pass, look for Water Pipits. In Shell Canyon, stop and take the nature trail to Shell Falls. Townsend's Solitaires abound in its vicinity.

2. Drive south from Sheridan on I 90 for 36 miles to Buffalo, then west on US 16 for 70 miles, over the Big Horns (Powder River Pass at 9,677 feet elevation) and down through Tensleep Canyon, to the community of Ten Sleep on the plains. On the way down the Canyon, just below the edge of coniferous forest where the terrain becomes rocky and barren, look for Chukars which occasionally cross the road to drink in the Creek.

On his way to Buffalo, the bird finder may wish to leave Sheridan on US 87 for 15 miles, then turn right on State 193, which passes through the village of Story and loops back to US 87. Once in Story, leave State 193 and proceed west up the paved road to the State Fish Hatchery, where hiking trails go into remote forests of ponderosa pine and deciduous thickets to open areas along streams—good for such species as the Dusky Flycatcher, Western Flycatcher, Black-capped Chickadee, both White-breasted and Red-breasted Nuthatches, Swainson's Thrush, Cedar Waxwing, Audubon's Warbler, Ovenbird, Western Tanager, and Red Crossbill. Yellow-bellied Sapsuckers are common.

Backtrack to State 193 and continue south to US 87. Just before reaching it and before crossing Piney Creek, turn right on Little

Piney Road (paved at first, then becomes gravel) and follow signs to Fort Phil Kearney and the site of the Wagon Box Fight. In open country along the Road and in both historic areas, Sharp-tailed Grouse and Gray Partridges are permanent residents and Bobolinks nest. In the Wagon Box Fight site, wildflowers bloom profusely in late spring and early summer, attracting Broad-tailed and Calliope Hummingbirds. Look for them on the barbed-wire fence surrounding the site.

Return to US 87, proceed south about 0.5 mile to the I 90 access ramp and continue south to Buffalo, then go west on US 16 as directed above. During the drive over the Big Horns on US 16, take one or more of the following side roads (all on the left and marked by signs; usually gravel and passable to conventional cars for the first few miles) which should yield good representations of mountain birds.

Crazy Woman Canyon Road, about 25 miles from Buffalo. MacGillivray's Warblers nest in streamside thickets near the Road, beginning about 3 miles from the entrance.

Next, *Hazelton Peak Road,* about 2 miles after the entrance to above. Continue on Hazelton Peak Road for a mile and turn left onto the road to Billy Creek. This passes meadows where Sprague's Pipits are found in summer, also stretches of conifers with some fallen timber, and patches of aspen and willow—breeding habitats for the Least Flycatcher, Western Pewee, Olive-sided Flycatcher, Gray Jay, Clark's Nutcracker, Mountain Chickadee, Red-breasted Nuthatch, Pine Siskin, and Oregon Junco. Watch for raptors—the Northern Goshawk, Sharp-shinned and Cooper's Hawks, and others.

Powder River-Webb Creek Road, about 7 miles after the entrance to Hazelton Peak Road. This passes meadows and sagebrush flats, deciduous trees and thickets along streams, and some conifers and clear-cut areas. In suitable habitats, look for the Ruffed Grouse in any season and the Vesper Sparrow in summer.

Canyon Creek Road, about 4 miles after the entrance to the Powder River-Webb Creek Road, leads through 'burns' for the Williamson's Sapsucker and Olive-sided Flycatcher, possibly the Northern Three-toed Woodpecker, open stands of conifers for the

Blue Grouse and Pine Grosbeak, and deciduous brush for the White-crowned Sparrow.

SUNDANCE
Bear Lodge Mountains | Devils Tower National Monument

In the northeastern corner of Wyoming the **Bear Lodge Mountains** represent a western extremity of the Black Hills (*see under* **Rapid City, South Dakota**). Like the Black Hills, they are forested principally with pine, except for their highest summits, the Warren Peaks (maximum elevation 6,673 feet), which are treeless and grass-covered. Aspens and birches thrive in ravines, or on slopes that were once burned over; scrub oaks predominate on several high ridges and in numerous draws on the outskirts of the range. Unlike the Black Hills, the Bear Lodge Mountains are without deep canyons and stands of spruce; hence they do not have as wide a variety of bird species.

One of the best sections for bird finding in the Bear Lodge Mountains may be reached by exiting from I 90 to Sundance and here proceeding north on the Reuter's Canyon Road (no route number) to Warren Peaks, 6 miles distant. After traversing range lands for the first 2.5 miles, the Road crosses a draw containing oaks and shrubs where such breeding birds reside as the Dusky Flycatcher, Blue Jay, Veery, Warbling Vireo, Black-headed Grosbeak, Lazuli Bunting, and Rufous-sided Towhee. Beyond the draw the Road begins to climb through Reuter's Canyon, a wide ravine with deciduous trees bordering a streambed and pines on the slopes. From the Road in the evening, Poor-wills may sometimes be heard calling. Among the other summer birds are the Western Flycatcher, Swainson's Thrush, Ovenbird, MacGillivray's Warbler, Western Tanager, Pine Siskin, and White-winged Junco. As the Road approaches the treeless heights of the Warren Peaks, Ruffed Grouse are fairly numerous and frequently may be heard drumming, even as late in the breeding season as the last of June.

The fire lookout tower on the highest of the Warren Peaks offers a fine view, which includes to the northwest one of the most extraordinary geologic formations on the continent—the Devils Tower. Commonly likened to 'a gigantic tree stump,' the Tower is in reality what is left of an ancient lava intrusion. Measuring 865 feet from bulging base to top and looming 1,280 feet above the nearby Belle Fourche River, it is an astonishing topographic feature from the distance; close by, it is even more amazing, for its sides are then seen to be composed of enormous columns of remarkable design and symmetry. The Tower and its adjacent land—altogether 1,194 acres—comprise the **Devils Tower National Monument.**

The rolling landscape in the immediate vicinity of the Tower is covered largely with ponderosa pine, mixed here and there with aspens, oaks, and other deciduous growth, including shrubby thickets. Some of the bird species in summer near the base of the Tower are the Hairy Woodpecker, Western Pewee, Pinyon Jay (occasionally), Black-capped Chickadee, Brown Thrasher, Warbling Vireo, Audubon's Warbler, Yellow-breasted Chat, Western Tanager, Pine Siskin, White-winged Junco, and Chipping Sparrow. Probably no birds nest among the ferns, grasses, and shrubs that grow in the acre and a half of surface on the top of the Tower, but a pair or more of Rock Wrens usually nest either in the low niches of its near-perpendicular sides or among the talus of broken columns below. In past years Prairie and/or Peregrine Falcons were reported nesting on sides of the Tower where there are situations apparently suitable for their aeries.

The Devils Tower National Monument may be reached from Sundance by driving west on US 14 for 21 miles, then turning north onto State 24 and going 6 miles to the entrance road, which passes around the south side of the Tower and up the hill to the visitor center near the west base. Soon after entering the Monument and crossing the Belle Fourche River, the road traverses an open grassy stretch occupied by a prairie dog 'town.' Lark Sparrows are common in its vicinity. A foot trail encircling the Tower and an additional 4.5 miles of nature trails give the bird finder every opportunity to take in the natural history of the Monument.

YELLOWSTONE NATIONAL PARK

Established in 1872, **Yellowstone National Park** is the oldest of all National Parks. It is also the largest, with a total area of 3,472 square miles. Though most of the Park is in the northwestern corner of Wyoming, its northern and western boundaries embrace small strips of the adjoining states of Montana and Idaho. Coupled with Yellowstone Park's supremacy in age and size is its singular standing as the most celebrated showplace among all American wildernesses. Considering the natural features that lie within its confines—the imposing array of mountains, lakes, rivers, canyons, and waterfalls, the amazing geysers and hot springs, and the exceptionally wide diversity of wildlife—the reputation is wholly justified.

Yellowstone Park occupies the high, volcanic Yellowstone Plateau (altitudes between 7,000 and 8,500 feet), partly ringed by mountain ranges that reach from 2,000 to 4,000 feet above the general level. The major ones are, on the northwest and north, the Madison and Gallatin Ranges; on the northeast and east, the rugged Absaroka Mountains; and, on the south, the majestic Tetons. In the Park loom a great many peaks, several of which exceed 10,000 feet, the highest being Eagle Peak, with an elevation of 11,360 feet.

Of the several rivers providing drainage, the chief one is the Yellowstone, whose headwaters, rising south of the Park, are soon interrupted in their northwest course by the natural dam that forms Yellowstone Lake, 139 square miles in surface area and 7,731 feet above sea level, the largest natural body of water at such an altitude on the continent. North of Yellowstone Lake, the River winds through Hayden Valley and then plunges 417 feet in two thundering cataracts, Upper and Lower Yellowstone Falls, into the Grand Canyon of the Yellowstone. For spectacular beauty, this enormous chasm has few competitors in the world. Its 1,000-foot walls from the blue-green torrents at the base to the deep-green conifers at the rim have a dazzling yellow-whiteness tinted with a multitude of pastel shades.

The 10,000 or so hot springs, geysers, steam vents, mud volcanoes, and 'paint pots,' nearly all in the west and central part of the Park, comprise an unrivaled aggregation of thermal phenomena. Of these, approximately 200 geysers are large enough to have names.

Over three-fourths of Yellowstone Park is forested, primarily with lodgepole pine, mixed with Douglas-fir in cool canyons. Deciduous trees, such as aspen, cottonwood, willow, and alder, line many of the watercourses; groves of aspen are prevalent on the floors of wide valleys or on slopes, where they appear as light-green patches among the dark pines. At higher altitudes, on mountains, Engelmann spruce and subalpine fir become increasingly common and, as the timber line is approached, are intermixed with whitebark pine, which, together with the spruce and fir, here exhibits the peculiar, dwarfed, matted, and one-sided condition characteristic of wind-beaten, subalpine trees. Natural open areas in Yellowstone Park range from lush meadows and dry, gray sagelands to bleak alpine meadows above the timber line.

Probably as great an attraction to Park visitors as the canyons and thermal phenomena is the variety of mammals, including black bear, moose, elk, mule deer, grizzly bear, bighorn sheep, pronghorn, and bison. No visitor can be in the Park very long before he will see red squirrels, ground squirrels, chipmunks, marmots, and, in certain localities, beavers and pikas. All the mammals are potentially dangerous in one way or another and must never be approached, molested, or fed.

Throughout the Park in summer the Northern Raven and Clark's Nutcracker are perhaps the most conspicuous of the passerine birds. The Northern Goshawk, Red-tailed and Swainson's Hawks, and American Kestrel are among the commoner raptors. Other birds that may be seen without much difficulty: Audubon's Warblers and Western Tanagers in large stands of pines; Rock Wrens in rocky outcrops; North American Dippers along swift streams; Gray Jays at campgrounds; Western Meadowlarks and Vesper Sparrows in grassy stretches; and Violet-green Swallows in canyons.

Automobile roads in Yellowstone Park are usually open from the first of May through October, except the road from the North Entrance to the Northeast Entrance which is kept open the year round. Bird finding in the Park is best during the height of the singing season from 15 June to 15 July, but there are many birds at all times of the year except in winter. When coming into the Park at any one of the five entrances, obtain a detailed map at the entrance station. This will locate the areas in the park mentioned below.

The road from the East Entrance, reached from Cody, Wyoming, via US 14, 16, and 20, skirts the northeastern shore of Yellowstone Lake and proceeds to the village of Lake, where it joins the Grand Loop Road—the 142-mile road system laid out in a figure 8. By turning left on the Loop Road, one may continue along the west side of Yellowstone Lake to Thumb and thereby obtain additional views of this great basin of water.

The shore of Yellowstone Lake is a fine vantage for seeing a number of waterbirds and waterfowl. Outstanding are the American White Pelicans, which nest on the Molly Islands in the Lake's Southeast Arm. At least a few of the adults belonging to this colony may be seen anywhere around the Lake, since they wander afar in search of food. California Gulls also have colonies on these islands and are likewise easily observed. Visiting the colonies is forbidden. Other birds likely in view anywhere on the Lake, though more often near its remote and undisturbed marshy edges, include the Eared Grebe, Common Pintail, Blue-winged Teal, Cinnamon Teal, American Wigeon, Barrow's Goldeneye, Ruddy Duck, and Common Merganser. Two or more pairs of Ospreys have aeries near the Lake.

The Loop Road from the village of Lake to the village of Canyon goes through Hayden Valley close to the Yellowstone River. Look for an occasional Bald Eagle along the way. The LeHardy Cascades, 4 miles north of Lake village, is an excellent place to watch for Harlequin Ducks. Where the Road crosses Trout Creek, Cliff Swallows nest beneath the bridge. At the Ranger station in Canyon, inquire about trail directions to Grebe Lake, to the northwest, where Trumpeter Swans breed.

Between Norris and Mammoth Hot Springs the Loop Road follows Obsidian Creek for a considerable distance and at the same time passes many grassy meadows, willow flats, and small lakes. Beaver Lake, on the west side of the Road, opposite Obsidian Cliff, has such birds nesting in its vicinity as Canada Geese, Mallards, and Green-winged Teal. Swan Lake, on the west side of the Road 5.5 miles north of Beaver Lake, has an occasional pair of Trumpeter Swans. Between Beaver Lake and Swan Lake are extensive willow flats, favorite haunts of the Traill's Flycatcher and Wilson's Warbler. Frequently these flats are visited by moose. Other birds to watch for, especially around lakes and meadows, are Sandhill Cranes (never common, but there is always a possibility of one or more), Common Yellowthroats, Brewer's Blackbirds (very common), and White-crowned and Lincoln's Sparrows. On nearby brushy slopes are a few Green-tailed Towhees.

A good opportunity to observe birds of the mountain slopes is by hiking to the summit of Mt. Washburn (10,243 feet elevation) on the old Mt. Washburn Road (closed to cars but accessible to bus service). This leaves the Loop Road at Dunraven Pass between Canyon and Tower Junction. The north side of the mountain is treeless, except for scattered groves of lodgepole pine, Douglas-fir, and quaking aspen on the lower slopes, and Engelmann spruce and whitebark pine on the higher slopes to the timber line at about 9,800 feet; the south side is more or less continuously forested with the same succession of trees from base to timber line. Among the birds to look or listen for during the climb are the following: Blue Grouse, Ruffed Grouse, Dusky Flycatcher, Steller's Jay, Mountain Chickadee, Red-breasted Nuthatch, Hermit Thrush, Townsend's Solitaire (higher slopes), Ruby-crowned Kinglet (higher slopes), Cassin's Finch (higher slopes), Pine Siskin, Oregon Junco, White-crowned Sparrow (at the timber line). Above the timber line, Water Pipits reside, and there is always a good chance of finding Black Rosy Finches or spotting a soaring Golden Eagle.

Yellowstone Park's various naturalist services, available during the Park's main season from 10 June to Labor Day, are centered at Mammoth Hot Springs, Norris Geyser Basin, Old Faithful, Grant Village, Fishing Bridge, and Canyon Village. Activities at most of these points include nature walks and interpretive programs. When entering the Park, obtain a full schedule of activities.

METRIC CONVERSION TABLE

Length						Area			
FEET	METERS	YARDS	METERS	MILES	KILOMETERS	ACRES	HECTARES	SQUARE MILES	SQUARE KILOMETERS
1	0.3048	1	0.9144	1	1.609	1	0.4047	1	2.590
2	0.6096	2	1.8288	2	3.219	2	0.8094	2	5.180
3	0.9144	3	2.7432	3	4.828	3	1.2141	3	7.770
4	1.2192	4	3.6576	4	6.437	4	1.6187	4	10.360
5	1.5240	5	4.5720	5	8.047	5	2.0234	5	12.950
6	1.8288	6	5.4864	6	9.656	6	2.4281	6	15.540
7	2.1336	7	6.4008	7	11.265	7	2.8328	7	18.130
8	2.4384	8	7.3152	8	12.875	8	3.2375	8	20.720
9	2.7432	9	8.2296	9	14.484	9	3.6422	9	23.310
10	3.0480	10	9.1440	10	16.093	10	4.0469	10	25.900
20	6.096	20	18.288	20	32.187	20	8.0937	20	51.800
30	9.144	30	27.432	30	48.280	30	12.1406	30	77.700
40	12.192	40	36.576	40	64.374	40	16.1874	40	103.600
50	15.240	50	45.720	50	80.467	50	20.2343	50	129.499
60	18.288	60	54.864	60	96.561	60	24.2812	60	155.399
70	21.336	70	64.008	70	112.654	70	28.3280	70	181.299
80	24.384	80	73.152	80	128.748	80	32.3749	80	207.199
90	27.432	90	82.296	90	144.841	90	36.4217	90	233.099
100	30.480	100	91.440	100	160.934	100	40.4686	100	258.999
200	60.960	200	182.880	200	321.869	200	80.937	200	517.998
300	91.440	300	274.320	300	482.803	300	121.406	300	776.996
400	121.920	400	365.760	400	643.738	400	161.874	400	1035.995
500	152.400	500	457.200	500	804.672	500	202.343	500	1294.994
600	182.880	600	548.640	600	965.606	600	242.812	600	1553.993
700	213.360	700	640.080	700	1126.541	700	283.280	700	1812.992
800	243.840	800	731.520	800	1287.475	800	323.749	800	2071.990
900	274.320	900	822.960	900	1448.410	900	364.217	900	2330.989
1000	304.800	1000	914.400	1000	1609.344	1000	404.686	1000	2589.988
2000	609.600	2000	1828.800	2000	3218.688	2000	809.372	2000	5179.980
3000	914.400	3000	2743.200	3000	4828.032	3000	1214.058	3000	7769.960
4000	1219.200	4000	3657.600	4000	6437.376	4000	1618.744	4000	10,359.950
5000	1524.000	5000	4572.000	5000	8046.720	5000	2023.430	5000	12,949.940
6000	1828.800	6000	5486.400	6000	9646.064	6000	2428.116	6000	15,539.930
7000	2133.600	7000	6408.000	7000	11,265.408	7000	2832.802	7000	18,129.920
8000	2438.400	8000	7315.200	8000	12,874.752	8000	3237.488	8000	20,719.900
9000	2743.200	9000	8229.600	9000	14,484.096	9000	3642.174	9000	23,309.890
10,000	3048.000	10,000	9144.000	10,000	16,093.440	10,000	4046.860	10,000	25,899.880

Index

Abbreviations used in the Index

AZ	Arizona	MN	Minnesota	OK	Oklahoma
AR	Arkansas	MO	Missouri	OR	Oregon
CA	California	MT	Montana	SD	South Dakota
CO	Colorado	NB	Nebraska	TX	Texas
ID	Idaho	NV	Nevada	UT	Utah
IA	Iowa	NM	New Mexico	WA	Washington
KS	Kansas	ND	North Dakota	WY	Wyoming
LA	Louisiana